The Evolution of Plants

K. J. Willis
The School of Geography and the Environment
University of Oxford

J. C. McElwain
The Field Museum, Department of Geology
Chicago

OXFORD
UNIVERSITY PRESS
Great Clarendon Street, Oxford OX2 6DP

Oxford University Press is a department of the University of Oxford.
It furthers the University's objective of excellence in research, scholarship,
and education by publishing worldwide in

Oxford New York

Athens Auckland Bangkok Bogotá Buenos Aires Cape Town
Chennai Dar es Salaam Delhi Florence Hong Kong Istanbul Karachi
Kolkata Kuala Lumpur Madrid Melbourne Mexico City Mumbai Nairobi
Paris São Paulo Shanghai Singapore Taipei Tokyo Toronto Warsaw

with associated companies in Berlin Ibadan

Oxford is a registered trade mark of Oxford University Press
in the UK and in certain other countries

Published in the United States
by Oxford University Press Inc., New York

First published 2002

Library of Congress Cataloging in Publication Data
(Data Applied for)

ISBN 0 19 850065 3

Typeset by Newgen Imaging Systems (P) Ltd., Chennai, India

Printed in Great Britain on acid-free paper by
Bath Press Ltd., Bath, Avon

Acknowledgements

There are a number of people that have been involved in the shaping of this book. In particular we are especially grateful to Keith Bennett and Janice Fuller for reading an early draft of the manuscript and providing invaluable feedback and discussion at a critical stage of the book. Nick Butterfield, Simon Crowhurst, Lindsey Gillson, Adam Kleczkowski, Peter Lang and Scott Lidgard also read various sections in detail. All these people provided excellent constructive criticism and lively debate that very much shaped the final product. Our warm and extended thanks also goes to Professor Bill Chaloner for thoroughly reviewing the book in its final stages and providing comments in such a positive light. For the meticulous line drawings and illustrations our thanks go to Simon Crowhurst of the Godwin Laboratory, University of Cambridge and for the illustrated legends to the biome maps to Marlene Donnally, Scientific Illustrator in the Department of Geology, The Field Museum, Chicago. Peter Lang is also thanked for graphically redrawing the base maps on which all of the biome illustrations were based. We would also like to thank John Weinstein of the Field Museum for photographing a number of the fossil plant images throughout the book and on the cover. Many people generously provided unpublished data and in particular we are especially grateful to Fred Zieglar and Allistair Rees, Department of Geosciences, University of Chicago for invaluable advice on biogeographical reconstructions and unpublished data on the extent of deserts through time and allowing us to reproduce a number of PaleoAtlas Projects paleographical maps, Karl Niklas from the Department of Plant Biology, Cornell University for use of his fossil plant database, and Chris Scotese from the Department of Geology, University of Texas for allowing us to reproduce a number of the paleogeographical reconstructions from the PaleoMap project and data in press on climatically sensitive sediments. Specialist advice was also freely given by Margaret Collinson on plant biogeography during the Cenozoic, Scott Lidgard on macroevolutionary theory, Paul Goldstein on cladistics, Nick Butterfield on early eukaryotic organisms, Paul Barrett on Cretaceous dinosaurs, Gary Upchurch on Cretaceous plant biogeography, Mark Tester on plant physiology and Tom Phillips on Carboniferous plant biogeography. We are immensely grateful to all of them. In the final stages of the book a number of people provided invaluable help including Linda Atkinson and Sue Bird, who using their librarian skills to the full in order to search out the most obscure of references, Rebekah Hines and Gina Wesley for proof reading and checking terms for the glossary, and Daniel Boyce for checking the index. We are also extremely grateful to Maja Andrič for her tireless work in helping to tie-up the final loose ends. At Oxford University Press we consider ourselves privileged to have been under the guidance of our editor Jonathan Crowe and production editor John Grandidge—their enthusiasm in this book at all stages was very much appreciated. Finally we would like to thank our families and most of all our long-suffering husbands—Andrew Gant and Peter Lang. Without their support this book would never have been completed and we dedicate it to them.

November 2001

K. J. W.
J. C. McE.

Contents

1 The evolutionary record and methods of reconstruction

2 Earliest forms of plant life

3 The colonization of land

4 The first forests

5 Major emergence of the seed plants

6 Flowering plant origins

7 The past 65 million years

8 Mass extinctions and persistent populations

9 Ancient DNA and the biomolecular record

10 Evolutionary theories and the plant fossil record

1 The evolutionary record and methods of reconstruction

1.1 Introduction

Historically the study of evolution has concentrated almost exclusively upon the animal record. If plants are mentioned it is only in passing and usually to reconstruct the environment in which to view the animal record. Yet from the earliest fossil records plants have provided an equally diverse and interesting picture. The earliest fossil photosynthesizers date back to approximately 3500 million years ago (Ma) (early Archaean) and some of the earliest eukaryotes were plants, namely red and green algae (Knoll, 1992; Butterfield *et al.*, 1994). Plants colonized terrestrial environments relatively early on in the late Silurian/early Devonian (~415 Ma), and in a matter of 75 million years had evolved from toe-brushing specimens to trees up to 60 m in height. The first seed-bearing plants evolved approximately 380 million years ago, but it was another 180 million years before the first flowers appeared, and grasses did not appear until 65 million years ago.

Interesting and dramatic evolutionary change has therefore occurred in the fossil plant record, but interpretation of processes and mechanisms of biological evolution have been based largely upon the fossil faunal record. In comparison to animals, plants have a limited list of basic needs (H_2O, CO_2, N, Mg, P, K, some trace elements, photons, plus various biochemical pathways necessary for photosynthesis) and this list has remained relatively unchanged throughout the evolutionary record (Knoll, 1984; Traverse, 1988a). Many plants also have in-built mechanisms for coping with environmental stress, including seed dormancy and leaf-shedding, mechanisms of die-back where the plant can perennate as underground stems or rhizomes, and the ability to regenerate vegetatively after trauma. In addition, plants are not, as a rule, nearly as sensitive as animals to population size dynamics. A 'deme' of vascular plants consisting of only a few individuals can persist far better than a small tetrapod animal deme (Knoll, 1984). Plants can also live for a very long time, with some species, such as the bristle cone pine (*Pinus longaeva*) for example, living for up to 11 000 years (Ingrouille, 1992). Unlike the animal record, there are also many extant plant families that have persisted, often in morphologically identical forms, for over 200 million years (Traverse, 1988a), including, for example, members of the ferns

(Filicopsids), the cycads (Cycadales), the ginkgos (Ginkgoales), and the conifers (Coniferales). Thus in both the study of phenotypic and genotypic evolution, plants offer a fascinating record.

Over the past 25 years a revolution in evolutionary thinking has occurred. In particular, the traditional view that evolution is a gradual process of change resulting from increasing minute adaptation to other species and the environment (Darwin, 1859) has been challenged. An alternative school of thought proposes that speciation is a sudden and rapid process interspersed with periods of stasis (punctuated equilibrium) (Eldredge and Gould, 1972). These ideas have been further developed to suggest that long-term environmental change, especially that brought about by the long-term cyclical changes in climate resulting from the position of the Earth in relation to the sun (orbital forcing), might provide the necessary pulses to bring about rapid speciation (Vrba, 1985, 1995; Bennett, 1990, 1997). But the majority of discussion has centred on the animal record and the same questions need to be asked of the plant record. For example, is there evidence for gradual evolutionary change or periods of stasis interspersed with pulses of rapid speciation? How do the patterns seen in the plant record relate to the animal record? Are there mass extinction events comparable to those seen in the animal record? Also, if the theory of punctuated equilibrium is applicable to the plant record, is there evidence to suggest that the mechanisms driving the pulses of speciation in the animal record are the same for plants?

Important advances in evolutionary studies have also occurred though an increased understanding of genotypic change. In particular, the ability to reconstruct phylogenies based on molecular characteristics and to calculate rates of genotypic evolution, are starting to provide yet another data set. These can be used to compare and contrast processes and mechanisms of long-term biological evolution. Although there is still a long way to go in relating evolutionary change seen in the genotypic record to changing morphology, one area of great potential must be in the extraction of ancient DNA from fossil material. Twenty-five years ago the concept of Jurassic Park and the reconstruction of dinosaurs from ancient DNA would have been placed in the realms of science fiction. However, there are now examples in the literature of extraction and amplification of plant DNA (up to 600 base pairs), being recovered from fossil leaves encased in 130 million-year-old amber (Poinar, 1994). Although there is still much controversy surrounding the ability of the bonds holding the DNA strands together to remain intact for such long periods of time (Lindahl, 1993), the potential of ancient DNA in providing calibration for molecular clocks and determining the relationship between phenotypic and genotypic evolution, cannot be underestimated.

The first seven chapters of this book aim to provide an overview of the main evolutionary changes that have occurred in the plant kingdom, from the first photosynthesizers to the present. They examine the predominant evolutionary patterns in the plant fossil record that have occurred over the past

430 million years, and consider them alongside the geological evidence for long-term global environmental change. Next the evidence for mass extinctions and persistent plant populations in the fossil record is examined. The state of current research on ancient DNA and other biomolecular markers in the plant fossil record is then reviewed in Chapter 9, and how this fits within the broad framework of current evolutionary thought on genotypic and phenotypic evolution. The final chapter examines the plant record in the light of different evolutionary theories; does the evolutionary plant record indicate a 'cone of increasing diversity' or decimation and diversification? Is there evidence in the plant record for gradual long-term change (phyletic gradualism) or periods of stasis interspersed by rapid change (punctuated evolution)? Are changes in the plant record similar to those seen in the animal record? And is it possible to detect processes and driving mechanisms of long-term plant evolution?

1.2 The geological timescale

The geological timescale subdivides the past 4700 million years (since the Earth's formation) into hierarchical divisions; eras, periods, and epochs (Figure 1.1) (Harland *et al.*, 1990). All the standard stratigraphic divisions are based on fossil stratigraphies (biostratigraphy). There are three major era boundaries, two of which are defined by mass extinctions in the animal record and the third by the rapid expansion of multicellular animals (Gould, 1989). The boundary between the Mesozoic and Cenozoic is dated at *c.* 65 million years ago (Ma), when dinosaurs died out along with 75% of marine invertebrates. The boundary between the Palaeozoic and Mesozoic is dated at *c.* 248 Ma, when there was the extinction of up to 96% of marine species in the geological record. The boundary between the Palaeozoic and Precambrian is placed at *c.* 543 Ma, and is characterized by a major increase in numbers of multicellular animals with hard parts soon after the boundary, known as the Cambrian explosion. At all levels of the hierarchy, however, it is important to note how many of the boundaries are based solely on changes in the animal record. As pointed out by Traverse (Traverse, 1988a), if the geological record had been based on changes in the plant record then a very different hierarchical structure would have resulted (Figure 1.2).

Throughout this book the traditional geological timescale (geochronological scale) will be used and major changes that have occurred in the plant fossil record are viewed within this framework. In order to provide accessibility to non-geologists, all changes will be referred to in terms of million years before present (Ma) based on the 1999 Geological Timescale of the Geological Society of America (http://www.ucmp.berkeley.edu/fosrec/TimeScale.html) as well as their geochronologic terms (Figure 1.1).

Era	Sub-era	Period	Epoch	Age (Ma)
Cenozoic	Quaternary	Neogene	Holocene	
			Pleistocene	1.8
	Tertiary		Pliocene	5.3
			Miocene	23.8
		Palaeogene	Oligocene	33.7
			Eocene	54.8
			Palaeocene	65
Mesozoic		Cretaceous	late	
			early	144
		Jurassic	late	
			middle	
			early	206
		Triassic	late	
			middle	
			early	248
Palaeozoic		Permian	late	
			early	290
		Carboniferous: Pennsylvanian	late	323
		Carboniferous: Mississippian	early	354
		Devonian	late	
			middle	
			early	417
		Silurian	late	
			early	443
		Ordovician	late	
			early	490
		Cambrian	late	
			middle	
			early	543
Precambrian				4600

Figure 1.1 The geological timescale (simplified); redrawn from The Geological Society of America, 1999 Geological timescale.

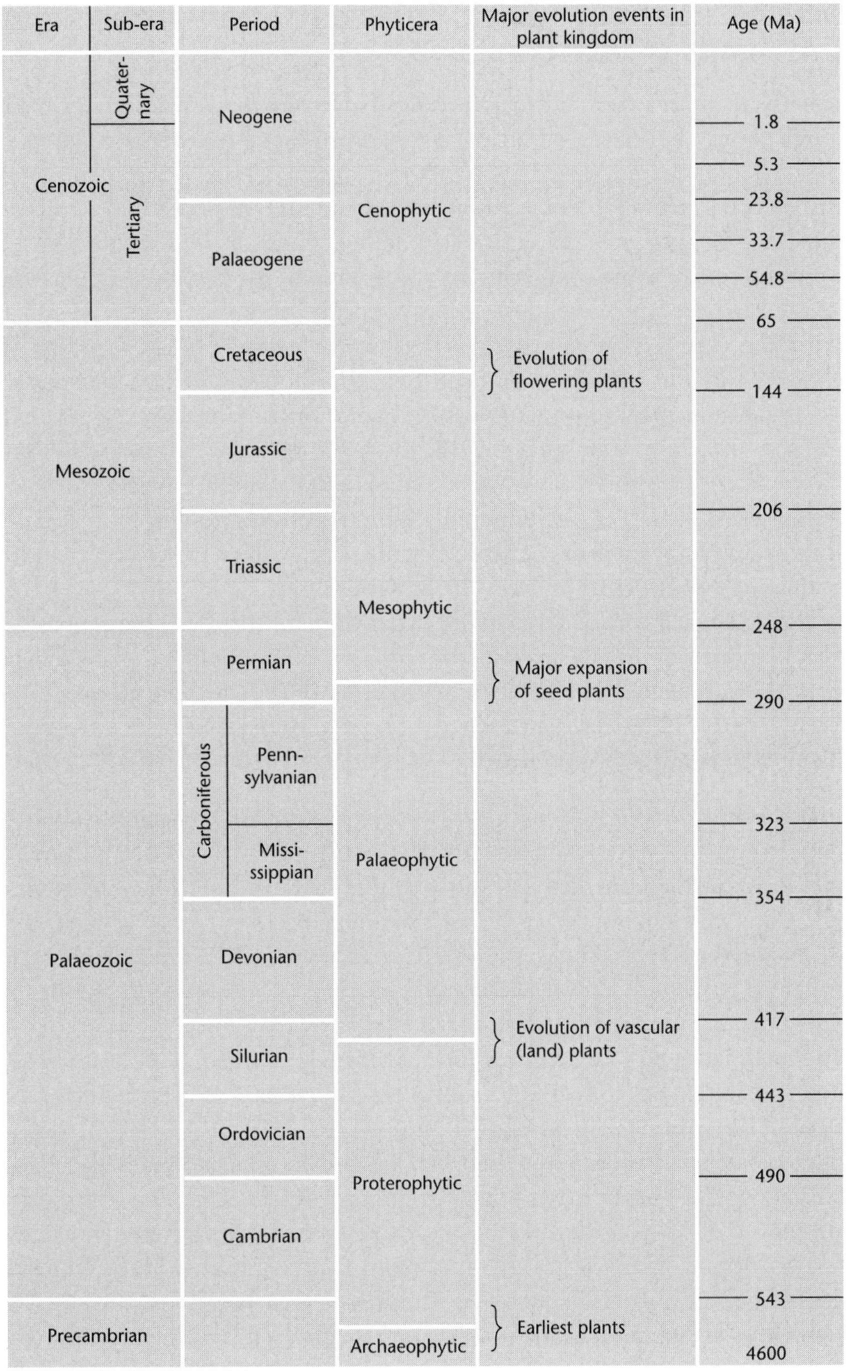

Era	Sub-era	Period	Phyticera	Major evolution events in plant kingdom	Age (Ma)
Cenozoic	Quater-nary	Neogene	Cenophytic		1.8
	Tertiary				5.3
					23.8
		Palaeogene			33.7
					54.8
					65
Mesozoic		Cretaceous		Evolution of flowering plants	144
		Jurassic	Mesophytic		206
		Triassic			248
Palaeozoic		Permian		Major expansion of seed plants	290
		Carboniferous: Penn-sylvanian	Palaeophytic		323
		Carboniferous: Missi-ssippian			354
		Devonian			417
		Silurian	Proterophytic	Evolution of vascular (land) plants	443
		Ordovician			490
		Cambrian			543
Precambrian			Archaeophytic	Earliest plants	4600

Figure 1.2 Alternative classification of the hierarchical structure in the geological timescale if the record had been constructed according to major evolutionary change seen in the plant fossil record (after Tr̀averse, 1988a).

1.3 Methods of reconstruction

Currently there are two approaches used to reconstruct evolutionary change through time. The first is based on the premise that 'the past is a key to the present'. This is the geological approach involving a detailed examination of the fossil record, including both whole organism and biomolecular studies to reconstruct a picture of evolutionary change through geological time. The second approach, which tends to be more favoured by ecologists and those interested in genotypic evolution, is based on the premise that 'the present is a key to the past'. This approach involves the examination of present-day organisms and, through the assessment of similar/dissimilar characteristics with other organisms, the inference of possible evolutionary pathways. Again studies for this approach include both those at a whole-organism and molecular level. Both methods have their advantages and shortfalls and are best used in conjunction with each other. Although much of the information given in this book will be based on the evidence provided by the fossil record, many examples from the study of extant plants will also be incorporated.

In the geological record three major types of plant fossil are preserved: compression/impression fossils, permineralized fossils, and unaltered plant remains. Each can be attributed to a distinctive preservational environment.

Compression/impression fossils

One of the most simple types of preservation for plants is as a compression fossil. Compression fossils are formed in a variety of freshwater and marine environments including lagoons, estuaries, rivers, and lakes or any other environment where plant parts are buried rapidly by a sedimentary layer. As subsequent layers accumulate, the water is squeezed out, the sediment becomes more compact, and the plant fragments within it become flattened. This will result in one of three types of compression fossil (Chaloner, 1999).

The most basic type of compression fossil is where internal material is obliterated and all that is left is a thin carbonaceous film that conforms to the original outline of the plant. These types of compressions are frequently found in shales above coal seams. Information that can be gained from such fossils includes details about the morphology of the part preserved, as well as some biochemical information. Work by Niklas (1976), for example, comparing the biochemical similarity of one of the earliest thalloid plant fossils, *Parka*, with genera of the Chlorophyta (green algae), involved scraping the carbonaceous film of compression fossil from the rock and putting it through a series of chemical processes to extract organic, soluble compounds. Gas chromatography and high-resolution mass spectroscopy were then carried out on the residue for identification of acid and ester mixtures. Using this method, it was possible to demonstrate a close biochemical similarity between fossil *Parka* (*c.* 400 Ma old) and present-day green algae (Niklas, 1976).

Another type of compression fossil results from the imprint of the fossil, thus in effect an 'impression' fossil which is a negative imprint of the organism. In this type of fossil formation, organic material has been removed and the only information retained will be features of the surface structure and details of shape and size.

When the compression occurs under anoxic (in the absence of oxygen) and acid conditions, a third type of fossil results where cell contents are often intact, providing much finer detail that in the above two examples. Plant cuticle, which remains structurally unaltered, is commonly preserved in this type of compression fossil (Taylor, 1999) (Figure 1.3). Aerial parts of all vascular land plants are covered with cuticle to prevent excessive water loss from the plant surface. Cuticle has a very stable chemical structure, providing both resistant outer protection to the plant and the ability to survive biological and geological processes through time (Mosle *et al.*, 1997).

The cuticle reflects the epidermal layer that it covers, so features such as stomata, papillae, and glands are preserved. Each plant genus has a very distinctive cuticular pattern (Figure 1.3) (the plant equivalent of the fingerprint), thus extraction and analysis can provide not only important information about the epidermal structure of the plant but also identification of disarticulated plant parts, e.g. flowers and seeds (Kerp and Krings, 1999).

Technically, coal is also a compression fossil (Scott, 1987; Taylor and Taylor, 1993), although in this case it represents a heterogeneous mix of plants originally growing in vast regions of swamp. Coal is classified according to the degree to which the deposit has been compressed and thermally altered. In the least-crushed material (lignite), many plant parts can be recognized and extracted. One of the main advantages of studying coal deposits is that they represent the community of plants growing in the ancient swamps.

Permineralized fossils

Permineralization results from the infiltration and permeation of the plant tissues by mineral-charged water. Over time, intracellular and interstitial

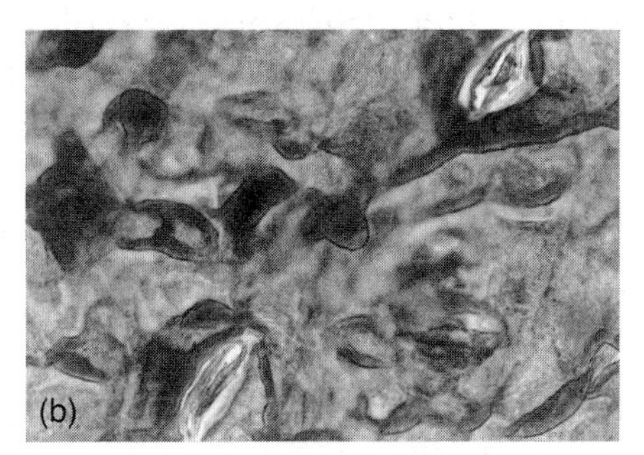

Figure 1.3 Compression fossil of (a) *Ginkgo huttonii* (~ 160 Ma) showing thick cuticle peeling of the rock (photograph W. G. Chaloner). (b) Detail of *Ginkgo huttonii* cuticle showing stomata and palillae (magnified ×400) (photograph J. McElwain).

precipitation of the dissolved mineral results in hardening within the plant and it effectively becomes solid rock. Thus, although the plant cell walls still consist of organic matter, they have been chemically altered and the intercellular spaces and cell lumens are filled with precipitated mineral. It is often stated that the process of permineralization is the most informative mode of preservation for the study of fossilized plants (Schopf, 1975). This is because permineralization tends to result in the three-dimensional fossilization of plants, with exquisitely preserved detail down to the cellular level.

To study the fine structure of permineralized fossils it is necessary to prepare thin sections of the fossil that will allow the passage of transmitted light (Figure 1.4). This enables examination of not only external morphological features of the plant, but also transverse and longitudinal cross-sections (Kenrick, 1999b).

A number of different types of permineralized fossils are found in the geological record (Figure 1.5). Pyritic and limonitic permineralization, for example, results from the infiltration and precipitation of the cells by iron sulphides (pyrites) and hydrated iron oxides (limonites), whereas siliceous

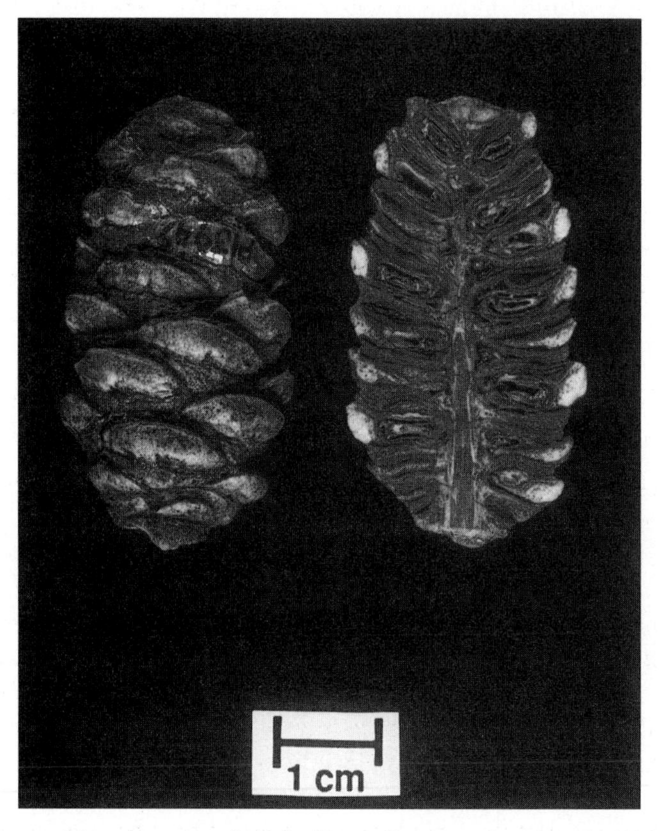

Figure 1.4 Cross section of permineralized fossil cone (*Proaraucaria patagonica*) Triassic, Argentina (photograph J. Weinstein, courtesy of The Field Museum, Chicago, neg # GEO86323).

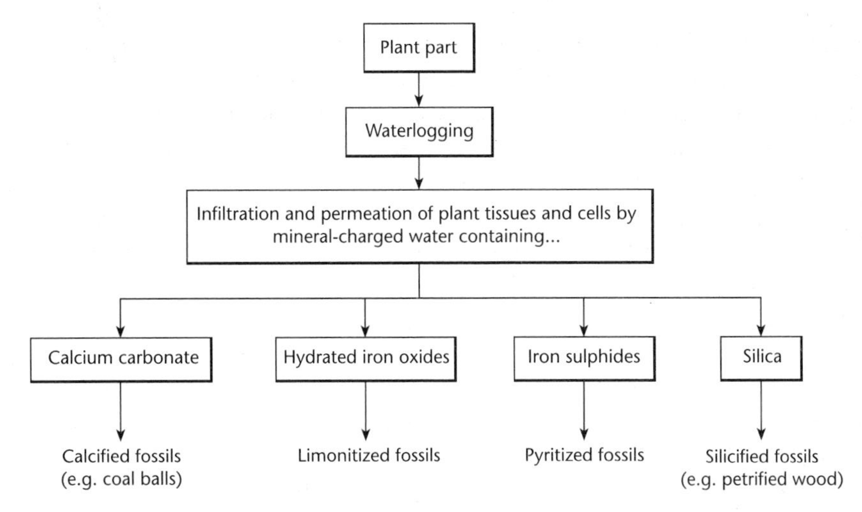

Figure 1.5 Most common types of permineralized plant fossils found in the geological record.

permineralization results from the deposition of microcrystalline colloidal silica. Whole plants (including trees) can be preserved this way, with famous examples including the petrified forests of Arizona which date back to the late Triassic (~230 Ma).

'Coal balls' are a distinctive and much cited form of permineralized fossil. These are balls of calcite, found exclusively in peat of Carboniferous age (~354–290 Ma) from deposits in Europe and North America, and of Permian age (~290 Ma) from Chinese deposits. They vary considerably in size, but are usually crudely spheroidal, *c.* 10–20 cm radius (Figure 1.6), and composed of numerous plant fragments incorporated within a calcium carbonate matrix. Similar to lignite, they contain a broad spectrum of plant fragments from the swamps in which they were formed (Phillips and DiMichele, 1999).

There are a number of theories as to how coal balls are formed (Scott and Rex, 1985), since it is not immediately apparent as to why such rounded calcium carbonate concretions should be found within swamp and peat deposits. One theory suggests that coal balls resulted from the transgression of marine water over the swamps and that calcium carbonate precipitated directly from this water into the organic layers beneath (Figure 1.7a). This theory is not entirely satisfactory because it does not explain why the material is concentrated in balls. Another marine theory, therefore, is that coal balls were only formed when storms brought mud rollers and carbonate into the swamp, and that it was these rollers that formed the coal balls (Figure 1.7b). A non-marine theory has also been proposed, suggesting that coal balls result from the carbonate-rich permineralizing water derived from a groundwater supply (Figure 1.7c) infiltrating up into the swamp layers.

Besides being in plentiful supply and providing a broad representation of plant life in the Carboniferous and Permian swamps, coal balls also have the

Figure 1.6 Fossil coal ball showing detail of fossil plant fragments (photograph J. Weinstein, courtesy of The Field Museum, Chicago, neg # GEO8626).

additional attraction of easy extraction of fossil material contained within them. The extraction process involves slicing the coal ball in half with a diamond saw, fine-abrasive polishing of the newly cut surface, and then placing the polished surface in a dilute solution of hydrochloric acid (Galtier and Phillips, 1999). Thus the calcium carbonate is etched away, leaving the plant material prominent on the surface. Acetone is then poured on this surface, and an acetate sheet applied. The acetone partially dissolves the lower layer of the acetate sheet, which moulds around the exposed plant fragments. Once the acetone evaporates, the acetate sheet once again hardens, and as it is pulled away, a detailed thin-section of the organic plant material contained within the coal ball is obtained on the peel. This can be examined using conventional microscopy.

Cryo-preservation is another form of permineralization, where tissue becomes permeated by microcrystalline ice. However, this form of permineralization tends to contain only relatively recent fossils; for example, the Pleistocene frozen mammoths found on the plains of Siberia and dating back to approximately 10 000 years ago (Roberts, 1998).

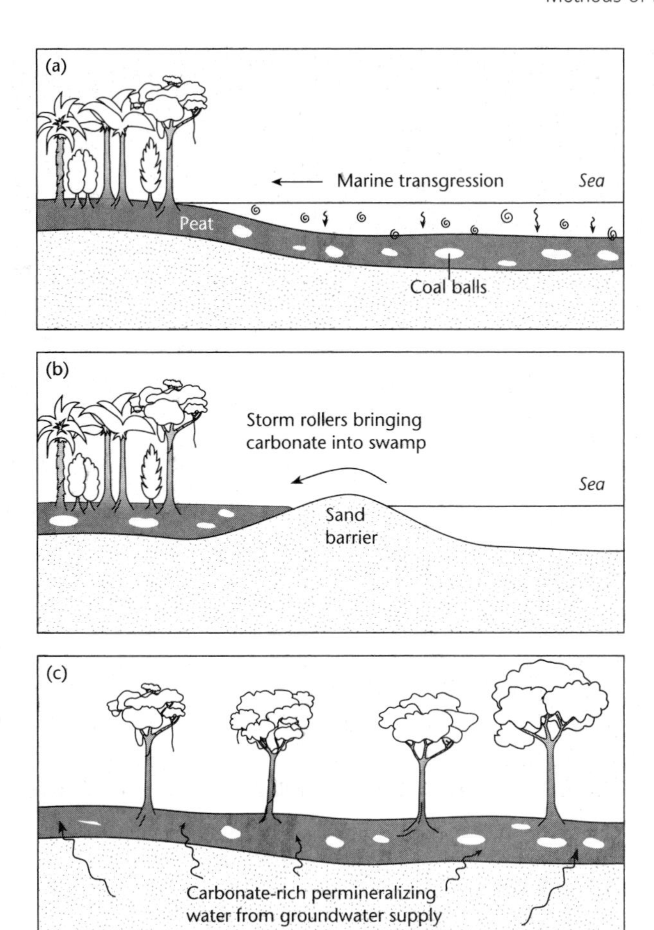

Figure 1.7 Three main theories accounting for the formation of coal balls; (a) marine-influenced coal balls; (b) mixed coal-ball formation; (c) non-marine coal-ball formation (after Scott and Rex, 1985).

Unaltered plant remains

The final plant 'fossilization' process to be discussed in this section is one that results in relatively unaltered plant material being preserved for millions of years. The most important factor in the preservation of unaltered plant material appears to be burial in an environment where microbial activity is inhibited, or at least severely restricted (Eglinton and Logan, 1991). Such situations include those where the organic material is desiccated or frozen, where there are environmental extremes of salinity, temperature, or pH, or in the absence of oxygen (anoxic environment). Three of the most common deposits from which unaltered plant remains have been recovered are lake sediments, amber, and packrat middens. Such an exceptional state of preservation exists that often, when first exposed, the leaves are still green and flowers identifiable (Figure 1.8). It is from this category of plant 'fossil' that DNA (Golenberg *et al.*, 1990;

Figure 1.8 Fossil flower from the Eocene-aged Green River formation (~50 Ma) (photograph J. McElwain of FM specimen P28978).

Golenberg, 1999) and other ancient biomolecules (Briggs and Eglinton, 1994) have been extracted.

Many exceptional plant fossils have been recovered from lake sediments, where burial has been rapid and the environment at the bottom of the lake anoxic. An example of this type of burial includes the Clarkia beds in Idaho, where thousands of early Miocene fossil leaves (~17–20 Ma) have been recovered. Some of these leaves are in such an exceptional state of preservation that it is possible to extract strands of ancient DNA (Golenberg *et al.*, 1990; Soltis *et al.*, 1992). In the Clarkia beds it has been proposed that, in addition to rapid burial, the leaves went through natural dehydration before abscission, resulting in a release of high concentrations of plant phenols. They were then rehydrated in tannin-rich waters (plants with normally high tannin contents are often found in the best state of preservation) when deposited in the lake (Niklas *et al.*, 1985), thus leading to exceptional preservation. Other ancient biomolecules to be recovered from the Clarkia beds include lipids, flavonoids, carotenoids, and chlorophyll derivatives (Rieseberg and Soltis, 1987).

The outer walls of pollen and spores, the exines, are also exceptionally well preserved in lake sediment, although, in this case, it is due both to the anoxic conditions in the lake and also the remarkable resilience of pollen (Coxon and Clayton, 1999). Exines are composed of a material called sporopollenin that has exceptionally strong chemical bonds. The strength of this material is demonstrated by the fact that extraction of pollen and spore exines from a sedimentary matrix uses hydrofluoric acid to dissolve the rock, yet the cell walls of the pollen and spores remain in a relatively unaltered state. One main advantage of using the pollen and spore record is that it provides a broader picture of the regional vegetation. Their inclusion in the fossil record is not reliant upon the plant growing in a very specific environment, such as a lowland swamp. This has a tendency to bias other forms of fossil preservation such as leaves and wood. It is not surprising, therefore, that some of the earliest plant fossil records are from pollen and spores, and notable features in the plant fossil record, such as the earliest gymnosperms, angiosperm origins, and the rise of the grasses, are first detected in the pollen record (Traverse, 1988).

Amber is another important preservation medium for 'unaltered' plant remains. Amber is a fossilized plant resin, most often formed by members of the conifer family Araucariaceae, and the angiosperm family Leguminosae (Poinar, 1994). It is made up of the sugars glucose, galactose, and arabinose, as well as various acids, alcohols, and esters. When the resin is formed it is very sticky, and any animal or plant fragments (e.g. small flowers, leaves, seeds, and insects) that fall into it are incorporated. Once covered by the resin, the sugars withdraw the moisture from the original tissue and initiate the process of inert dehydration (Lindahl, 1993; Poinar, 1994). This process is rapid and effectively the plant fragments become encased in a dehydrated state. Oligocene fossil leaves contained in amber ($\sim$24–34 Ma) have been recovered from a number of localities, and ancient DNA has been extracted successfully from the extinct plant *Hymenaea protera* contained in fossil amber from the La Toca mines, Dominican Republic (Poinar, 1994).

The most unusual preservation medium for unaltered fossil plants, however, must be in the packrat, porcupine and hyrax middens found in the arid areas of south-western North America, the Middle East and South Africa (Betancourt *et al.*, 1990). These middens, which provide shelter/nests for the animals, are constructed of plant material that they collected and which became encased in crystallized urine. The animals appear to sample species richness adequately within their foraging area. The crystallized urine (amberat), combined with the extremely dry environment, provides an exceptional preservation medium, and although the plant material within the midden represents a time-slice of probably no more than 5 years, middens dating back to beyond 40 000 years have been discovered. Extraction of plant material from the middens is simple since the amberat dissolves in water, releasing beautifully preserved plant fragments with ease. Similar to the pollen record, middens have provided invaluable information on vegetation from regions where conditions are such that other forms of preservation are unlikely (even for pollen-collecting sources such as lakes). The added advantage of midden deposits are that plant fragments contained in them are usually large enough to be identified down to a generic level. Again, the main use of the information has been to describe past plant distributions and climate change. However, the molecular potential of these deposits is also being recognized and preliminary work to extract DNA and other ancient biomolecules from these fossils has demonstrated encouraging results (Rogers and Bendich, 1985; Lowenstein and Scheuenstuhl, 1991).

1.4 Dating methods

In order to provide a firm chronological framework in which to place plant fossils, a number of dating methods can be employed. The method chosen will be determined by both the approximate age range of the material to be dated and the sedimentary matrix in which they are contained. Although radiocarbon dating is probably the most commonly cited dating technique, for the geological record it is of limited use, since it is only appropriate where the sample, or the matrix that it is contained within, is still in an organic state (i.e. not permineralized). The sample must also be less than approximately 40 000 years old (Aitkin, 1990). Methods more commonly used in the dating of the geological record include radiometric dating, magnetostratigraphy, thermoluminescence dating, and biostratigraphy.

Radiometric dating

Radiometric dating uses the spontaneous decay rate of long-lived, naturally occurring radioactive isotopes (Dalrymple, 1991) contained in various minerals that make up a rock (Aitken, 1998). Certain radioactive isotopes start to decay to a stable form as soon as the mineral is formed (for example, ^{40}K decays to ^{40} Ar) regardless of changes in the physical or chemical environment. If the decay rate is known (expressed as the half-life), which in the case of ^{40}K to ^{40}Ar is 1.25 billion years, then the ratio of $^{40}K/^{40}Ar$ in the sample will give an indication of the time that has elapsed since the whole rock or mineral sample was formed. Other isotopes used in this type of age measurement include, for example, the decay of ^{238}U to ^{206}Pb and ^{40}Ar to ^{39}Ar (Table 1.1). Different isotopes have different rates of decay, thus affecting their potential as a dating tool. Uranium

Table 1.1 Decay rates of various naturally occurring radioactive isotopes used to determine the ages of rocks and minerals (after Dalrymple, 1991)

Parent isotope (radioactive)	Daughter isotope (stable)	Half-life (Ma)	Decay constant (year)
^{40}K	^{40}Ar	1250	5.81×10^{-11}
^{87}Rb	^{87}Sr	48 800	1.42×10^{-11}
^{147}Sm	^{143}Nd	106 000	6.54×10^{-12}
^{176}Lu	^{176}Hf	35 900	1.93×10^{-11}
^{187}Re	^{187}Os	43 000	1.612×10^{-11}
^{232}Th	^{208}Pb	14 000	4.948×10^{-11}
^{235}U	^{207}Pb	704	9.8485×10^{-10}
^{238}U	^{206}Pb	4470	1.55125×10^{-10}

series dating, for example, has an age range from the present to approximately 10 million to 4.6 billion years ago (Dalrymple, 1991).

Although it is difficult to date minerals younger than 100 000 years old using radiometric dating (Williams *et al.*, 1993), the main disadvantage in using this technique lies in the fact that these minerals are most common in igneous and metamorphic rocks, whereas plant fossils are most common in sedimentary rocks. However, the date of sedimentary rocks can be established when they are overlain or intruded by igneous rock, and there are many examples where fossil assemblages have been dated this way.

Geomagnetic polarity timescale

The chronological position of the sample according to its palaeomagnetic signature is often used in conjunction with radiometric dating. Throughout geological time there have been variations in the Earth's geomagnetic field, including polarity reversals where the magnetic North Pole has become the South Pole and vice versa (Jacobs, 1994). Reversal records are preserved in both terrestrial and oceanic sediments, and the dating of these sedimentary sequences using the radiometric methods described above (especially K/Ar dating) has enabled the development of a detailed geomagnetic polarity timescale, extending back to at least 300 Ma (Figure 1.9) (Cande and Kent, 1992; Dunlop and Özdemir, 1997). Magnetostratigraphy has been employed successfully as a dating tool for a number of sedimentary sequences containing fossil assemblages (Jacobs, 1994), although this method still relies upon other dating techniques to calibrate the sequence.

Luminescence dating

An alternative method for dating fossils contained within sedimentary sequences is luminescence dating. This has an age range of between 1000 and 8 million years. Luminescence dating measures the amount of radioactive 'damage' to mineral grains such as quartz and feldspar in a sedimentary matrix. This damage results from bombardment with alpha, beta, and gamma radiation from radioactive isotopes such as uranium, thorium, and potassium that are also contained within the sediment. The process starts as soon as the sediment is buried and the result is damage to the crystal lattice of the mineral grains and displacement of electrons, which are subsequently trapped in another part of the lattice (Aitken, 1985; Stokes, 1999). Thus due to the cumulative effect, the number of electrons trapped represents the length of time since the sediment was buried. Exposure to light, or heating to greater than 500 °C, vibrates the crystal lattice and releases the trapped electrons, resetting the 'clock' to zero (Lowe and Walker, 1997). It is therefore possible, either through the use of heat or light, to stimulate a time-dependent signal from mineral grains. To stimulate a signal from heat (known as thermoluminescence dating, TL) the sample must not be

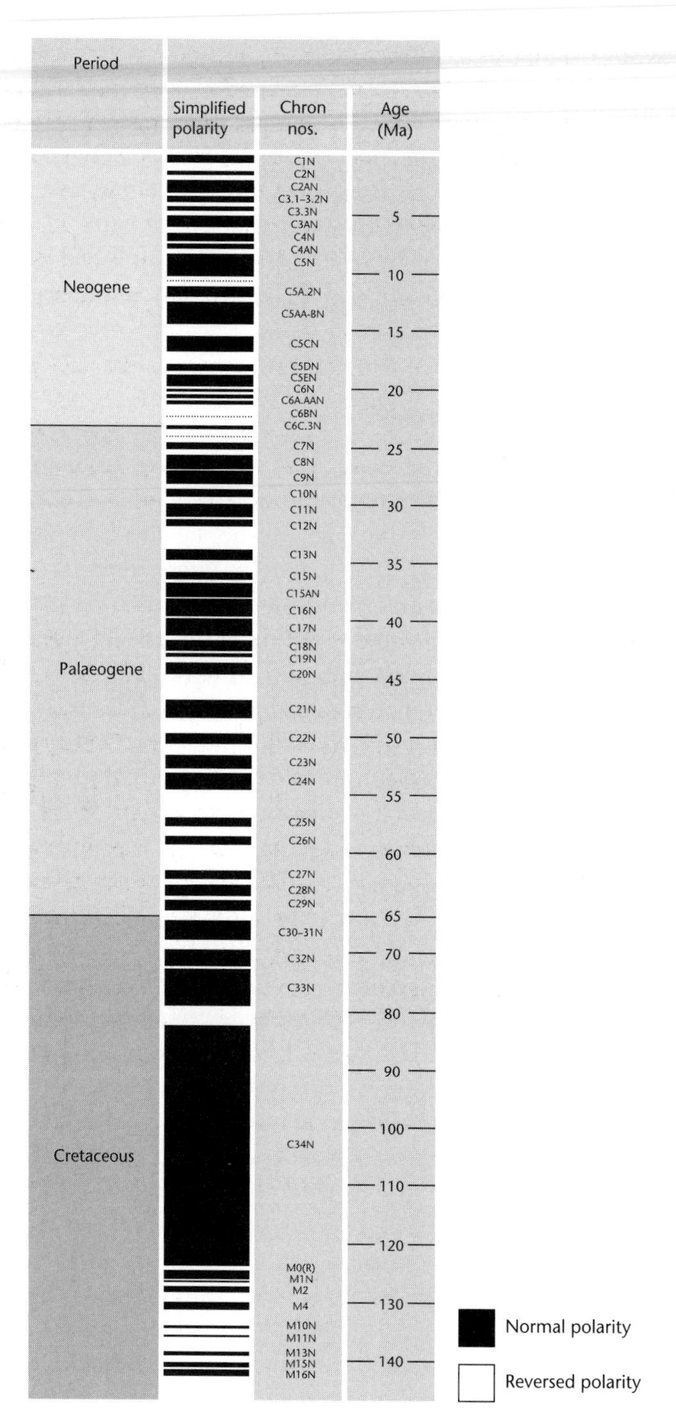

Figure 1.9 Palaeomagnetic timescale (after Harland *et al.*, 1990; Cande and Kent, 1992).

exposed to light for more than a few seconds before being placed in an opaque container. In the laboratory, the sample is heated to over 500 °C and the TL measured by photoelectronically counting the photons emitted (Aitken, 1998). To stimulate a signal from light (known as optically stimulated luminescence dating, OSL), the minerals are exposed to differing wavelengths of light, in particular green and infrared light. The eviction of charge from traps is observed as a rapidly depleting signal when expressed as a function of light exposure (Stokes, 2000). Both methods enable a measurement of the radiation damage, known as the palaeodose.

Luminescence dating has proved to be very successful in the dating of mineral grains from terrestrial sediments such as buried dunes, loess, and other sedimentary sequences. It has not been so successful in the dating of waterlain sediments because of variable sedimentation rates, attenuation of ultraviolet rays, and periods of flooding and drying out (Williams *et al.*, 1998).

Biological correlation

Biological correlation, or biostratigraphy, is a method of relative (rather than absolute) dating and involves the use of presence/absence of index fossils within a sequence to provide an approximation of its age. These index fossils should be easily identifiable, have existed through a relatively short period of geologic time (less than 1 million years), be abundant, widely distributed geologically, and have lived in a range of different sedimentary environments. If the date of the index fossil is known in one region, then the assumption is made that its occurrence in another locality indicates a deposit of similar age. Obviously there is great potential for circularity of argument in this method, because it will never be known if the index fossil did exist in different periods, if it is always assumed to be of one age. However, this method was extensively used before many of the above dating techniques were fully developed, and subsequent dating has indicated that many of the early correlations developed using this method were correct (Harland *et al.*, 1990).

2 Earliest forms of plant life

The Earth was formed approximately 4600 million years ago and within 1800 million years cellular life had evolved. Palaeoenvironmental reconstructions suggest that global temperature and the earliest composition of the atmosphere, ocean, and land would have provided a challenging combination of environmental extremes for the onset of biological evolution. However, evolutionary change was such that by 540 million years ago (Cambrian) an array of multicellular organisms, both plant and animal, had evolved. This chapter outlines the patterns of change in these early environments (3500–540 million years ago) and the processes leading to the development of the first forms of multicellular plant life.

2.1 The earliest environments

Geological evidence indicates that the continental crust started to form by ~4200 Ma and that by 1900 Ma a large, single, lens-shaped body had amassed (Goodwin, 1991; Condie, 1997). This supercontinent has been named 'Rodinia' and is thought to have been located around the equatorial belt, in roughly the same position as present-day Africa (Briggs, 1995). Initial splitting up of Rodinia occurred from approximately 1000 Ma (Figure 2.1), resulting in three major parts, Laurasia (consisting of North America, Greenland, Baltica and Siberia), East Gondwana (southern Africa, India, most of Australia, and East Antarctica) and West Gondwana (principally parts of South America and West Africa) (Rogers, 1996; Mallard and Rogers, 1997) (Figure 2.2).

The early continental crust would have been thin and extremely hot due to heat flow from the mantle. An estimate of Earth's heat production in the early Archaean (~4000–3500 Ma), suggests that it was two or three times greater than present (Goodwin, 1991). However, there is also metamorphic evidence to indicate that by 3000 Ma, the heat had subsided sufficiently to allow the development of a continental crust, up to 40 km in thickness. This crust was composed of igneous rock, weathered sediments (including volcanic sands), and meteoritic compounds resulting from intense cometary bombardment (Anders, 1989).

Although the continental crust was hot, it has been calculated that endogenic heat supplied only an extremely small fraction ($<0.001\%$) to the Earth's

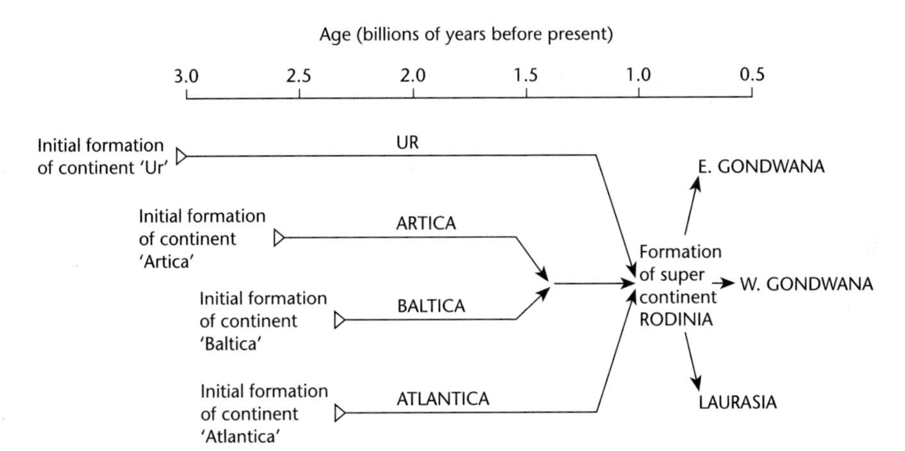

Figure 2.1 Diagrammatic representation of major continental development and movement in the Earth's early history (redrawn from Rogers, 1996).

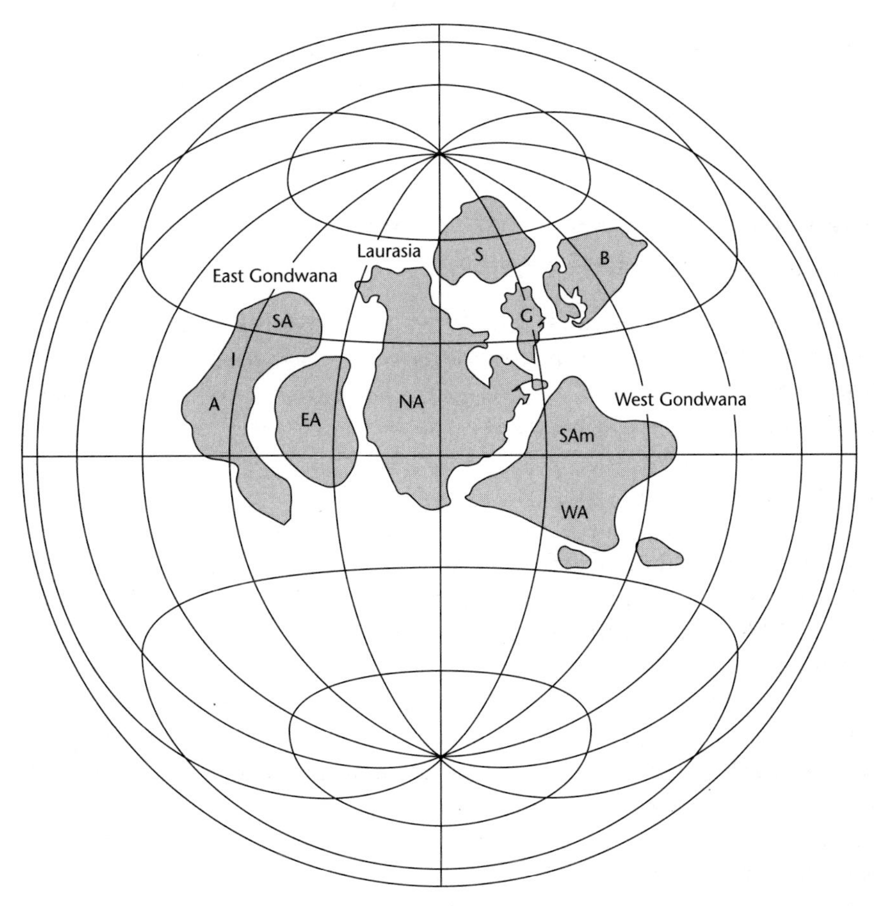

Figure 2.2 Splitting up of supercontinent Rodinia from approximately 1000 million years ago resulting in Laurasia (NA, North America; G, Greenland; B, Baltica; S, Siberia), East Gondwana (SA, South Africa; I, India; A, most of Australia; EA, East Antarctica), and West Gondwana (SAm: parts of South America; WA, West Africa) (redrawn from Rogers, 1996; Mallard and Rogers, 1997).

energy budget, and that this was not an important factor in controlling surface air temperatures (Worsley *et al.*, 1986). Rather, surface air temperatures would have been strongly influenced by the high levels of atmospheric methane (CH_4) and carbon dioxide (CO_2) from volcanic degassing. Therefore, even though the sun was considerably less luminous than today (solar luminosity has increased by approximately 25% since the origin of the solar system (Condie and Sloan, 1998), high levels of atmospheric CO_2 and CH_4 would have blocked outgoing long-wave radiation, promoting greenhouse warming (Kasting *et al.*, 1983; Kasting, 1988). Climatic modelling suggests that CO_2 greenhouse warming could have been responsible for global temperatures between 30 and 50 °C in these early environments (3500–3200 million years ago) (Lowe, 1994).

In contrast to the high levels of carbon dioxide, early atmospheric oxygen (O_2) was extremely low. Some estimates suggest that before 2200 Ma (early Proterozoic) the O_2 content of the air accounted for 1% of total atmospheric composition, and was probably a major limiting factor for organic evolution (Holland, 1994). One of the most restrictive factors associated with extremely low levels of oxygen would have been the lack of an ozone layer, and therefore no protection for terrestrial organisms from incoming solar radiation. Between 2200 and 1900 Ma (middle Proterozoic), however, various sedimentary deposits that require free oxygen for their formation become apparent in the geological record. These include red-beds (sandstones and shales with red iron oxide) (Eriksson and Cheney, 1992) and highly oxidized palaeosols (Holland and Beukes, 1990), and are taken to be indicative of increasing amounts of atmospheric O_2.

Other gases contributing to the overall composition of the early atmosphere included hydrogen, hydrogen cyanide, formaldehyde, and hydrogen sulphide. These would have been present in trace amounts, resulting from volcanic out-gassing, photochemical reactions in the atmosphere, and cometary infall (Deamer, 1993; Chang, 1994; Hayes, 1994).

It is suggested that oceans formed early in Earth history (between approximately 4400 and 3900 million years ago) from the condensation of atmospheric water vapour (Chang, 1994; Condie and Sloan, 1998). Composition and temperature of the early oceans would have been strongly influenced by mantle outgassing (Holland, 1984), with some estimates suggesting that they were warm (between 80 and 100 °C) and contained high concentrations of ferrous irons, dissolved CO_2 and bicarbonate, with perhaps a pH as low as 6.0 (Lowe, 1994).

2.2 Accumulation of organic material and formation of the first cell

A precondition to any form of biological evolution requires the accumulation of organic material and formation of the cell. There are at least two contrasting theories as to how organic matter accumulated in the prebiotic environment: terrestrial accretion and cometary infall.

The oldest and most widely cited theory is that of terrestrial accretion. This theory, first proposed by Miller in 1953, is based on the principle that, under laboratory conditions, when gases such as hydrogen, carbon dioxide, methane, and ammonia are heated with water, and energized by electrical discharge for at least 24 hours, approximately half the carbon originally present in the methane gas will be converted into amino acids, sugars, and other organic molecules, including purines and pyrimidines (required to make nucleotides). Amino acids and nucleotides can associate to form polymers, including polypeptides (proteins) and polynucleotides (RNA and DNA) (Miller, 1992) thus forming the first building blocks of the cell.

An alternative theory is that the first organic material on Earth came from other planets and entered the Earth's atmosphere via meteoritic input. This idea was originally viewed as better suited to the realms of science fiction. However, there is increasing evidence to support the theory that meteoritic and cometary infall provided the first organic material on Earth (e.g. McKay *et al.*, 1996). Analysis of the present-day meteorites on the ice beds of the Antarctic (carbonaceous chrondrites), for example, indicate that they typically contain between 1 and 4% carbon, mainly as graphite but also as much as 1% organic molecules (Dalrymple, 1991). Organic compounds contained within the carbonaceous chrondrites include hydrocarbons, amino acids, carbon, hydrogen cyanide and amphiphilic molecules (Figure 2.3).

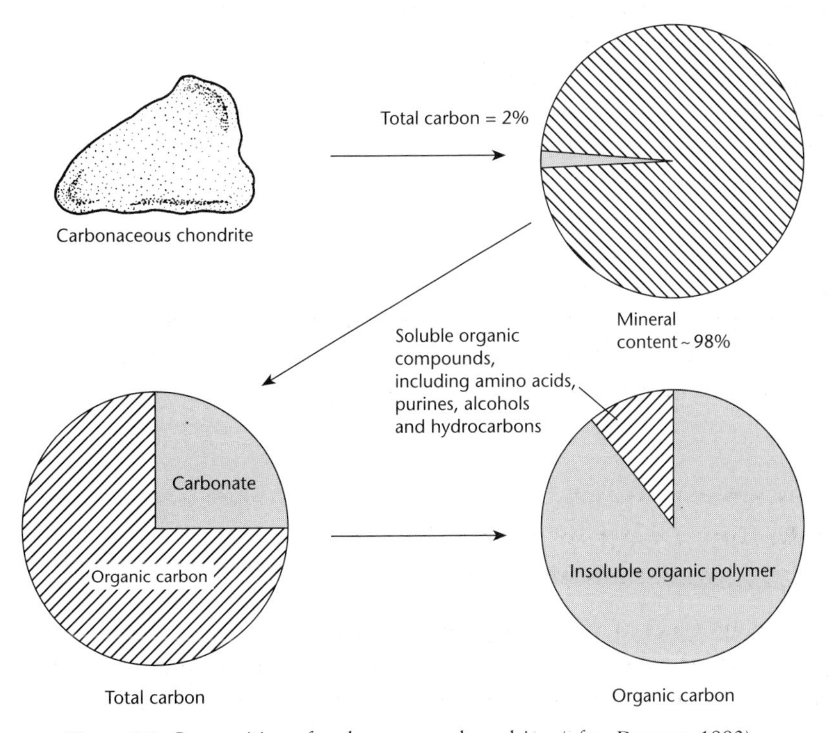

Figure 2.3 Composition of carbonaceous chrondrites (after Deamer, 1993).

In the literature, there is considerable debate as to which process is more probable. Critics of the terrestrial accretion mechanism suggest that although such processes may occur under laboratory conditions, there is no conceivable environmental equivalent of a closed glass flask in which such reactions could occur. On the other hand, critics of the meteoritic theory argue that the extremely high atmospheric pressures (as a consequence of the high levels of atmospheric carbon dioxide) would have resulted in organic materials in these rocks burning up before reaching the surface of the globe. Recent calculations, however, indicate that although this might be the case for large meteorites, very small micrometeorites, in the size range 10^{-6} to 10^{-9} g, lose relatively little of their mass during atmospheric entry and could have withstood the predicted early atmospheric pressures (Deamer, 1993). Estimates suggest that in the early environment (~4700–3500 Ma), when there would have been a heavy meteoritic bombardment, organics accumulated at a rate of 10^6–10^7 kg/year (Chyba *et al.*, 1990), resulting in 'a layer of organic material several centimetres deep spread over the surface of the earth' (Deamer, 1993, p. 15).

Another crucial event in the formation of the first cells would have been development of the outer cell membrane, since without this there would be no means of simple containment of the macromolecules within the cell. There is also the question of how organic compounds became sufficiently concentrated to form more complex molecules. An extremely important environment for both processes to occur would have been shallow tidal pools, where cycles of solar heating and cooling, combined with tidal wetting and drying, would have provided the necessary conditions for concentration and molecular self-assembly (Figure 2.4).

Containment of the macromolecules probably resulted from the presence of amphiphilic molecules. These molecules have one part that is hydrophobic (i.e. water insoluble) and the other part hydrophilic (i.e. water soluble). In water, amphiphilic molecules tend to aggregate to form bilayers (with the water-soluble parts aligning), creating small, closed vesicles, the aqueous contents of which are isolated from the external medium. On drying, these bilayers fuse into multilayers and any other molecules present become trapped between the layers and incorporated into the vesicles on rehydration (Deamer *et al.*, 1994). It is envisaged therefore that wetting and drying cycles in the prebiotic tidal pools may well have provided an ideal environment for formation of an early protocell (Figure 2.4).

A more recent, and just as controversial, third theory is closely associated with the terrestrial accretion theory, in that it suggests that the evolution of the first cell took place on another planet, possibly Mars. These cells were then carried to Earth via meteoritic bombardment (Sleep and Zahnle, 1998). Again it is acknowledged that most organic material would have perished in space, but it is also noted that 'it only takes one cell to infect a planet' (Nisbet and Sleep, 2001).

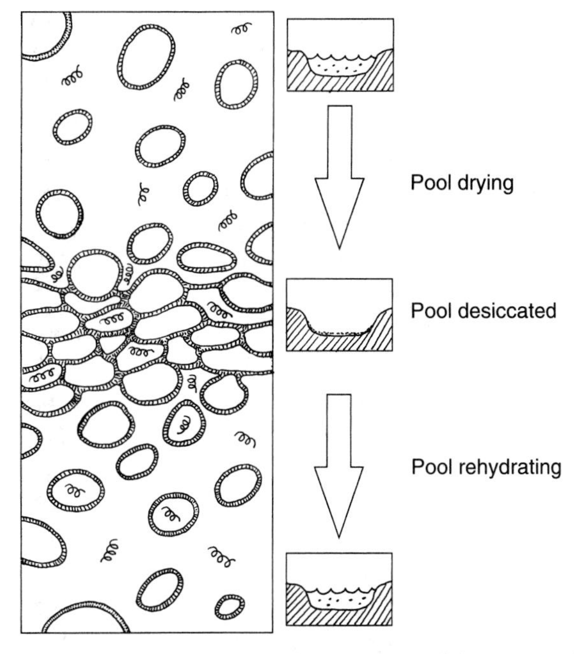

Figure 2.4 Formation of outer cell membrane and containment of the macromolecules within the cell. It is suggested that an extremely important environment for development of the outer cell membrane and the concentration of organic compounds to form more complex molecules would have been shallow tidal pools. Containment of the macromolecules probably resulted from the presence of amphiphilic molecules. These molecules have one part that is hydrophobic (i.e. water insoluble) and the other part hydrophilic (i.e. water soluble). In water, amphiphilic molecules tend to aggregate to form bilayers (with the water-soluble parts aligning) creating small, closed vesicles, the aqueous contents of which are isolated from the external medium. On drying, these bilayers fuse into multilayers and any other molecules present become trapped between the layers and incorporated into the vesicles on rehydration (Deamer *et al.*, 1994). It is envisaged therefore that cycles of solar heating and cooling, combined with tidal wetting and drying, would have provided the necessary conditions for concentration and molecular self-assembly, leading to the formation of the first cell (after Deamer, 1993).

2.3 The first prokaryotes: the geological evidence

Of all the organisms presently on the Earth, prokaryotes are the simplest in structure, smallest physically, and most abundant in terms of number of individuals (Raven *et al.*, 1992). They range between approximately 1 and 10 μm in size, are single celled (although many types have cells joined together within a mucilaginous sheath), lack an organized nucleus surrounded by a nuclear envelope, and reproduce by binary fission (i.e. each cell increases in size and divides into two). Prokaryotes also tend to have one of three basic cell shapes (Figure 2.5): a straight rod shape (bacilli); a spherical shape (cocci); or a long, coiled shape (spirilli).

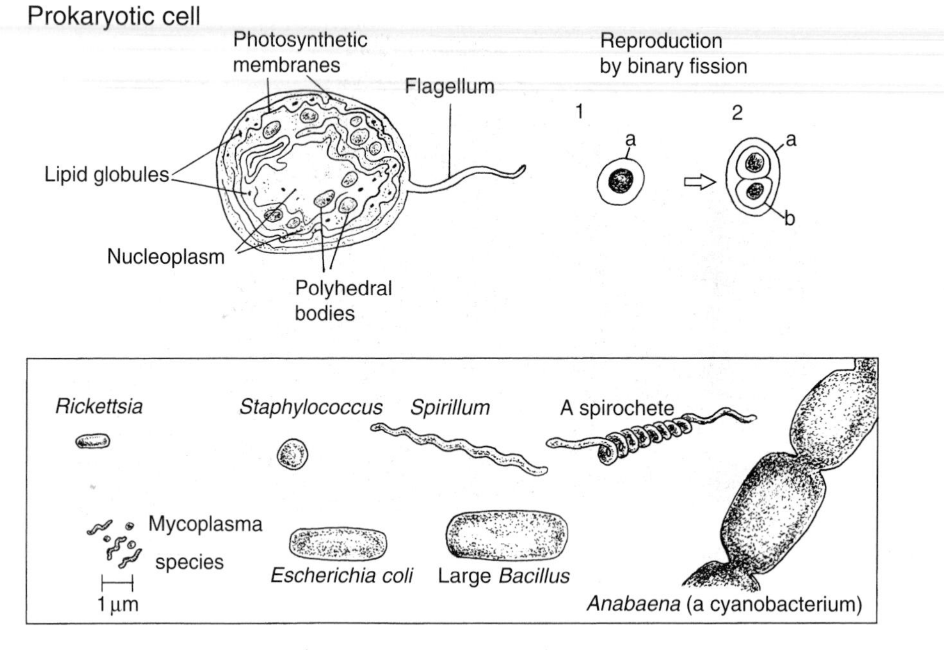

Figure 2.5 The most common cell shapes of prokaryotes.

Evidence for some or all of these characteristic features (shape, size, arrangement of organelles, and method of reproduction) is apparent in the earliest fossil records, suggesting that by 3500 million years ago, a complex system of prokaryotes had developed on planet Earth (Knoll, 1992; Nisbet and Sleep, 2001).

Evidence from stromatolites

Stromatolites, some dating as far back as 3500 Ma (early Archaean), are found in localities ranging from Spitsbergen and central East Greenland to South Africa, Australia, and parts of Antarctica (Schopf and Packer, 1987; Walsh, 1992; Ramussen, 2000). They are composed of numerous alternating light and dark layers of sediment (Figure 2.6) and are thought to represent the trace fossils of ancient microbial mat communities (algal/bacterial mats) (Golubic, 1976; Walter, 1994). One of the most widely cited examples to explain stromatolite formation are the stromatolite beds presently accreting in Shark Bay on the west coast of Australia. Here, laminated cushion-like structures can be seen in the intertidal zone (Figure 2.7), composed of alternating layers of calcium carbonate and filamentous and coccoid micro-organisms (algal mats). These layers are formed when algal mats spread over the ocean substrate and trap and bind a layer of sediment particles. A layer of calcium carbonate forms on top due to microbially mediated precipitation, and the light-requiring micro-organisms respond by growing upwards. A subsequent layer of sediment is trapped by the

Figure 2.6 Composite fossil stromatolite beds (made up of very many columns, each approximately 5 cm in diameter), Victoria Island, arctic Canada (Neoproterozoic Shaler Supergroup) (photograph N. Butterfield).

Figure 2.7 Stromatolite beds growing in Shark Bay on the west coast of Australia. The columns are approximately 20–30 cm diameter (photograph N. Butterfield).

newly formed algal mat, and thus through time, a layered structure will develop (Figure 2.8). This is, however, just one example, and research indicates that stromatolite organisms are extremely diverse and colonize both hard and loose substrates, preparing a broad spectrum of benthic environments for stromatolite construction (Golubic, 1976). Such structures appear to form today only in hypersaline environments which exclude the grazing invertebrates that now

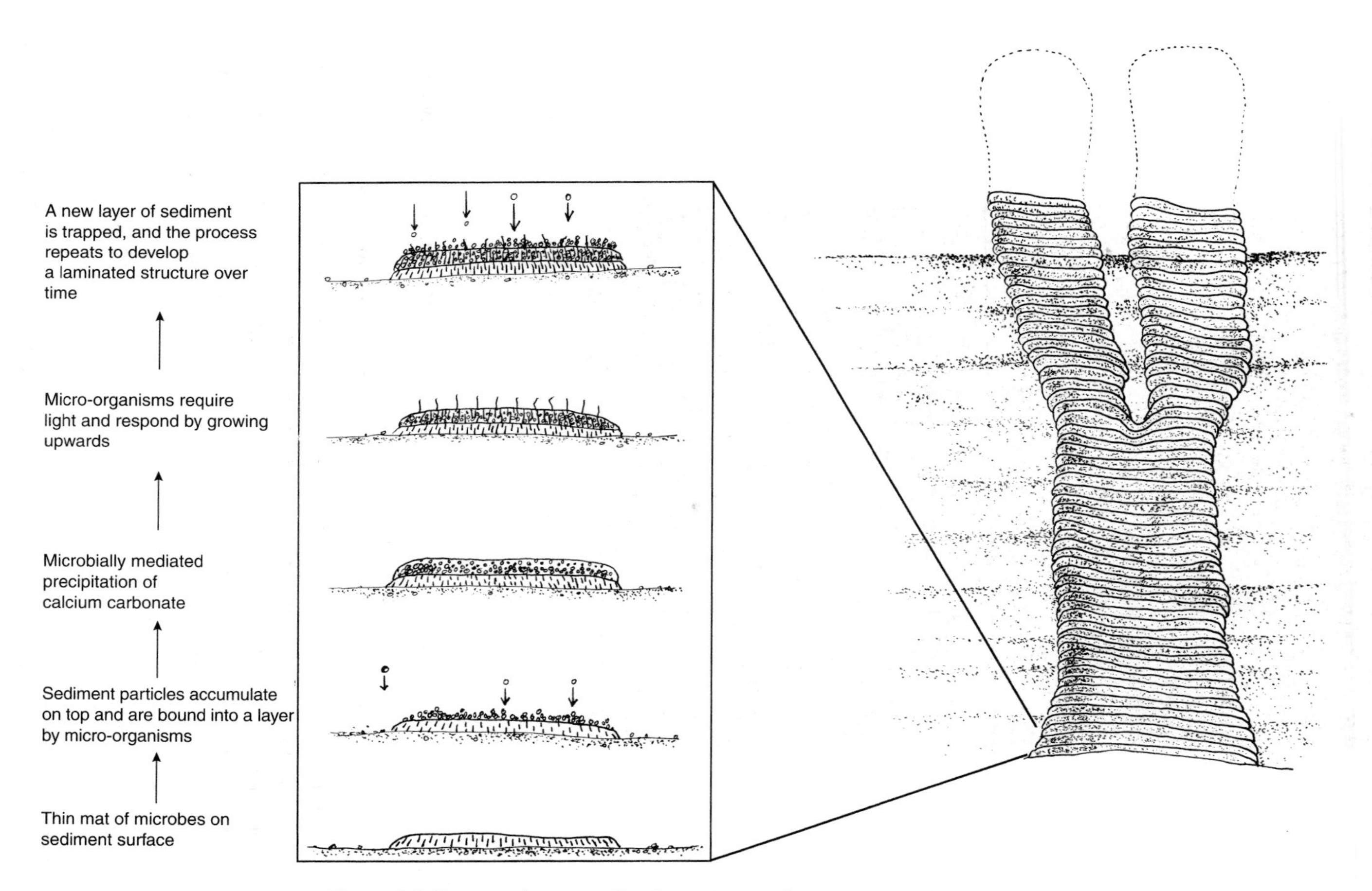

Figure 2.8 Process of stromatolite formation on the west coast of Australia.

Figure 2.9 Fossil Conophyton beds in the Mesoproterozoic Bangemall Group, Western Australia; scale bar in cm (photograph N. Butterfield).

inhibit their formation in other marine settings. Their Precambrian equivalents had no such grazers to contend with.

The distribution and composition of stromatolite communities are strongly determined by ecological factors, including temperature, pH, salinity and water potential, light intensity, and the ability for anaerobic growth. Subtle differences in these environmental factors determine the microbial communities present and this, in turn, affects the shape of the stromatolite. For example, many cone-shaped stromatolites (Conophyton) have been found in fossil deposits dating as far back as 3500 Ma (Figure 2.9). Broadly similar structures have been found forming in the hot springs of the Yellowstone National Park in Wyoming, USA (Walter, 1977), where the thermal waters have temperatures of between 32 and 59 °C and pH values of 7–9. The predominant micro-organisms responsible for the formation of these 'conophytons' in the Yellowstone springs have been identified as filamentous cyanophytes (blue-green algae), but other photosynthetic filamentous bacteria are also present (e.g. *Chloroflexus*).

The presence and shape of the fossil stromatolite beds dating back to the Proterozoic have therefore been used to infer a number of features. These include the early presence of prokaryotes, the probable types of microbial communities represented, evidence of early prokaryotic metabolism (especially photosynthesis), and the type of environment in which they were formed (Dyer and Obar, 1994). Until recently, information from the fossil stromatolite beds was always inferential since they rarely contained direct evidence of the micro-organisms responsible for their accretion. However, ancient biomolecules (lipids; see Chapter 9) have recently been extracted from 2700 million-year-old rocks that are characteristic of cyanobacteria, thus confirming the early presence of organisms capable of oxygenic photosynthesis (Summons *et al.*, 1999). Most geologists would still agree, however, that they are a better indication of the long-term evolution of the Earth's environment rather than the microbial communities associated with them (Grotzinger and Knoll, 1999).

Evidence from cherts and shales

Fossil micro-organisms have also been extracted from other early sedimentary deposits, including cherts and shales. Some of the earliest examples are found in the Warrwoona Group in Western Australia, where filamentous and colonial fossil structures have been reported from carbonaceous cherts dated between 3500 and 3300 Ma (Schopf and Packer, 1987; Schopf, 1992, 1993). These fossils, which are predominantly small spheres and filaments, indicate close morphological comparison with extant prokaryotes, based on both cell size and also structure (i.e. whether single cells or in colonies, whether the colonies are sheath enclosed, etc.). The micro-organisms contained within the Warrwoona Group sediments have therefore been 'classified' as photoautotrophic cyanobateria (oxygen-producing blue-green algae) and taken to indicate the presence of organisms capable of oxygen-producing photosynthesis as far back as 3500 million years ago (early Archaean). Cell size and structure have not been the only factors used to identify these early micro-organisms as prokaryotes; evidence of their reproductive mode has also been discovered. There are now many examples in the early fossil record of cells that have been preserved while in a state of cell enlargement and division, suggesting binary fission (e.g. Knoll and Barghoorn, 1977; Knoll, 1985).

Younger Proterozoic deposits (between 2500 and 540 Ma) indicate an increasing morphological diversity of prokaryotic life. It is estimated that by the Cambrian (590 Ma) there were at least 122 taxa present in approximately 40 different biotas (Knoll, 1992). One of the best examples of increased morphological diversity is from the Gunflint Formation in southern Ontario, Canada. Here a number of micro-organisms have been identified in thin sections from black carbonaceous chert dating to 1900 Ma (Barghoorn and Tyler, 1965). From these sediments at least 12 species have been recognized, including a number that are comparable to extant prokaryotes. These include: a filamentous form named *Animikiea*, that in structure and cell organization is nearly morphologically identical to the extant filamentous cyanobacterium *Lyngbya* or *Oscillatoria*; a type named *Gunflintia*, the general morphological make-up (cell contents and structure) of which is similar to that of extant species of the chemosynthetic, iron-forming bacterium, *Crenothrix* (Stewart and Rothwell, 1993); and a multi-filament micro-organism named *Eoastrion*, which is thought to be related to an extant, metal-oxidizing, budding bacterium.

Thus many different types of micro-organisms, morphologically similar to extant prokaryotes, are recognizable extremely early on in the fossil record. Characteristic features of these early prokaryotes probably included an ability to ferment, to withstand hot temperatures (thermophily), and to respire without the use of oxygen (anaerobic respiration) (Dyer and Obar, 1994). All would have been extremely important requisites in the earliest environments. From a plant evolutionary viewpoint, however, the most interesting of these numerous examples of early micro-organisms must be those resembling photosynthetic bacteria and, in particular, the cyanobacteria (blue-green algae). Structurally extant photosynthetic cyanobacteria possess chlorophyll *a*, together with carotenoids and phycobilins, layers of photosynthetic thylakoids, and ribosomes (Raven *et al.*, 1992). It is therefore probable that the early cyanobacterial ancestors had a similar suite of organelles and were capable of photosynthesis. There are various hypotheses as to the evolution of the photosynthetic pathway (for reviews see Nisbet *et al.*, 1995; Xiong *et al.*, 2000). One of the most plausible is that the evolution of photosynthesis paralleled the evolution of the structural mats (stromatolites) (Nisbet and Fowler, 1999; Nisbet and Sleep, 2001). Thus, as organisms grew in the more productive but more dangerous uppermost layers of the microbial mats, new forms evolved that had the biochemical pathways to utilize the energy of the sunlight and the electrons from the water to convert atmospheric CO_2 into organic compounds (i.e. photosynthesis). The splitting of the water molecule in this reaction would have resulted in the release of O_2 into the atmosphere—a process that was essential for the evolution of aerobic life on Earth.

2.4 Evolution of the eukaryotes

Eukaryotes differ from prokaryotes in that they have a membrane-bound nucleus in which the DNA is contained, organelles including mitochondria, integrated multicellularity, and sexual reproduction (sometimes) (Figure 2.10). They constitute the three major groups of multicellular organisms (plants, animals, and fungi), along with many groups of the Protista, including species of red, green, and brown algae (Lipps, 1993a). Because of the diversity and importance of eukaryotic organisms to life on Earth, it is often stated that the evolution of eukaryotes was one of the most important events in the history of life (Knoll, 1992).

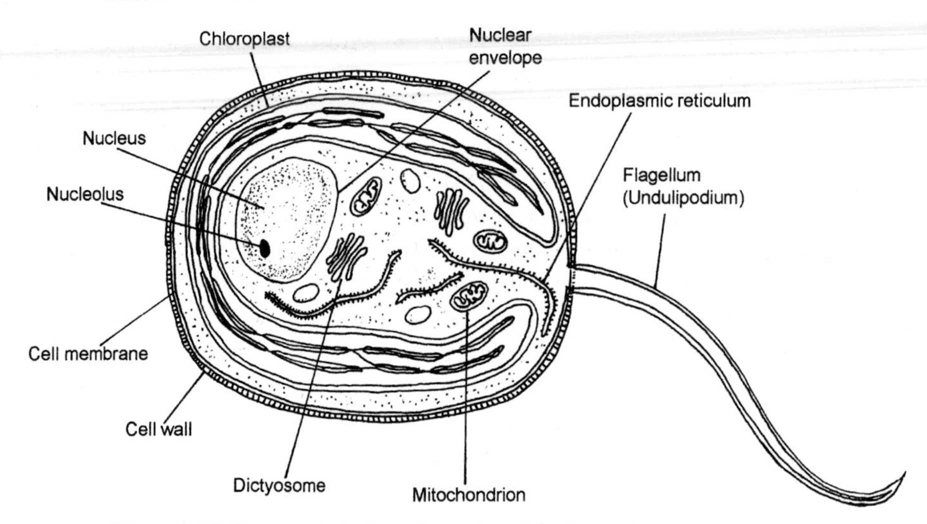

Figure 2.10 Cross-section of a eukaryotic cell (redrawn from Lipps, 1993b).

Geological evidence

Evidence for eukaryotes (from fossil organisms and ancient biomolecules) suggests that they were probably present on Earth from as early as 2700 Ma (Brocks *et al.*, 1999; Summons *et al.*, 1999). Three records of early eukaryotes of particular interest to the evolution of plants include the oldest recorded alga *Grypania*, dated to 2100 Ma (Han and Runnegar, 1992), fossil bangiacean red algae, dated to approximately 1200 Ma (Butterfield *et al.*, 1990), and cladophoralean green algae, dated to between 800 and 700 Ma (Butterfield *et al.*, 1994).

Grypania

The oldest recorded alga is *Grypania*, which was discovered in the banded iron formations of northern Michigan and dated to approximately 2100 Ma (Han and Runnegar, 1992). This fossil, which is a coiled cylindrical organism, 0.5 m in length and 2 mm in diameter, has a number of morphological characteristics, including its large size, morphological complexity, and structural rigidity, to suggest that it represents a giant unicellular alga. A suggested modern-day analogue is *Acetabularia*, which is a dasycladalean (green algae) and a photosynthetic autotroph (Runnegar, 1994).

Bangiophyte red algae

Some of the earliest evidence for eukaryotic organisms bearing a close resemblance to extant species of red algae comes from the silicified carbonate rocks of the Huntington Formation in arctic Canada (Butterfield *et al.*, 1990; Butterfield, 2000). These rocks, dated to approximately 1200 Ma, contain fossils that are extremely close in morphological detail to the extant red algal genus *Bangia*. Because of the exceptional preservation of the fossils in this formation (they are permineralized), it has been possible to cut thin sections both transverse and perpendicular to the bedding plane. Perpendicular cross-sections have revealed unbranched, uniserate filaments, 15–45 mm in diameter and up to 2 cm long (Figure 2.11). The filaments appear to consist of stacked disc-shaped cells enclosed in a relatively transparent enveloping sheath and, in some sheaths, constrictions are apparent (Figure 2.12) (Butterfield *et al.*, 1990; Butterfield, 2000). Examination of these filaments in transverse section has revealed that they are composed of up to eight wedge-shaped cells arranged axially around a central core (Figure 2.13a). With this level of morphological detail, the case for arguing that these fossils are closely related to the extant *Bangia* is extremely convincing. Extant *Bangia* have a filamentous form, a similar size, and are composed of stacked cells within an almost transparent sheath (Figure 2.13b). Restrictions are often apparent in the sheath, and in transverse cross-section of the *Bangia* filament there are eight wedge-shaped cells.

Cladophoralean green algae

Early multicellular green algae have been found in 800–700 million-year-old shale deposits from the Svanbergfjellet Formation on Spitsbergen (Butterfield *et al.*, 1994). In these deposits there is strong morphological evidence to suggest

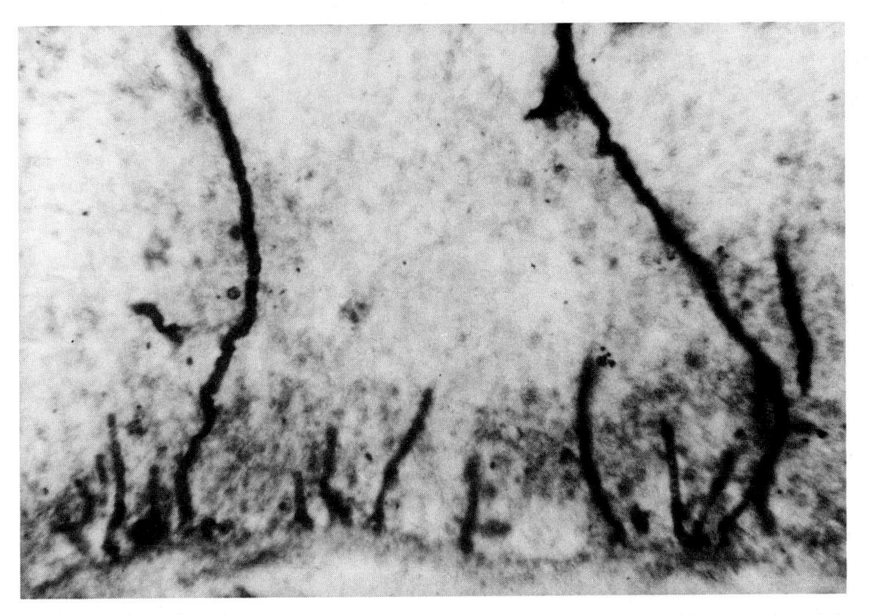

Figure 2.11 Perpendicular cross-section of unbranched, uniserate filaments of bangiacean red algae from the Hunting Formation in arctic Canada (~2100 Ma) (photograph N. Butterfield, 1990). The longest filament is approximately 75 μm in length.

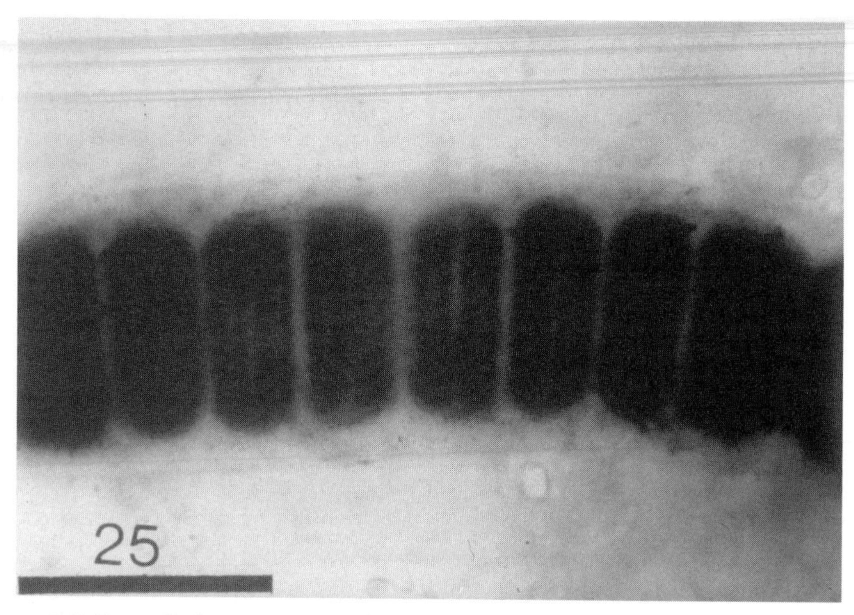

Figure 2.12 Perpendicular cross-section of these fossil filaments has revealed stacked disc-shaped cells in a relatively transparent enveloping sheath (scale bar, 25 μm), indicating close morphological similarity to *Bangiomorpha pubescens* (photograph N. Butterfield).

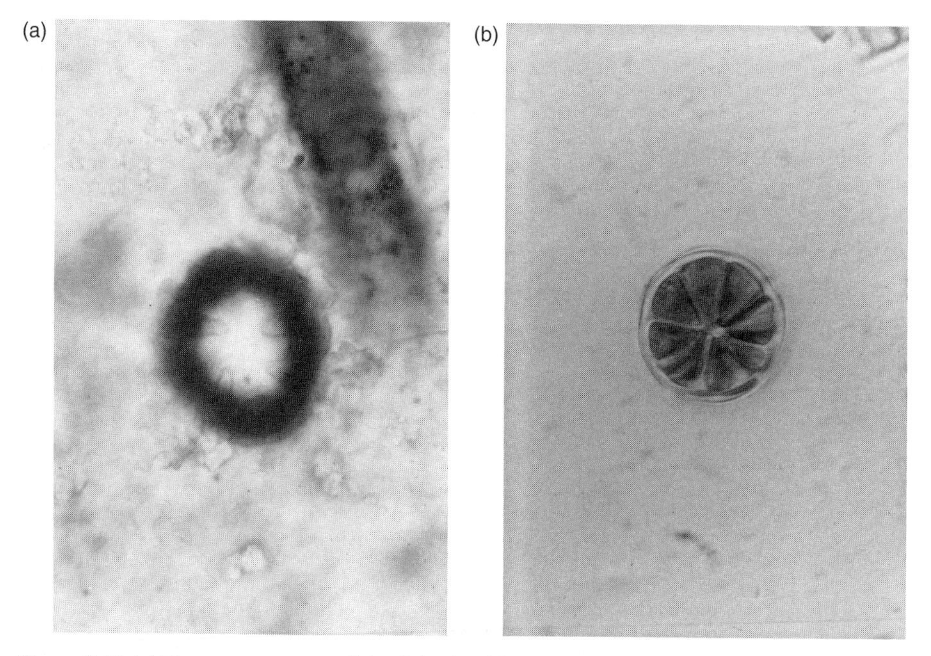

Figure 2.13 (a) Transverse cross-section of the fossil bangiophyte showing wedge-shaped cells arranged axially around a central core (photograph N. Butterfield). (b) Transverse cross-section of an extant species of *Bangia*, showing eight wedge-shaped cells (photograph N. Butterfield).

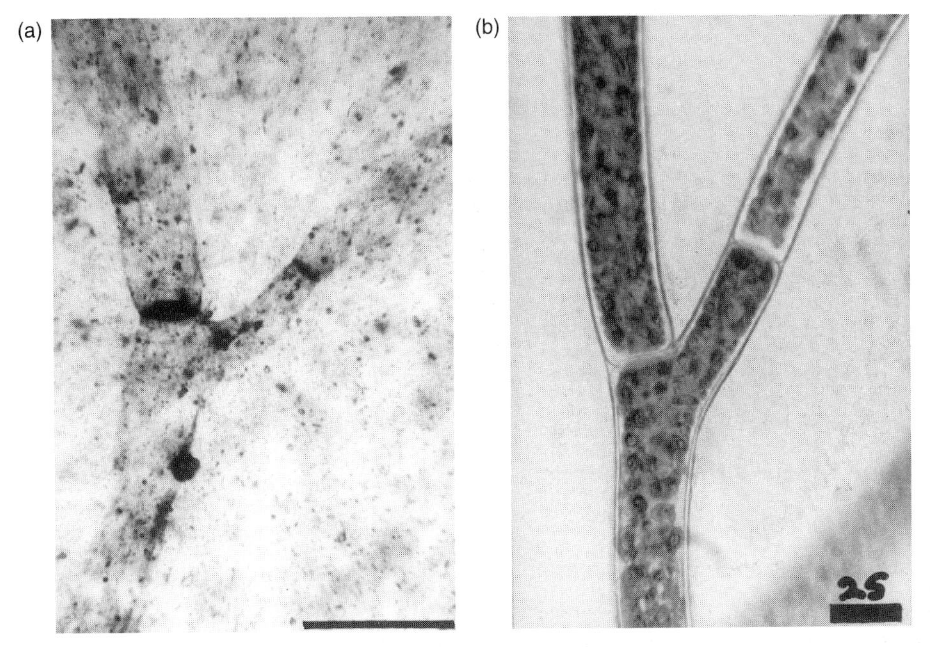

Figure 2.14 (a) Fossil green alga from Spitzbergen shales, Svanbergfjellet Formation. (b) Extant cladophoralean green alga, *Cladophora* sp. (photographs N. Butterfield).

the early presence of green algae comparable to the extant Cladophorales. The morphological evidence to link them to this group includes extensive branching filamentous thalli (up to 1 cm in height) composed of large cylindrical cells (50–800 mm in diameter), that are attached to adjacent cells by thickened septal plates (Figure 2.14a) (Butterfield *et al.*, 1988). When compared to extant species of cladophoralean green algae (e.g. *Cladophora*, Figure 2.14b) there is a remarkable similarity.

It is interesting to note that present-day classifications divide the green algae (over 7000 species) into five classes (Charophyceae, Chlorophyceae, Pleurastrophyceae, Prasinophyceae, and Ulvophyceae). Each is thought to represent a separate evolutionary line (Raven *et al.*, 1992). The Ulvophyceae, which include the cladophoralean green algae, are predominantly marine (and include the green seaweeds). They contain a number of species that form 'algal mats' in tropical reef areas.

Geological evidence has therefore provided many pieces of the jigsaw to understanding the earliest evolution of life on Earth (Figure 2.15). It gives a first estimation as to what was around and when. However, further insights into the evolutionary relationship between the early prokaryotes and the eukaryotes have been gained not by looking at the fossil record, but by examining extant forms at both a structural and a molecular level.

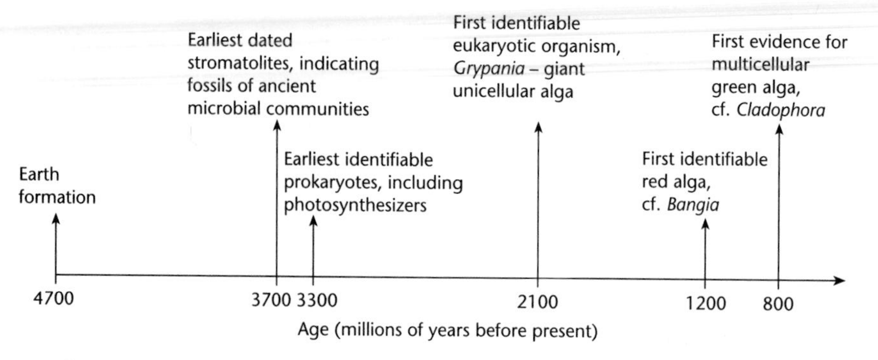

Figure 2.15 Major events recorded in the geological record of early life on Earth.

Comparison with extant forms

Structural comparisons

From a structural viewpoint, extant forms of the fossil prokaryotes, resembling the cyanobacteria (blue-green algae), are some of the most interesting. Certain species of extant photosynthetic cyanobacteria bear a remarkable similarity, both in terms of intracellular structure and molecular make-up, to chloroplasts (Cattolico, 1986). These similarities have led to the generally accepted theory that the early cyanobacteria may have been instrumental in evolution of the first eukaryotic cells. It is suggested that the cyanobacterium became incorporated into a pre-existing eukaryotic cell as a result of endosymbiosis. By transferral of genetic material to the nucleus, chloroplasts became an integral part of the eukaryotic cell (Lipps, 1993) (Figure 2.16).

Although purple non-sulphur bacteria have not been identified in the earliest fossil record, it is highly probable that they were also present from approximately the same time as the cyanobacteria. In extant purple non-sulphur bacteria, the morphological, biochemical, and metabolic features indicate a strong resemblance to mitochondria. Endosymbiosis of a purple non-sulphur bacterium is therefore suggested as another probable process that led to the formation of an integrated and functioning eukaryotic cell (Deamer, 1993; Deamer *et al.*, 1994; Dyer and Obar, 1994) (Figure 2.16).

Molecular comparison

As species diverge from a common ancestor, they accumulate genetic mutations (usually substitutions in the genetic code), resulting in species becoming more genetically different over time (Lewin, 1997). Thus by comparing the genetic difference between two groups, it is possible not only to relate them to a common ancestor but also to calculate the amount of time that has elapsed since they diverged from that ancestor. Using evidence attained from the geological record as to which groups to examine (i.e. those first present), a number of molecular studies have been carried out in recent years to determine both the probable

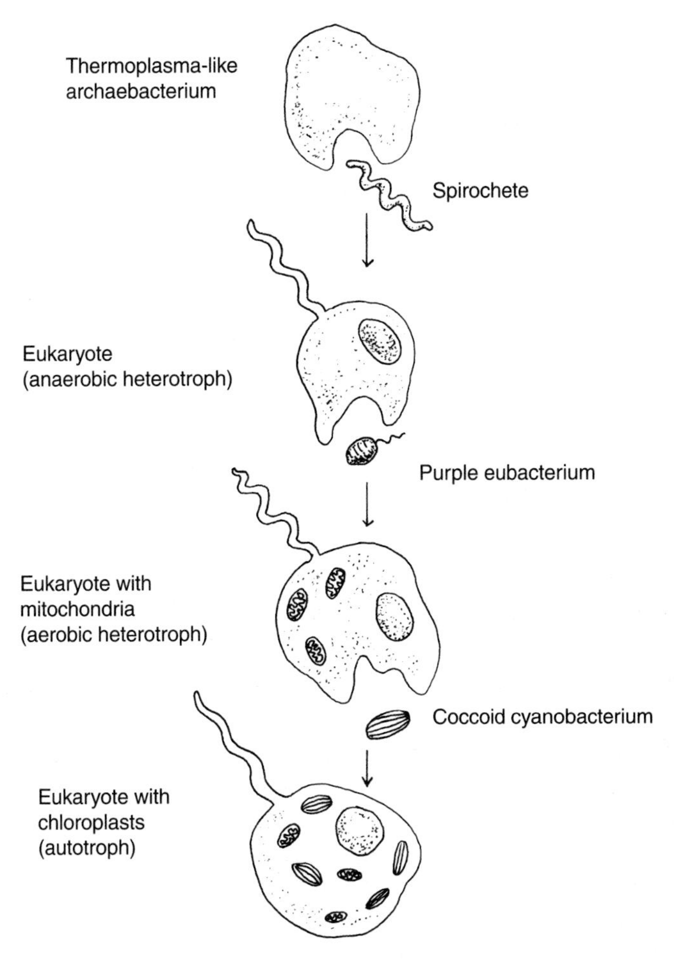

Figure 2.16 Proposed endosymbiosis of cyanobacteria and purple non-sulphur bacteria, leading to the acquisition of chloroplasts and mitochondria in the eukaryotic cell (after Lipps, 1993b).

evolutionary pathway to, and time of divergence of, the earliest eukaryotes (e.g. Woese *et al.*, 1990; Schlegel, 1994; Sidow and Thomas, 1994; Woese, 1994). By comparing base sequences of ribosomal RNA (rRNA) among a large selection of bacteria (more than 1000 species have been analysed), as well as a number of protein sequences, a number of interesting molecular relationships have emerged. These have upset the traditional two-kingdom classification of eukaryotes and prokaryotes. The main conclusions drawn from these molecular studies to date are as follows (Dyer and Obar, 1994) (Figure 2.17).

Prokaryotes have a deep phylogenetic split between archaebacteria and eubacteria Molecular evidence has established that there are two distinct groups of bacteria that are as genetically distinct from each other as each is from the eukaryotes (Woese *et al.*, 1990). These two groups have been named the archaebacteria (ancient bacteria) and eubacteria (true bacteria). The

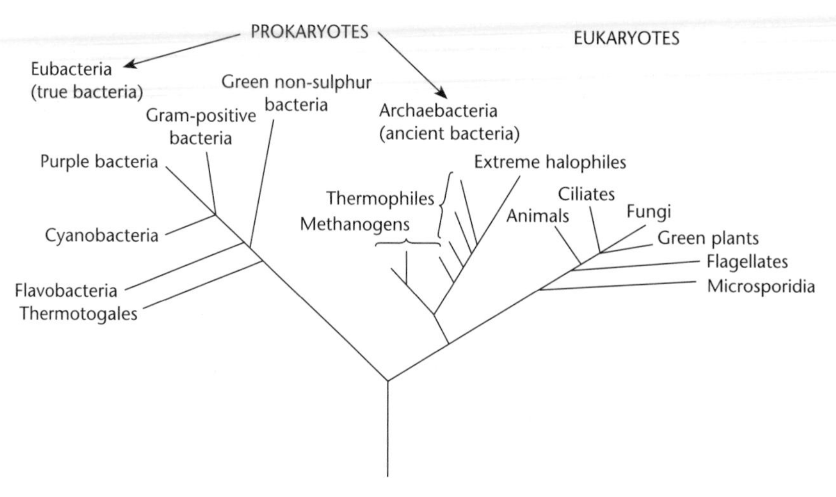

Figure 2.17 Molecular phylogenetic split of the domains: the primary kingdoms (after Woese and Fox, 1977; Woese *et al.*, 1990; Dyer and Obar, 1994).

archaebacteria include the extreme halophiles (salt-tolerant bacteria), the methanogens, and the extreme thermophilic (heat tolerant) sulphur-metabolizing bacteria (eocytes). The eubacteria include cyanobacteria and purple sulphur bacteria (Figure 2.17). To date there is direct evidence in the geological record for numerous cyanobacteria (eubacteria) but not, as yet, any archaebacteria. However, carbon isotope signatures from ancient sedimentary carbonates indicate that the onset of methane cycling occurred from at least as early as approximately 2800 million years ago (Hayes, 1994), thus providing indirect evidence for the early presence of at least the methanogens.

The molecular result that prokaryotes have a deep phylogenetic split between archaebacteria and eubacteria thus has significant implications in understanding eukaryotic evolution and, in particular, suggests that the traditionally cited evolutionary pathway of the eukaryotes evolving from the prokaryotes may be too simplistic.

Eukaryotes bear a specific phylogenetic relationship to archaebacteria Molecular evidence also indicates that the eukaryotes are phylogenetically closer to the archaebacteria than the eubacteria (Figure 2.17). That is, the archaebacteria either separated early from the eubacterial lineage and/or the eukaryotic lineage branched from the archaebacteria after they separated from the eubacteria (Lipps, 1993b). Further molecular phylogenetic work on the archaebacteria has also revealed that it is probably the eocytes (that is the extreme thermophilic, mostly sulphur-metabolizing bacteria) that are the closest bacterial relatives to eukaryotes (Rivera and Lake, 1992) (Figure 2.18). Thus, rather than the eukaryotes sharing a common ancestor with all three main groups in the archaebacteria (the halophiles, the methanogens, and the eocytes), the eukaryotes only share a most recent and common ancestor with the eocytes.

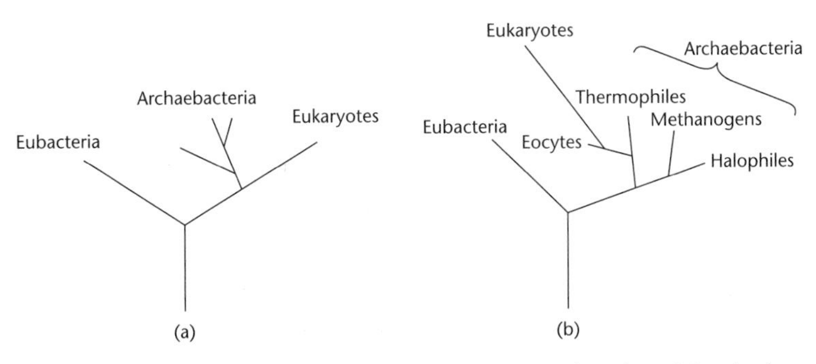

Figure 2.18 Two theories based on molecular phylogenies to explain the origin of eukaryotes: (a) molecular phylogeny proposed by Woese *et al.* (1990) (see also Figure 2.17), and (b) the new phylogeny indicating that the eukaryotes share a most recent common ancestor with only the eocytes (Rivera and Lake, 1992).

Eukaryotes are an ancient group almost as old as the prokaryotes Using a molecular clock based on DNA sequences, dating of the split between the three major lineages (eukaryotes, eubacteria, and archaebacteria) suggests that diversification occurred at around 3500 million years ago (or earlier) (Knoll, 1992; Dyer and Obar, 1994). The earliest eukaryotes evolved at, or very shortly following, the split of the archaebacteria and eubacteria. However, these 'early' eukaryotes, including those that are in the lowermost branches of the phylogenetic tree and therefore most similar genetically to the archaebacteria, are simple microorganisms that have no mitochondria and are confined to oxygen-free environments. The middle branches of the tree (i.e. the next most similar), which have a calculated molecular age of between 2800 and 2400 million years ago, include protists that contain mitochondria but no chloroplasts. It is not until a calculated molecular age of 1000 million years ago that major radiation of multicellular organisms, especially photosynthesizers, is believed to have occurred.

The pattern of this preliminary phylogeny raises some interesting questions as to the rate and process of evolution of life on Earth. In particular, the fact that the major clades (animals, fungi, plants) appear to branch near a common point implies a rapid burst of evolution. This, along with other lines of evidence, has led to the suggestion that the major epochs in eukaryotic evolution are in some way causally linked to significant geological events in the early period of Earth's history (e.g. Schopf, 1994).

2.5 Possible triggering mechanisms of eukaryotic evolution

A number of researchers have suggested that the combined geological/molecular record demonstrates a general trend of episodic increases in biological diversity (Knoll, 1992) through the Archaean and Proterozoic (~3500–540 Ma) (Knoll and Lipps, 1993; Dyer and Obar, 1994; Knoll, 1994a; Riding, 1994; Nisbet and Sleep, 2001). Six major biological events, which are specific to the plant fossil record are recognized, including:

(i) the origin of life on Earth and diversification of anaerobic archaebacteria and eubacteria at or around 3500 Ma;
(ii) origin of photosynthetic organisms around 3300 Ma;
(iii) the appearance of organisms capable of aerobic metabolism and cellular acquisition of the organelles mitochondria and, later, chloroplasts, between 2800 and 2400 Ma;
(iv) the appearance of complex eukaryotic organisms in the geological record from approximately 2100 Ma;
(v) the large increase in diversity of organisms and radiation of acritarchs—a group of organic-walled microfossils, the majority of which are thought to represent reproductive cysts of green algae or alage cysts (Mendelson, 1993), from approximately 1000 Ma (Knoll, 1994b);
(vi) the large increase in diversity of planktonic algae from approximately 540 Ma (Knoll, 1992, 1994b).

The question therefore arises as to whether it is possible to relate these periods of evolutionary change to evidence in the geological record for environmental change. A number of tentative links have been suggested for events (1) and (2), including decreasing meteoritic bombardment between approximately 4000 and 3800 Ma and rapid accretion of continental crust at around 2700 Ma. However, more detailed consideration of the environmental conditions associated with tectonic activity and also climatic conditions associated with periods of glaciations (Knoll, 1994a) have been given for the changes apparent from 1000 Ma (in particular rising levels of atmospheric oxygen resulting from increased burial of organic carbon).

Rising levels of atmospheric oxygen

It has long been suggested that a rise in atmospheric oxygen could have been one of the triggering mechanisms responsible for the appearance of complex eukaryotes (Knoll 1994a,b, 1996).

In order to examine a possible causal link, it is necessary to have a means of measuring levels of atmospheric oxygen other than by increases in the number of aerobic organisms, which involves a circularity of argument. Two geochemical methods that provide independent records are the measurement of the isotopic fractionation of carbon and sulphur in ancient sedimentary rocks.

Isotopic analysis of carbonate carbon in sedimentary rocks can give an indication of times of burial of organic carbon. When organic carbon is buried, it effectively becomes shielded from oxidation, causing the oxygen that was hitherto bound to it as CO_2 to become 'free' to enrich the atmosphere (Knoll,

1996). Results suggest that episodic increases in the burial of organic carbon occurred between ~2200 and 1800 Ma and 1100 and 700 Ma (Des Marais *et al.*, 1992) (Figure 2.19). In contrast, measurement of the isotopic composition of biogenic sedimentary sulphides can be used to detect the process of sulphide production and whether it occurred in conditions with or without the presence of oxygen. Comparison of deposits dated between 1050 and 640 Ma with younger deposits (i.e. < 540 million years ago) indicate a change in the process of sulphide production, from one driven within anaerobic conditions to one within aerobic conditions (Canfield and Teske, 1996). Both lines of evidence therefore point to increasing availability of atmospheric oxygen between ~2200 and 540 Ma, and the former to episodic bursts of increase.

When the proposed times of increase in atmospheric oxygen (measured through carbon burial rates) are compared to geological evidence for periods of global rifting and orogeny (Figure 2.19), it becomes apparent that there might be some causal link between the two. Thus increased atmospheric oxygen may be related to periods of tectonic activity. Such a process is therefore proposed for the earliest environments, whereby during periods of major global tectonic

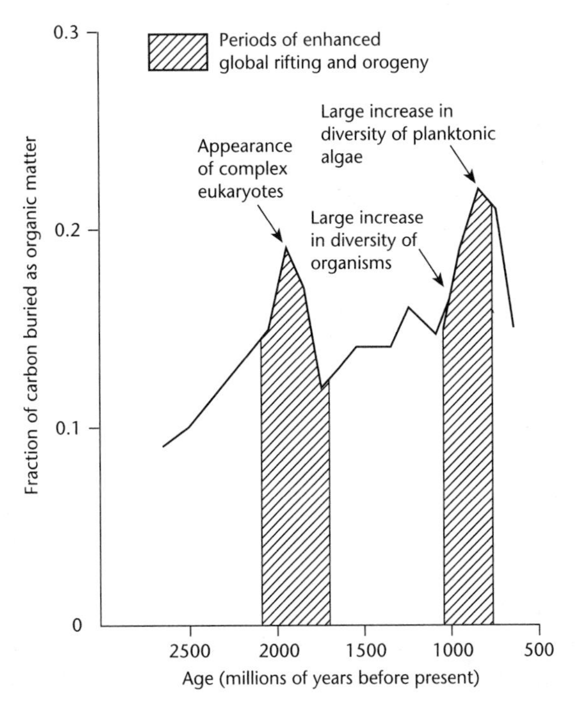

Figure 2.19 Organic carbon burial between 3000 and 500 million years ago measured through the isotopic analysis of carbonate carbon in sedimentary rocks (redrawn from Des Marais *et al.*, 1992). When organic carbon is buried, it effectively becomes shielded from oxidation, causing the oxygen that was previously bound to it as CO_2 to become 'free' to enrich the atmosphere. This has been used as a proxy record of levels of atmospheric oxygen between ~2600 and 600 Ma. Also indicated are periods of enhanced global rifting and orogeny and the timing of major events in eukaryotic evolution.

activity, atmospheric oxygen increased, and this was in some way causally linked to eukaryotic evolution (Knoll, 1994a).

Changing environmental conditions associated with glaciations

Geological evidence suggests that as many as four major glacial periods occurred in the late Proterozoic (~1000–540 Ma), the final one, the Varanger ice age, taking place between approximately 610 and 590 Ma (Harland *et al.*, 1990; Briggs, 1995). During the Varanger ice age, more acritarch extinctions occurred than originations, resulting in a decline in overall species number (75% reduction). Following the ice age, there was a large increase in species number and many new and highly ornamented acritarchs appeared. A tentative link has therefore been suggested between the acritarchs and environmental/oceanographic changes (e.g. changing patterns of marine sedimentation) that accompanied the end of the Varanger glaciation (Knoll and Lipps, 1993; Knoll, 1994a,b).

Although just two examples are provided above, there are many other suggestions in the literature of bursts of evolutionary activity occurring at or around the same time as significant geological events in this early period of Earth history (e.g. Knoll, 1992; Knoll and Lipps, 1993; Schopf, 1994). However, it is also highly probable that in these early environments there would have been spiralling ecological relationships (Knoll, 1991, 1994). Without a doubt, these would also have had a major impact on species diversity and number, resulting in the development of a complex marine ecosystem of plants and animals by the beginning of the Cambrian period, 590 Ma.

Summary

1. Continental crust started to form on Earth from approximately 4200 Ma, and by 1900 million years ago a large landmass called Rodinia had developed.

2. The earliest terrestrial environments were harsh, with global temperatures estimated at between 30 and 50 °C and levels of atmospheric oxygen at only 1%, compared with the present-day value of 21%. As a result, there would have been no ozone layer and high levels of atmospheric CO_2 and CH_4 would have created a strong greenhouse effect.

3. Oceans formed early in Earth history (between 4400 and 3900 Ma) by condensation of atmospheric water vapour. Estimates suggest that the earliest oceans were warm (between 80 and 100 °C) and fairly acidic.

4. Prokaryotic organisms are present in the fossil record from as early as 3500 Ma and eukaryotic organisms from approximately 2700 million years ago.

5. Three geological records of early eukaryotes of particular interest to the evolution of plants are *Grypania*, fossil bangiacean red algae, and cladophoralean green algae.

6. Molecular and structural examination of extant forms of the earliest types of prokaryotes and eukaryotes apparent in the fossil record indicate a possible evolutionary pathway between the two.

7. Eukaryotic cells (containing a membrane-bound nucleus and organelles, including mitochondria and chloroplasts) were probably created by endosymbiosis of certain prokaryotes. Blue-green algae (photosynthetic cyanobacteria) and purple non-sulphur bacteria bear close structural and molecular similarities to chloroplasts and mitochondria, respectively.

8. Evidence from both the geological and molecular records suggest that both prokaryotic and eukaryotic evolution occurred in relatively rapid bursts. Tentative links have been implied between major epochs in early eukaryotic evolution and times of major climatic and environmental changes, such as during times of increasing atmospheric O_2 and during glaciations.

3 The colonization of land

The colonization of the land by plants was one of the most significant evolutionary events in the history of the planet. Before colonization could occur, however, major changes were necessary, both to the environment and to early plants, in order to enable growth outside of an aquatic environment. By the early Silurian (~430 Ma) plants that were permanently adapted to a terrestrial or water-deficient habitat (land plants) had evolved. This chapter examines the environmental changes leading up to and during terrestrialization, the fossil evidence for changes in plant morphology that enabled land dwelling, and the possible evolutionary pathways from the earliest aquatic plants to land plants. It concludes with a consideration of the global biogeographical distribution of the earliest land plants and the factors influencing their distribution.

3.1 Environmental changes during the Cambrian and Ordovician (543–443 Ma)

Land surfaces were available for colonization by plants from an early point in the history of multicellular life (see Section 2.1). However, other essential environmental prerequisites were still developing. These included the formation of sizeable and stable near-shore environments, the development of soils, and the amelioration of atmospheric and climatic conditions suitable for terrestrial plant survival.

Formation of new sizeable and stable near-shore environments

The Cambrian and Ordovician were periods of relatively intense tectonic activity, resulting in reorganization of the continental plates. The supercontinent Rodinia fragmented and there was a collision of East Gondwana with West Gondwana. This was followed by the rotation and collision of West Gondwana relative to Laurasia (Rogers, 1996; Condie and Sloan, 1998), resulting in the assembly of the supercontinent, Pangea, by approximately 300 million years ago (Figure 3.1).

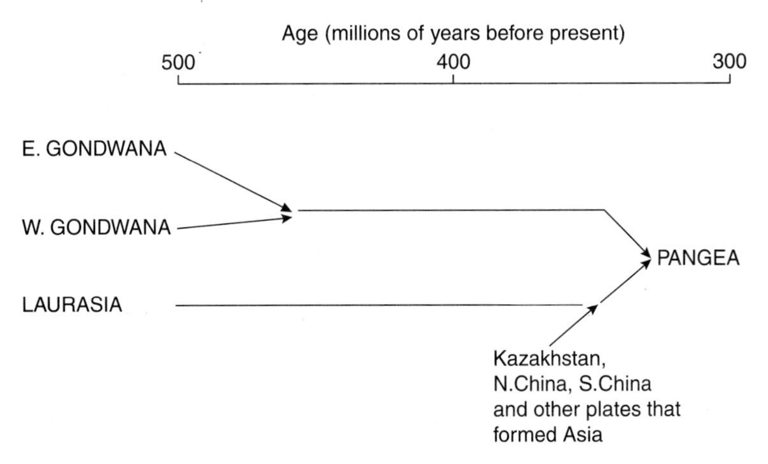

Figure 3.1 Diagrammatic representation of major continental development and movement in Earth's history during the Cambrian and Ordovician (~543–443 Ma) (redrawn from Rogers, 1996).

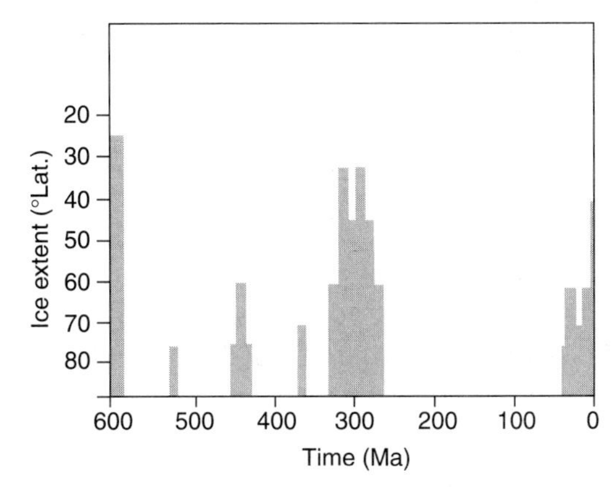

Figure 3.2 Latitudinal extent of periods of glaciation during the past 600 Ma (redrawn from Crowley, 1998).

During this time interval there were also dramatic sea-level changes. Initially sea levels were rising in response both to mantle upwelling, associated with the tectonic activity, and the melting of ice sheets that had developed during the Varanger ice age in the late Proterozoic (~650–590 Ma) (Figure 3.2). Widespread flooding of the continental plates during this time created large areas of shallow continental shelf (Briggs, 1995), with estimates suggesting that more than two-thirds of North America was covered by a shallow sea (Crowley and North, 1991). However, another period of glaciation from approximately 440 million years ago (end of the Ordovician) (Figure 3.2), led to a dramatic reduction in sea level, indicating a drop of as much as 70 m. In the marine animal record a severe extinction event lasting 1–2 million years coincides with this glacial event (Briggs, 1995). In the plant record it is around this time that there is the first unequivocal evidence for the colonization of the land by plants (see Section 3.2).

Formation of soils

The bare surfaces of the earliest terrestrial environments would have had no humic material or 'biologically available' mineral elements (e.g. N, P, Fe, S)—a factor that would have posed serious limitations to plant colonization. However, geological evidence indicates that by 440 Ma (end of the Ordovician) well-established soil profiles, with evidence of *in situ* oxidation of organic matter and deep-burrowing organisms, had developed (Retallack, 1985, 1986; Yapp and Poths, 1992). Various processes (biological and non-biological) are thought to have played important roles in the formation of these first soils, including atmospheric elemental input and weathering by acid rain, and organic acids produced by early microbial organisms and lichens.

One of the most important mechanisms responsible for enriching these early soils with Fe and P was probably weathering of parent material: a process in which the early prokaryotic and eukaryotic organisms would have been extremely important. In extant systems, organic acids secreted from microbes, including cyanobacteria, non-photosynthetic bacteria, and eukaryotic algae, play a significant role in the process of breaking down rock material and releasing Fe and P. For example, the chelation of Fe(III) from rocks is carried out by specific organic molecules called siderophores. Siderophores are secreted by prokaryotic and eukaryotic organisms. These siderophores, containing the Fe(III), are then taken up by other organisms and their breakdown results in the release of Fe (Raven, 1995). Thus even though true vascular plants were not yet around to supply organic acids for weathering in these early environments, prokaryotic and eukaryotic organisms could have operated the full range of Fe and P acquisition processes necessary for soil formation.

High levels of atmospheric CO_2 in these early environments would also have played an important role in the soil formation process. Levels of atmospheric CO_2 in the Cambrian/Ordovician period (~543–443 Ma) are estimated to have been up to 18 times higher than present (Berner, 1991; Yapp and Poths, 1992; Berner, 1993) (Figure 3.3). There are at least three mechanisms by which higher CO_2 could have contributed to soil formation. First, CO_2 enrichment would have favoured increased growth of all photosynthetically active organisms (thus enhancing the processes described above), leading to the growth of microbial mats (Graham, 1993; Raven, 1995). The subsequent decomposition and breakdown of these microbial mats would have released inorganic nutrients back into the system, to be utilized by the next generation of organisms. Secondly, high levels of atmospheric CO_2 would have made precipitation more acidic. Acid rain in these early environments would have increased chemical decomposition of the rocks, promoting the physical disintegration of the bare mineral surfaces and initiation of soil formation. Thirdly, it is interesting to note

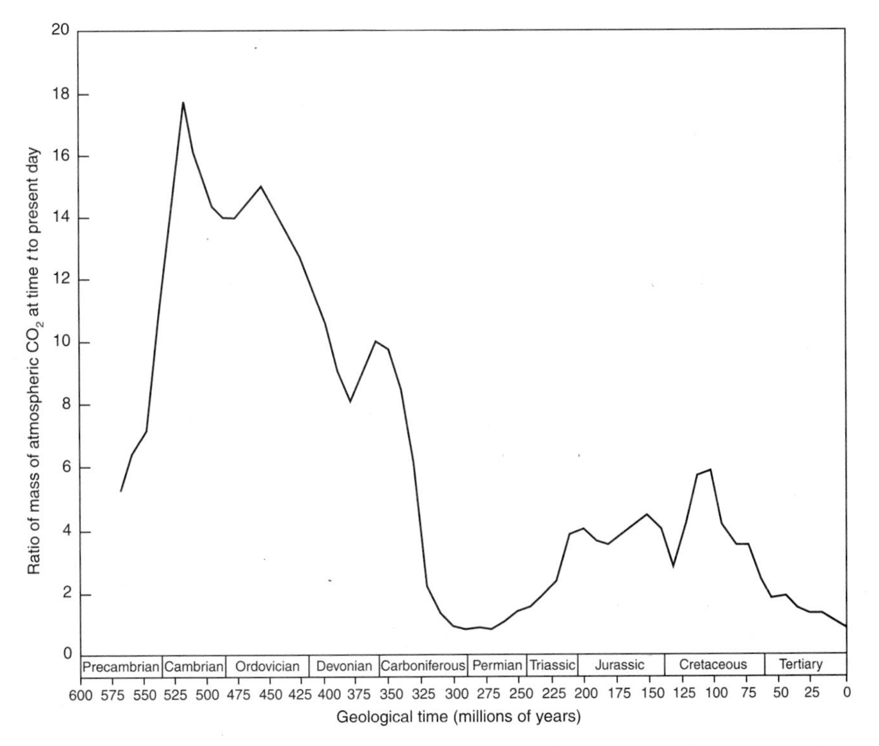

Figure 3.3 Estimated variations in atmospheric CO_2 during the past 600 million years (redrawn from Berner, 1991: GEOCARB I). Values are plotted against ratio of mass of atmospheric CO_2 at time *t* to present day (where present is taken at the pre-industrial value of 300 p.p.m.v.). (Note that the last 1.8 Ma represents the Quaternary—not shown on the timescale.)

that modern ecosystem studies demonstrate increased activity of burrowing organisms (cast activity) when CO_2 concentrations are elevated experimentally. Increased cast activity results in higher water content of soils, leading to increased chemical weathering and aiding the soil formation process. Such a mechanism may have occurred in these early environments.

Another process that probably contributed to early soil development is related to fossil evidence for an early appearance of lichens. Lichens, which are a symbiotic association between a fungus and a green alga or cyanobacterium, have now been found in deposits dating back to at least 400 million years old (Taylor *et al.*, 1995). The importance of these early lichens must be viewed, as in extant groups, in their ability to colonize and weather rock, thus contributing to the formation of soils. Although there are few unequivocal examples of fossil lichens dating back to this period, it has recently been suggested that some of the widespread Ediacaran fossils found in late Proterozoic deposits (~650–550 Ma) and classified as Vendobionta (worms and jellyfish) may, in fact, be fossil lichens (Retallack, 1994). Although this suggestion remains highly controversial, if some of these Edicarian 'fauna' are in fact 'flora', then the impact of lichens on early soil formation processes might have been significant.

Despite the contribution to soil formation from these processes, estimates suggest that the demand for elements in these early environments would have far exceeded those released by weathering (Andreae and Jaeschke, 1992). Thus other mechanisms of elemental input must also be examined. Another process for the addition of S and N to the soils in these early environments might have been via atmospheric transfer of volatile products from the metabolism of marine organisms (Raven, 1995). In extant phytoplankton, for example, when the organism dies, SO_2 is volatized and oxidized and then transferred to the terrestrial environment by dry deposition or in rain. A similar process is envisaged for the early environments. In addition, the fixing of atmospheric N into biologically usable forms through lightening strikes is thought to have played an important role. It has been suggested that up to 1 kg N/ha/year could have originated from the electrical storms in these early environments (Raven and Sprent, 1989).

Development of suitable climatic and atmospheric conditions

Although high atmospheric CO_2 might be viewed as being advantageous to early soil development, it would have been the opposite for the establishment of plants outside of an aquatic environment. In fact, one of the most serious limitations to the plant terrestrialization process was probably due to very high atmospheric CO_2 concentrations (Figure 3.3) and associated high global temperatures. It is estimated that during the Cambrian and Ordovician an increased greenhouse effect resulted in global summer temperatures as high as 40 °C (Crowley and North, 1991). Geological deposits such as evaporites and redbeds dating to the Cambrian support these results and indicate that there were widespread arid to semiarid environments in regions within 45° of the palaeoequator (Condie and Sloan, 1998).

However, by the late Ordovician (~458–443 Ma) there is evidence to suggest that global climates had become much more variable, and that certain regions were becoming cool and moist. Two reasons are thought to have contributed to these changes. First, modelling (Figure 3.3) indicates there was a gradual reduction in atmospheric CO_2, probably resulting from a combination of factors, including a decrease in volcanic outgassing and increased burial of organic carbon (Berner, 1991, 1998). The latter process is thought to have occurred as a consequence of phytoplankton fixing the CO_2 then subsequently becoming buried in marine sediments, thus reducing the amount of organic carbon readily available to form CO_2. Thus, the influence of greenhouse gases on global temperatures would have been reduced. Secondly, geological evidence suggests that glaciers were forming at the South Pole from approximately 445 Ma (late Ordovician glaciation). This would appear to contradict the predicted levels of atmospheric CO_2 because, even though they were dropping, estimates do not suggest temperatures low enough for the formation of ice. Computer-simulated climate modelling to examine the processes that may have been responsible for this apparent paradox (Crowley *et al.*, 1987) indicate that the coastal position of the South Pole at this time was of fundamental importance (Figure 3.4). The proximity of coastal water to the pole would have resulted in a small seasonal cycle (because the large heat capacity of the water tends to suppress the magnitude of warming on adjacent land areas), with summer temperatures not getting above freezing even with high atmospheric CO_2. Thus a steep climatic gradient would have developed between the South Pole (estimated January temperature −2 °C) and the low latitudes (estimated January temperature of 34 °C). The model also indicates that arid conditions occurred in mid-continent areas and monsoon conditions in coastal areas (Crowley *et al.*, 1987; Crowley and North, 1991; Graham, 1993). This pattern of global temperature and precipitation thus suggests that in some areas of the globe more favourable weather conditions were developing for colonization of the land.

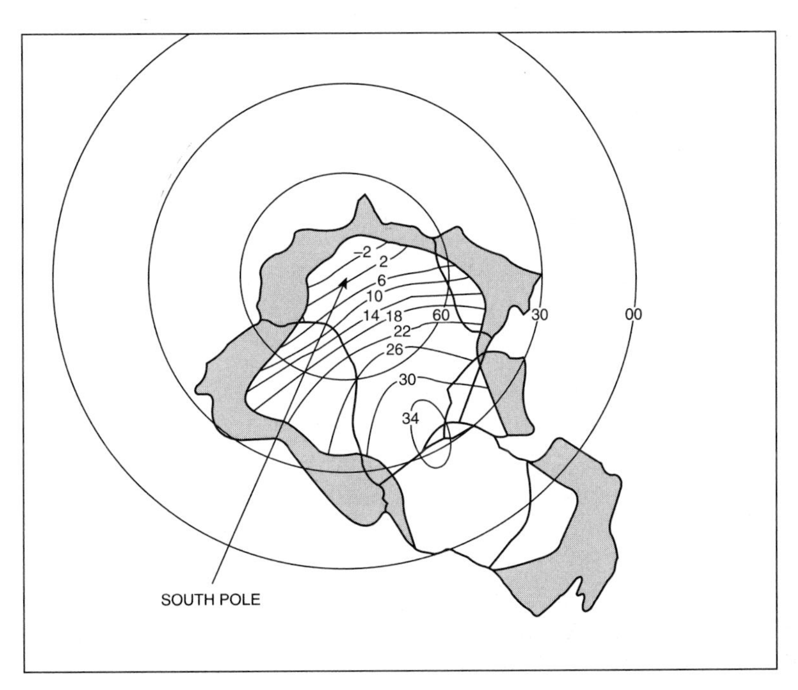

Figure 3.4 Reconstructed January temperatures for the late Ordovician (~450 Ma) of Gondwanaland (redrawn from Crowley *et al.*, 1987). Circular contours represent lines of latitude, with the coastal position of the South Pole indicated. Contours across the Gondwana landmass represent reconstructed January temperatures (°C) and indicate a steep climatic gradient between the South Pole and low palaeolatitudes at this time. Shaded areas represent continental margins flooded by shallow seas.

3.2 Fossil evidence for plant terrestrialization

In parallel to changing environmental conditions, major evolutionary changes in plant structure, shape, and reproduction were necessary in order for plants to exist in a terrestrial environment. From the middle Ordovician to early Silurian (~470–430 Ma) evidence starts to emerge in the plant fossil record for the development of specialized cells for water and nutrient transport, various measures to protect against desiccation, mechanical support, and a reproductive mode that did not depend predominantly upon external water. These morphological changes were essential prerequisites for land dwelling by plants.

Reduction of dependence on water for reproduction

Some of the earliest spores visible in the geological record date back to the late Ordovician (~450 Ma) and are in a tetrahedral arrangement (Gray *et al.*, 1982). Others tending to be found in slightly younger deposits from the early Silurian (Llandoverian, ~430 Ma) appear as single spores, but with a distinctive trilete mark (Gray and Shear, 1992; Taylor and Taylor, 1993) (Figure 3.5). There is strong morphological evidence to suggest that both forms are indicative of

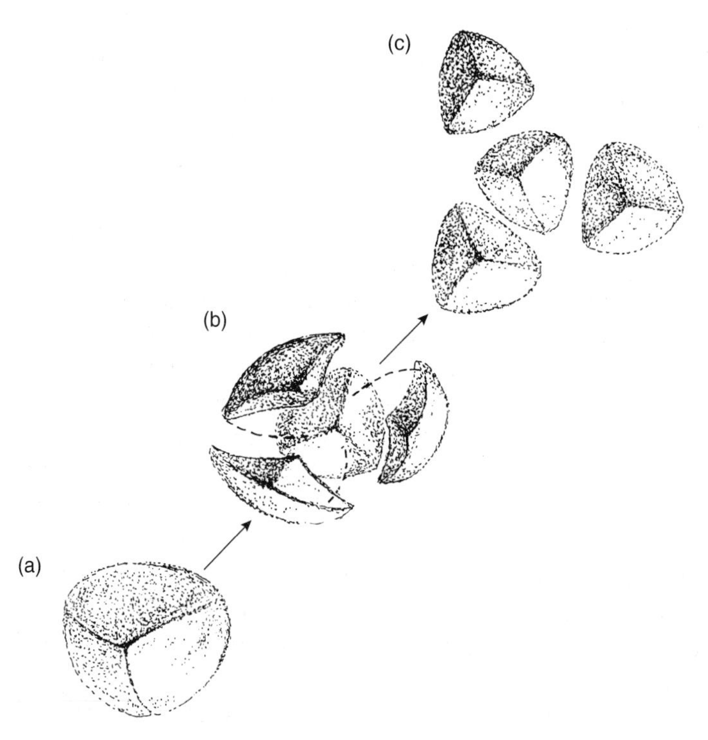

Figure 3.5 (a) Spores in tetrahedral arrangement; (b) composition of the tetrahedral spore; (c) single spores with distinctive trilete Y-shaped marking.

meiotic division, since when a diploid cell divides by meiosis, it produces four spores. The tetrahedral arrangement in the earliest spores is thus thought to demonstrate this process. Similarly, the trilete marking on the slightly younger spore also demonstrates meiotic division. This pattern results from a Y-shaped ridged surface on the spore wall (Traverse, 1988) and is seen in the spores of extant land plants, such as bracken (and also those of some bryophytes). This marks the scar left from the time when four spores were attached to each other in a pyramidal arrangement. When they came apart to form single spores, a trilete mark was created at their central joining position (Figure 3.5).

The significance of these early tetrad spores in the fossil record is that many of these, especially those later in the record (from the early Silurian, ~430 Ma), probably represent the first evidence for the elaboration of the sporophyte phase (spore producing) in the plant life cycle. However, to fully appreciate the importance of the development of a more complex sporophyte phase to land dwelling, it is necessary to first examine the life cycle of extant aquatic plants, algae in particular.

Many algal groups are known to reproduce both asexually and sexually. The latter involves an alternation of two phases or generations, namely a diploid sporophyte phase and a haploid gametophyte phase (Thomas and Spicer, 1987) (Figure 3.6). In the gametophyte phase, male and female gametes (each gamete containing a single set of chromosomes) are released from the gametophyte. The male gametes swim to the female and fuse, producing diploid zygotes which contain double sets of chromosomes. Upon germination the zygotes form diploid 'plants' called sporophytes. When mature, the sporophyte undergoes meiotic division to form haploid spores and these are released from the plant to grow into gametophytes. Thus the alternating cycle (gametophyte/sporophyte) is complete (Figure 3.6a).

In different species of extant algae, different parts of the cycle are emphasized. This feature can also be seen in the early land plants. Some plants acquired an amplified gametophyte generation (resulting in a sporophyte generation nutritionally dependent on the gametophyte phase) (Figure 3.6b), whereas others developed an amplified sporophyte generation (resulting in a nutritionally independent sporophyte) (Figure 3.6c). Elaboration of the gametophyte stage would, if anything, have increased the need for water, since water is essential for the survival of the gametophyte, the transfer of the sperm to egg, and the initial growth of the sporophyte embryo (Niklas, 1997). It can be suggested, therefore, that in continually moist environments with little desiccation stress the gametophyte generation would have been emphasized or selected for. In contrast, elaboration of the sporophyte stage would have decreased the necessity for water, since neither the production nor dissemination of spores is reliant upon water. Thus, desiccating environments would have imposed intense selective pressure in favour of such amplification of the sporophyte generation.

This differentiation of amplification in the gametophyte or sporophyte generation has persisted to the present day. In extant vascular plants

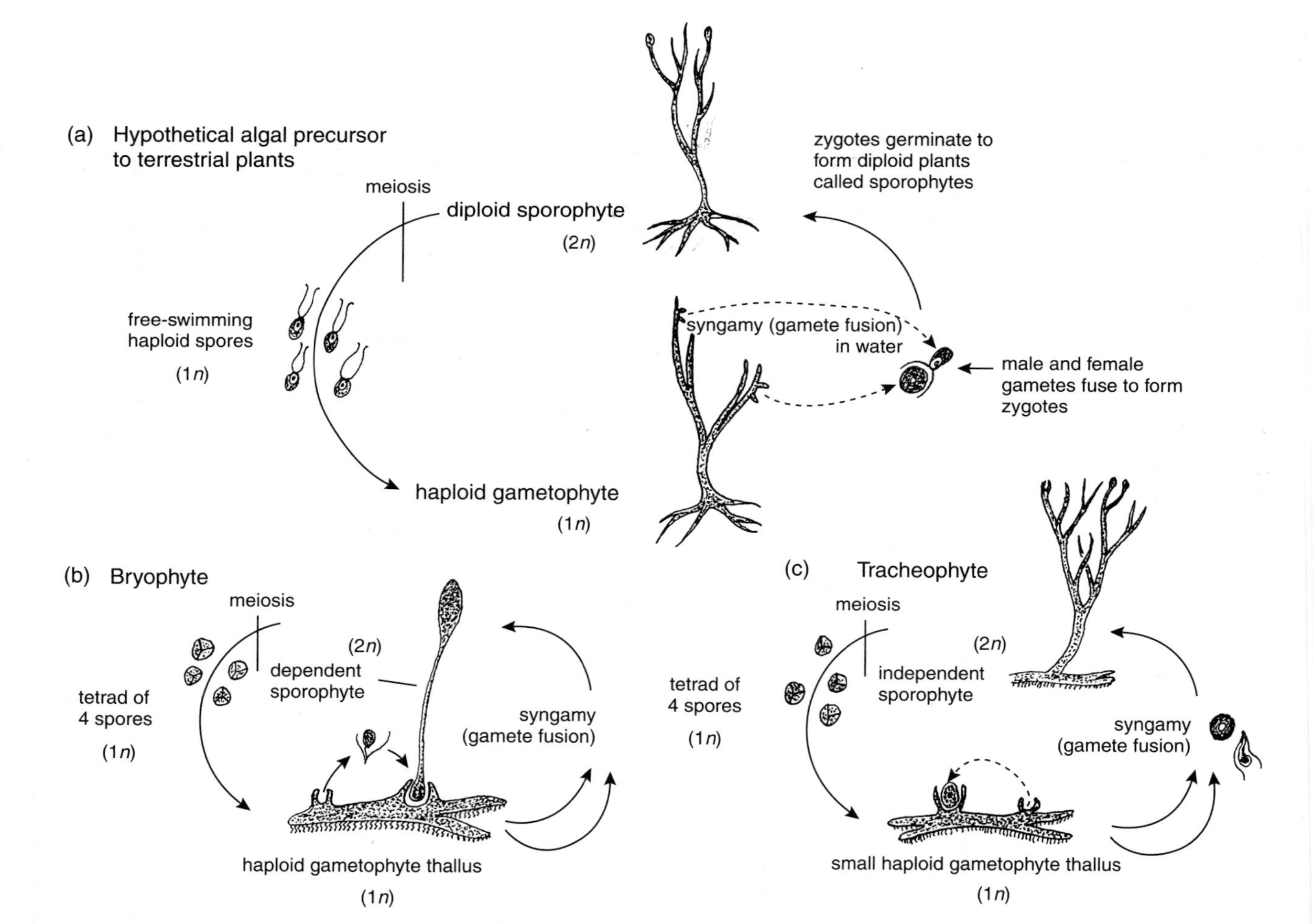

Figure 3.6 Diagram of simplified plant life cycle, showing alteration of generation phases. (a) Simplified diagram to indicate the alteration of two phases of generation in algal reproduction; (b) the cycle with an amplified gametophyte generation—most bryophytes follow this mode of reproduction; and (c) the cycle with an amplified sporophyte generation—all vascular plants follow this mode of reproduction (redrawn from Price, 1996).

(tracheophytes) the vegetative sporophyte is the 'visible' plant body, whereas in non-vascular groups (bryophytes) the vegetative gametophyte is the 'visible' plant body (Figure 3.6b,c). This differentiation may well explain why, throughout the geological record, bryophytes have remained relatively small (less that 1 m in height) and restricted to damp, moist regions, whereas tracheophytes have evolved not only to some of the largest plants on this planet, but also to fill almost every ecological niche available.

Protection against desiccation

It has been suggested that the appearance and gradual increase in spores in the fossil record is not only indicative of elaboration of the sporophyte phase but also the development of spores that were resistant to decay. In particular, it is suggested that spores with walls containing sporopollenin were developing (Graham, 1993). Sporopollenin is found in the walls of most extant pollen and spores, in both vascular and non-vascular plants and also in some algae, and is a complex polymer that provides desiccation resistance, robustness, and protection from ultraviolet radiation (Kenrick and Crane, 1997a,b). The evolution of desiccation-resistant spores was extremely important to the terrestrialization process. It would have enabled long-distance spore dispersal by wind and therefore a means of establishment of isolated communities far into the continental interiors (Taylor and Taylor, 1993). It is proposed that desiccant-resistant spores may have been produced initially by aquatic or semiaquatic green algae in response to periodic drying (Thomas and Spicer, 1987).

However, it was not only spores that needed protection from desiccation, both the vegetative sporophyte and vegetative gametophytes would also have been extremely susceptible without special adaptations. This is in direct contrast to the situation that would have occurred in aquatic environments. Here, any barrier preventing the free flow of water, solutes, soluble CO_2 and waste products between the plant and the nutrient solution in which it was bathed, would have been disadvantageous (Raven, 1993; Taylor and Taylor, 1993).

Cuticle is a layer of wax and insoluble lipid polymers that covers and impregnates the walls of the epidermal cells to reduce water loss, and is present on the aerial parts of every living land plant. Fossil evidence for sheets of cuticle appear in the geological record from the early Silurian (~430 Ma) (Edwards and Fanning, 1985; Gray and Shear, 1992). Although this early cuticle is 'disarticulated', i.e. not attached to whole plant bodies, its presence in the fossil record suggests land colonization by the earliest Silurian (~430 Ma). In comparison to other lines of evidence for terrestrialization (e.g. spores with trilete markings, described above), the evidence for cuticles appears comparatively late in the fossil record (Gray, 1993; Kenrick and Crane, 1997). It has been suggested that this may be due to the fact that the fatty acids of which the cuticle is composed saponify in water (Thomas and Spicer, 1987). However, another more likely interpretation is that the various features providing protection against desiccation, were acquired sequentially.

Although formation of the cuticle would have protected against desiccation, microbial attack, abrasion, and mechanical injury, it would also have prevented the flow of gases (in particular CO_2 and O_2) into and out of the plant. The earliest evolutionary solution to this dilemma appears to have been the development of an imperforate but thin cuticle, which would have allowed the movement of gases by diffusion (Thomas and Spicer, 1987). This facilitation of gas diffusion is one that occurs in some extant mosses. However, a more refined feature for regulation of gases and water came with the evolution of stomata, from approximately 408 million years ago in the latest Silurian/earliest Devonian.

There is increasing evidence to suggest that these early fossil stomatal densities were closely related to levels of atmospheric CO_2 (McElwain and Chaloner, 1995; Chaloner and McElwain, 1997; McElwain, 1998). In the early Devonian, for example, fossil stomatal densities were extremely low in relation to other periods in the geological record. When these fossil stomatal densities are compared to estimated changes in global atmospheric CO_2 through time (Berner, 1991, 1993, 1997) there appears to be a close correspondence. High stomatal densities occur at times of low atmospheric CO_2 and vice versa (Figure 3.7). This evidence therefore suggests that even the earliest land plants were in tune physiologically with concentrations of atmospheric CO_2 (McElwain, 1998).

Development of specialized cells for water and nutrient uptake

When plants are located in an aquatic environment there is little requirement for a specialized system to distribute water, solutes, and photosynthetic products, since all cells are a short distance from the source of water and nutrients

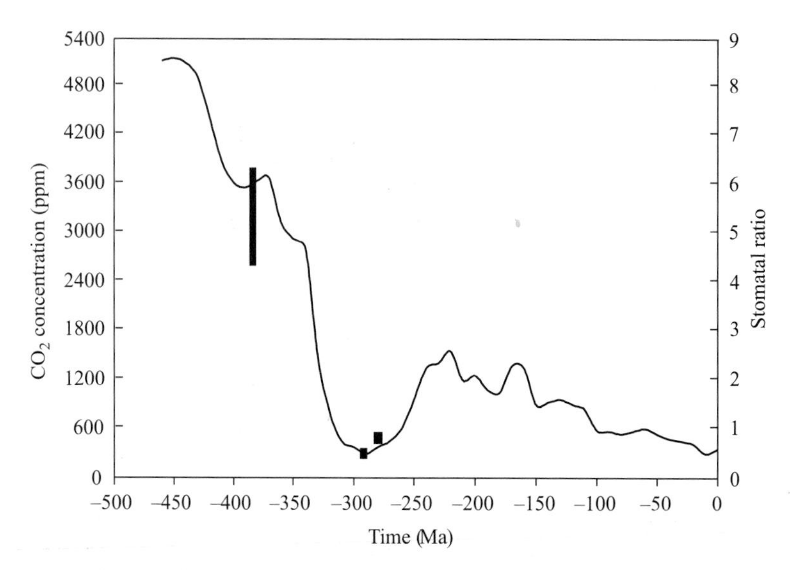

Figure 3.7 Fossil stomatal ratios (a ratio of stomatal numbers of a modern plant to that of a fossil), superimposed on estimated changes in CO_2 concentration from a long-term carbon cycle model (Geocarb II) of Berner (1993) (redrawn from McElwain and Chaloner, 1995).

(Thomas and Spicer, 1987). With terrestrialization, early land plants would therefore have required specialized conducting tissue to transport food and water about the plant, especially if vertical growth was involved. Calculations have indicated, for example, that even a modest height of 2 cm requires specialized conducting tissues (Niklas, 1997).

From approximately 430 million years ago (early Silurian) various tubes and tracheids (tubes with lignified-type thickening) appear in the fossil record (Figure 3.8). These structures are taken to be indicative of the evolution of an early conducting system (Gray and Shear, 1992). The tubes can be divided into two groups according to size and structure. There are those that are narrow (approximately 8–20 μm in diameter and 50 μm long) and frequently smooth, and those that are longer (up to 200 μm long) with an annular–helical thickening

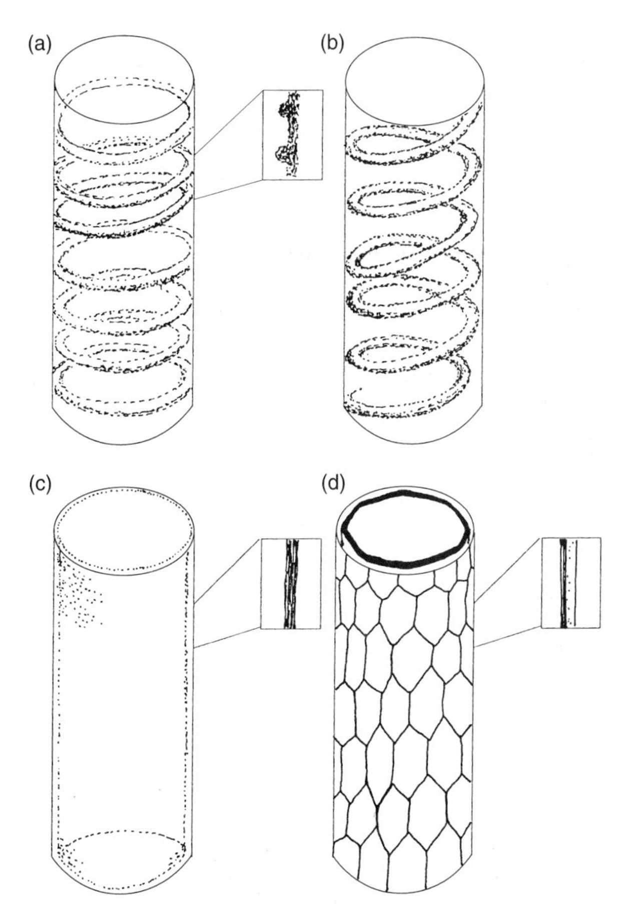

Figure 3.8 Structure of early fossil records of tubes and tracheids and comparison to extant plants (redrawn from Taylor and Taylor, 1993). (a) Tracheid of extant vascular plant. (b) Silurian tubes (*c.* 440 Ma) with helical thickenings on the inner surface. These tubes have measured up to 200 μm in length. (c) Extant moss hydroid with smooth inner surface. (d) Tube from fossil *Aglayphton major* (up to 50 μm in length).

on the inner surface (Taylor and Taylor, 1993) (Figure 3.8). The former have been likened to those found in many extant bryophytes and the latter to those of vegetative sporophytes (Edwards, 1993).

In extant vascular plants, water-conducting cells have differentially thickened cell walls (provided by lignin) which mechanically reinforce the cell wall from collapse as a consequence of negative internal pressure created when fluid moves rapidly through a tube. The greater the demand for water (and the higher the plant may grow), the more important the thickening, since there is a 'critical size' (*sensu* Niklas, 1997) at which the threshold of vertical height will impose selection favouring conducting tissue. It has therefore been suggested that many of these early tubes with thickenings probably contained lignin (and should therefore be classified as tracheids), although as yet this has been almost impossible to demonstrate biochemically from the early fossil record (Thomas, 1986). It is interesting to note here that the ability to synthesize lignin is apparently highly dependent on levels of atmospheric O_2, with estimates suggesting that lignin formation could not have proceeded until the Cambrian ($\sim$540 million years ago) (Chapman, 1985).

More recent work on early Devonian water-conducting cells ($\sim$417–391 Ma) contained within early whole-plant bodies (rather than disarticulated tubes, etc.) has identified three distinct tracheid types (Kenrick and Crane, 1991, 1997a,b). These have been distinguished according to their cell wall construction, as follows (Figure 3.9):

(1) G-type cell wall with annular–reticulate thickenings;
(2) S-type cell wall with helical thickenings that appear to combine certain features of tracheids (e.g. lignified thickenings) and moss hydroids; and
(3) P-type cell wall with scalariform pittings (pits in the secondary cell wall are elongated transversely and parallel to each other).

Evidence from the fossil record therefore suggests that by the earliest Devonian a system had evolved to transport water and nutrients around the plant body.

Mechanical support

With the loss of support provided by water, another adaptation necessary in the process of terrestrialization was a means of staying upright. Although the mechanical strength of lignin would have provided some support (Taylor and Taylor, 1993), calculations on stem diameter suggest that structural organization was also important (Niklas, 1986, 1994, 1997).

Calculations on the biomechanics of early terrestrial vegetation have indicated that although there is no single 'optimal' plant morphology, the simplest means of attaining maximum surface area to volume ratios for light interception, gas exchange, and indeterminate growth is the flat dorsiventral shape. This is the condition in which the plant does not have any vertical growth but instead grows

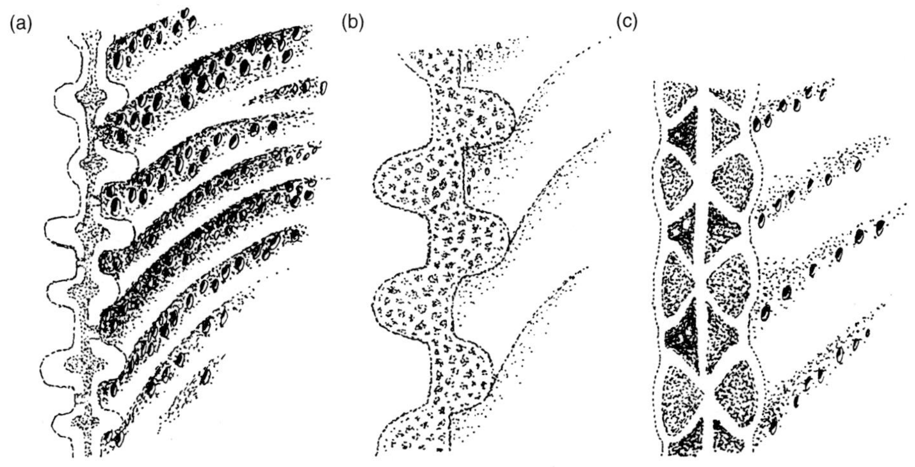

Figure 3.9 Cell wall structure of early tubes and tracheids in the fossil record (redrawn from Kenrick and Crane, 1997a). Three types of tracheid have been distinguished according to their cell wall construction, as follows: (a) G-type cell wall with annular–reticulate thickenings; (b) S-type cell wall with helical thickenings that appear to combine certain features of tracheids (e.g. lignified thickenings) and moss hydroids; and (c) P-type cell wall with scalariform pittings (pits in the secondary cell wall that are elongated transversely and parallel to each other).

horizontally across the rock or substrate on which it is situated (Niklas, 1986). Many extant liverworts and some green algae (e.g. *Coleochaete*) conform to this shape, and in the fossil record there is evidence to suggest that one of the most convincing precursors to vascular plants, *Parka decipiens* (see p. 72), possessed this biomechanical design.

For attainment of height, the next best biomechanical solution is to have a stem that allows a plant to continue to grow longitudinally with only a minor reduction in volume to surface area ratio, resulting in a large, wide, flat top and a small stem. It has been calculated that the best shape for such a stem is a cylinder (Niklas, 1986). The maximum height of such a cylindrical stem before buckling will depend upon its 'flexural rigidity', which combines the material properties of the structure with the cross-sectional geometry. Calculations suggest that if flexural rigidity is to be maintained/maximized, stems lacking secondary growth (i.e. wood) must be solid and/or thick walled and as broad as possible. For example, calculations suggest that a hollow, thickened walled cylinder will grow approximately 26% taller than a solid cylinder (Niklas, 1986). The fossil record appears to confirm these calculations in that the early stems appear to become wider, longer, and with a greater internal diameter with decreasing age of deposit (Niklas, 1997). However, a note of caution must be added here since thickening of the stem wall in the early land plants would have created potentially conflicting requirements between mechanical strength and photosynthesis. In the absence of leaves, the optimal position for photosynthetic tissue would be just below the surface of the stem (Niklas, 1987), but this would also be the optimal position for the bulk of the strengthening material (cell wall) in the stem. In early plants, however, this problem was overcome by the fact that the cells that provided support functioned hydrostatically (that is when inflated with water these cells became stiff and rigid). The optimum location for hydrostatic cells, since they are water conducting, is in the middle of the stem. Evidence from the fossil record indicates, therefore, that the earliest land plants had short, wide stems with a photosynthetic 'rind' just beneath the surface and hydrostatic tissue (parenchyma) in the centre (Niklas, 1997).

Calculations also suggest, however, that to gain greater height than a few centimetres, it would have been necessary to have some specialized mechanical tissue. The next best mechanical solution, therefore, is to have mechanical tissue (outer cortical tissue, collenchyma) between the photosynthetic rind and the inflatable hydrostatic core (Niklas, 1997) (Figure 3.10). A number of early vascular plants demonstrate this morphology. However, it is calculated that eventually the thickening required to attain greater heights would have presented a physical barrier to photosynthesis and the solution would have been to compartmentalize the functions of the plant body into those for photosynthesis and those for mechanical support (see Chapter 4).

Anchoring mechanisms

Standing upright was not the only structural problem associated with the terrestrialization process. Early land plants also needed to acquire a system to anchor them to the ground and enable them to obtain the mineral elements necessary for nutrition and water from soils.

The earliest unequivocal fossil evidence for rooting systems comes from plant and root traces preserved in palaeosols dating from around 408 million years ago (early Devonian). These traces indicate dichotomous roots between 0.5 and 2 cm in diameter and up to 90 cm in length, thus providing evidence that early vascular plants were capable of rooting to nearly 1 m in depth (Elick *et al.*, 1998).

Early rooting structures were relatively undifferentiated from the aerial parts of the plant (Taylor and Taylor, 1993). It is envisaged that as the lower part of the plant became covered with mud, it eventually became specialized into a rooting system.

In summary, evidence from the plant fossil record suggests that the period between the late Ordovician and early Silurian was a time of major innovation. The terrestrialization process included the evolution of a reproductive system that was not primarily dependent on water and various mechanisms to enable plant growth outside of an aquatic medium. All of these features appeared between approximately 470 and 430 million years (Figure 3.11), followed closely by the first evidence of entire land plant organs, such as stems and reproductive structures, from the early Devonian (~408 Ma).

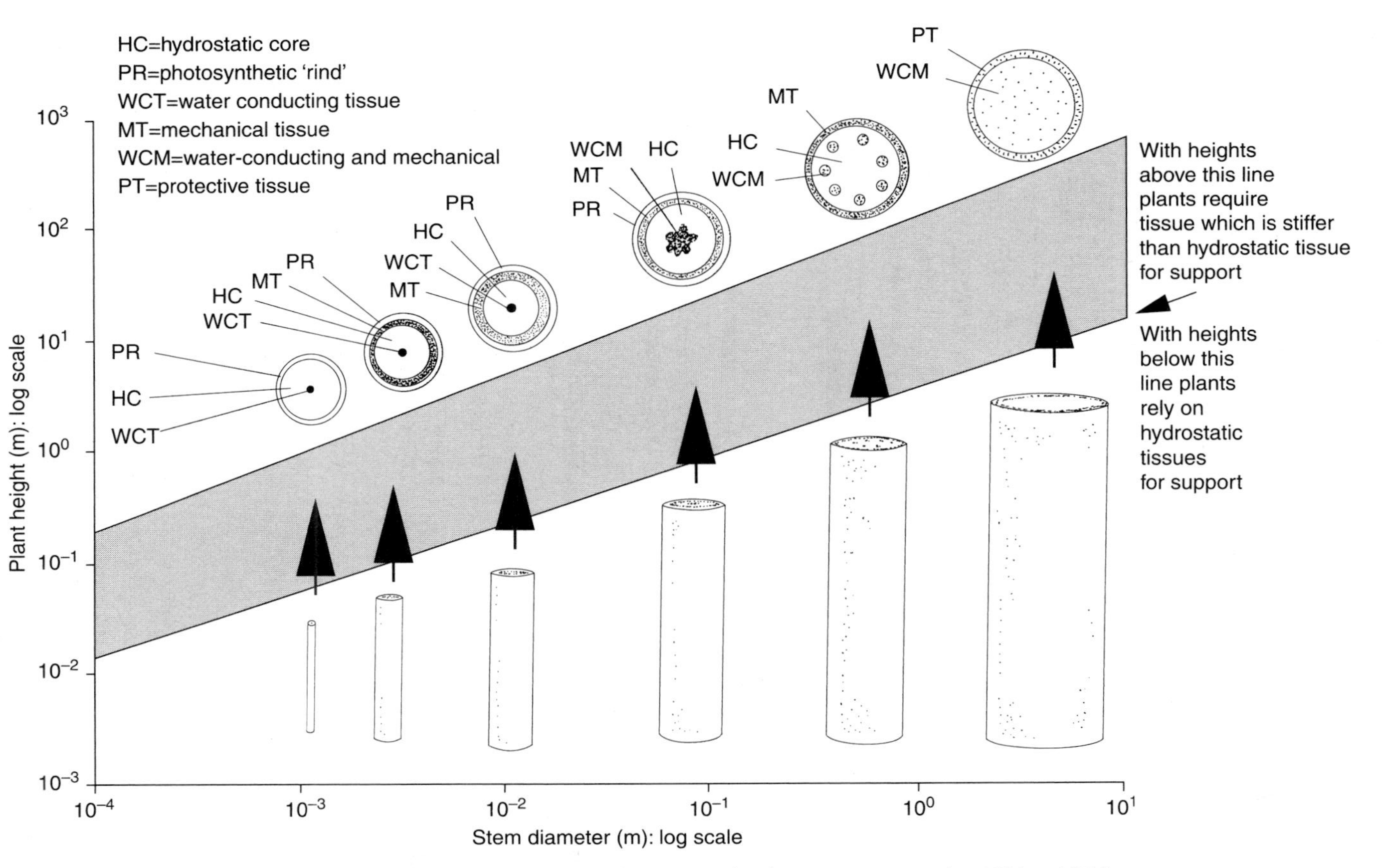

Figure 3.10 Relationship between stem diameter and stelar construction (after Niklas, 1997).

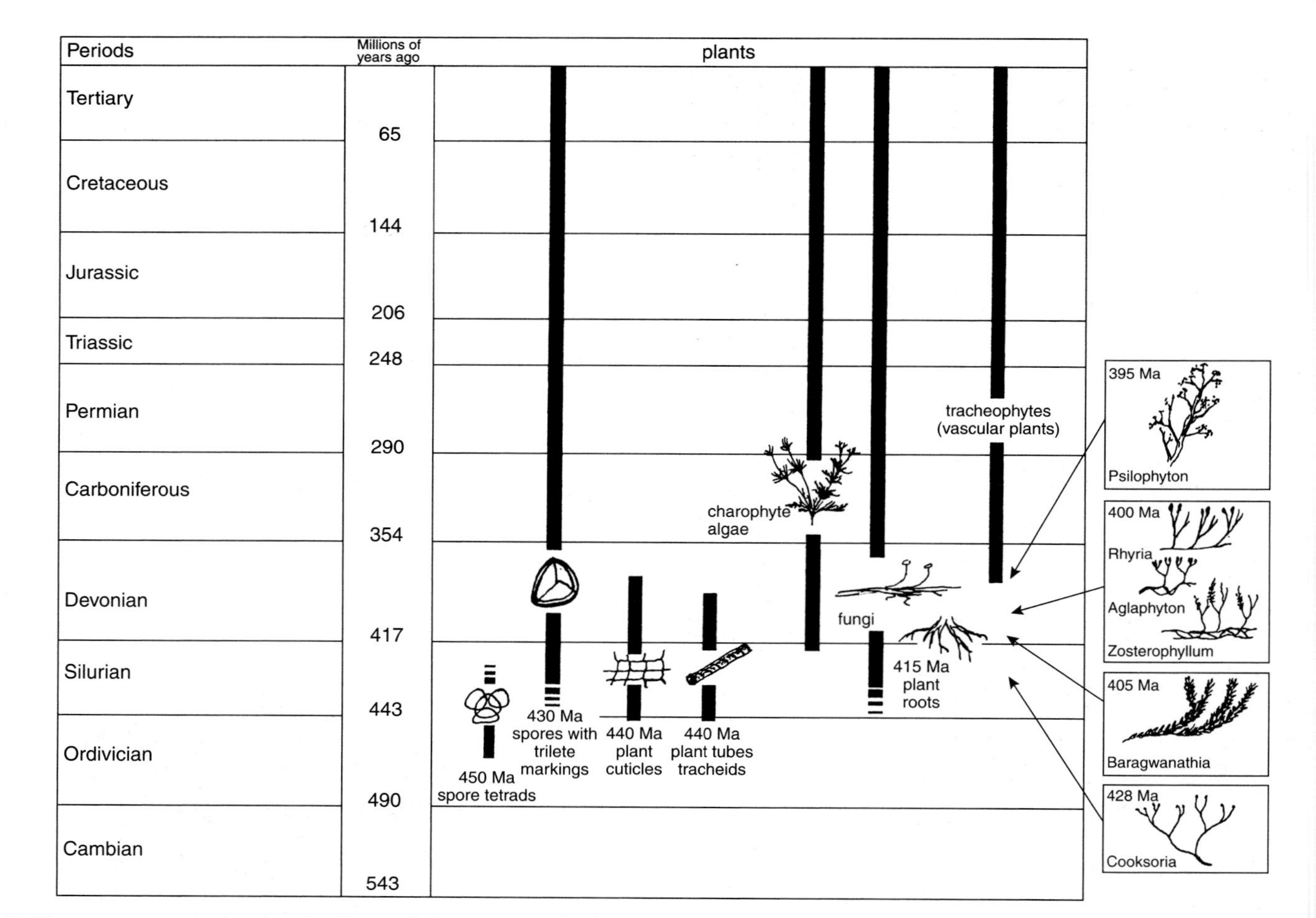

Figure 3.11 First appearance in the plant fossil record of innovations leading to the colonization of land by plants. Approximate time of first appearance of spore tetrads, spores with trilete markings, plant cuticles, plant tubes and tracheids, and plant roots (redrawn from Gray and Shear, 1992).

3.3 Examples of earliest land plants in the fossil record

Evidence for entire vascular land plant organs, such as stems and reproductive structures, start to appear in the fossil record in the early Devonian (~408 Ma). These macrofossils (rather than the microfossils of single tubes, tracheids, spores, etc.) have been identified from enough fossil localities to suggest that a vegetation of small-stature vascular plants had become established on many continents (Edwards and Selden, 1991; Algeo and Scheckler, 1999). Taxonomic descriptions of early land plants in the literature (e.g. Edwards, 1982; Edwards and Fanning, 1985; Gensel and Andrews, 1987; White, 1990; Kenrick and Crane, 1997a,b) also suggest that by the early Devonian, there was some diversity in the terrestrial plant world. It is beyond the scope of this book to describe all these types in detail, but six examples of the most common early land plants found in the fossil record are given below.

Cooksonia

Early plants that are classified as *Cooksonia* (Figure 3.12) occur in large numbers of fossil localities in North America and Europe. The earliest examples of *Cooksonia* have been found in deposits from Ireland dating to early Silurian age (Wenlockian, ~428 Ma) (Edwards *et al.*, 1983). They consist of a simple

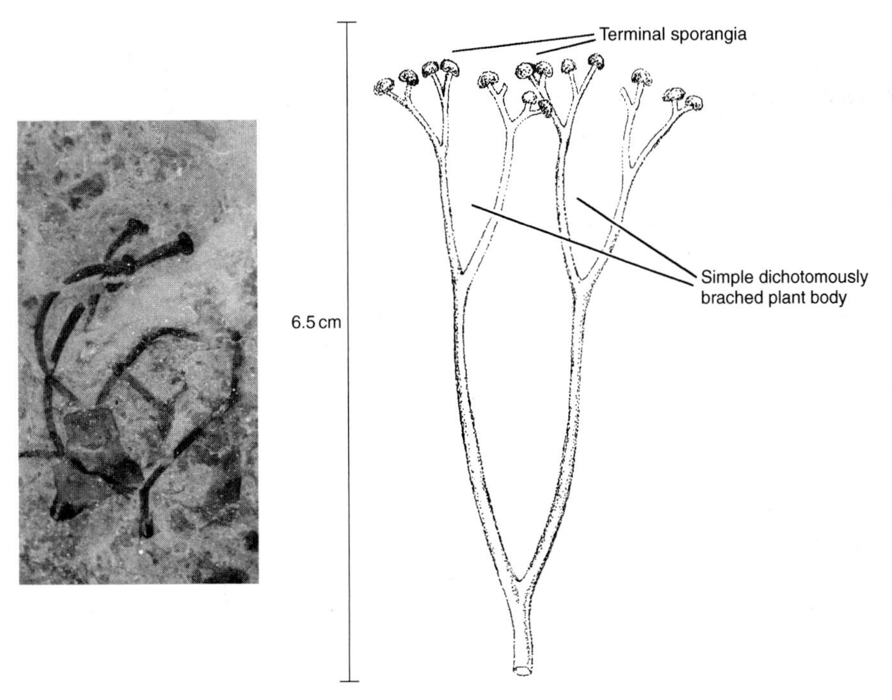

Figure 3.12 Photograph of fossil *Cooksonia* alongside sketch to indicate main features (photograph by H. Barks courtesy of P. Gesel; sketch redrawn from Bell, 1992).

dichotomously branched plant body with sporangia (the spore-bearing structures) at the end of each branch tip (known as terminal sporangia). It would appear that these plants grew to no more that 6.5 cm tall and probably formed swards in swampy places. Stomates and some details of internal anatomy have been described in some of the earliest *Cooksonia* (e.g. *Cooksonia pertonii*; Kenrick and Crane, 1997a). In many examples, however, no internal anatomy or cuticular features have been observed, and the suggestion has been made that they took up CO_2 through their underground organs (Bell, 1992).

Aglaophyton major

The only record of *Aglaophyton major* (Figure 3.13) is found in the Rhynie Chert, Scotland, and is dated to approximately 400 Ma (early Devonian, Emsian) (Rice, 1994). It was morphologically more complex than *Cooksonia*, and consisted of branched aerial stems that grew up to 20 cm tall, rising from a horizontal rhizome (Edwards, 1986; Bell, 1992). As in *Cooksonia*, the sporangia were located at the tips of the stems, and there was a well-defined cuticle with stomata in the aerial parts of the plant. A rooting system was provided by prostrate rhizomes bearing small rhizoids that branched off from the main body of the plant (Figure 3.13). It has recently been suggested that the presence of unornamented water-conducting cells in this species associates it with a level of anatomical differentiation closer to that of moss sporophytes than vascular plants (Edwards *et al.*, 1995).

Rhynia gwynne-vaughanii

Numerous permineralized specimens of *Rhynia gwynne-vaughanii* (Figure 3.14) (Edwards, 1980, 1986) dating back to approximately 400 million years ago (Rice, 1994) have been recovered from the Rhynie Chert in Scotland. These fossils record a plant growing to approximately 18 cm tall with simple dichotomizing stems arising from a rhizome bearing thread-like rhizoids (Figure 3.14). The stems were 2–3 cm in diameter (Taylor and Taylor, 1993) and were vascularized with helical S-type elements (Kenrick and Crane, 1991). Sporangia were borne singly at the tip of the main stem and produced spores in tetrahedral tetrads. Each spore was approximately 40 μm in diameter, covered in closely spaced spines.

Zosterophyllum divaricatum

Zosterophyllum divaricatum (Figure 3.15) is one of a number of early fossil plants (dating back to approximately 400 million years ago) showing the presence of sporangia borne laterally along the stem, or attached to it by a short branch (Gensel, 1982; Gensel and Andrews, 1987). These sporangia were borne on short stalks and usually in two rows orientated to one side of the axis

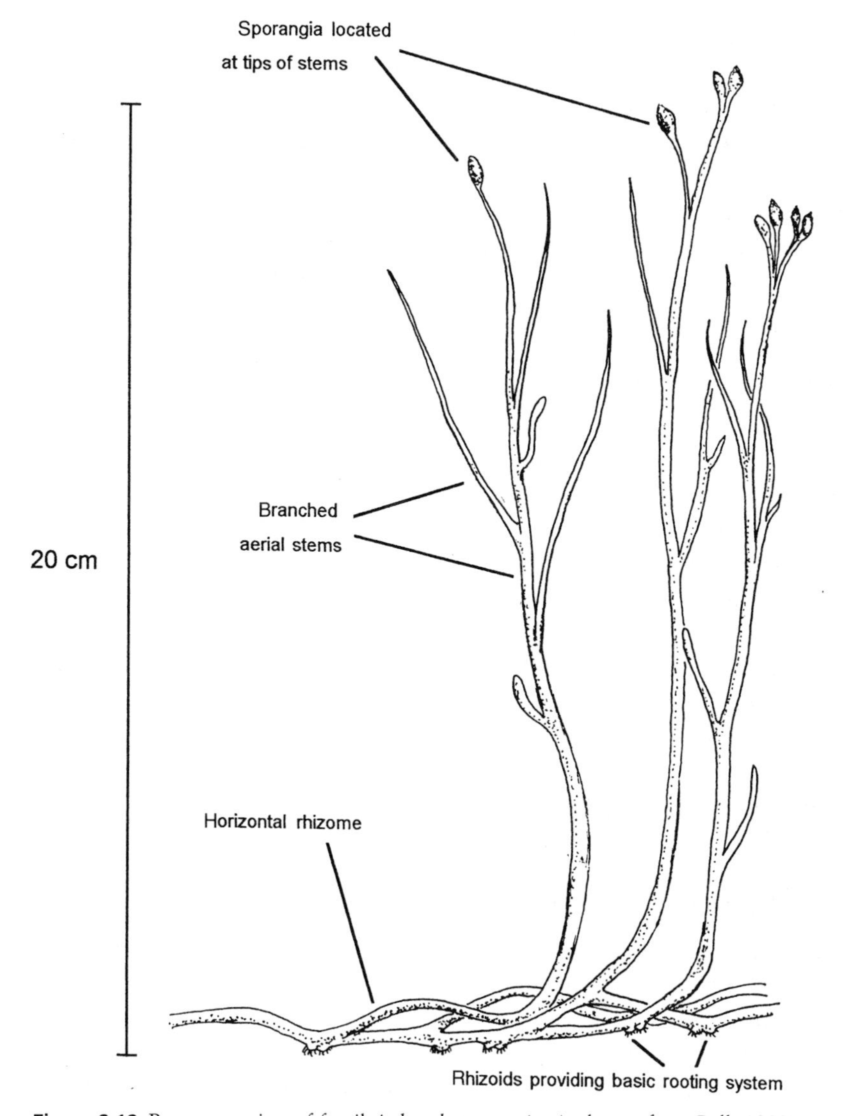

Figure 3.13 Reconstruction of fossil *Aglaophyton major* (redrawn from Bell, 1992).

(Figure 3.15). The spores were approximately 55–85 μm in diameter, circular to sub-triangular, smooth, and bore a distinctive trilete marking (Kenrick and Crane, 1997a). The plant grew to an overall height of approximately 30 cm and consisted of branches that grew up and out from a rhizome at an acute angle (sometimes described as H-type branching, Figure 3.15). These branches were between 1 and 4 mm in diameter and covered with numerous tapered spines (Taylor and Taylor, 1993). Inside the stem was a solid core of tracheids, interpreted possibly as G-type cells (see above; Kenrick and Crane, 1991).

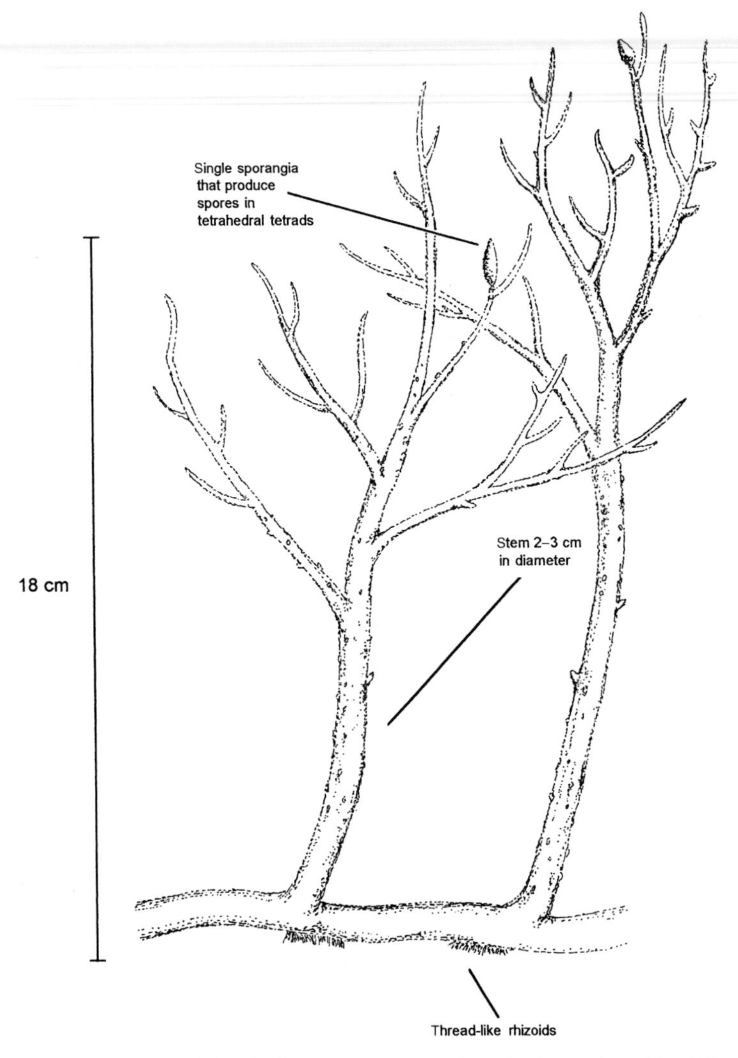

Figure 3.14 Reconstruction of fossil *Rhynia gwynne-vaughanii* (redrawn from Kenrick and Crane, 1997a).

Baragwanathia longifolia

Baragwanathia longifolia (Figure 3.16) has been identified from a number of Australian sites which was part of the Gondwanaland supercontinent (Figure 3.16). They were composed of a robust stem that was approximately 1–2 cm in diameter and thickly covered with long, slender leaves (Gensel and Andrews, 1987). Fossil evidence shows that the stems were extremely long, extending out from a rhizomatous base up to 1 m. The stems contained tracheids with annular or helical thickenings, some with a G-type appearance, and were covered with closely arranged leaves up to 4 cm in length (Figure 3.16).

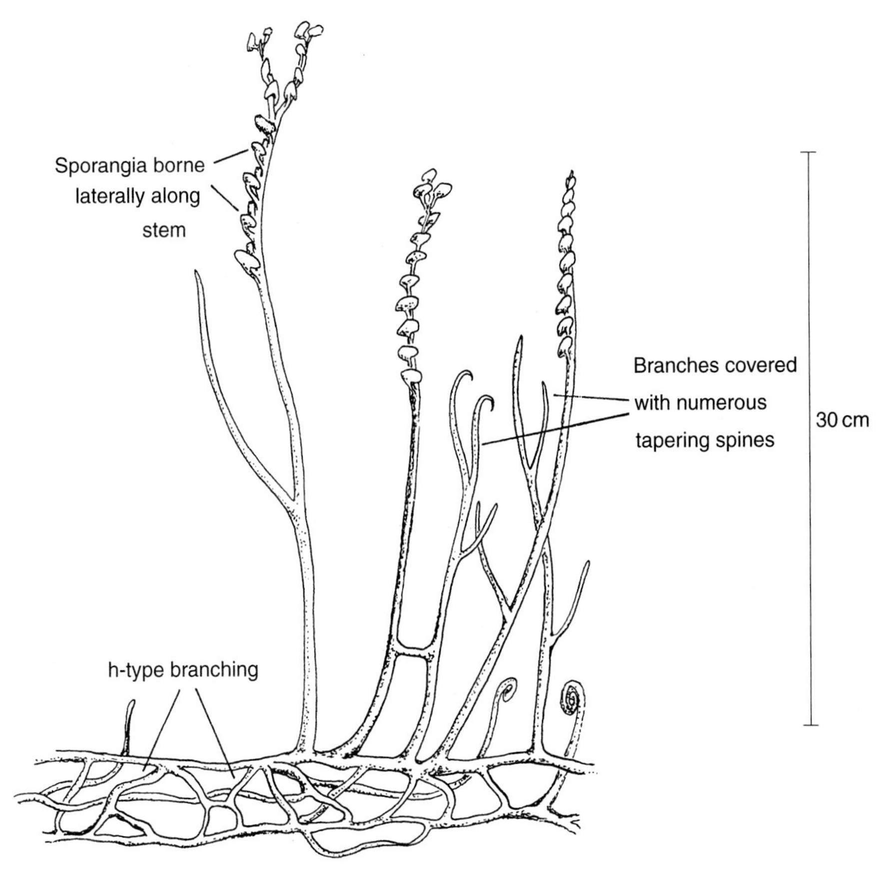

Figure 3.15 Reconstruction of fossil *Zosterophyllum divaricatum* (redrawn from Gensel and Andrews, 1987; Bell, 1992; White, 1990).

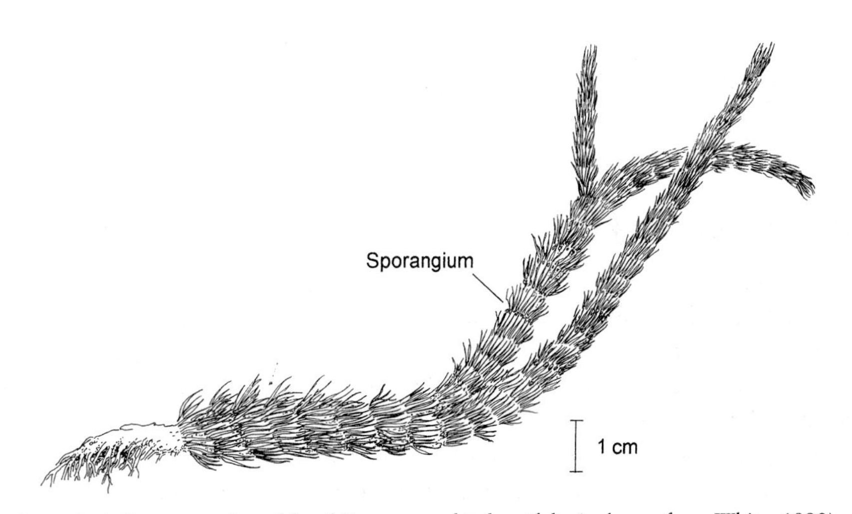

Figure 3.16 Reconstruction of fossil *Baragwanathia longifolia* (redrawn from White, 1990).

Sporangia were arranged along the stem, although it is unclear whether these were actually attached to the stem itself or to the base of the leaves. The spores were smooth, had trilete markings, and were up to 50 μm in diameter.

The age of *Baragwanathia longifolia* has been the subject of heated discussion. Originally it was described as late Silurian (~420 Ma) and then lower Devonian (~410 Ma) (Lang and Cookson, 1935). Its co-occurrence with well-constrained, late Silurian invertebrate fossils appeared to support the suggestion of a considerably earlier first appearance (~420 Ma), at least on Gondwanaland (Garratt *et al.*, 1984; Tims and Chambers, 1984). If this earlier age is accepted, it has important evolutionary implications since *Baragwanathia longifolia* was morphologically more complex, large, and had leaves (Garratt *et al.*, 1984), yet was present on the land before all but *Cooksonia* (Gensel and Andrews, 1987). More recent research, however, has once again questioned the age of these deposits (for discussion, see Hueber, 1992; Kenrick and Crane, 1997a) and again it is suggested that they are more likely to be of early Devonian age.

Psilophyton dawsonii

Psilophyton dawsonii (Figure 3.17) has been identified at a number of fossil localities dating back to approximately 395 million years ago (Banks *et al.*, 1975; Edwards and Fanning, 1985). It is known to have grown to a height of approximately 60 cm, with greatly branched fertile and vegetative lateral branches growing out from a central stem. The fertile branches typically dichotomized up to six times before terminating in clusters of approximately 32 sporangia (Figure 3.17) (Bell, 1992; Taylor and Taylor, 1993). This plant had a sophisticated vascular system made up of radially aligned tracheids (scalariform pitted P-type cells, see Figure 3.9c) (Kenrick and Crane, 1997a,b).

Since the first discovery of early land plants in the fossil record, many have attempted to establish a phylogenetic relationship between these earliest plant fossils and extant species. However, a problem encountered in many of these studies has been in determining whether early land plants such as *Aglaophyton* (described above) are vascular (i.e. tracheophytes: plants that contain lignified tracheids) or non-vascular (i.e. bryophytes: plants that do not contain lignified tracheids—mosses, liverworts, and hornworts) (Edwards, 1986). Recent cladistic studies (e.g. Kenrick and Crane, 1997a,b) have therefore included a basal clade, polysporangiophytes (see below), which incorporates both vascular and non-vascular plants. According to these cladistic studies (based on morphological traits), the earliest land plants can be divided into three nested groups: the polysporangiophytes, the tracheophytes, and the eutracheophytes (Figure 3.18). These are classified as follows.

1. Polysporangiophytes: the defining features of plants in this group include branching sporophytes that bear multiple sporangia (Kenrick and Crane, 1991) and independence of the gametophyte from the sporophyte

Figure 3.17 Photograph of *Psilophyton forbesi* alongside sketch of *Psilophyton dawsonii* to indicate main features (photograph R. Lupia, PP33630; sketch redrawn from Kenrick and Crane, 1997a).

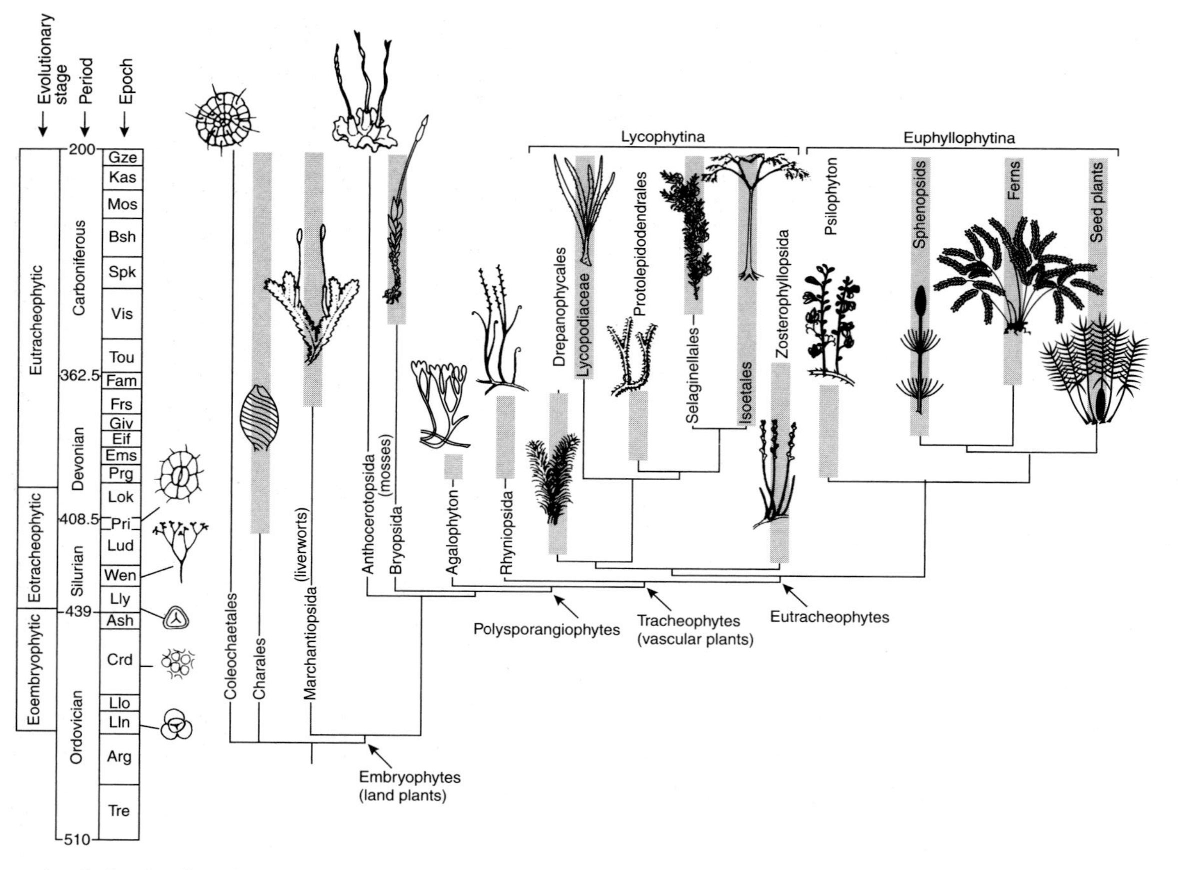

Figure 3.18 Phylogenetic relationship between extinct and extant early plants (based on cladistic analyses of morphological traits) (redrawn from Kenrick and Crane, 1997b).

(independent alteration of generations). Nested within this group are the tracheophytes and a separate clade consisting of the genus *Aglaophyton*.

2. Tracheophytes: this term is used to describe early and extant plants that have well-defined tracheids (i.e. water-conducting cells with helical, annular, or other more complex thickenings). This group is further divided into two clades, Rhyniopsida and Eutracheophytes (see Figure 3.18). The Rhyniopsida (which includes *Rhynia*, described above) are defined by a distinctive type of adventitious branching, an abscission layer at the base of the sporangium, and the distinctive S-type tracheid. The Eutracheophytes include two distinct groups (for a detailed description, see Kenrick and Crane 1997a and 3 below).
3. Eutracheophytes: the eutracheophytes contain all extant vascular land plants and most vascular plant fossils. This group is divided into two major clades—the Lycophytina and the Euphyllophytina (see Figure 3.18). The Lycophytina are characterized by a number of features, including sporangia borne laterally along the stem or attached to it by a short stalk with marked thickening on the sporangial line of dehiscence (i.e. where the sporangium opens to release the spores). Fossil examples in this clade include both *Zosterophyllum* and *Baragwanathia longifolia* (described above). The Euphyllophytina clade includes all other land plants, including angiosperms (comprising over 99% of extant vascular plant species), *Equisetum*, seed plants, and many fossil taxa related to extant ferns. As a group, it is characterized by monopodial branching (i.e. a central stem from which other stems branch), a helical arrangement of branches, sporangia in pairs and grouped into terminal trusses (i.e. clustered at the ends of fertile branches), and multicellular appendages (spines). A fossil member of this clade includes *Psilophyton dawsonii* (described above).

The phylogenetic relationship between these early vascular plant groups and extant vascular plants will be described in subsequent chapters. It is interesting to note here, however, that over 99% of extant vascular plants can be included in the Euphyllophytina clade, with most of the remaining 1% in the Lycophytina clade. It would therefore appear that many of the characteristics determining our present-day flora were first established at least 400 million years ago.

3.4 Evolutionary trends: green algae to land plants?

Geological evidence suggests that by 400 million years ago plant evolution had progressed from multicellular eukaryotic organisms that were reliant upon an aquatic medium for survival (e.g. green algae), to vascular land plants permanently adapted to a water-deficient habitat. However, what is not so clear from the geological record is the evolutionary pathway that led from the algae to

Table 3.1 Classification of the division Chlorophyta (green algae) in the subkingdom Algae (from Bell, 1992)

Class/Subclass	Order
Prasinophyceae	Pedinomonadales
	Pyraminodales
	Pterospermatales
Chlorophyceae	Volvocales
	Tetrasporales
	Chlorococcales
	Chlorosphaerales
	Chaetophorales
	Oedogoniales
	Sphaeropleales
Ulvophyceae	Ulotrichales
	Ulvales
	Prasiolales
	Cladophorales
	Siphonocladales
	Codiales
	Caulerpales
	Dichotomosiphonales
	Dasyclydales
Charophyceae	Klebsormidiales
	Mesotaeniales
	Desmidiales
	Zygnematales
	Coleochaetales
	Charales
Pleurastrophyceae	Tetraselmidales
	Pleurastrales
	Trentepohliales

the first land plants. In particular, it is unclear whether there was a single precursor to land plants. One of the predominant evolutionary theories for the development of land plants (vascular and non-vascular) is that they evolved from the green algae (Chlorophyta) and, in particular, the Charophyceae (Table 3.1). This is based on evidence from biochemical, morphological, and molecular analyses of extant groups, in combination with evidence from the fossil record.

Biochemical evidence

When comparing the biochemical similarities between land plants and green algae it is apparent that their metabolism and photosynthetic pigments have much in common (Bell, 1992). In general, as in vascular and non-vascular land plants, all green algae contain chlorophyll *a* and *b*, true starch, and cellulose in their cell walls. However, particular groups of green algae, such as the Charophyceae (Table 3.1) have an even closer biochemical affinity.

A study of the enzymes involved in the non-mitochondrial part of respiration in groups of extant green algae and land plants has demonstrated that, in the Charophyceae, the enzyme glycolate oxidase, characteristic of land plants, replaces glycolate dehydrogenase which is present in the other groups of Chlorophyta (Frederick *et al.*, 1973; Bell, 1992). Both enzymes are important as scavengers of carbon that would otherwise be lost by excretion from photosynthetic cells. However, these enzymes differ in cellular location and in the process of retrieving carbon. It has been demonstrated that the presence of glycolate oxidase would have been particularly advantageous to early land-dwelling plants (Graham, 1993).

Another enzyme present in the Charophyceae algae and land plants, but not other green algae, is an enzyme (copper/zinc superoxide dismutase) that eliminates damaging oxygen radicals from plant cells (deJesus *et al.*, 1989; Graham, 1993). Again the advantages of this enzyme to land dwelling cannot be overestimated and it is interesting to note its presence in this group of green algae.

The presence of sporopollenin in spore walls and cuticles in the early fossil record is often taken to indicate an important stage in the terrestrialization process. It is therefore of some significance that certain Charophycean green algae, including, for example, *Spirogyra, Chara*, and *Coleochaete*, all contain sporopollenin in their zygote cell walls (DeVries *et al.*, 1983; Blackmore and Barnes, 1987; Delwiche *et al.*, 1989; Graham, 1990). In addition, a resistant surface similar in structure to cuticle has been recognized on the thalli of certain *Coleochaete* species (Graham and Delwiche, 1992; Graham, 1993).

Molecular evidence

Molecular systematics also appears to support the suggestion that Charophyceae are the closest in evolutionary terms to land plants (e.g. Manhart and Palmer, 1990; Chapman and Buchheim, 1991). One example of this type of approach is work based on an examination of group II introns (non-coding DNA that interrupts short sections of coding DNA) found in the tRNA genes of all land-plant chloroplast DNA (Manhart and Palmer, 1990). Previous work had indicated that all algae and eubacteria had uninterrupted tRNA, demonstrating both their antiquity and also their distance (in evolutionary terms) from land plants. However, more recent examination of the tRNA genes in Chlorophyta (Manhart and Palmer, 1990) has found the distribution of group II introns in *Coleochaete, Nitella*, and *Spirogyra* (all Charophycean algae), again suggesting a closer affinity of this group than other green algae to

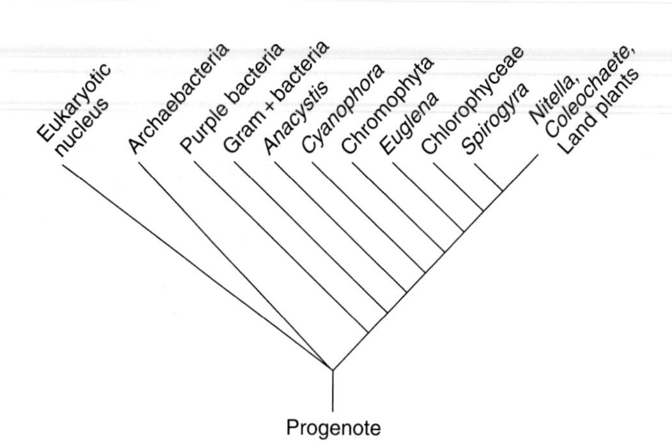

Figure 3.19 A proposed phylogenetic relationship based on molecular evidence for an evolutionary link between Charophyceae and land plants (from Manhart and Palmer, 1990).

land plants (Figure 3.19). Furthermore, the intron data place *Coleochaete* and *Nitella* closer to land plants than *Spirogyra*. It is suggested that these introns were probably acquired by the chloroplast genome between 400 and 500 million years ago.

Morphological evidence

Certain morphological similarities between some species of green algae and vascular plants have also been used to suggest a different evolutionary pathway between these two groups of organisms. For example, at the whole-organism level it has been recognized that various groups of algae are structurally similar to land plants, consisting of a multicellular plant body (thallus) composed of both prostrate and erect components (Stewart and Rothwell, 1993; Taylor and Taylor, 1993). One example that has been cited many times as the possible progenitor to land plants is *Fritschiella* (Figure 3.20), which is classified with the Chlorophyceae group (Table 3.1). This morphologically complex terrestrial green algae has a prostrate system, buried deep in mud, producing a nodule of cells which serves as a perennating organ and an erect system that has a multicellular thallus (Bell, 1992). However, more detailed analysis at the ultrastructural level has now revealed that other groups of algae have an even closer morphological affinity to land plants (Graham, 1993).

Ultrastructural examination indicates many similar characteristics between certain green algae and land plants, especially relating to cell division and reproduction. For example, the reproductive cell ultrastructures of zoospores, meiospores, and gametes in the freshwater green algae, the Charophyceae (Table 3.1), are remarkably similar to those of land plants (both vascular and non-vascular) (Marchant *et al.*, 1973; Pickett-Heaps, 1975). It has also been

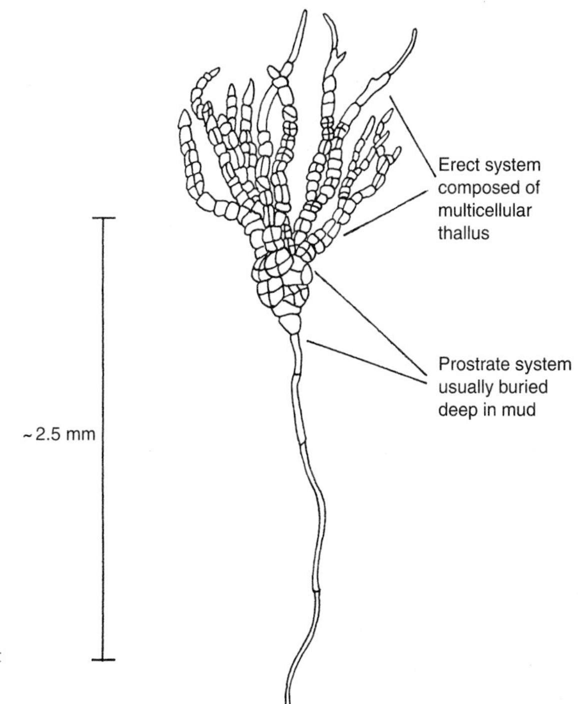

Figure 3.20 Sketch of extant *Fritschiella* (redrawn from Stewart and Rothwell, 1992).

demonstrated that the Charophytes *Coleochaete* and *Chara*, plus a few others (*Spirogyra, Stichococcus, Raphidonema, Klebsormidium*, and *Chlorokybus*), demonstrate a process of cytoplasmic division in mitosis closely resembling that occurring in land plants (Graham, 1993).

Using ultrastructural information in combination with the biochemical evidence, a possible evolutionary pathway between different groups of the green algae and land plants has been proposed that places the Charophycean algae as the nearest relative to land plants (Mattox and Stewart, 1984) (Figure 3.21).

Ecological evidence

As a group, green algae have a wide ecological tolerance, with the ability to exist in a large range of habitats, extending from the subtidal zones of seas, to fresh water, damp soils, the surfaces of leaves, and, in symbiosis with fungi, to the harshest environments on Earth. Before the advent of ultrastructural or molecular techniques, it was these ecological tolerances that were most often cited as evidence for an evolutionary link to land plants. These observations still hold true but, with increasing evidence from biochemical, structural, and molecular

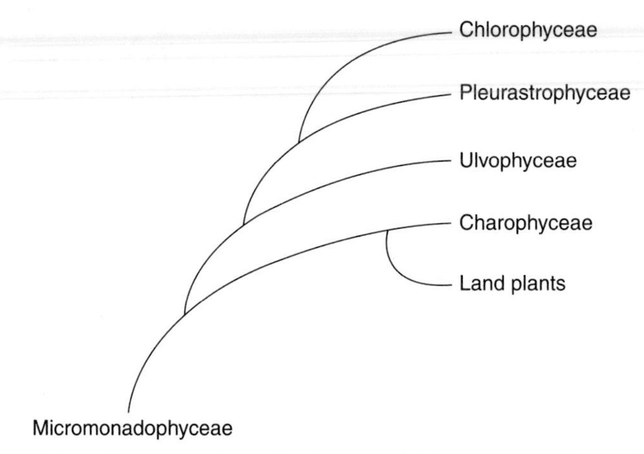

Figure 3.21 Possible evolutionary relationship between different groups of green algae and land plants, based on ultrastructual and biochemical features (from Mattox and Stewart, 1984; Graham, 1993).

studies linking Charophycean algae to land plants, it is more fruitful to examine the ecological tolerances of this particular group (Graham, 1993).

Although various members of the Charophyceae, such as *Chlorokybus* and *Klebsormidium*, occupy terrestrial habitats such as limestone walls, bark, wood, and rooftops, a genus of special interest is *Coleochaete*, because species within it exhibit a range of morphological variation depending upon the environmental conditions (Graham *et al.*, 1992). For example, thalloid species tend to occur in shallow water whereas filamentous species are more common at greater water depths (Figure 3.22). It is suggested that this range of morphological variation may illustrate the kind of morphological transformations that occurred when the charophytes invaded the terrestrial environment (Graham, 1993). These transformations possibly gave them a competitive advantage over other species of green algae.

Fossil evidence: *Coleochaete/Parka*

Fossil evidence would also appear to support the theory that members of the Charophyceae may have provided an evolutionary link between green algae and land plants. In particular there is evidence for a late Silurian/early Devonian terrestrial thalloid plant (~415–395 Ma) that bears a remarkable morphological similarity to the thalloid form of *Coleochaete*. This fossil plant, called *Parka decipiens* (Figure 3.23), was between 0.5 and 7.0 cm in diameter and composed of cells arranged in a pseudoparenchymous organization, spread horizontally across the surface of the substrate. Disc-shaped structures found on the surface of the thallus are thought to be fossil remains of the sporangia (Taylor and Taylor, 1993) and many of them contain small, compressed bodies that are thought to be spores.

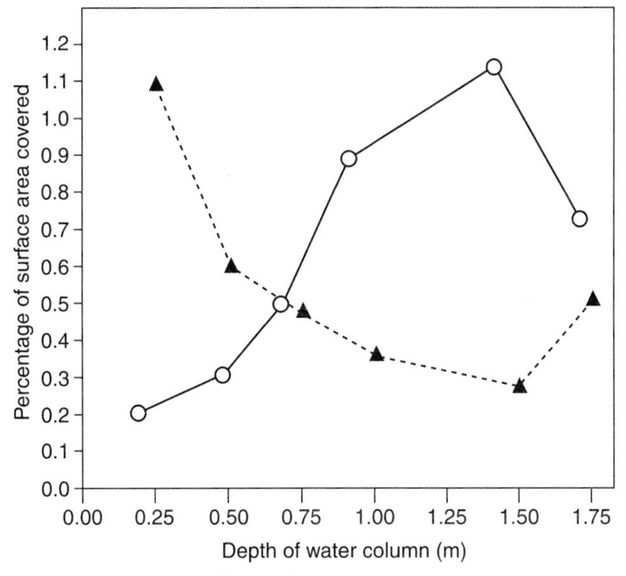

Figure 3.22 Relationship between water column depth and type of extant *Coleochaete* species present (from Graham, 1993).

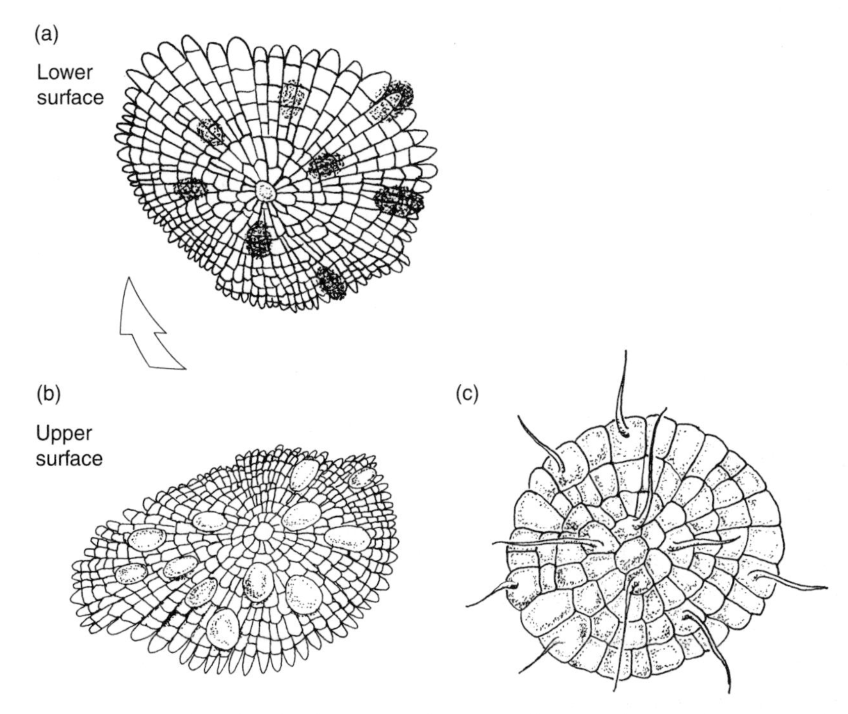

Figure 3.23 Sketch of fossil *Parka decipiens*, indicating (a) lower surface and (b) upper surface. (c) Extant *Coleochaete* algae. Specimens range from 0.5 to 7.0 cm in diameter (redrawn from Taylor and Taylor, 1993).

Chemical analysis of the fossil remains of *Parka decipiens* supports the suggestion that this fossil species was possibly an evolutionary link between green algae and land plants. Starch, cellulose, and a range of hydrocarbon compositions similar to those seen in present-day green algae have been extracted from *Parka* fossil remains. Further chemical analysis revealed that it also contained two steranes that are only present today in land plants and Chlorophyta (Niklas, 1976). From the fossil chemical evidence, therefore, the conclusion is drawn that *Parka* is a fossil representative of the green algae but defined in the broadest sense, i.e. a green plant that is classified neither as a bryophyte nor a tracheophyte. (A note of caution must be added here, though, since the fossil record also suggests that *Parka* grew right next to *Zosterophyllum*, which is a tracheophyte.)

3.5 Evolutionary trends: non-vascular to vascular plants?

As more fossil evidence comes to light, it is becoming apparent that the record of the earliest plants is highly biased towards vascular plants (tracheophytes). There are few macrofossil records of early non-vascular land plants (bryophytes), although, as described above, there are 'disarticulated' microfossils to suggest their presence, such as unlignified tubes, rhizoids, and tetrahedral spores. This biased record is certainly due in part to the poor preservation potential of non-vascular plants (with less cuticle and other thickened parts). It is therefore usually assumed that the earliest terrestrial flora contained vascular (tracheophytes) and non-vascular plants (bryophytes), and plants that had characteristics of both groups (Graham, 1993; Bateman *et al.*, 1998).

A number of interesting questions regarding the evolutionary relationship between the vascular plants, the non-vascular plants, and green algae therefore emerge. For example, did vascular plants evolve from non-vascular plants, which in turn evolved from green algae? Or did the two groups (vascular and non-vascular) evolve separately from a common ancestor? There is also the question of whether a closer evolutionary relationship exists between one branch of the bryophytes (for example the mosses) and vascular plants, than other groups (such as the liverworts or hornworts). Similar to tracing the possible evolutionary links between green algae and land plants, a number of different lines of evidence have been addressed, including geological, morphological, and molecular evidence and cladistic studies.

Geological evidence

There is increasing geological evidence to support the notion that the earliest land plants were non-vascular, that is they were bryophytes (mosses, liverworts, and hornworts), and that vascular plants (tracheophytes) evolved later. Although the geological record is probably biased due to poor preservation, most evidence for the earliest land plants based on microfossils is closer in affinity to extant bryophytes, in particular the liverworts, than tracheophytes (Edwards *et al.*, 1995; Taylor *et al.*, 1995; Kenrick and Crane, 1997). For example, in the geological record the earliest spores with a tetrahedral arrangement bear a close morphological resemblance to the spores of modern liverworts (Gray, 1985). It is also suggested that some of the earliest land plants bear a stronger morphological resemblance to the simple spore-producing phase of mosses and liverworts than to vascular plants (e.g. Edwards, 1982; Mishler and Churchill, 1985; Kenrick and Crane, 1997). Many of the earliest tubes and tracheids also bear a close morphological resemblance to those found in extant mosses and liverworts (Edwards *et al.*, 1995; Kodner and Graham, 2001).

Morphological evidence

Extant groups of bryophytes and tracheophytes share some developmental features, including an alternating life cycle, development of an embryo when the eggs of the gametophyte are fertilized by the sperm, and production of non-motile spores with sporopollenin in their walls. This has been used as evidence that they evolved from a common ancestor (Gray and Shear, 1992). However, phylogenetic studies based on these extant features suggest that although they share a common ancestor in the green algae, mosses and tracheophytes are most closely related, followed by the hornworts. Liverworts are a separate basal lineage (Bateman *et al.*, 1998 and references cited therein).

Molecular evidence

Increasingly, molecular studies also support the suggestion that liverworts are the basal group and that either mosses or hornworts are the living sister group to tracheophytes (Hiesel *et al.*, 1994; Kranz *et al.*, 1995; Kranz and Huss, 1996). In a recent study, for example, 353 extant diverse land plants were examined for the presence of three mitochondrial group II introns (Qiu *et al.*, 1998). Results indicate that they are present in mosses, hornworts, and all major lineages of vascular plants, but entirely absent from liverworts and green algae. These results therefore imply that liverworts were the earliest land plants and that the three introns were acquired in a common ancestor of all other land plants (Figure 3.24).

In summary, therefore, recent geological, morphological, and molecular evidence appears to suggest that all land plants (vascular and non-vascular) evolved from a common ancestor, probably the green algal group, Charophyta.

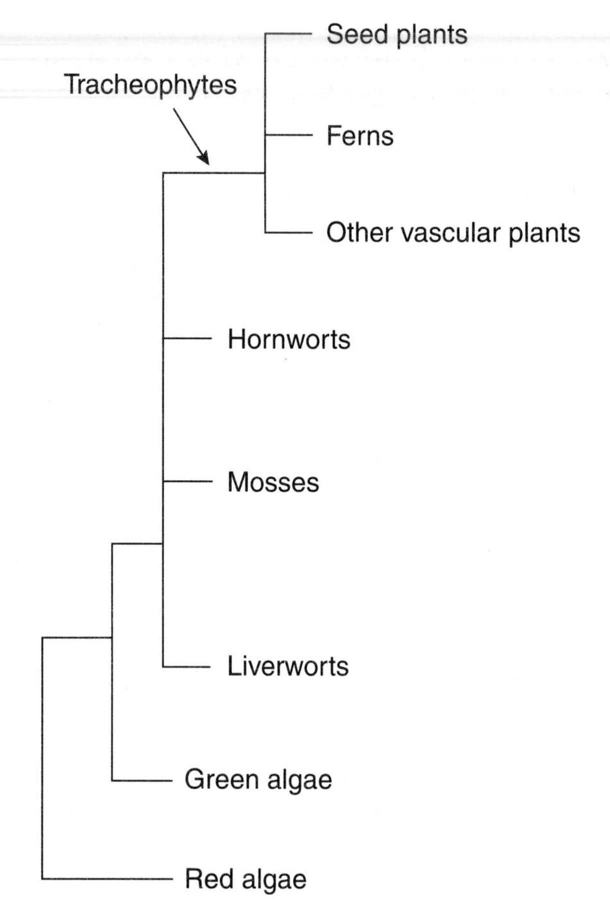

Figure 3.24 Suggested molecular lineage between bryophtes and tracheophytes (after Qiu *et al.*, 1998).

In evolutionary terms, the land-plant lineage then split between the liverworts and all other land plants (Kenrick and Crane, 1997). A second split then occurred between the hornwort lineage and all other land plants, and finally a split between the mosses and the two tracheophyte clades (lycophytes and the euphyllophytes). This interpretation therefore suggests that mosses should bear the closest evolutionary affinity to vascular land plants. Many of the early land plants have morphological features found in both mosses and vascular plants. In *Aglaophyton major*, for example (described in Section 3.3), there is evidence for a branched, nutritionally independent sporophyte which is typical of extant vascular plants, but also an absence of leaves, roots (they have rhizoids), and well-defined thickening in the tracheids, typical of extant mosses (Kenrick and Crane, 1991, 1997a; Edwards *et al.*, 1995). In a sense, therefore, such fossils provide an important missing link to interpreting the evolution of vascular land plants.

3.6 Biogeographical distribution of the earliest land plants in the late Silurian and early Devonian (~430–390 Ma)

A framework detailing the various stages towards the colonization of land by plants (Gray, 1993; Kenrick and Crane, 1997a,b; Figure 3.18) suggests that three epochs can be recognized in the spore record. The first is termed the Eoembryophytic epoch (i.e. emphasizing the role of the bryophytes and tracheophytes in the early stages of terrestrialization). It is dated from mid-Ordovician to early Silurian (~476–432 Ma) and represents the time in the fossil record when there is evidence for spore tetrads with a decay-resistant wall and tetrahedral configuration. The second epoch, the Eotracheophytic epoch, is dated to between early Silurian and early Devonian (~432–402 Ma). It is marked by an increase in the presence of single spores with trilete markings, and also an increase in elaboration of spores, along with the first unequivocal microfossil evidence for land plant fossils (e.g. tubes, tracheids, cuticle, stomata). Macrofossil evidence includes axes and reproductive structures indicating the first land plants, all below 10 cm in height. Thirdly, the Eutracheophytic epoch, dated to between mid and late Devonian (~389–356 Ma), represents a time when there was a widespread increase in spores and megafossils, a substantial increase in vascular plant diversity, and the appearance and diversification of many important sporophyte groups (discussed in Chapter 4).

It was originally assumed that the biogeographical distribution of plants in the Eotracheophytic epoch (432–402 Ma) was uniform and cosmopolitan (reviewed by Wnuk, 1996). There is, however, increasing geological evidence to suggest that provinciality was well developed from the first appearance of vascular plants (Raymond *et al.*, 1985; Raymond, 1987; Edwards and Berry, 1991, Wnuk, 1996).

At the broadest level, five biogeographic units have been recognized for this period (late Silurian–early Devonian) (Raymond, 1985, 1987; Edwards and Berry, 1991; Wnuk, 1996), a Siberian–North Laurussian region, a Kazakhstan region, a South Laurussian–China region an Australian region (Figure 3.25) and up to two Gondwanan regions (northern and southern), of which only the northern is shown in Figure 3.25. Each had a distinctive flora, although many of the localities also shared common genera. Cluster analysis of the fossil assemblages covering this time period have also revealed that the same biogeographic units are determined if morphological traits (such as clustering of sporangia) are analysed instead of generic names (Raymond, 1987). A tenuous link has therefore been made between these traits and climatic conditions. For example, in the case of clustered sporangia, it is suggested that this arrangement would have provided protection of the reproductive organs from dry or cold periods, by reducing the area of sporangia exposed to drying or freezing winds. This has led to the suggestion that even in the earliest stage of terrestrialization, environmental factors were a strong determinant in the global distribution of plants.

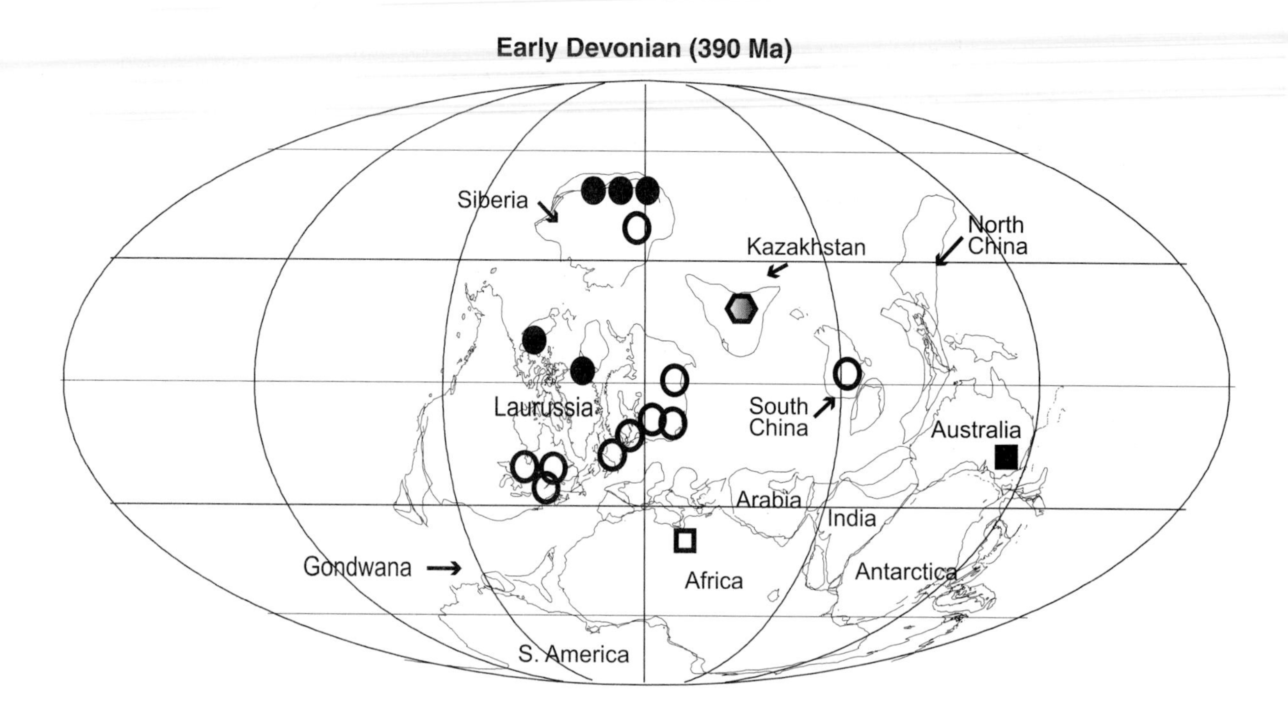

Summary

1. Land surfaces were available for colonization by plants from an early point in Earth's history. However, other essential environmental prerequisites were still developing. These included the formation of sizeable and stable near-shore environments, the formation of soils, and the development of suitable climatic and atmospheric conditions.

2. During the Cambrian and Ordovician (543–443 Ma) a combination of climate change and changing continental configurations resulted in widespread flooding of the continental plates. This was then followed by a period of glaciation at the end of the Ordovician (440 Ma) which led to a dramatic reduction in sea level and exposure of large areas of continental shelf.

3. Processes important for the early development of soils would have included atmospheric elemental input, and weathering by acid rain and organic acids produced by early microbial organisms and lichens. Geological evidence indicates that by the end of the Ordovician (~440 Ma) well-established soil profiles had developed.

4. By the late Ordovician (458–443 Ma) global climates were becoming much more variable and certain regions were becoming cool and moist. These climatic changes are attributed to two main factors: a gradual reduction in atmospheric CO_2, leading to a reduced greenhouse effect, and the formation of glaciers at the South Pole.

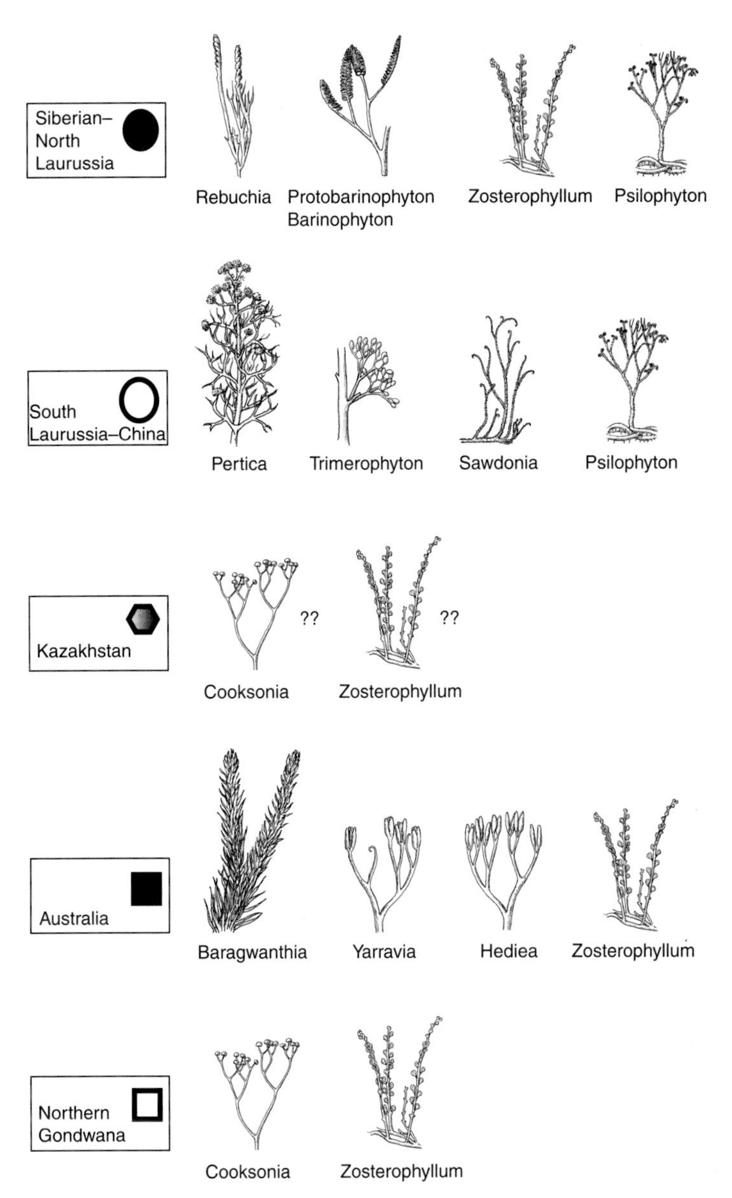

Figure 3.25 Biogeographical distribution of the earliest land plants between approximately 417 and 390 Ma (early Devonian) (from Raymond, 1987), with representatives of the most characteristic fossil plant taxa of each biogeographical region shown. The distributions have been plotted on a global palaeogeographic reconstruction for the Emsian (390 Ma) (courtesy of C. R. Scotese, PaleoMap Project).

5. From the middle Ordovician to early Silurian (~470–430 Ma) evidence starts to emerge in the plant fossil record for the development of specialized cells for water and nutrient transport, various measures to protect against desiccation, mechanical support, and a reproductive mode that did not depend predominantly upon external water. These morphological changes were essential prerequisites for land dwelling by plants.

6. Evidence for whole vascular land plant organs, such as stems and reproductive structures, start to appear in the fossil record in the early Devonian (~408 Ma). These macrofossils (rather than the microfossils of single tubes, tracheids, spores, etc.) have been identified from enough fossil localities to suggest that a vegetation of small-stature vascular plants had become established on many continents by this time.

7. Six examples of the most common early land plants found in the fossil record are: *Cooksonia, Aglaophyton major, Rhynia gwynne-vaughanii, Zosterophyllum divaricatum, Baragwanathia longifolia*, and *Psilophyton dawsonii*.

8. Cladistic studies indicate that the early land plants can be divided broadly into three nested groups, the polysporangiophytes, the tracheophytes, and the eutracheophytes. The latter group contains all extant vascular land plants and most vascular plant fossils, and it would appear from these studies that many of the characteristics that determine our present-day flora were first established at least 400 million years ago.

9. Various lines of evidence, including biochemical, morphological, and molecular analyses of extant groups, in combination with evidence from the fossil record, indicate that land plants evolved from green algae (Chlorophyta) and in particular the Charophyceae.

10. The record of the earliest land plants is highly biased towards vascular plants. This bias is certainly due in part to the poor preservation potential of non-vascular plants (with less cuticle and other thickened parts). It is therefore usually assumed that the earliest terrestrial flora contained vascular plants (tracheophytes), non-vascular plants (bryophytes), and plants that had characteristics of both groups.

11. Tracing the evolutionary links between vascular and non-vascular plants through geological, morphological, and molecular evidence suggests that all land plants (vascular and non-vascular) evolved from a common ancestor, probably the green algal group, Charophyta. In evolutionary terms, the land plant lineage then split between the liverworts and all other land plants. A second split then occurred between the hornwort lineage and all other land plants, and finally a split between the mosses and the two vascular clades (lycophytes and the euphyllophytes). This interpretation therefore suggests that liverworts (i.e. non-vascular) were the earliest land plants.

12. Five biogeographic regions can be recognized in fossil floras by the late Silurian–early Devonian (~430–400 Ma).

4 The first forests

Between 395 and 286 million years ago (early Devonian to late Carboniferous) the terrestrial flora evolved from one composed of small vascular and non-vascular plants to a vegetation that included trees towering to over 35 m. During this time significant changes were also occurring to the global environment. This chapter examines the evidence for these environmental and evolutionary changes and discusses the possible evolutionary pathways from the earliest vascular plants to trees. It concludes with a consideration of the biogeographical distribution of the first multi-storied forests and the factors influencing their distribution.

4.1 Environmental changes spanning the mid-Devonian to late Carboniferous (~395–290 Ma)

The period between 395 and 290 Ma was one of active plate movement. It was also a time of dramatic global climatic change. The continental blocks that had formed the Gondwana and Laurussian groups during the Silurian (Figure 3.1) moved northwards, resulting in the formation of the supercontinent Pangea by 300 Ma (Figure 4.1) (Rowley *et al.*, 1985) and global climates altered from warm, humid, and ice-free, to cooler, drier climates with extensive glaciation in the high latitudes of the southern hemisphere (Crowley and North, 1991; Frakes *et al.*, 1992). Various links between the changing continental configurations and this period of dramatic global climate change have been suggested.

During the early to middle Devonian (~395–360 Ma) the South Pole was situated either over Central America or South Africa (Figure 4.2). This would have resulted in extensive warming and retention of heat at the pole (Crowley *et al.*, 1987). Estimated January temperatures at the South Pole, for example, were between 20 and 24° C and at the equator up to 32° C. From the late Devonian (~360 Ma), however, the South Pole became much closer to the coast and modelling suggests that this would have caused significant global cooling (Figure 4.2). The larger heat capacity of the water would have suppressed the magnitude of warming on the adjacent land masses, and resulting summer temperatures might therefore not have reached above freezing in the high

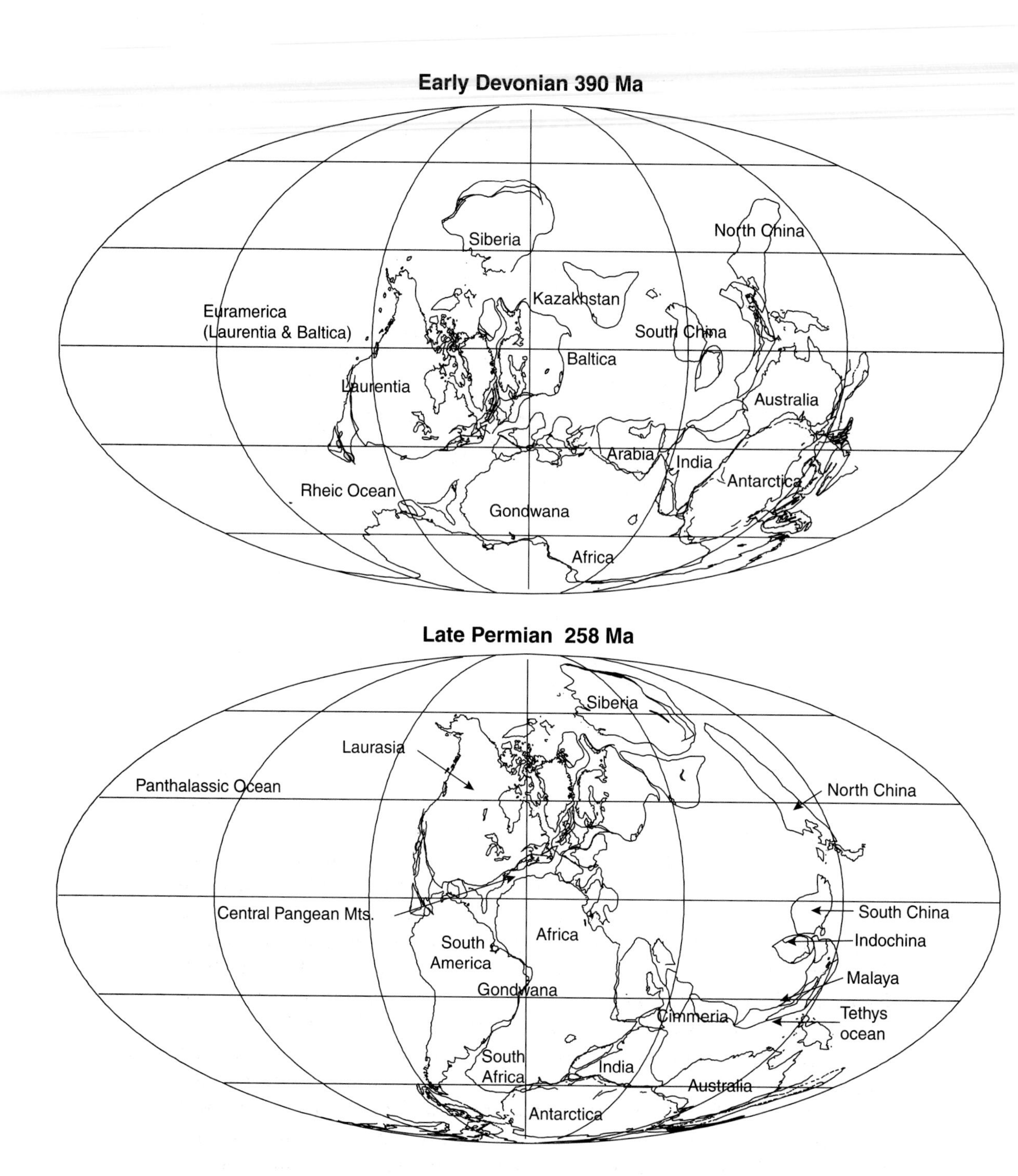

Figure 4.1 Continental plate changes between 390 and 258 million years ago, showing assembly of Pangea (courtesy of Scotese and McKerrow, 1990).

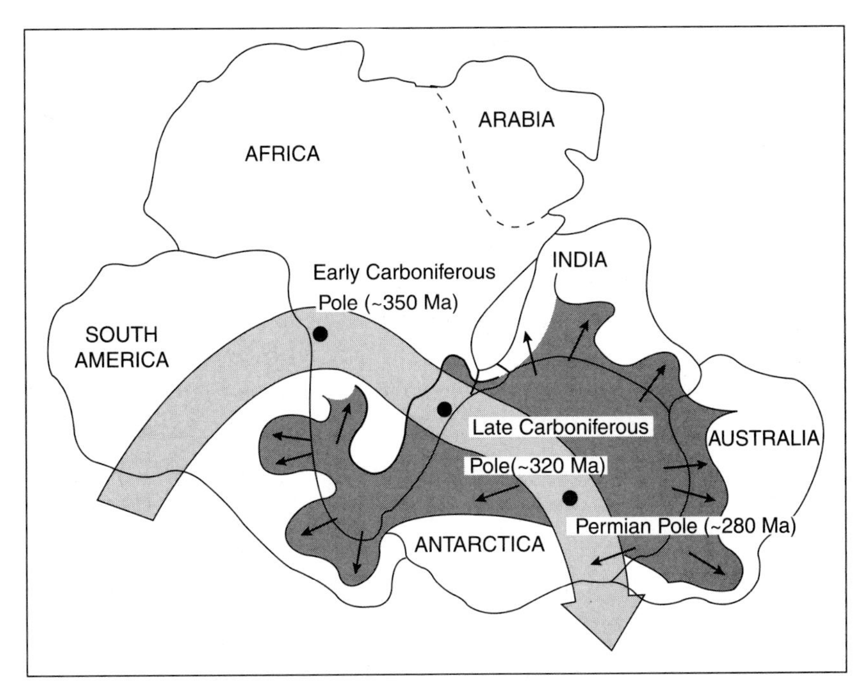

Figure 4.2 Postulated polar wander across Gondwana during the Carboniferous and Permian (~350–248 Ma) (redrawn from Crowley *et al.*, 1987; Condie and Sloan, 1998). The large arrow indicates the direction of wander. Also indicated (stippled) are major glacial centres during this time and ice flow directions (small arrows) reconstructed from glacial pavement studies.

latitudes of the southern hemisphere (Crowley and North, 1991). This would have led to the formation of a large southern hemisphere ice sheet, with increasingly cool and arid conditions in continental interiors.

At least four periods of southern hemisphere glaciation have been recognized between approximately 360 and 290 Ma (Bambach *et al.*, 1980; Crowley and North, 1991) (Figure 3.2). Initial areas of glaciation were in South America and Africa, but then it became more widespread, with evidence for continental glaciation also occurring in Antarctica and Australia (Crowley and North, 1991) (Figure 4.2). During times of glaciation, sea levels were lowered by an estimated 100–200 m and there was aridity in the high latitudes (Aitken, 1985).

Although continental climates became generally drier, fossil evidence also suggests that precipitation became enhanced in a narrow equatorial belt (Raymond, 1985, 1987). High year-round wetness in this tropical belt is thought to be partially responsible for the development of extensive lowland swamps, resulting in the formation of coal deposits in eastern North America, western Europe, and parts of Russia during the Carboniferous (Ziegler *et al.*, 1987; Chaloner and Creber, 1988, 1990). The effect of the continental plate movements is also thought to have been responsible for this. Extensive mountain-building episodes resulted from the continental plate collisions. The formation of the Central Pangean Moutains at tropical latitudes, at approximately 305 Ma (Westphalian), for example, resulted in a mountain range over 3 km in altitude (Condie and Sloan, 1998). Modelling indicates that this uplift would have had a significant impact on global circulation and precipitation (Otto-Bliesner, 1998). In particular, there would have been enhanced local precipitation in palaeoequatorial latitudes, thus accounting for the formation of extensive coal deposits at this time.

Another factor affecting the global climates during this period (395–290 Ma) was less directly related to the continental plate movements and was a result of the widespread colonization of the land by plants. Estimates of atmospheric CO_2 in the early Devonian indicate that levels were up to 8–9 times higher than those of the present-day (Figure 4.3), in fact at its highest level in the past

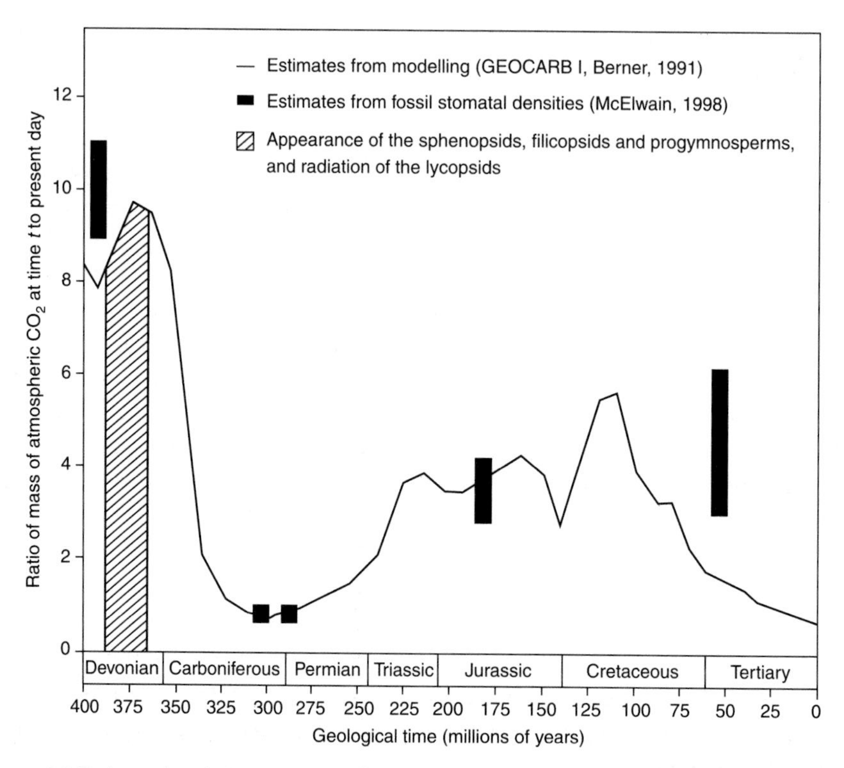

Figure 4.3 Estimated variations in atmospheric CO_2 during the past 400 million years (redrawn from Berner, 1991, GEOCARB I). The general direction of the trends are supported by independently derived results from carbon isotope analysis of palaeosols (e.g. Mora *et al.*, 1996) and stomatal densities on fossil leaves (e.g. McElwain, 1998). For comparison the latter are indicated on the graph. Both are plotted as ratio of mass of atmospheric CO_2 at time *t* to present day (where present is taken at the pre-industrial value of 300 p.p.m.v). The early to mid-Devonian rise in atmospheric CO_2 is highlighted, along with the major changes apparent in the plant fossil record during this time. Note that the Quaternary (past 1.8 Ma) is not marked on the geological timescale.

400 million years (Berner, 1991, 1998; Han and Runnegar, 1992; McElwain and Chaloner, 1995) (Figure 4.3). This would have had a considerable effect on global warmth through an enhanced greenhouse effect. However, between 360 and 286 Ma modelling estimates that atmospheric CO_2 levels plummeted from 3600 p.p.m. to 300 p.p.m., such that by the late Carboniferous, levels of atmospheric CO_2 were comparable to those of the present day (Berner, 1993, 1997) (Figure 4.3). This declining atmospheric CO_2 concentration would also have made a significant contribution to the pattern of global cooling evident from approximately 360 million years onwards. One of the main processes thought to be responsible for this rapid decline was the global expansion of vascular plants and the effect that they would have had on the acceleration of silicate rock weathering and the production of bacterially resistant organic matter (e.g. lignin) (Schidlowski, 1983; Berner, 1998). Both of these processes are thought to have contributed significantly to a decline in atmospheric CO_2.

It is envisaged that with the widespread colonization of land surfaces by vascular plants and their associated release of organic acids (e.g. released through root mycorrhizae, humic and fluvic acids) would have greatly increased chemical weathering of silicate rocks. This process is one that uses atmospheric CO_2 ($CO_2 + CaSiO_3 \leftrightarrow CaCO_3 + SiO_2$), resulting in a transfer of carbon from the atmosphere to carbonate minerals. These would then be washed away and eventually deposited on the ocean floor where they would subsequently become buried and therefore removed from the atmospheric pool (Berner, 1997, 1998). In addition, it is suggested that the large increases in plant organic matter composed of organic polymers such as lignin, that were highly resistant to decay, would have exceeded the numbers of primary decomposers (e.g. various fungi). This would have led to a scenario where rates of accumulation would have exceeded those of decomposition (Robinson, 1991). Burial of these vast quantities of organic material would again have resulted in the removal of carbon from the atmosphere, but this time by locking it away in long-term carbon sinks, namely coal.

4.2 Major changes and innovations in the plant fossil record during the mid-Devonian to late Carboniferous (~395–290 Ma)

During this period of dramatic global environmental change, major changes and innovations were occurring in the terrestrial vegetation. The period spanning the middle to late Devonian (~390–365 Ma) was a time of emergence of new plant groups, and relatively rapid increase of species numbers (Niklas *et al.*, 1985) (Figure 4.4). Estimates suggest, for example, that within a matter of 20 million years, the numbers of spore-producing plants had increased threefold (Knoll,

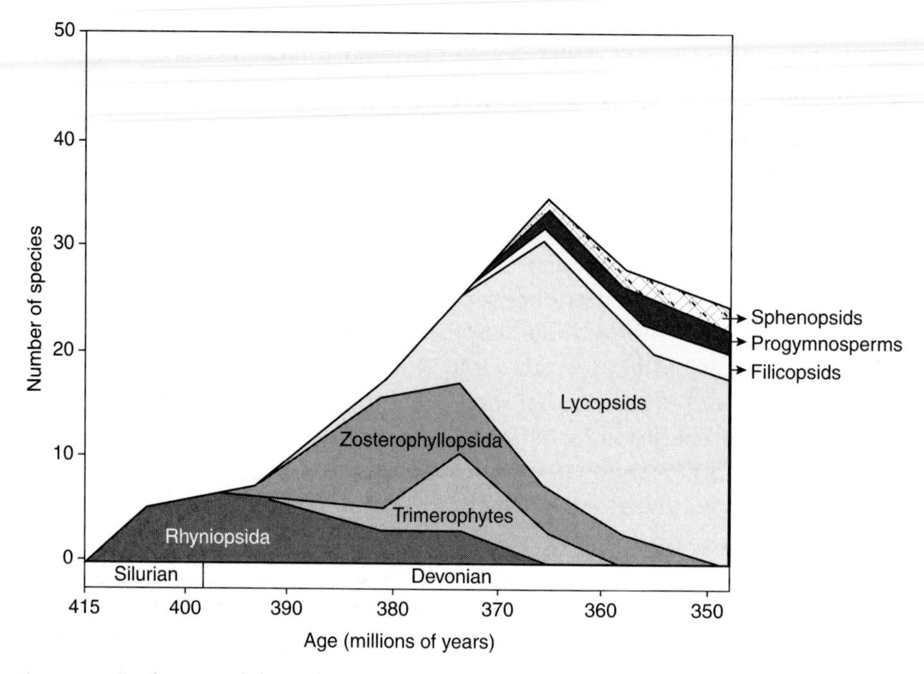

Figure 4.4 Evolution and diversification of vascular land plants during the mid to late Devonian (~390–365 Ma) (redrawn from Niklas *et al.*, 1985a). Data are taken from a compilation of approximately 18 000 fossil plant species citations.

1984; Edwards and Davies, 1990; Raymond and Metz, 1995; Niklas, 1997). This pattern of increasing diversity was not to last, however, and by the late Devonian (~360 Ma), numbers of new species started to plateau such that there were less originations than extinctions, leading to a slight drop in overall species number (Raymond and Metz, 1995). This reduction in overall species number occurred at a time (late Devonian, Frasnian/Famennian) when there was a mass extinction event in the animal record.

Throughout this period (~390–365 Ma) the plant fossil record also indicates that major innovations were occurring in land plant morphology (Chaloner, 1967; Chaloner and Sheerin, 1979; Gensel and Andrews, 1987; Niklas, 1997; Algeo and Scheckler, 1999). In particular, terrestrial plants were getting much bigger and acquiring more refined reproductive structures. There is also evidence for the evolution of a more advanced vascular system for the transport of water and nutrients, additional supporting mechanisms, advanced rooting systems, and leaves. Evolution of these features would have facilitated the large increases in vascular plant size and structure. The fossil evidence for these morphological changes will therefore be discussed first, followed by a description of some of the earliest trees on Earth and their biogeographical distribution.

4.3 Evidence of further plant adaptations to land dwelling between early Devonian and late Carboniferous (395–290 Ma)

Advanced vascular system (stelar evolution)

In order to attain greater height, one of the most important adaptations would have been the development of the central conducting cylinder (the stele) for the effective transport of water and nutrients around the plant (Stewart and Rothwell, 1993; Taylor and Taylor, 1993; Niklas, 1997). Evidence from the geological record suggests that the stele of plants became progressively more complex with time and that by the late Devonian (~374 Ma) at least three different stele types were apparent.

Protostele

The earliest stele found in the fossil record is the protostele, found in most of the earliest vascular plants (Chapter 3) and also some early lycopsids and sphenopsids (see Section 4.5). This stele (Figure 4.5a) is composed of a solid strand of vascular tissue in which phloem (food-conducting tissues) either surrounds the xylem (water and mineral conducting tissues) or is interspersed within it.

Siphonostele

A more complex stele, termed the siphonostele (Figure 4.5b), became apparent in the geological record in the Emsian (~395 Ma). This stele was made up of xylem

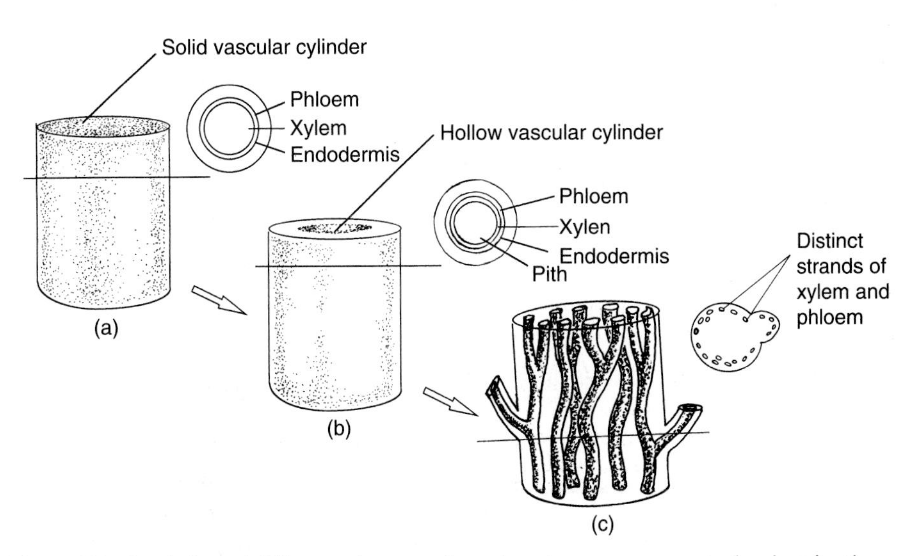

Figure 4.5 Morphological differences between the early stele types apparent in the plant fossil record: (a) protostele, apparent in the earliest vascular plants from ~420 Ma; (b) siphonostele, apparent in the fossil record from ~395 Ma; (c) eustele, apparent in the fossil record from ~380 Ma.

and phloem forming a cylinder around a central core filled with pith, and has been found in fossil filicopsids, lycopsids, and some sphenopsids (see Section 4.5).

Eustele

A third stele type, termed the eustele (Figure 4.5c), first emerged in the fossil record in the Givetian (~380 Ma). This stele appears to be the most complex of the three structures in that it was composed of distinct strands of phloem and xylem, separated by parenchymal tissue. These strands were arranged either in a circle or throughout the ground tissue (Taylor and Taylor, 1993). The eustele first occurred in the progymnosperms (the precursor to the seed plants).

Evidence from the geological record thus suggests that the stele of plants became progressively more complex with time. It is also notable that with increased stelar complexity there was an increase in stem diameter and height. As shown in the previous chapter, the earliest vascular land plants were small and narrow-stemmed, and the transport of water and nutrients in the plants occurred through specialized tracheids. The central stem therefore contained the tissue for support, transport of water/nutrients, and also photosynthesis. Calculations indicate that the best arrangement of tissue for these three functions would have been a photosynthetic 'rind' of tissue just beneath the surface of the stem and a central water-conducting strand running through an inner core of non-photosynthetic tissue (Niklas, 1997). This type of arrangement is seen in the protostele, where the xylem cells would have provided mechanical support and the phloem cells would have transported sugars and nutrients. These two layers were then surrounded by an outer layer of photosynthetic tissue (Figure 4.5a).

However, calculations indicate that in order to attain greater heights the required thickness of the xylem cells (to provide mechanical support) would eventually present a barrier between the photosynthetic and water-conducting tissues. Therefore, beyond a certain size it would have been necessary to compartmentalize functions of the plant body into organs devoted to photosynthesis and organs devoted to mechanical support and water conduction (Niklas, 1986, 1997). Increasing complexity of the stele (Figure 4.5) in combination with innovations in supporting mechanisms and the evolution of true leaves (discussed in the next section) are therefore regarded as fundamental to the evolution of arborescence (a woody habit) in plants (Speck and Vogellehner, 1988; Algeo and Scheckler, 1999). Ultimately this would lead to the appearance of well-stratified forested ecosystems on Earth.

Additional supporting mechanisms

Arborescent forms in the extant flora display a variety of different supporting mechanisms, ranging from thickening of the trunk with wood and bark, to above-ground root mantles and extensive underground rooting systems. Fossil evidence suggests that many of these supporting mechanisms developed from as early as 380 Ma (Givetian). Probably the most common form of support for extant trees is achieved by thickening of the trunk through the growth of wood. Wood results from the growth of secondary xylem and phloem from lateral meristematic zones known as the vascular cambium (Raven *et al.*, 1992). Evidence from the fossil record suggests that the formation of secondary xylem and pholem first occurred in the middle Devonian (~380 Ma) and enabled some of the earliest known trees (e.g. *Archaeopteris*, see p. 110) to achieve heights of up to 30 m (Meyer-Berthaud *et al.*, 1999).

However, palaeobotanical evidence indicates that in other early trees, thickening of the trunk for increased stature was achieved in other ways. These included the development of large quantities of inner bark (usually referred to as secondary cortex) (Thomas and Spicer, 1987; Taylor and Taylor, 1993) and the development of greatly thickened stems as a result of numerous leaf bases (e.g. *Lepidodendron*, see p. 102). Another mechanism for support appears to have been the development of abundant mantles of roots, that were thickest at the base of the trunk. These would have acted as guy-ropes, tethering the tree to the ground and enabling much greater stature to be achieved (e.g. *Psaronius*, see p. 108).

In addition to above-ground root mantles, evidence from palaeosols (fossil soils) indicates the development of at least seven morphologically distinguishable below-ground rooting systems during the Devonian and Carboniferous (Bockelie, 1994) (Figure 4.6). They ranged from highly branched systems extending up to 1 m into the substrate to small annulated-segmented roots from which long root hairs grew at the point of segmentation. These rooting systems would have been vital, for both support and the increased uptake of water and nutrients. They would also have had a significant impact on weathering rates and processes, increasing the depth of mature soil profiles (Algeo and Scheckler, 1999). The variation between different root types is thought to reflect the need for anchorage and support, and also the different environmental conditions in which the plants grew (Thomas and Spicer, 1987).

Leaves

The earliest land plants possessed stomata and cuticles in their stems and were therefore photosynthesizing from these axes. For at least the first 40 million years of their existence land plants were therefore leafless or had only small, spine-like appendages (Kenrick, 2001). However, leaf-like structures and true leaves, primarily devoted and optimized for the photosynthetic process, started to appear in the fossil record from the mid to late Devonian (~390–354 Ma), respectively. There were two early forms of leaf type—microphylls and megaphylls—and the differences between them can still be seen in plant groups today.

Microphylls

These are generally relatively small leaves (although there are exceptions) that grow out directly from the stem. They contain only a single strand of

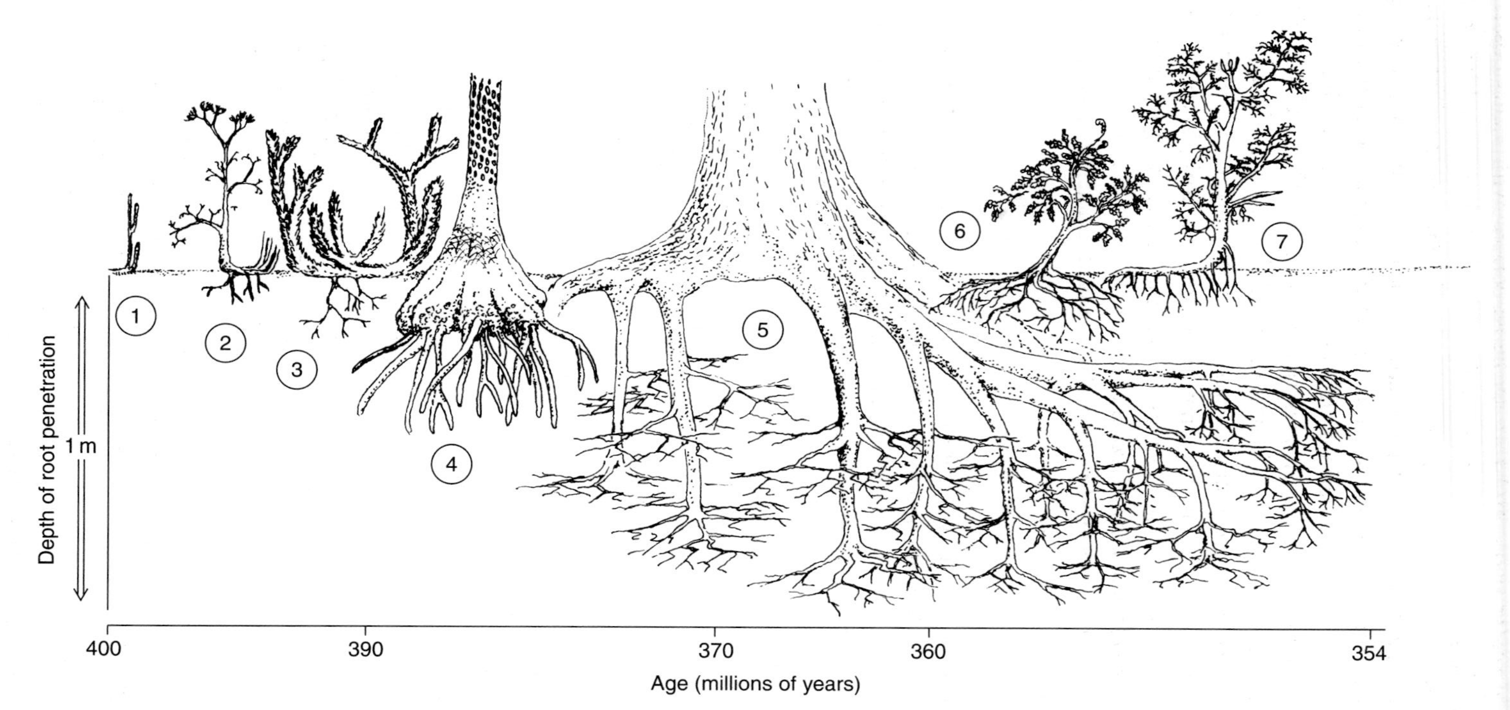

Figure 4.6 Relative sizes, shapes, and penetration depths of root systems during the early, middle, and late Devonian (~400–354 Ma; age axis is not drawn to scale) (redrawn from Algeo and Scheckler, 1999). Representative plant types are as follows: 1, Rhyniopsida, e.g. *Aglaophyton*; 2, trimerophytes, e.g. *Psilophyton*; 3, early herbaceous lycopsids, e.g. *Asteroxylon*; 4, early tree lycopsids, e.g. *Lepidodendron*; 5, progymnosperm, e.g. *Archaeopteris*; 6, early gymnosperm, e.g. *Elkinsia*; and 7, filicopsid, e.g. *Rhacophyton*.

unbranching vascular tissue and tend to be associated with stems possessing protosteles. It is proposed that they evolved from one of two hypothetical pathways, termed the Enation theory and the Telome theroy. The 'Enation theory' (Bower, 1935) proposed that microphyll leaves evolved from the spiny stems of the earliest land plants, such as *Sawdonia*. Lateral outgrowth of vascular tissue into the spines (enations) resulted in the spines 'fleshing out' and the formation of microphyllous leaves (Figure 4.7a(i)). The 'Telome theory' (Zimmerman, 1952), on the other hand, suggested that microphylls represented the end product of an evolutionary series of reduction from once-forked side branches, which became aligned (planation) (Figure 4.7a(ii)). Such 'stem-hugging' leaves, with a single vascular strand and no petiole, are typical of extant species of lycopsids such as *Lycopodium* (see Figure 4.17) and were the characteristic leaf-type of the earliest, but now extinct, lycopsids, such as *Baragwanthia* (Figure 3.16).

Megaphylls

These are leaves associated with stems that have either a siphonostele or eustele, and are attached to the stem by a petiole (Figure 4.7b). Their evolution is thought to be closely linked to the three-dimensional vegetative branching pattern (i.e. branches with no sporangia on them) of the earliest vascular plants

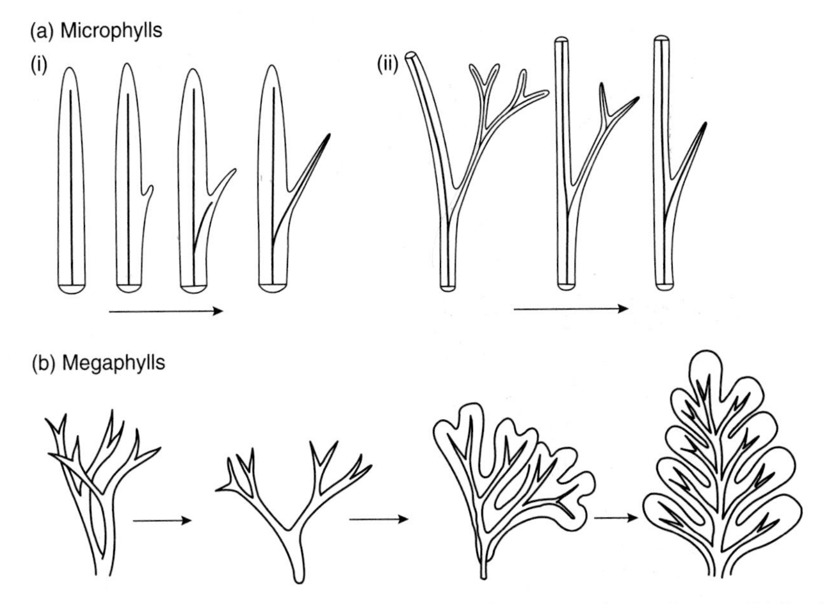

Figure 4.7 (a) Formation of microphylls according to (i) the 'Enation theory' (Bower, 1935), which proposed that these 'stem-hugging leaves' evolved from the outgrowth of vascular tissue into the spines on the stems, and (ii) the 'Telome theory', which suggested that they were the end product of an evolutionary series of reduction from once-forked side branches. (b) Formation of megaphylls according to 'Telome' theory' (Zimmerman, 1952), which proposed that these leaves evolved by a fusion of non-fertile branches, with undifferentiated tissue forming webbing between the branches. (Redrawn from Stewart and Rothwell, 1993.)

(e.g. *Psilophyton dawsonii*, Figure 3.17). According to the Telome theory (Zimmerman, 1952), megaphylls are thought to have evolved by a fusion of these non-fertile branches, with undifferentiated tissue forming webbing between the branches. The former branches would thus provide vascular tissue into the megaphyll leaf, resulting in the evolution of a broad leaf containing many vascular strands (Figure 4.7b). Early and extant filicopsids (ferns), a number of extinct sphenopsids (horsetails), and all flowering plants possess megaphylls.

Recent studies based on morphological observations and modelling of the biophysical principles of plant physiology suggest that evolution of the megaphyll leaf may have occurred in response to the massive reduction in atmospheric CO_2 during the late Devonian and early Carboniferous (Robinson, 1991; Beerling *et al.*, 2001). It is suggested that in the high CO_2 world of the early and mid Devonian (~417–370 Ma), the evolution of a megaphyll leaf would have been detrimental to plant survival, as leaf temperatures would have exceeded lethal threshold limits. It was not, therefore, until the late Devoninan/early Carboniferous, when CO_2 levels plummetted, that increased leaf size would have been advantageous to land plants (Beerling *et al.*, 2001).

4.4 Further adaptations to the plant life cycle

Alongside the morphological innovations that were taking place in plant vegetative structures, the period between 395 and 286 Ma (early Devonian to late Carboniferous) saw the development of the seed, which marked a major step in plant reproductive strategy. Evolution of the seed habit revolutionized the plant life cycle by freeing plants from the necessity of external water for sexual reproduction and by affording protection and providing nutrients to the developing embryo. These three functions enabled plants with the seed habit to expand their geographical extent beyond the confines of stream sides and sources of water, and to exploit drier, upland, and primary successional environments (Algeo and Scheckler, 1999).

Homospory to heterospory

Evidence from the fossil plant record indicates that plants were producing sporangia yielding two kinds of spores (Figure 4.8) from the early Devonian (Emsian, ~400 Ma) onwards. These include megaspores that were between approximately 150 and 200 μm in diameter and microspores that were usually <50 μm (Traverse, 1988). The transition from plants that were homosporous (one spore size) to heterosporous (two spores sizes) is considered one of the most important evolutionary trends in the development of seed-bearing plants. It is postulated that the larger spores of heterosporous plants were the precursor of ovules, and the small spores, the precursor to pollen (Figure 4.9).

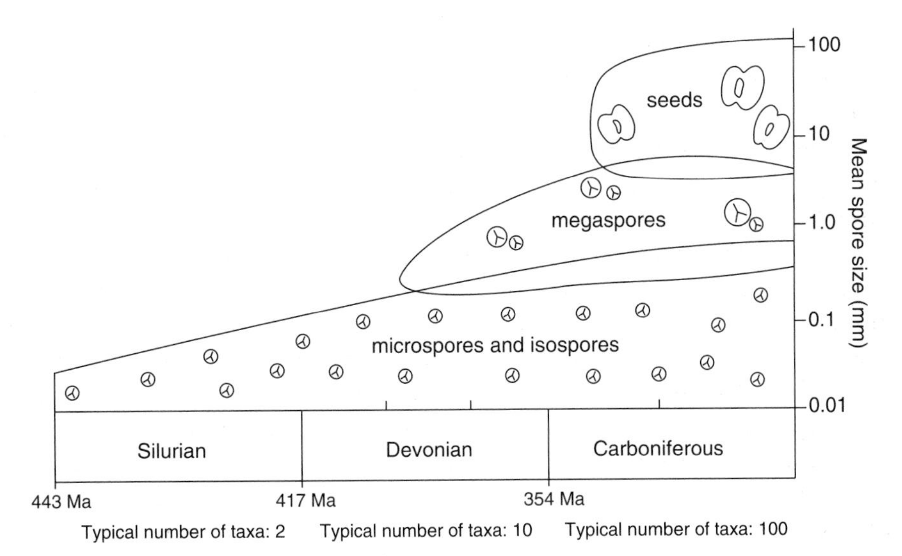

Figure 4.8 Spore sizes in the geological record (redrawn from Traverse, 1988b) indicating that a transition from one to two spore sizes occurred from the early Devonian (~400 Ma).

There is much discussion in the literature as to the formation of megaspores. The most widely accepted theory is that mutation resulted to two spore sizes developing in a sporangium (Thomas and Spicer, 1987). Evidence from the geological record would appear to support this theory in that records of dispersed spores of two size classes (anispory) indicate a gradual increase in occurrence of heterospory through the Devonian (Chaloner, 1967). An early example in the fossil record of a heterosporus plant is *Chaleuria* (Gensel and Andrews, 1984, 1987). Specimens of this late Devonian plant have been shown to contain two types of sporangia, those that produce spores between 30 and 48 μm and others that produce spores between 60 and 156 μm.

Ovule evolution after the onset of heterospory

Evidence from the plant fossil record indicates that by the end of the Devonian heterospory had progressed to the point that development of megaspores involved the abortion of three out of the four spores in a sporangium, with all the energy going into the remaining spore to form a single functional megaspore (Figure 4.10). Retention of the megaspore in the megasporangium was the first step in the direction towards evolution of the ovule. The second step would have been the development of a seed coat.

Development of a seed coat (integument)

Analysis of late Devonian (370–354 Ma) fossil megaspores has revealed an outer protective coating. This would have protected the megasporangium from

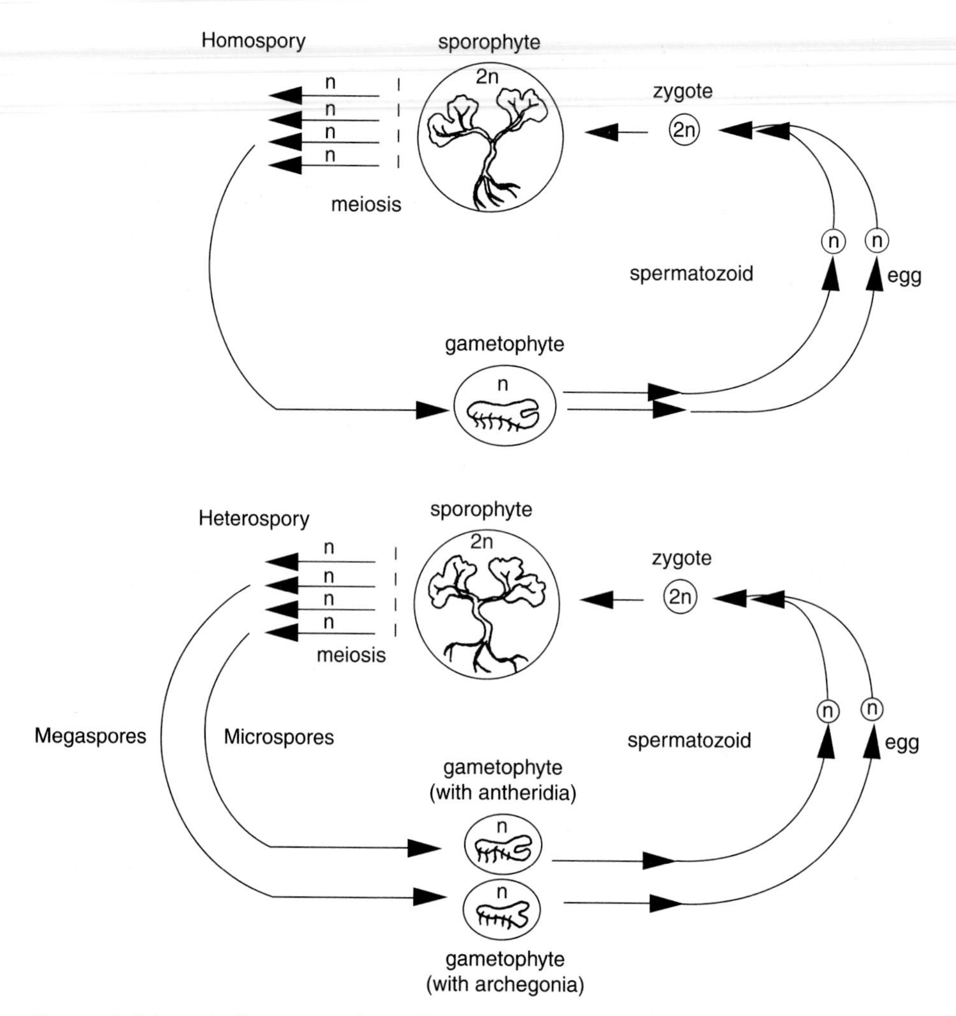

Figure 4.9 Schematic diagram to indicate the proposed transition from homospory to heterospory in the vascular plant life cycle, leading to the development of seed-bearing plants.

desiccation and attack. According to taxonomic nomeclature, once an outer coat or integument has formed around a megasporangium, it technically becomes an ovule, since in structure it now represents an unfertilized seed. Seed coats probably resulted from the envelopment of the megasporangium by sterile telome trusses (another use in addition to the formation of megaphyllous leaves, see Section 4.2), which fused around the megasporangium to encase it, so forming an ovule (Figure 4.11) (Thomas and Spicer, 1987; Taylor and Taylor, 1993). Many of the early ovules were further enveloped by stalked cupules, with up to four ovules within a cupule. Again, the cupule was thought to have developed from fusion of a vegetative branching system.

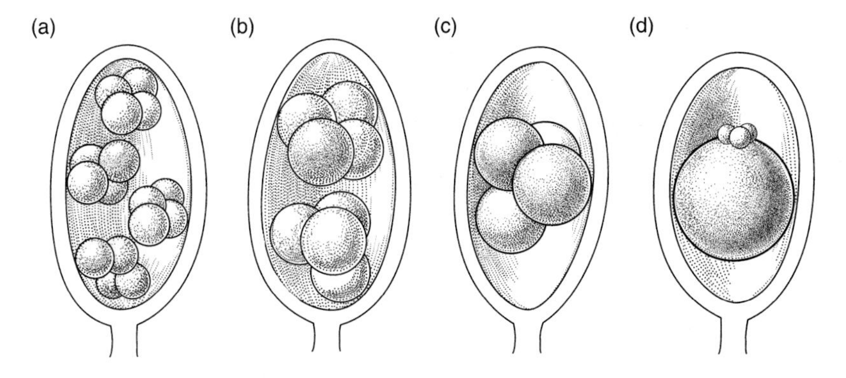

Figure 4.10 Schematic diagram to indicate the probable development of a single functional megaspore within the megasporangium (redrawn from Niklas, 1997).

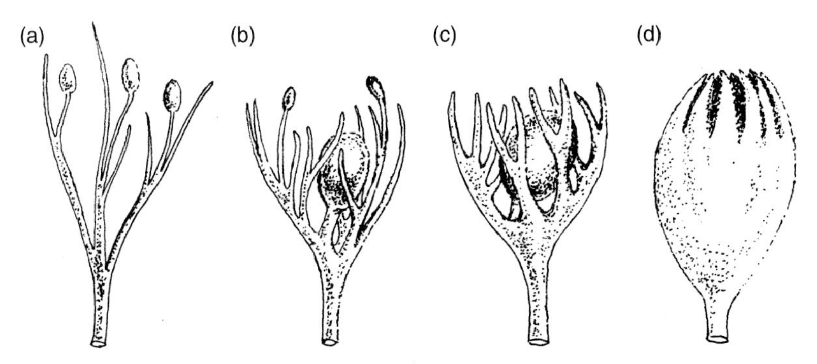

Figure 4.11 Development of the seed coat from sterile telome trusses (redrawn from Andrews, 1961; Stewart and Rothwell, 1993).

Evolution of pollen grains

In association with the development of the ovule (from the megasporangium), palaeobotanical evidence indicates the evolution of pollen from approximately 364 million years ago (Famennian). Evolution of pollen represents adaptation of the male part of the plant life cycle, since prior to this time only spores had been found in the fossil record. Pollen is distinguishable from spores in both structure and the method by which gametes are released (Traverse, 1988b). In exant heterosporous plant groups, microspores liberate flagellated gametes from the proximal end of the spore (i.e. from the trilete aperture), which then swim to the archegonia for fertilization. Pollen, in comparison, produces a pollen tube from the distal end, through which the gametes are transferred directly into the ovule (Figure 4.12).

The earliest pollen grains evident in the fossil record (termed 'prepollen') are halfway between pollen and spores. They contain morphological features of spores, such as the trilete scar, but evidence suggests that germination occurred on or very near to the opening of the megasporangium (Traverse, 1988b). By the

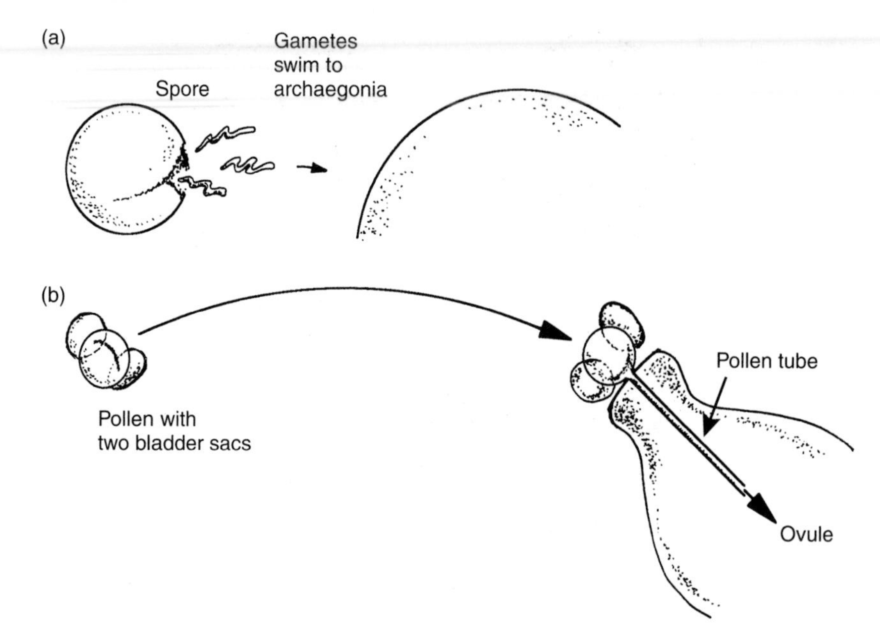

Figure 4.12 Schematic diagram to illustrate the release of gametes from (a) spores and (b) pollen. In exant heterosporous plant groups (a), micro-spores liberate flagellated gametes from the proximal end of the spore (i.e. from the trilete aperture) which then swim to the archegonia. (b) Pollen, in comparison, produces a pollen tube from the distal end, through which the gametes are transferred into the ovule.

late Carboniferous (~310 Ma), however, at least two morphologically distinct pollen types were present in the fossil record (Traverse, 1988b). One kind was saccate (i.e. it had either one or two bladder sacs attached to the grain) and germination occurred from its distal end (i.e. away from the centre of the original tetrad). This type of pollen is morphologically similar to extant gymnosperm pollen (Figure 4.13). The other type consisted of a single grain with a long suture running down the distal edge, presumably from where germination occurred. This morphological type of pollen grain is known as monolete. Both monosaccate and monolete pollen in the early fossil record are attributed to members of the pteridosperms and cordaites (see Section 4.5).

Development of pollen reception mechanisms

With the encasement of the ovule in a seed coat, it would have been necessary to develop a mechanism by which the pollen could reach the egg contained within the ovule, for germination. In the earliest seeds, where the integumentary envelope was only fused at the base (see p. 93), there is evidence for evolution of various elaborate structures at the neck of the megasporanium. In effect, these acted as a funnel to channel the pollen towards the ovule (Figure 4.14a). These pollen-receiving structures, in combination with the lobes of the seed coat and

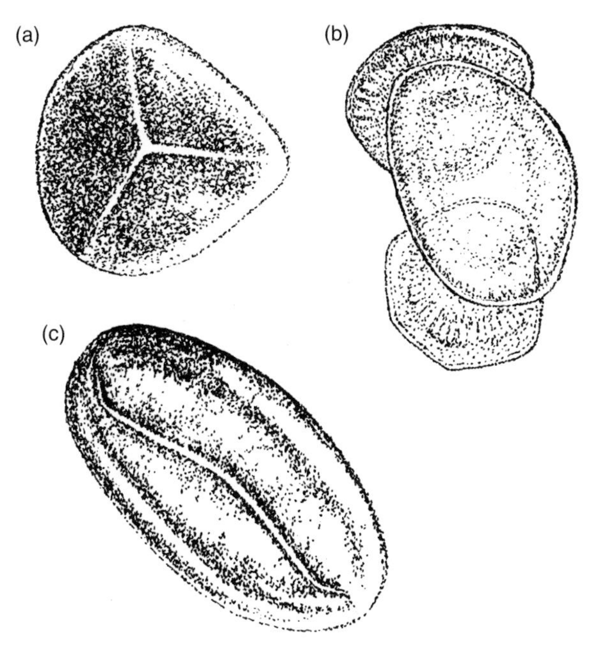

Figure 4.13 (a) Earliest spores with trilete scar; (b) saccate pollen, with two bladder sacs attached to the grain; (c) monolete pollen, with a long suture running down the distal edge. All grains ~20–60 μm in diameter.

cupules, would have been highly effective in trapping wind-borne pollen (Taylor and Millay, 1979). Wind tunnel experiments attempting to reconstruct the aerodynamics of these structures (Niklas, 1983) have indicated, for example, that lobed integuments would have offered a large surface area to capture air-borne pollen (Niklas, 1997) (Figure 4.14a).

However, some of the earliest ovules had much simpler mechanisms for attracting pollen, including hairs that lined both the inside and outside of the integumentary lobes (Figure 4.14b) and pollination droplets (Figure 4.14c). Pollination droplets were composed of a sticky material that exuded from tentacle-like projections situated on the outside of the integument, to which pollen became stuck (Thomas and Spicer, 1987; Taylor and Taylor, 1993).

In extant seed plants, the seed coat is fully enclosed, resulting from the fusion of the integumentary lobes with each other and the nucellus. There is a clearly defined hole in the top of the seed (a terminal pore) called the micropyle, and it is through this that pollen or the pollen tube extends to deposit the male gametes near or in the archegonia. In extant gymnosperms, for example, a sugary 'pollination drop' exudes from the micropyle in which wind-blown pollen becomes trapped. As the pollination drop is withdrawn back into the nucleus, the pollen is taken in with it (Bell, 1992).

Evidence in the fossil record suggests that development of an enclosed integument and associated micropyle occurred in the late Devonian. The advantages

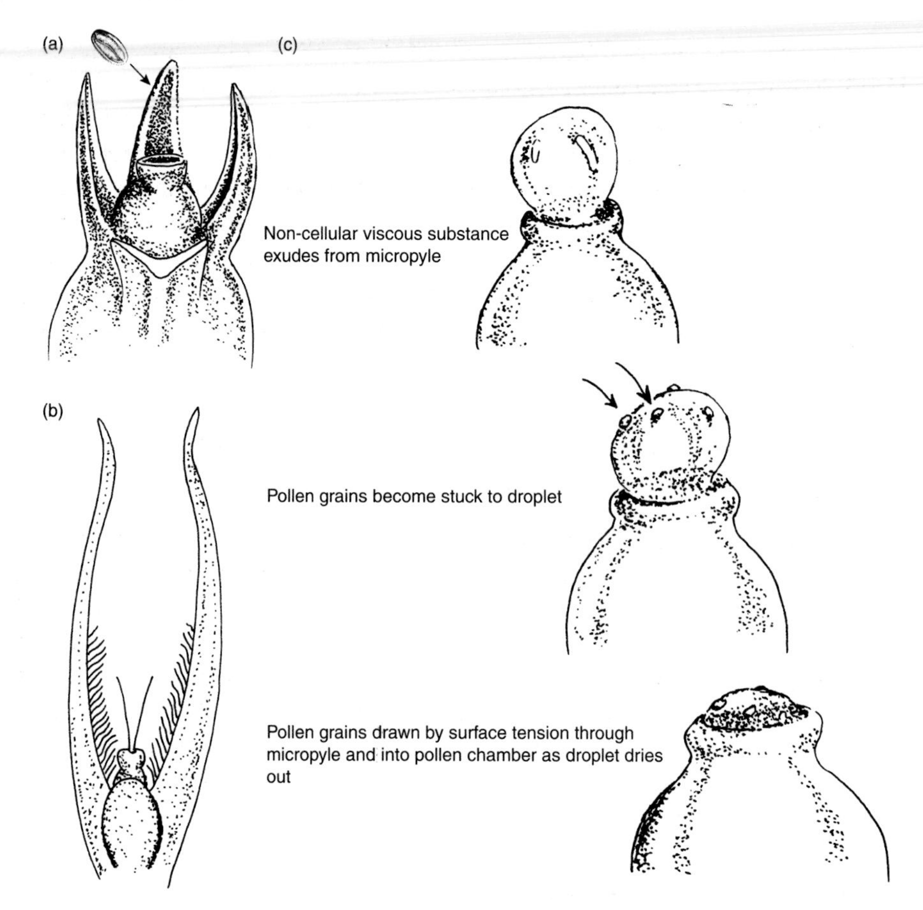

Figure 4.14 Development of pollen reception mechanisms visible in the fossil record: (a) adaptation of funnels to channel pollen towards the ovule; (b) hairs (both redrawn from Stewart and Rothwell, 1993); (c) schematic diagram to indicate the 'pollination drop' mechanism.

of an enclosed integument were threefold (Niklas, 1997). First, it would have provided much greater protection to the megasporangium against predators and desiccation. Secondly, it would have increased the efficiency of wind pollination; a more fused morphology and reduced lobes would have conferred aerodynamic efficiency for deflecting windborne pollen into the pollen drop, and thence into the micropyle. Thirdly, it would have aided the dispersal of mature seeds. The evolution of true seeds would therefore have enabled much greater chance of lateral dispersal away from the parent plant (Chaloner, 1967; Niklas, 1997) and greater potential for survival, once dispersed, in the newly created, drier, primary successional environments (Algeo and Scheckler, 1999).

In summary, therefore, it is apparent that throughout the early Devonian, major innovations were occurring in vascular plant morphology. These would not only have enabled greatly increased size and stature but also the ability to reproduce and survive in more arid environments over a wider geographic distribution (Chaloner and Sheerin, 1979; Niklas, 1997; Algeo and Scheckler, 1999). It is interesting to note that these innovations occurred during a dramatic transitionary phase in global climates, moving from a period of high global temperatures and high atmospheric CO_2 to one of cold glacial climates, with a rapid and major decline in atmospheric CO_2. However, these climate changes did not merely provide a background to vascular plant evolution but, rather, vascular plant evolution played a significant role in altering the global climates as the terrestrialization of the earth resulted in a draw-down of atmospheric CO_2 which in turn contributed to global climatic cooling (Robinson, 1991; Berner, 1997, 1998).

4.5 Earliest trees in the fossil record

The term 'earliest trees' must be used fairly loosely since the record of when and where the first trees were located is strongly influenced by the fact that certain regions provided better preservation potential, such as in areas of lowland swamps, and therefore the record is somewhat biased. However, two irrefutable facts emerge about the vegetation of this time. First, that the earliest arborescent forms of plants (trees) appeared in the middle Devonian (~380 Ma). Secondly, from this time on until the late Carboniferous, forests composed of spore-producing trees (sporophytes) plus two groups of early seed-producing trees dominated large regions of the continents.

Earliest spore-producing trees

The spore-producing trees, shrubs, and herbs that evolved during this interval (~380–290 Ma) can be classified into four main groups, namely lycopsids, sphenopsids, filicopsids, and progymnosperms (Table 4.1). The progymnosperms are extinct, but the other three groups all have extant members, indicating the remarkable length of the fossil record of some groups. It is beyond the scope of this book to describe in detail all fossil members of each group (for full descriptions see, for example, Taylor and Taylor, 1993; Stewart and Rothwell, 1993). However, the features of one common fossil type from each group are described in the following section to provide a preliminary picture of the composition of vegetation in the earliest forests.

Lycopsids (lycopods)

Although there is evidence in the geological record for small, herbaceous, homosporous lycopsids from as early as 410 million years ago (e.g. *Baragwanathia longifolia*), and many subsequent forms (see Gensel and Andrews, 1987; Thomas and Spicer, 1987), lycopsid trees were first present in the Frasnian (370 Ma). One of the most common arborescent lycopsids in the fossil record was the giant *Lepidodendron* tree, which achieved heights of

Table 4.1 Cladistic classification of land plants (adapted from Kenrick and Crane, 1997a)

Infrakingdom embryobiotes (land plants)
Superdivision Marchantiomorpha (liverworts)
Division Marchantiophyta
Class Marchantiopsida
Superdivision Anthoceromorpha (hornworts)
Division Anthocerophyta
Class Anthocerotopsida
Superdivision Bryomorpha (mosses)
Division Bryophyta
Class Bryopsida
Superdivision Polysporangiomorpha
Class Horneophytopsida†
Aglaophyton major†
Division Tracheophyta (vascular plants)
Class Rhyniopsida†
Subdivision Lycophytina
Species *Zosterophyllum myretonianum incertae sedis*†
Class Lycopsida (lycopsids)
Order Drepanophycales†
Order 'Lycopodiales'
Order Protolepidodendrales†
Order Selaginellales
Order Isoetales
Class Zosterophyllopsida†
Order Sawdoniales*†
Family Sawdoniaceae†
Family Barinophytaceae†
Family 'Gosslingiaceae'†
Family Hauaceae†
Subdivision Euphyllophytina (fern–equisetum–seed plant clade)
Eophyllophyton bellum†
Psilophyton dawsonii†
Infradivision Moniliformopses*
Class 'Cladoxylopsida'*† (early fern–like plants)
Subclass Cladoxylidae†
Subclass Stauropteridae†
Subclass Zygopteridae†
Class Equisetopsida (sphenopsids)
Class Filicopsida* (ferns)
Subclass Ophioglossidae

Table 4.1 (*Continued*)

Infrakingdom embryobiotes (land plants)
Subclass Psilotidae
Subclass Marattiidae
Subclass Polypodiidae
Infradivision Radiatopses
Pertica varia†
Supercohort Lignophytia*
Order 'Aneurophytales'† (progymnosperms)
Order 'Archaeopteridales'† (progymnosperms)
Order 'Protopityales'† (progymnosperms)
Cohort Spermatophyta† (seed plants)
Family 'Calamopityaceae'† (pteridosperms)
Family 'Hydraspermaceae'† (pteridosperms)
Family 'Lyginopteridaceae'† (pteridosperms)
Family Medullosaceae† (pteridosperms)
Subcohort Euspermatoclides*
Infrascohort Cycadatae (cycads)
Family callistophytaceae† (pteridosperms)
Infracohort Coniferophytatae
Superclass Cordaitidra† (cordaites)
Superclass Coniferidra (conifers)
Family Glossopteridaceae† (glossopterids)
Infracohort Ginkgoatae (ginkgos)
Family 'Peltaspermaceae'† (pteridosperms)
Family 'Corystospermaceae'† (pteridosperms)
Family Caytoniaceae† (pteridosperms)
Infracohort Anthophytatae*
Order Pentoxylales†
Order Bennettitales† (bennettites)
Superclass Gnetidra
Superclass Magnolidra (angiosperms)
Class 'Magnoliopsida'
Class Lilliopsida
Class Harnamelidopsida
Subclass Ranunculidae
Subclass Hamamelididae
Infraclass Caryophyllidna
Infraclass 'Rosidna'
Infraclass 'Dilleniidna'
Infraclass Lamiidna
Infraclass Asteridna

*, Ambiguous relationships between members of the group;
'', Paraphyletic group (i.e. a group containing an ancestral species plus some, but not all, of its descendents);
†, extinct group.

between 10 and 35 m and had a trunk up to 1 m in diameter (Figure 4.15a). The upper part of the trunk was branched many times to form a dense crown covered in simple microphyll leaves that grew directly out from the stem (i.e. without a petiole), leaving the swollen leaf base or 'leaf cushion' when shed. Leaves of *Lepidodendron* trees were up to 1 m in length, triangular in cross-section and arranged in regular spirals (Bell, 1992) (Figure 4.15b). The area of attachment of the leaf to the branch/trunk (the leaf base) remained when the leaf dropped off, resulting in a diamond-shape pattern. Fossil *Lepidodendron* bark indicates this distinctive pattern (Figure 4.16). Furthermore, in many examples the area where the vascular bundle and two channels of parenchyma tissue entered the leaf from the cortex is visible as three dot-like impressions (Figure 4.15d) (Taylor and Taylor, 1993).

The stem of *Lepidodendron* consisted of either a protostele or a siphonostele surrounded by zones of xylem and sometimes secondary xylem (Figure 4.15c). However, the main tissue contributing to trunk diameter and providing support was bark (periderm), in extensive layers (Eggert, 1961; Gensel and Andrews, 1987).

The rooting system of the *Lepidodendron* tree was in many ways as impressive as the above-ground part of the plant (Figure 4.15e). It consisted of four or more radiating arms (axes) that were extensively branched and, although not deeply penetrating into the substrate, extended up to 12 m in length (Bell, 1992; Taylor and Taylor, 1993). These axes, referred to as stigmaria, were composed of large amounts of secondary xylem and cortical tissues, and bore spirally arranged lateral appendages, which were, in effect, the roots. These 'roots' became abscised from the older axes, leaving a circular scar (Thomas and Spicer, 1987).

Lepidodendron was heterosporous (unlike earlier plants in the lycopsid group) and the sporophylls were clustered together in cones borne on the ends of the branches. These cones were composed of a central axis with helically arranged sporophylls (Taylor and Taylor, 1993) and were between 1 and 3.5 cm in width and 5 and 40 cm in length. From these cones the megaspores and microspores were shed.

Following their evolutionary radiation, arborescent lycopsids greatly increased in number to become a dominant component of the world flora. They formed a significant part of the organic material deposited in the coal measures of the late Carboniferous (~315–290 Ma), leading to estimates that they accounted for over two-thirds of the earliest global forests (Gensel and Andrews, 1987). There were also numerous herbaceous and shrubby forms of lycopsids (for detailed descriptions see Taylor and Taylor, 1993 and references therein).

Extant lycopsids include a number of genera, including *Lycopodium* (clubmoss), *Selaginella*, and *Isoetes* (Figure 4.17). All three are herbaceous plants with microphyllous leaves, ranging from small epiphytes to large climbing plants (Bell, 1992). However, no extant members of this group are arborescent.

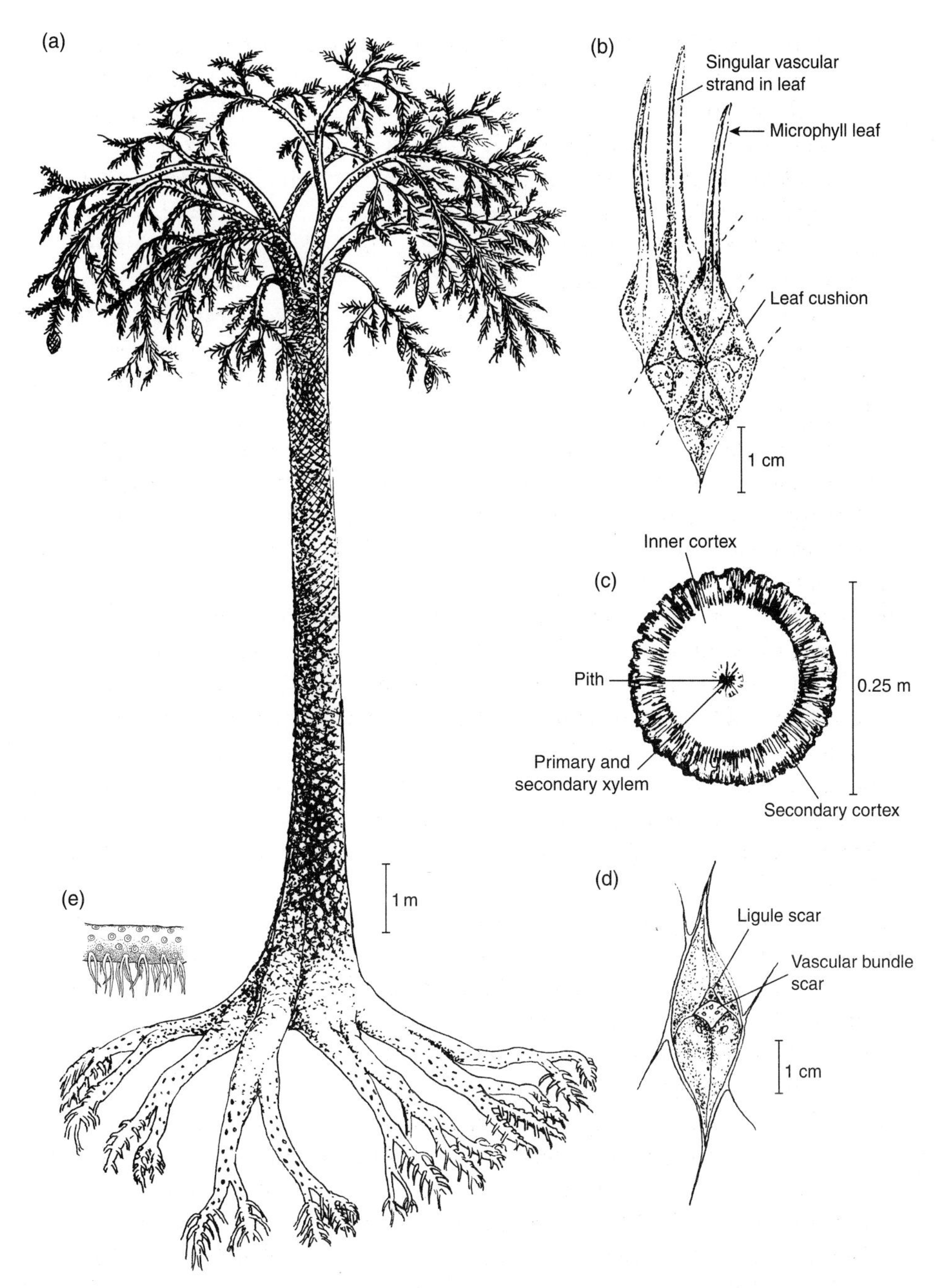

Figure 4.15 *Lepidodendron* tree: (a) habit; (b) bark pattern with leaves; (c) cross-section of stem with cortical tissue; (d) details of leaf scar; (e) roots (redrawn from Bell, 1992; Stewart and Rothwell, 1993; Taylor and Taylor, 1993).

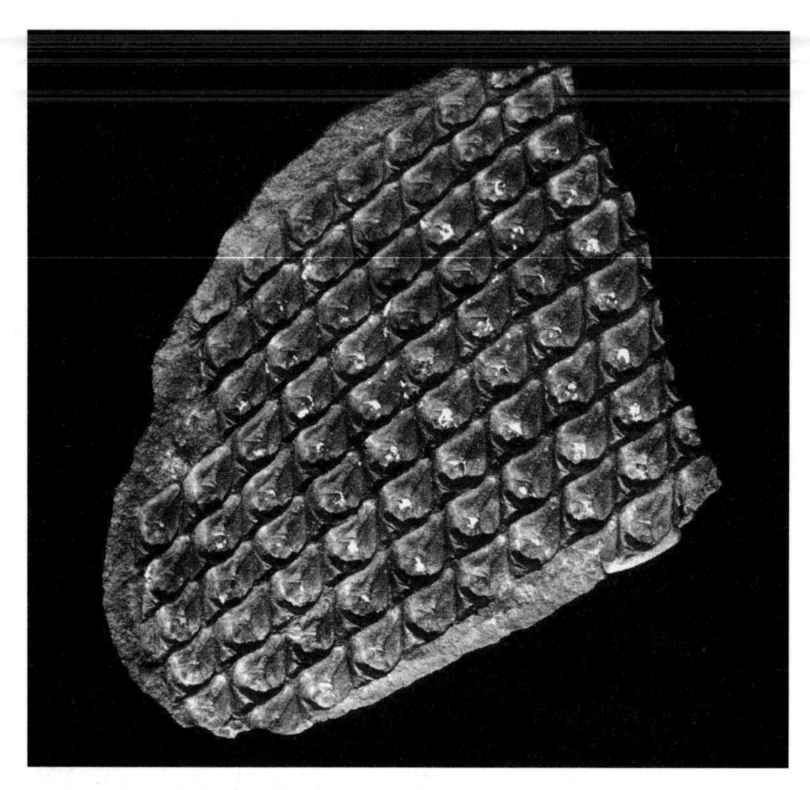

Figure 4.16 Fossil *Lepidodendron* bark, indicating diamond-shaped leaf bases (photograph by Ron Testa, courtesy of The Field Museum, neg # GEO84974–NC).

Equisetopsida (sphenopsids or giant horsetails)

Presently there are approximately 20 extant species of sphenopsids, all of which are herbaceous and belong to a single extant genus. They are widely distributed in temperate and tropical regions of the world (Figure 4.18). All have a distinctive morphology, comprised of whorled microphyllous leaves and branches borne as lateral appendages on a central stem. These distinctive morphological features have enabled ready identification of extinct sphenopsids in the plant fossil record from the early Carboniferous (~354 Ma).

Many fossil species of sphenopsids have been identified from the Carboniferous and Permian (~354–248 Ma) (Taylor and Taylor, 1993), including several arborescent forms. The largest of these was the giant horsetail tree, *Calamites*. These trees grew up to 18 m in height and had a creeping rhizome from which massive aerial stems arose, bearing whorls of branches and leaves (Figure 4.19a).

The leaves of the *Calamites* tree had a single vien running down the middle of the leaf. Two type of leaves have been identified in the fossil record: needle-like leaves borne in whorls of 4–40 and up to 3.0 mm long, and lanceolate leaves up to 8.0 mm long, which were borne in clusters of 5–32 (Figure 4.19b).

Calamites stems were horizontally dissected into many nodes from which branches and whorls of leaves sprouted (Figure 4.19a). The upper branches also

Figure 4.17 Extant *Lycopodium*, indicating the characteristic microphyllous leaves on the stem (photograph courtesy of The Field Museum, Chicago).

bore whorls of leaves (Figure 4.19b). In structure the stem was a siphonostele with a central pith. This was surrounded by vascular bundles of primary phloem and xylem, which were then surrounded by secondary xylem (Figure 4.19c). Fossil evidence suggests that in many examples this secondary thickening was at least 12 cm thick (Taylor and Taylor, 1993).

Calamites possessed a complex root system composed of large underground rhizomes, which were segmented, and from which arose the aerial stems, root branches, and root hairs (Bockelie, 1994) (Figure 4.19d). Some *Calamites* were homosporous but there is strong evidence for heterospory in late Carboniferous *Calamites* (Stewart and Rothwell, 1993). Similar to *Lepidodendron* trees, the spore-bearing structures were aggregated into cones. Location of the cones varied greatly, with evidence suggesting single cones, clusters at nodes, or several cones on fertile appendages at the end of lateral branches (Figure 4.19e). The evolution of heterospory in the sphenopsids appears to have paralleled that of *Lepidodendron*, with evidence for megasporangia dating back to late

Figure 4.18 Model of extant horsetail (*Equisetum*), indicating its distinctive morphology (photograph courtesy of The Field Museum, Chicago, neg # B83077c).

Carboniferous (Westphalian, ~300 Ma) (Bell, 1992; Stewart and Rothwell, 1993). The life cycle of the homosporous *Calamites* is thought to have most closely resembled that of extant *Equisetum*.

Similar to the lycopsids, following their radiation, arborescent forms of sphenopsids rapidly increased in number and diversity. These trees flourished in

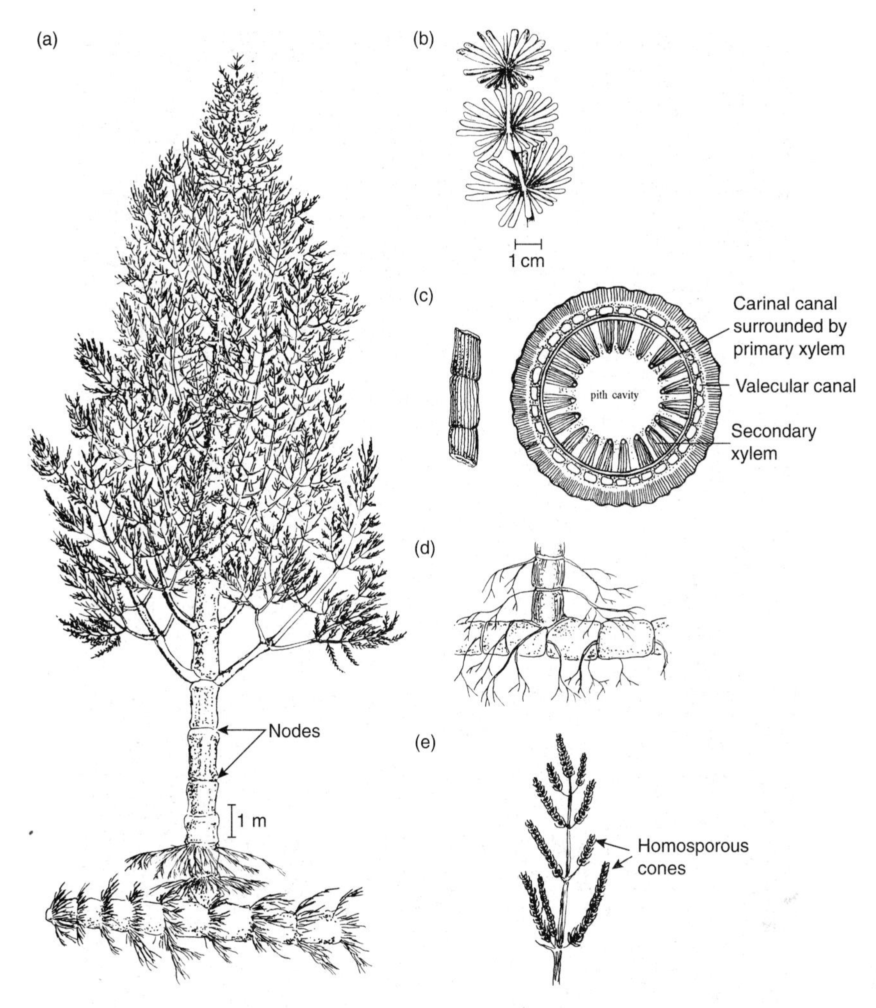

Figure 4.19 *Calamites* tree: (a) habit; (b) leaves; (c) stem and cross-section of stem; (d) roots; (e) reproductive structures (redrawn from Bell, 1992; Taylor and Taylor, 1993; Stewart and Rothwell, 1993).

wet, swampy conditions and their large underground rooting rhizomes would have provided an ideal base from which aerial stems could force their way through the swamp. They probably did not form an understorey to the lycopsid trees, but rather grew in clearings or on the waterside edges of forests (Thomas and Spicer, 1987). Arboresecent forms of sphenopsids formed a significant fraction of the organic material present in the late Carboniferous coal measures deposited between 315 and 290 million years ago.

Filicopsids (ferns)

Fossil ferns have a record dating back to the early Carboniferous (~360 Ma) and many of these bear remarkable similarity to extant forms. There are numerous examples in the fossil record of the tree fern *Psaronius* (Figure 4.20a), which grew up to 10 m tall and was the largest of the ferns found in the coal-measure swamps (Thomas and Spicer, 1987). It is suggested that their gigantic form would certainly have enabled them to exploit light efficiently in the *Lepidodendron* swamp forests, where they occupied the drier areas.

The leaves of *Psaronius* were megaphylls consisting of large pinnately compound fronds. Some of these fronds were fertile in that they bore sporangia on the lower side of the pinnules, similar to many extant ferns (Thomas and Spicer, 1987) (Fig. 4.20b).

The trunk of *Psaronius* was unbranched and the fronds developed at the top of the main axis forming an apical crown (Bell, 1992). When the leaves abscised

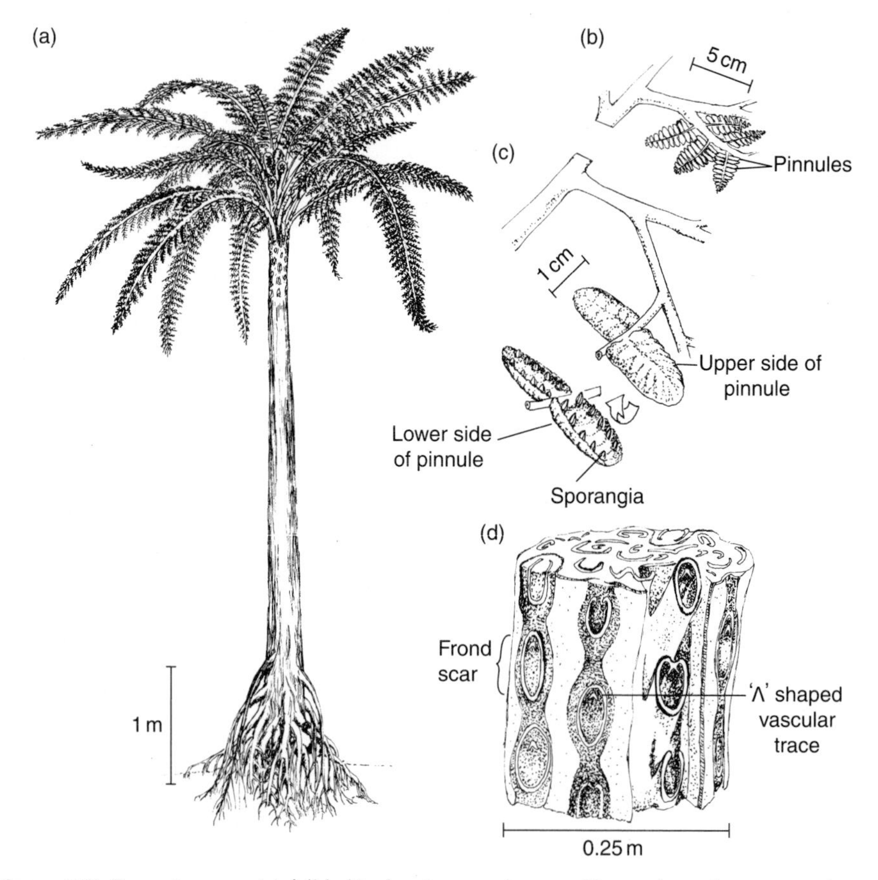

Figure 4.20 *Psaronius* tree—(a) full habit showing attachment of large pinnately compound fronds forming an apical crown. Also illustrated are an abundant mantle of roots which grew out as lateral appendages from the main trunk, and then became orientated vertically downwards parallel to the main stem; (b) detail of frond showing individual pinnules; (c) detail of reproductive structures (synangia) on the lower surfaces of two individual pinnules (enlarged), 10 individual sporangia are present on the lower surface of each pinnule; (d) portion of trunk showing distinctive circular scars from frond abscission and trace of vascular strand (redrawn from Bell, 1992; Taylor and Hickey, 1992; Stewart and Rothwell, 1993).

from the trunk, they left a distinctive circular leaf scar, which is diagnostic of different species. *Psaronius* possessed a complex stelar structure with a small protostelic stem at the base surrounded by an abundant mantle of roots. The roots of the mantle grew out as lateral appendages from the main trunk, and then became orientated vertically downwards parallel to the main stem to act, in effect, as guy-ropes (Figure 4.20a).

Towards the top of the trunk, the stem diameter became greater while the layer of adventitous roots became proportionately narrower. The stem was thus composed of an inverted stelar cone surrounded by adventitious roots (Taylor and Taylor, 1993). The construction of the stele was also complicated in the fact that each of the leaf bases received an extensive leaf trace consisting of several strands of vascular tissue (Figure 4.20d). This resulted in numerous concentric leaf gaps visible in cross-sections of the stele (Bell, 1992). There was no secondary vascular tissue (wood) in the trunk. Rather, their tree-like stature was made possible by the supporting strength of adventitious roots as in all extant tree ferns, which share the same structure. Palaeobotanical evidence suggests that in some specimens, the root mantle reached 1 m in diameter at the base of the stem (Taylor and Taylor, 1993).

The majority of extinct ferns, including *Psaronius*, were homosporous. In *Psaronius* the sporangia were large and arranged in fused clusters, called synangia, on the lower surface of the pinules of fertile fronds. It is suggested that this arrangement probably evolved, when webbing, linked fertile branches to form megaphylls, incorporating the sporangia on the underside of the frond (Figure 4.21).

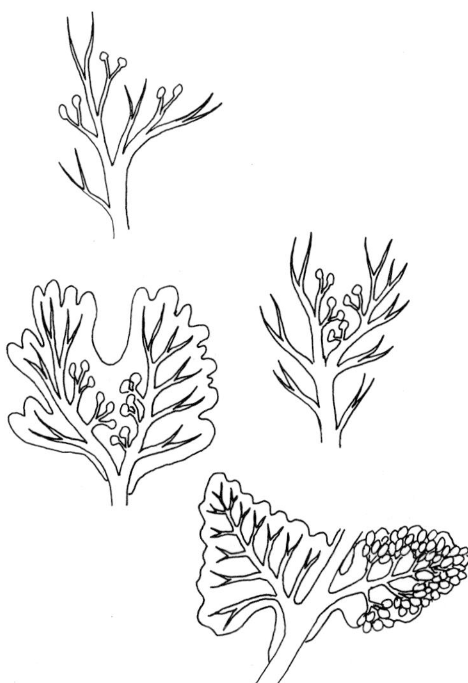

Figure 4.21 Formation of fertile fronds (from Stewart and Rothwell, 1993).

The life cycle of *Psaronius* is thought to have been similar, if not identical, to extant Marattiales, with the spores being released from the plant to germinate and form free-living, photosynthetic gametophytes (Taylor and Taylor, 1993).

Although the Marattiales are given as the example group, it must also be noted that there were a number of other filicopsid groups evolving arborescent, shrubby, and herbaceous forms during this period. In fact, up to 12 different groups of ferns are recognized in the fossil record by the lower Permian (~280 Ma), some of which have extant relatives and others that are now extinct (Stewart and Rothwell, 1993; Taylor and Taylor, 1993).

Progymnosperms (Aneurophytales, Archaeopteridales, Protopityales)

The progymnosperms are well represented in the fossil record, with 15–20 fossil genera. There are no extant members of this group, but they are extremely important in plant evolutionary history because they are thought to be the group from which all modern seed plants evolved. They are remarkable in having a gymnosperm-like stem anatomy but reproduce by spores, like in ferns and lycopods. Fossil evidence suggests that all progymnosperms were trees and many grew to up to 8 m in height with stems that reached diameters of over 1.5 m. They have a fossil record that extends from mid-Devonian to early Carboniferous (~390–340 Ma).

One of the first fossil trees to be documented as a progymnosperm was *Archaeopteris* (Beck, 1962; Gensel and Andrews, 1987) (Figure 4.22). Helically arranged, deciduous branches grew out laterally from the upper part of the main trunk, on which grew laminate leaves (Taylor and Taylor, 1993) (Figure 4.22b). Progymnosperm foliage is not dissimilar to the fossil fern fronds of *Psaronius* described previously, and for many years these leafy branches were regarded as large pinnate fronds (compound leaves). It was only when fossil examples of these 'fronds' were found attached to trunks, with clear evidence of thick secondary xylem, that it became apparent that they were part of a different evolutionary group (Beck, 1960). Recent fossil finds in Morocco of *Archaeopteris* trunks have indicated that the architecture of these trees was greatly advanced and much closer in evolutionary terms to modern trees than that of the other sporophyte trees present in the early forests (Meyer-Berthaud *et al.*, 1999).

The stem of *Archaeopteris* was eustelic in structure and composed of a central pith surrounded by primary xylem (Figure 4.22c). Outside of this layer was a thick secondary xylem (up to 1 m diameter in some examples) made up of tracheids and narrow vascular rays. In some examples there is also evidence of growth rings. The morphology of this wood suggests that it was dense and it bears a close similarity to the trunks of extant conifers (Bell, 1992).

The underground root system of *Archaeopteris* was extensive with a main axis and numerous lateral roots. It is presumed that the rooting system looked rather similar to that of modern conifers (Figure 4.22d), although it has been demonstrated that structurally they were very different (Thomas and Spicer, 1987). In particular, lateral roots were formed by branching of the main root axis.

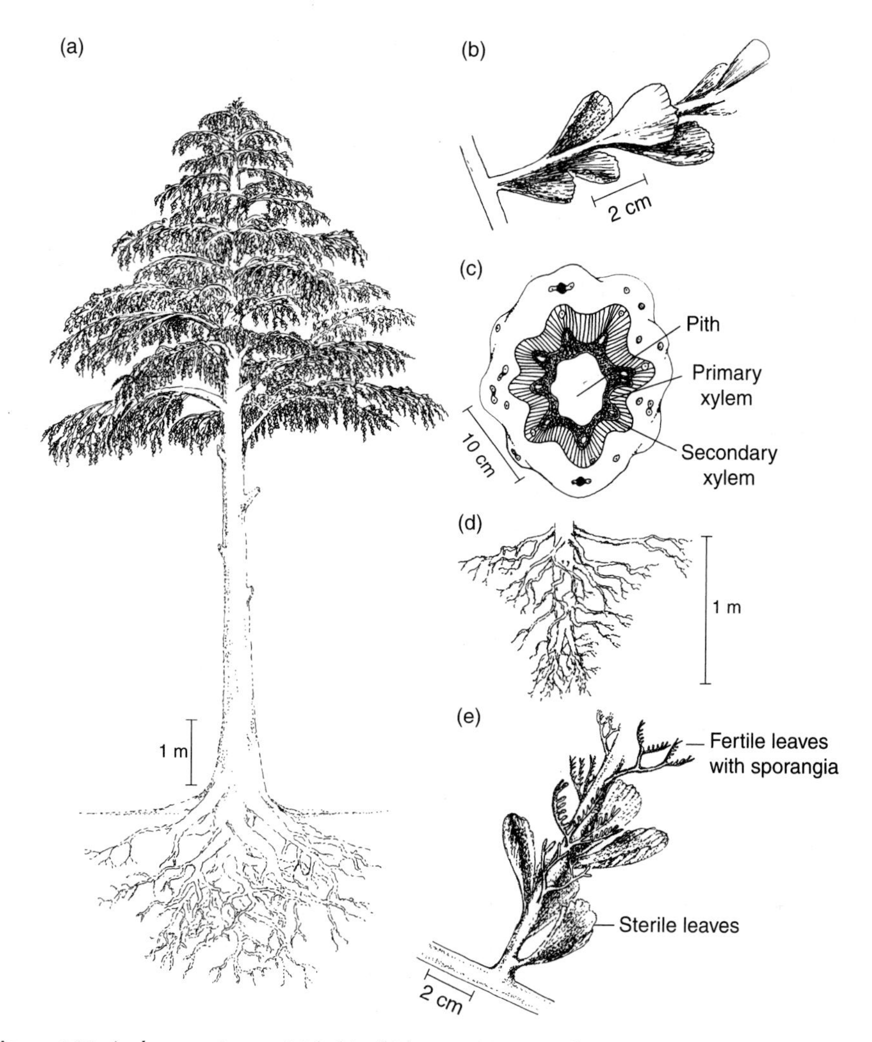

Figure 4.22 *Archaeopteris* tree: (a) habit; (b) leaves; (c) stem; (d) rooting system; (e) branch with sporangia and leaves (redrawn from Beck, 1962; Gensel and Andrews, 1987).

Archaeopteris had both fertile (with sporangia) and sterile (leafed) branches, as well as some branches that contained both sterile leaves and sporangia (Figure 4.22e). The sporangia were spindle-shaped and up to 3.5 mm in length (Bell, 1992). There is evidence in the fossil record that *Archaeopteris* was probably heterosporous (Taylor and Taylor, 1993).

In comparison to the other tree types in the early forests, *Archaeopteris* had a combination of key morphological characteristics that it is argued would have conferred a competitive advantage. In particular, these included perennial lateral branches, much deeper rooting structures, and megaphyllous leaves. These would all have enabled growth beyond the lowland swamps and probably contributed to the almost worldwide dominance of *Archaeopteris* trees in late Devonian floodplain environments (Algeo and Scheckler, 1999).

Earliest seed-producing trees

Along with the spore-producing giant horsetails, lycopsids, filicopsids, and progymnosperms, two groups of early seed plants, namely the pteridosperms and cordaites, were present in the forests of the early and late Carboniferous (~354–290 Ma).

Palaeozoic pteridosperms (seed ferns)

The seed-ferns are so called because although they had fern-like foliage, they reproduced by seeds. There is fossil evidence for at least six families of seed ferns during the Carboniferous and Permian (~354–248 Ma) (Table 4.1). Some had a prostrate vine-like habit, but most were arborescent forms and similar in general appearance to modern tree ferns (Stewart and Rothwell, 1993). The largest seed ferns in the fossil record were in the Medullosaceae family and included arborescent forms up to 10 m in height. One of the predominant genera in this family was *Medullosa noei* (Figure 4.23). The leaves of *Medullosa* were large megaphyllous fronds, pinnately divided and dichotomously branched (Figure 4.23b). They were usually spirally arranged on the stem and had petioles up to 20 cm in diameter. Individual pinnules were usually approximately 2–4 cm in length and had a distinctive venation pattern, often with a prominent mid-vein and secondary veins at angles from this (Taylor and Taylor, 1993).

Medullosa trees (e.g. *Medullosa noei*) grew to up to 10 m in height and had a wide-based stem with diameters of up to 0.5 m. They had a complex stelar structure composed of a single stele divided into many vascular segments (Figure 4.23c). Each vascular segment was made up of a central core of primary xylem and parenchyma cells surrounded by a cylinder of secondary xylem. The secondary xylem contained very large tracheids and was rather unusual in that it was thickest towards the centre of the stem (Thomas and Spicer, 1987). Thus the secondary xylem was formed towards the inside of the primary xylem, unlike the situation in living plants. Surrounding the secondary xylem was the vascular cambium, secondary phloem, and cortical tissues.

On the external part of the stem, the lower portion was covered with adventitious roots (Figure 4.23a), whereas higher up there were spirally arranged leaf bases (Stewart and Rothwell, 1993). The adventitious roots, together with a single stele that became divided at intervals along the trunk, provided support. Fossil evidence suggests that these roots could be up to 2.5 cm in diameter, and were abundant, with secondary tissues.

Medullosa ovules are common in the fossil record and at least 14 morphologically distinctive types have been recognized. They range in size between

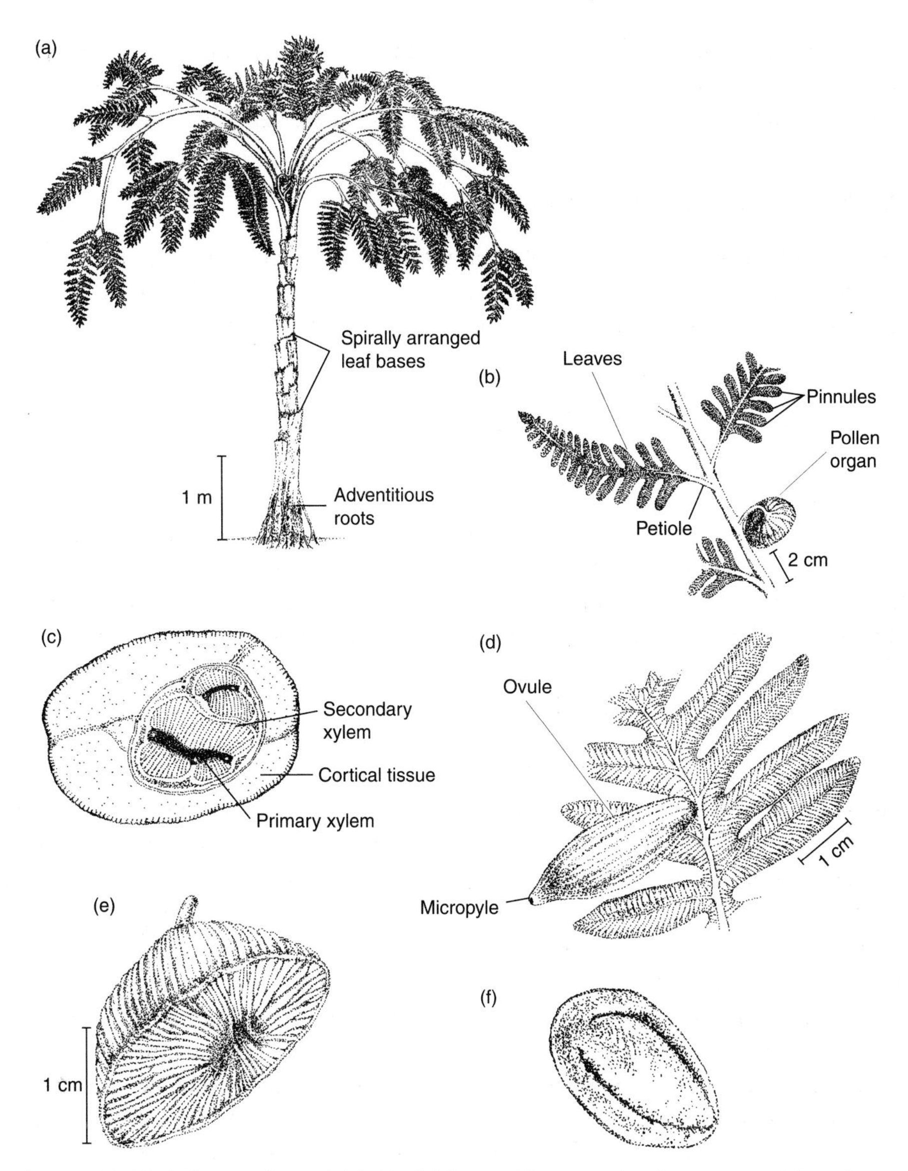

Figure 4.23 *Medullosa noei* tree: (a) habit; (b) leaves; (c) stem and stelar structure; (d) ovules; (e) pollen organ; (f) monolete pollen (up to 600 μm in diameter) (redrawn from Bell, 1992; Stewart and Rothwell, 1993; Taylor and Taylor, 1993).

1 and 11 cm and have a three-layered enclosed integument with a micropyle and simple pollen chamber (Figure 4.23d). These ovules share a close morphological similarity to cycad ovules, supporting the suggestion that the two are closely related (see Chapter 5). However, very few fossil *Medullosa* ovule specimens have been found attached to the fronds, and as such their phylogenetic position remains uncertain (Cleal and Thomas, 1999).

The prepollen grains of the *Medullosa* trees were located in pollen organs borne on fertile branches in clusters, or in place of the pinnae on fronds (Ramanujan *et al.*, 1994). These were large structures (2–3 cm in diameter) resulting from numerous fused sporangia (Taylor and Taylor, 1993) (Figure 4.23e), containing monolete pollen grains (with a single furrow) between 100 and 600 μm in length (Figure 4.23f).

Cordaites

Unequivocal fossil evidence for cordaites suggests that they were present from the lower Carboniferous into the Permian (~330–250 Ma), although there is some discussion as to whether unattributed, but substantial, woody remains, dating as far back as 380 million years ago, may also be associated with this group (Bell, 1992). All evidence obtained so far suggests that the cordaites were predominantly arborescent (trees and shrubs) and occupied a range of different habitats from mangrove-type habitats to drier uplands (Cleal and Thomas, 1999). With fossil trunks measuring up to 30 m in height and 1m in diameter, these plants may have represented some of the tallest trees in the late Carboniferous and Permian forests (Figure 4.24). Trees in the cordaites group all had a main stem with single branches radiating out towards the top of the axis (Figure 4.24a). In cross-section the trunk was composed of large amounts of secondary xylem surrounding a medullated primary stele (Figure 4.24b). Cordaitean leaves were variable, but usually helically arranged on long, slender branches. Many species in the fossil record bear a close morphological similarity to leaves of the exant genus *Agathis* (Araucariaceae). The leaves themselves were strap-shaped or tongue-shaped, and up to 1 m in length and 15 cm in width. The tip of the leaf was bluntly rounded with a broad leaf base attaching it to the branch (Figure 4.24f). Most leaves were morphologically distinctive, with long parallel veins but no central vein (midrib) running down its length (Taylor and Taylor, 1993). However, some cordaitean leaves were more needle-like in appearance, with only a single vein running down each leaf (Stewart and Rothwell, 1993).

Rooting systems of the cordaites were variable in structure. Fossil evidence suggests that some groups had extensive rooting systems, with lateral roots radiating from large branched primary roots (Bockelie, 1994), whereas others consisted of lateral roots forming in clusters on only one side of the main root. This latter type is typical of extant plants that grow in mangrove swamps and possibly indicates the environmental conditions in which these trees grew (Thomas and Spicer, 1987).

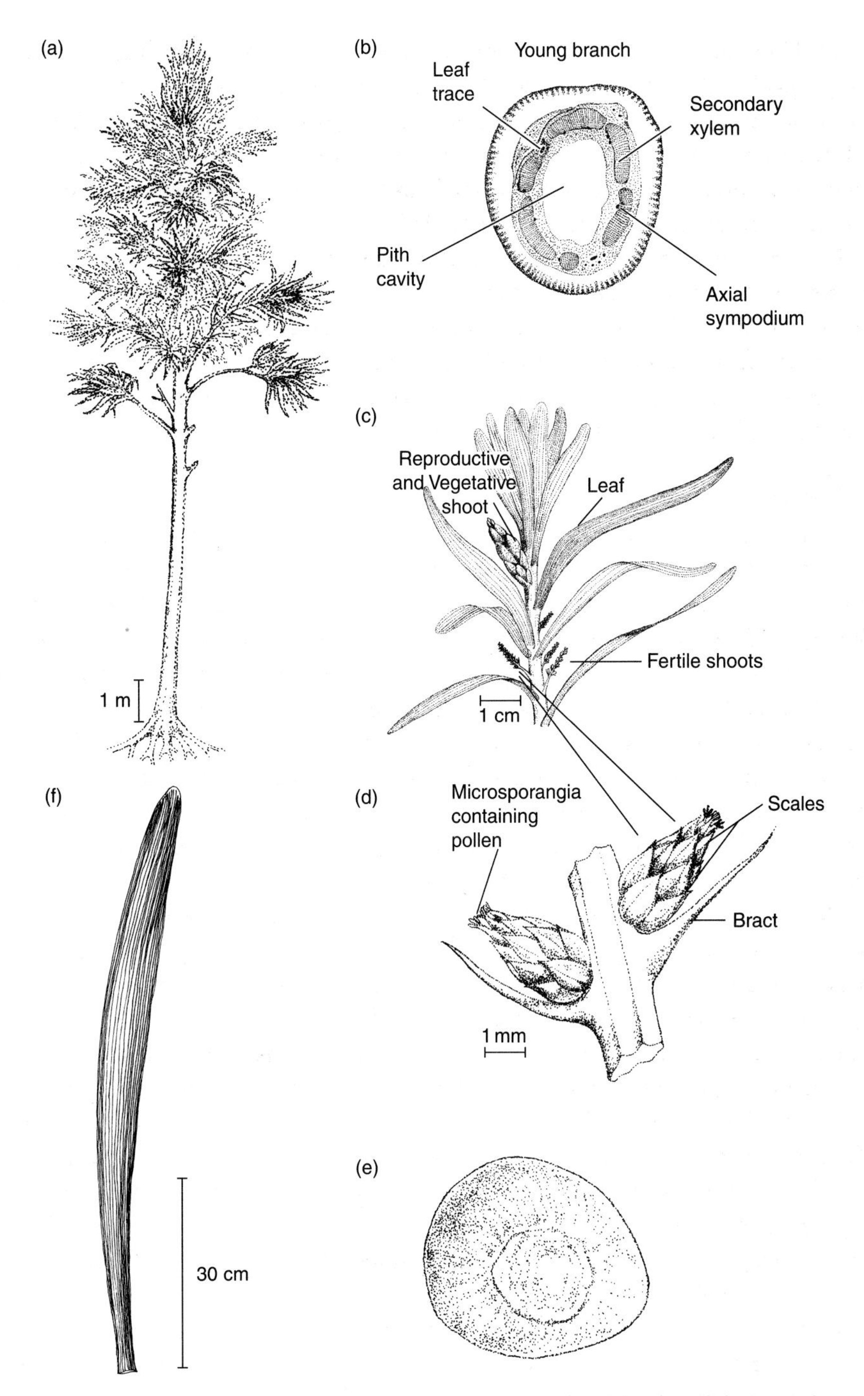

Figure 4.24 Cordaitales tree: (a) habit; (b) stem stelar structure; (c) branch with leaves and reproductive shoots; (d) reproductive structures; (e) pollen (45–65 μm in diameter); (f) leaf (redrawn from Bell, 1992; Stewart and Rothwell, 1993; Taylor and Taylor, 1993).

Reproductive structures were borne on the same branches as the leaves. Male and female organs were separate but probably located on the same tree (Bell, 1992) (Figure 4.24d). The male organs were located on shoots up to 1 cm long that contained a number of scales arranged around the shoot axis in a spiral, forming a structure similar to a conifer cone. The lower scales were sterile but the upper scales terminated in cylindrical pollen sacs (up to 1 mm in length), which contained numerous monosaccate pollen grains, each between 45 and 65 μm in diameter (Figure 4.24e).

The female reproductive organs were also located on the branches and bore some similarity to the male organs in that they were cone-shaped and composed of fertile and infertile scales. However, in the female organs, the fertile scales terminated in ovules rather than pollen sacs (Bell, 1992). Each seed was surrounded by three integumentary layers, forming an enclosed seed coat with micropyle. The margin of the seed extended as a wing (platyspermic).

4.6 Evolutionary trends: earliest vascular plants to trees

Ever since the first plant fossils were discovered, theories have been put forward to provide an evolutionary link between the earliest land plants, such as *Cooksonia* and *Aglaophyton*, to later groups, such as the progymnosperms and sphenopsids, and to our extant flora.

Recent cladistic analyses based on both morphological and biomolecular analyses of extant and extinct plant groups (e.g. Bell, 1992; Raubeson and Jansen, 1992; Hiesel *et al.*, 1994; Kranz and Huss, 1996; Crane and Kenrick, 1997; Kenrick and Crane, 1997a,b; Niklas, 1997) suggest that all vascular plants can be divided into two well-defined clades—the Rhyniopsida and Eutracheophytes. Two basal lineages emerged from the Eutracheophytina—the Lycophytina and Euphyllophytina (fern–*Equisetum*–seed-plant clade) (Figure 4.25)

The Lycophytina

The Lycophytina share a number of characteristics (known as synapomorphies) including kidney-shaped sporangia which develop on short laterally inserted stalks and exarch xylem. The lineage split early on in evolutionary terms to form two major clades, the Zosterophyllopsida (zosterphyllophytes), which bore lateral sporangia direct on their leafless stems, and the Lycopsida (lycopsids) which have microphyll leaves, with sporangia in their axils:

1. The zosterophyllophytes include only extinct members (e.g. *Zosterophyllum*, Figure 3.15) and no further evolution of this clade is thought to have occurred beyond the Devonian (~360 Ma).
2. The lycopsids, in comparison, include both extinct and extant members and indicate a phylogenetic relationship between the earliest vascular plants, such

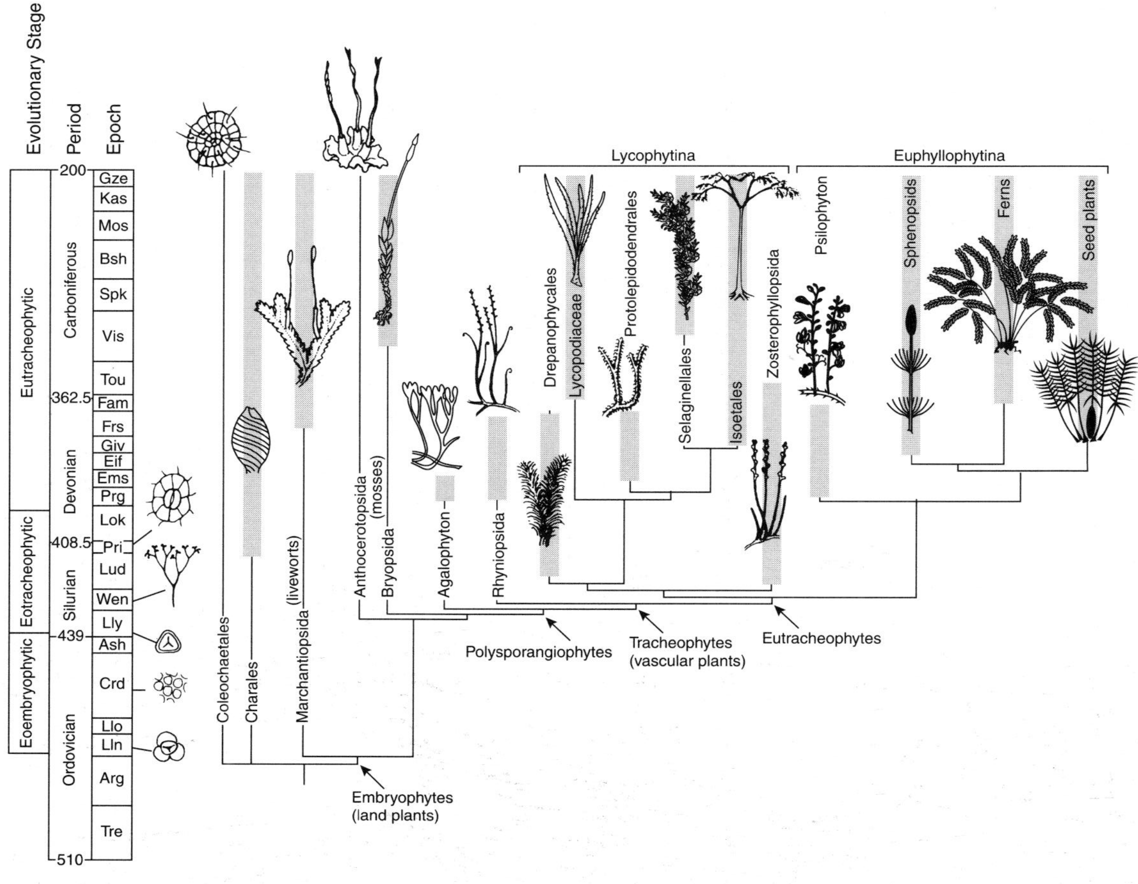

Figure 4.25 Phylogenetic relationship between extinct and extant early plants (redrawn from Kenrick and Crane, 1997b).

as *Baragwanathia* and the arborescent *Lepidodendron*, and extant families such as Lycopodiaceae and Selaginellaceae (Table 4.1) (Kenrick and Crane, 1997).

The Euphyllophytina

The Euphylophytina are all characterized by a helical arrangement of branches, tracheids with bordered pits in the form of a ladder and often by sperm which is multiflagellate. The Euphyllophytina also split early in evolutionary terms to form two clades, one containing the extinct genus *Psilophyton dawsonii* (Figure 3.17) and the other including the ferns (filicopsids), horsetails (sphenopsids), and seed plants (spermatophytes) (Figure 4.25).

From these cladistic analyses it would also appear that the spermatophytes (which will be described in Chapter 5) are a sister group to the sphenopsids and filicopsids, thus suggesting a close phylogenetic relationship between members of the Euphyllophytina.

4.7 Biogeographical distribution of global vegetation during the early and late Carboniferous (354–290 Ma)

The ecology, structure, and diversity of some of the earliest forests of the middle to late Devonian (380–360 Ma) are well defined (Scheckler, 1985, 1986). However, it is not until the early Carboniferous that there are enough fossil floras to get a detailed global picture of plant distribution and biogeography (Rowley *et al.*, 1985; Wnuk, 1996). The aim of the next section and throughout the book is to provide a general overview of the major biogeographical regions or 'biomes' and how they have changed in composition and spatial structure through time. First, however, plant biogeographical concepts and classification schemes are reviewed briefly, in order to outline some of the general limitations associated with global biogeographical analysis, and the extrapolation of modern biogeographical systems, such as the 'biome' concept, to the fossil record.

Plant biogeographical concepts and classification scheme

Plant biogeography has played an extremely important role in the reconstruction of continental positions in the past, particularly at times of high floral provincialism (e.g. during the Carboniferous). Many methods have been used to divide up global vegetation into recognizable units. These include taxonomic and/or ecological characteristics and measures of plant diversity (Table 4.2). Similarly, many classification schemes have been employed to define these plant

Table 4.2 Plant biogeograpical classification schemes (from Wnuk, 1996).

Classification based on taxonomic criteria	Classification based on ecological criteria
Realm Defined on the basis of the (palaeo) geographic distribution of distinctive suites of taxa. Each suite includes a significant proportion of endemic families	Realm Defined on the basis of the (palaeo) geographic distribution of distinctive suites of taxa. Each suite includes a significant proportion of endemic families
Region Defined on the basis of the (palaeo) geographic distribution of distinctive suites of taxa. Each suite includes a significant proportion of endemic genera and/or species	**Biome** Social groupings on an intercontinental scale composed of a characteristic set of taxa and has a characteristic overall physiognomy (such as plant stature, habit, leaf size, shape) and taxonomic diversity
Area Defined by differences in the proportions of the taxa comprising the suite and by lesser differences in the proportions of endemic species and genera	Biocenosis Social groupings of taxa representing integrated functional systems. The taxa comprising a biocenosis require a narrowly constrained range of environmental and edaphic conditions
Association Defined on the basis of a consistent co-occurrence of specific taxa	Community A biocenosis composed of a specific suite of taxa

biogeographical units and this has led to a long, and often confusing, list of nomenclature.

All of the various methods employed are limited by the incompleteness of the fossil record. In addition, the effect of differing preservation (taphonomy) and the fact that whole plants are rarely preserved, leading to the counting of the individual organs as separate species (which can result in overestimation of fossil plant diversity), can lead to difficulties in extrapolation of widely spaced floras (Scott and Galtier, 1996). Keeping these limitations in mind however, the use of 'climatically sensitive' sediments (e.g. coals, salt deposits) in addition to taxonomic, ecological, and diversity patterns in global vegetation have added greatly to our understanding of the delineation of fossil plant biomes and how they have changed spatially through time (Ziegler, 1990; Rees *et al.*, 2000). For instance, the presence of evaporites, which occur in environments where net evaporation exceeds net precipitation, helps to delimit the extent of past deserts (Parrish, 1998).

Table 4.3 Climatic and latitudinal parameters for the present vegetation biomes (modified from Ziegler, 1990)

Biome No.	Palaeo-biome shading	Present vegetation	Climate description	Subdivision	Climatic constraints of biome	Polward latitude (°)
10		None	Glacial	None	No month when $t > 0°C$	90°
9		Tundra (treeless)	Arctic	None	No month when $t > 0°C$	*c.* 70°–60°
8		Boreal coniferous forests (taiga)	Cold temperate	None	<1 month when $t \geq 10°C$	*c.* 66°–58°
7		Steppe to desert	Cold temperate	None	<4 months when $t \geq 10°C$	$51° \pm 2$
6		Nemoral broadleaf deciduous forest	Cool temperate	Western	<4 months when $t \geq 10°C$	*c.* 58°
				Eastern	<4 months when $t \geq 10°C$	$44° \pm 2$
5		Temperate evergreen forests	Warm temperate	Western	Winter $t < 0°C$	$46° \pm 14$
				Eastern	Winter $t < 0°C$	$36° \pm 2$

4		Sclerophyllous woody	Winterwet	None	≥ 10 months ppt > 20 mm	$38° \pm 5$
3		Subtropical desert vegetation	Subtropical desert	Coastal	Limit of winter rains	$32° \pm 4$
				Inland	Winter $t < 0°C$	*c.* 33°
2		Tropical deciduous forest or savannas	Summerwet	Subtropical	<3 months ppt > 20 mm	Up to 25°
				Tropical	<3 months ppt > 20 mm	$15° \pm 5$
1		Evergreen tropical rainforest	Everwet	Tropical	> 11 months ppt > 20 mm	Up to 25°
				Equatorial	> 11 months ppt > 20 mm	$10° \pm 7$

Where t is mean temperature and ppt is precipitation.

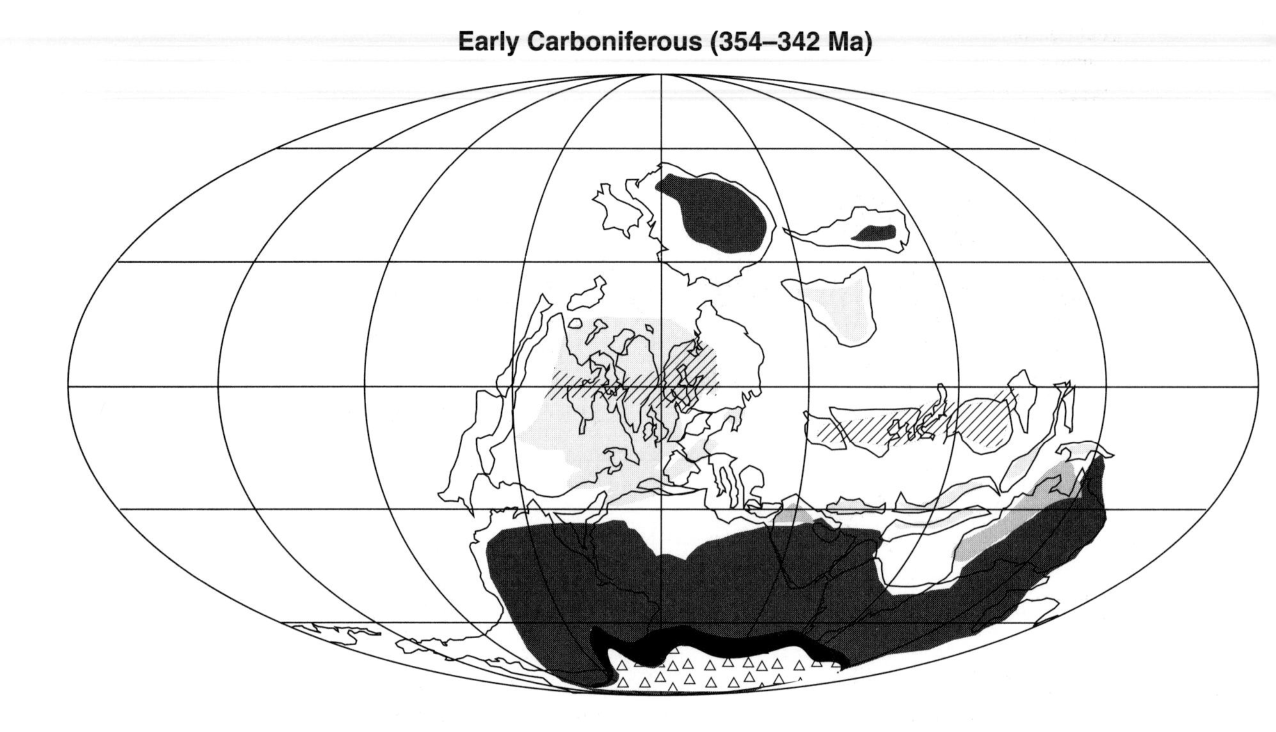

All of the biogeographical maps presented throughout this book have been standardized using the biome scheme of Ziegler (Ziegler, 1990; Table 4.3). They incorporate both fossil plant and lithological data in the delimitation of the fossil biomes, which have been defined according to our understanding of the climatic, edaphic, and plant physiognomic features which characterize the ten major biomes recognized in our world vegetation today as laid out in Table 4.3.

Biogeographical distribution of global vegetation 354–342 Ma (early Carboniferous)

Six distinct biomes are recognized for the early Carboniferous (~354–342 Ma) using the biome scheme of Ziegler (1990; Table 4.3), on the basis of detailed analyses of fossil floras (Rowley *et al.*, 1985; Cleal and Thomas, 1991, 1999; Wnuk, 1996 and references therein) and from evidence of climatically sensitive sediments for this time (PaleoAtlas project http://pgap.uchicago.edu/) (Figure 4.26).

Tropical everwet biome

The tropical everwet biome encompassed a narrow equatorial belt including present-day China, Scandinavia, and parts of Greenland and North America. It was characterized by fossil floras composed of arborescent lycopsids (e.g.

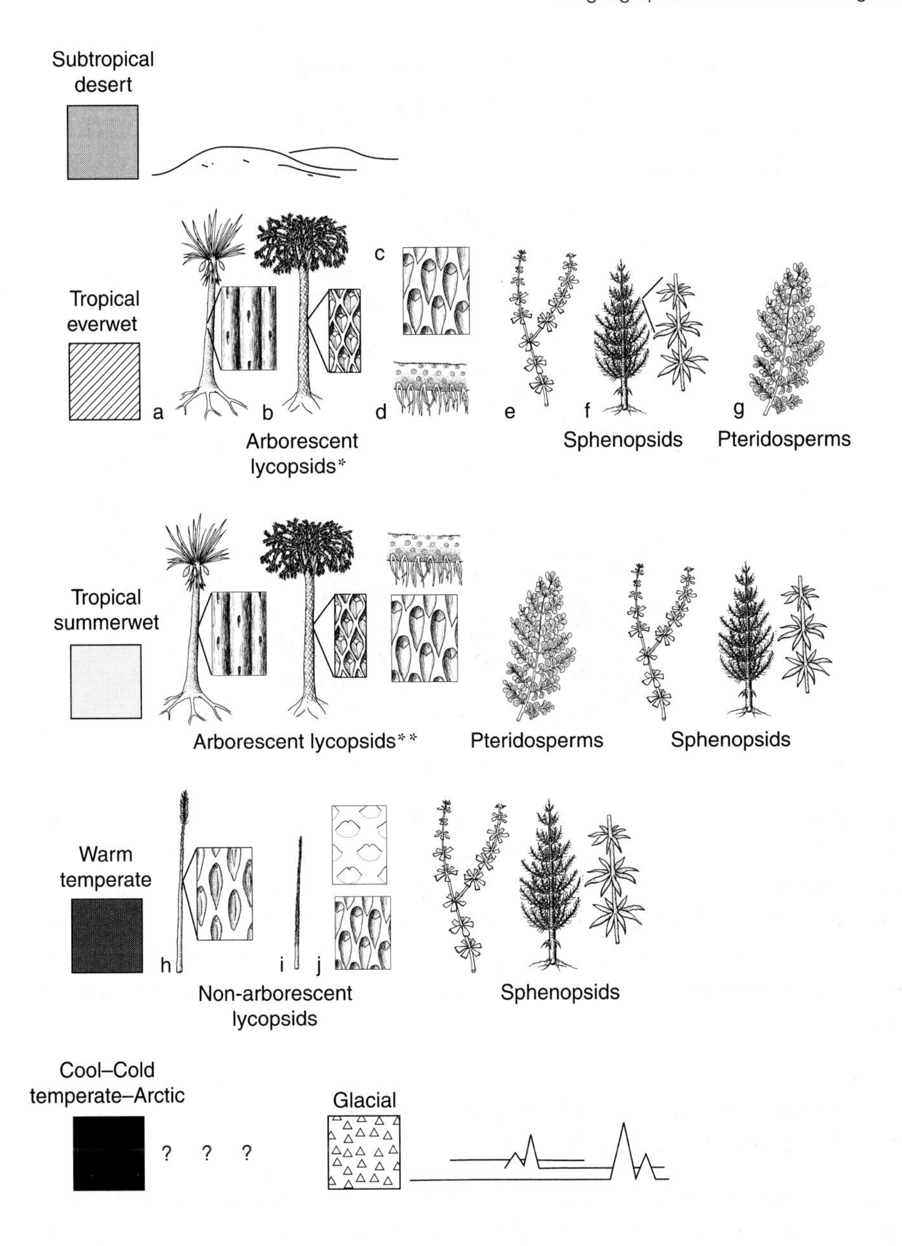

Figure 4.26 Suggested biomes for the early Carboniferous (354–342 Ma), with representatives of the most abundant and/or dominant fossil plant taxa shown. The biomes are superimposed on a global palaeogeographic reconstruction for the Visean (~340 Ma) (courtesy of C. R. Scotese and the PaleoMap Project). (a) *Sigillaria* tree with details of fossil bark; (b) *Lepidodendron* tree with details of fossil bark; (c) *Stigmaria* (lycopsid root); (d) details of fossil bark of *Lepidodendropsis*; (e) *Sphenophyllum*; (f) *Calamites* tree with details of leaves (*Asterophyllites*); (g) stylized pteridosperm frond; (h) *Tomiodendron* tree with details of fossil bark; (i) juvenile lycopsid; (j) *Ursodendron* bark. See Appendix 1 for sources of plant reconstructions and line drawings. * *Stigmaria*-type roots mainly present or ** mainly absent.

Lepidodendron and *Stigmaria*), sphenopsids (e.g. *Sphenophyllum*), and pteridosperms (Cleal and Thomas, 1991). The South China region had a similar 'broad equatorial' flora dominated by lycopsids with some sphenopsids and ferns, although endemic form-genera that made this region distinctive were also present. This biome is defined by others as the North Laurussian region (reviewed by Wnuk, 1996). The presence of abundant coals, which are indicative of wet or everwet climates, and a low palaeolatitude support the assignment of this region to the tropical everwet biome.

Summerwet (tropical) biome

A seasonally wet or summerwet biome encompassed the remainder of Euramerica from palaeolatitude 5–30°, and Kazakhstan. The vegetation of the swamp environments was composed predominantly of arborescent lycopsids and pteridosperms, whereas sphenopsids were less abundant than in the tropical everwet biome. The assignment of a summerwet biome to these areas is supported by evidence from palaeoecology for a seasonally wet climate (Knaus *et al.*, 2000; Falcon-Lang, 1999). Tropical desert belts, indicated by the presence of evaporites (Ziegler and Rees, personal communication), help to define the southerly limit of the summerwet biome. The southern hemisphere portion of this biome is defined by others as the Acadian–Laurussian region (Raymond, 1985; Wnuk, 1996).

Kazakhstan, which was a separate continent (Figure 4.1) was characterized by a flora composed of approximately 45% endemic form-genera (Raymond, 1985; Cleal and Thomas, 1991), including species of lycopsids and seed ferns (Meyen, 1987). Although Wnuk (1996) attributes a warm temperate climate to the Kazakhstan region during the Tournaisian, the presence of evaporites, the general short stature of the vegetation, and lack of the rhizophorous rooting structures known as stigmaria indicate that this flora was also part of the seasonally wet biome.

Subtropical desert biome

The subtropical desert biome was very restricted during the Tournaisian, with evidence of evaporites indicating its presence in western Australia and northwestern Saudi Arabia. Fossil floras are absent from this biome.

Warm temperate biome

The higher palaeolatitudes between 30–70° make up the warm temperate biome. In Siberia the floras were dominated by small lycopsid trees, distinguishable by their shorter stature than those of the tropical everwet biome, and sphenopsids. Fewer pteridosperms were present compared to the tropical everwet and summerwet biomes. This biome in the northern hemisphere is referred to by others as the Angaran region (Siberia and Kazakhstan) (reviewed by Wnuk, 1996). In the southern hemisphere, classified as the Gondwanan region, the vegetation was dominated by arborescent lycopsids, sphenopsids, and early seed ferns. Although the composition of the vegetation was very similar to that of the lower latitude biomes the diversity was considerably lower. Furthermore, the lycopsids of the Gondwanan region possessed similar morphological characteristic to those in the warm temperate biome of the northern hemisphere, in that they were shorter in stature and lacked stigmaria-type rooting structures, indicating cooler temperatures. The morphological similarities of these high-latitude floras, together with their lower general diversity than those of the equatorial biome and the presence of coals, are indicative of warm, wet climates typical of the warm temperate biome.

Cool–cold temperate, arctic and glacial biomes

A gradation of climatic conditions and also biomes most likely existed between the warm temperature biome covering most of Gondwana and the glacial biome (ice sheet) to the south. However, as no fossil floras of early Carboniferous age are known south of 60° it is not possible, as yet, to indicate the limits of these biomes with confidence, or to document the composition of the vegetation that was present. The northerly limits of the glacial biome are based on the presence of glacial deposits (tillites) and glacial pavement studies (see Figure 4.2).

Biogeographical distribution of global vegetation 300–286 Ma (late Carboniferous)

By late Carboniferous time (~300–286 Ma) 6 to 8 biomes are recognized based on the same criteria as those used for the early Carboniferous (p. 122) (Figure 4.27) and on biogeographical studies of Rowley *et al.* (1985), Ziegler *et al.* (1981), Cleal and Thomas (1991), and Wnuk (1996):

Tropical everwet biome

The tropical everwet biome encompassed present-day eastern North America, western Europe, China and north Africa. The vegetation of this biome was dominated by forests composed of pteridosperms, lycopsids, and tree ferns (filicopsids), as well as other early seed plants such as cordaites. Lycopsid diversity and abundance in all of these low-latitude forests was greatly diminished (with the exception of those in the western Appalachian region of North America), compared to the dominant position they held in nearly all of the major biomes during the early Carboniferous.

Summerwet (tropical) biome

The summerwet biome contained a higher proportion of more drought-resistant conifers, such as *Lebachia* and *Ernestiodendron*, with more reduced leaves compared to those of the tropical everwet biome. Sphenopsids such as *Phylotheca* were also present. This biome covered a narrow band immediately to the north and south of the tropical everwet biome (Figure 4.27). Cuneo (1989) suggests that floral differentiation was also apparent on the western coast of

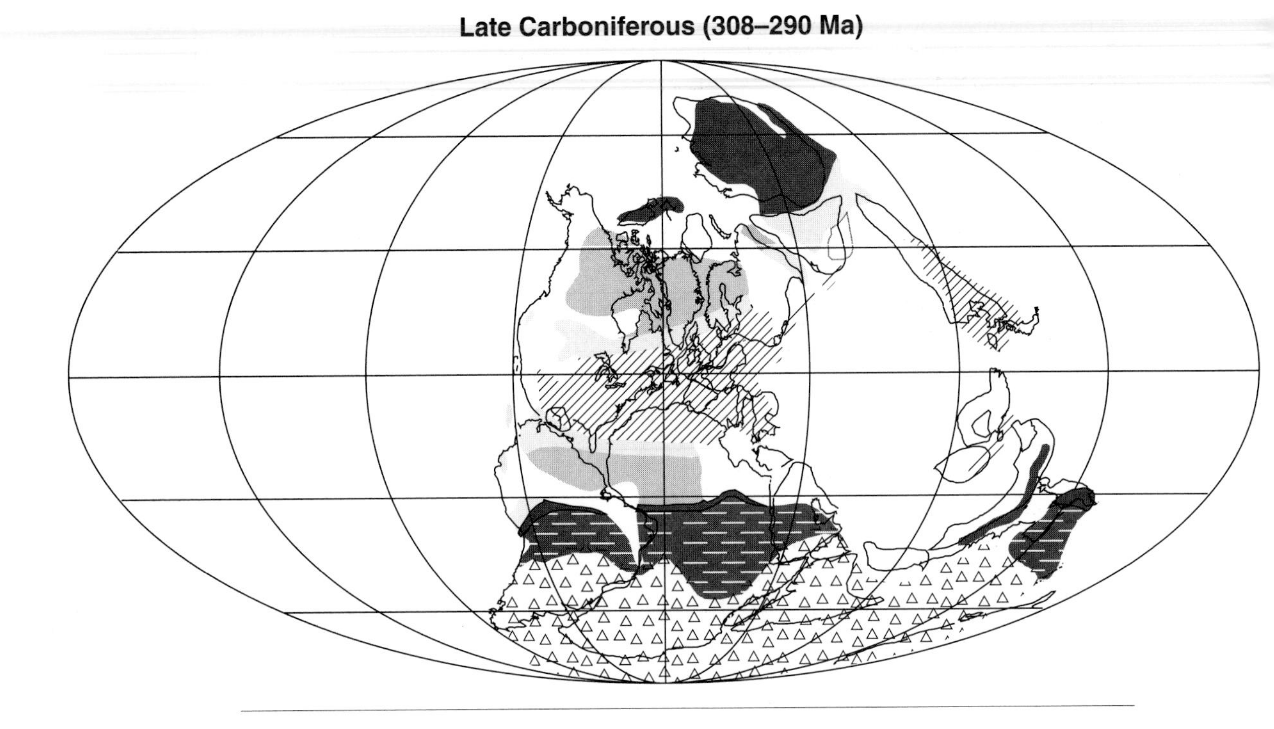

South America, with a distinctive subtropical flora in what is now western Argentina at this time.

Subtropical desert biome

The frequent occurrence of evaporites (Ziegler, personal communication) and almost complete absence of fossil plants characterized this biome. It spanned an area including most of Greenland and Scandinavia and northern Canada in the northern hemisphere and parts of Brazil and western Africa in the southern hemisphere.

Warm temperate biome

By the late Carboniferous (~300 Ma) Kazakhstan had collided with Siberia (see Section 4.1), thereby increasing the aerial extent of the warm temperate biome (part of the Angaran biogeographical realm) in the northern hemisphere (Figure 4.27). This biome spanned the entire area of present-day Kazakhstan and south-eastern Siberia (also known as the sub-Angaran realm) in the northern hemisphere and the north coastal fringes of Gondwana in the southern hemisphere. Fossil floras of this biome were abundant, of high diversity, and well mixed. They were dominated by sphenopsids (including *Calamites*), pteridosperms, and cordaites (Meyen, 1982; Cleal and Thomas, 1991).

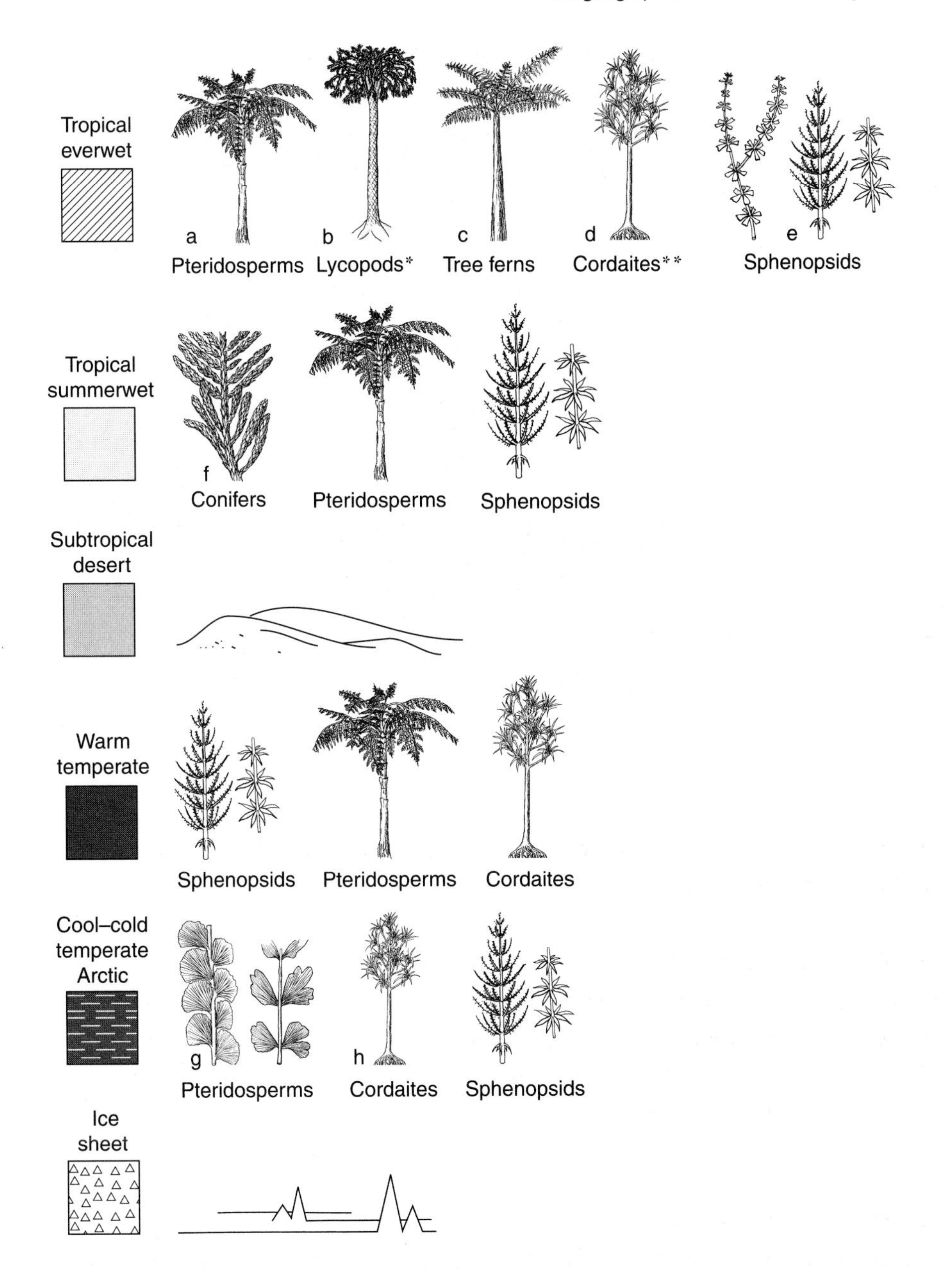

Figure 4.27 Suggested biomes for the late Carboniferous (300–286 Ma) with representatives of the most abundant and/or dominant fossil plant taxa shown. The biomes are superimposed on a global palaeogeographic reconstruction for the Wesphalian (~306 Ma) (courtesy of C. R. Scotese and the PaleoMap Project). (a) *Medullosa noei*; (b) *Lepidodendron* tree; (c) *Psaronius*; (d) cordaites tree; (e) *Calamites* tree with details of leaves (*Annularia*); (f) fertile branch of *Lebachia piniformis*; (g) *Botrychiopsis plantiana* frond; (h) rhacopterid frond. See Appendix 2 for sources of plant reconstructions and line drawings. * Abundant in China and regionally in eastern North America only; ** very abundant regionally in western North America.

Cool to cold temperate and Arctic biomes

The cool/cold temperate biome in both hemispheres covered present-day north-eastern Siberia in the northern hemisphere and the entire Gondwana supercontinent with the exception of the coastal regions and those covered by ice. It was characterized by significantly lower species diversity than that found in the warm temperate biome (Retallack, 1980; Meyen, 1987; Durante, 1996). The dominant vegetation included pteridosperms, cordaites, and significantly fewer sphenopsids.

In comparison to the wide diversity of taxa present in the warm temperate biome, vegetation became increasingly open, particularly in the southern hemisphere (Figure 4.27), by the late Carboniferous (300 Ma), and dominated by shrubby and herbaceous forms of lycopsids, sphenopsids, and the pteridosperm, *Botrychiopsis* (Archangelsky, 1990; Wnuk, 1996). The distinctly 'primitive' nature of the cold temperate floras of the southern hemisphere, compared with those of the tropical everwet and warm temperate biomes, have been interpreted by Retallack (1980) as a 'tundra'-type vegetation. It is possible therefore that an arctic biome existed between the cold temperate and glacial biomes, in areas close to the ice sheet. The assignment of these regions to the cool/cold temperate biomes is supported by the presence of both tillites and coals, which, when found in association, indicate cool temperate conditions.

Comparison of the composition of vegetation within each biome between the early (Figure 4.26) and late Carboniferous (Figure 4.27) reveals a number of striking differences, indicating major ecological turnover among a number of fossil plant groups. In the tropical everwet biome, for instance, the lycopsid and sphenopsid composition of wetland swamps declined (Cross and Phillips, 1990; Phillips *et al.*, 1985). In the tropical summerwet biome there was a coeval rise in the sporadic appearance of conifers (DiMichele *et al.*, 1992). Furthermore, the geographical extent of the subtropical desert biome expanded. All of these changes are thought to reflect a general aridification of late Carboniferous climates.

Summary

1. During the period spanning the early Devonian to late Carboniferous (395–290 Ma), global vegetation evolved from one dominated by small, weedy plants, most less than 1 m in height, to fully forested ecosystems with trees towering up to 35 m.

2. Global climates during this time can be divided into two distinct intervals. The first, spanning the period 390–360 Ma, had warm, humid climates with high levels of atmospheric CO_2. From 360 Ma onwards, global climates became increasingly cool and arid, with evidence for at least three periods of glaciation occurring between 360 and 290 Ma.

3. The transformation in global climates from ~360 Ma was due to a number of factors, including position of the South Pole in relation to the Gondwana landmass, mountain building episodes, and large-scale draw-down of atmospheric CO_2 associated with colonization of the land by plants.

4. Further adaptations to land dwelling that became apparent between 390 and 360 Ma included a more advanced vascular system, additional supporting mechanisms (roots, wood, bark), and leaves.

5. The earliest arborescent forms of plants (trees) appeared in the middle Devonian (~380 Ma).

6. Fossil evidence suggests that the plant life cycle also became more refined between 380 and 360 Ma, with the evolution of sporangia producing two spore sizes (megaspores and microspores), and the evolution of ovules with seed coats (seeds), pollen grains, and pollen reception mechanisms.

7. By the early Carboniferous (350 Ma) evolution and expansion of four major groups of spore-producing plants, namely the sphenopsids (horsetails), lycopsids (clubmosses), filicopsids (ferns), progymnosperms, and two groups of seed-producing plants (cordaites and pteridosperms) had occurred. Forests composed of these groups had widespread global distribution.

8. Broadscale biogeographic reconstruction for the early Carboniferous (354–342 Ma) indicates that six biomes are recognizible: tropical everwet, tropical summerwet, subtropical desert, warm temperate, cool temperate and glacial.

9. A slightly later biogeographic reconstruction (~300–286 Ma) indicates that provinciality of the global vegetation increased markedly through the Carboniferous, and that by this time at least six and possibly eight biomes can be recognized: tropical everwet, tropical summerwet, subtropical desert, warm temperate, cool to cold temperate, arctic and glacial, probably reflecting the general aridification of the late Carboniferous climates.

5 Major emergence of the seed plants

Evidence from the plant fossil record suggests that two groups of seed plants, the Cordaitales and pteridosperms, were established by the early Carboniferous (~360 Ma) (see Chapter 4). However, it was not until the Permian, some 50 to 80 million years later, that seed plants became dominant in the world flora. This major floral transition involved not only the expansion of these two early seed plant groups but also the emergence of new groups including the Cycadales (cycads), Ginkgoales (ginkgos), Bennettitales (bennettites), and Glossopteridaceae (glossopterids). By the upper Permian (~260 Ma) over 60% of the global flora was composed of gymnosperms (seed plants) and the arborescent lycopsids and sphenopsids, which had dominated early Carboniferous floras, were in decline (Figure 5.1). This chapter outlines the emergence and radiation of the seed plants and the environmental and climatic conditions that existed prior to and during their radiation.

5.1 Environmental changes during the Permian (290–248 Ma)

The series of continental collisions that occurred since the Carboniferous (see Chapter 4) finally culminated in the joining of Laurasia with Gondwana, Laurasia with Kazakhstania, and Siberia with Kazakhstania, resulting in the formation of the supercontinent, Pangea, by the early Permian (~300 Ma) (see Figure 4.1).

The impact of this supercontinent was far reaching in terms of global climate, affecting both atmospheric and oceanic circulation (Parrish, 1993). In the early Permian, Gondwanan continents were locked in deep glaciation (see Chapter 4) but by the mid-Permian, both geological and palaeontological evidence suggest that warmer climatic conditions were developing. By the late Permian (~256 Ma), it has been estimated that global climate was characterized by widespread aridity within continental interiors, high seasonal temperature fluctuations, equatorial aridity (in contrast to the present situation), and monsoonal conditions prevailing in both hemispheres (Erwin, 1993). Although the proposed mechanisms behind this dramatic change from a global 'icehouse' to a

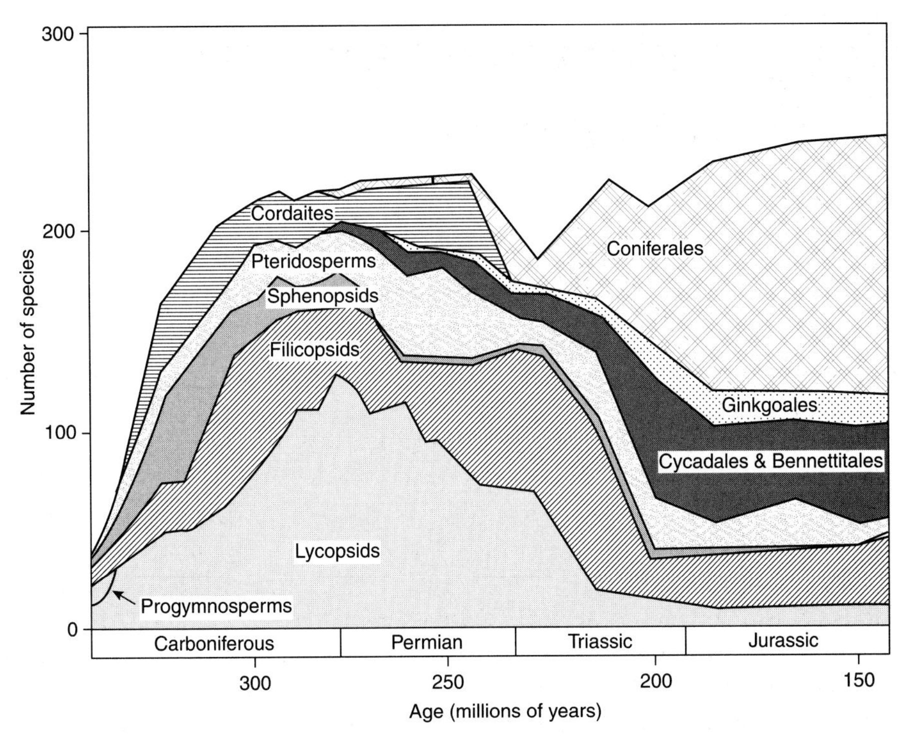

Figure 5.1 Evolution and diversification of the seed plants during the Permian (redrawn from Niklas *et al.*, 1985a). Data are taken from a compilation of approximately 18 000 fossil plant species citations.

'greenhouse' are extremely complex (for a full review see Erwin, 1993), three main processes are thought to have been primarily responsible. First, the northward movement of the southern supercontinent, Gondwana, to coalesce with Laurasia resulted in the movement of Pangea into the tropical latitudes (Figure 5.2). Secondly, it is suggested that formation of the large landmass of Pangea greatly increased continentality (Erwin, 1993). Thirdly, creation of extensive uplands as a consequence of folding and thrust faulting resulting from these plate collisions would have further increased continentality. Geological evidence suggests, for example, that uplifted mountain belts formed at the zone of division between Russia and the Siberian platform (the Eurasian Orogen) and between Laurasia and Gondwana (the Hercynian orogeny), producing mountains as high as 4 km (Hay *et al.*, 1982). These uplands would have acted as a barrier to the flow of air between the oceans and continents (Crowley, 1998), like the present-day Himalayas and Tibetan Plateau.

Climate modelling predicts that during the northern hemisphere winter, a very strong high pressure cell would have developed over Pangea, whereas during the summer a strong low pressure cell would have developed, drawing warm, moist air from the oceans (Parrish *et al.*, 1986; Kutzbach and Gallimore, 1989; Crowley and North, 1991). This seasonal contrast in both atmospheric and

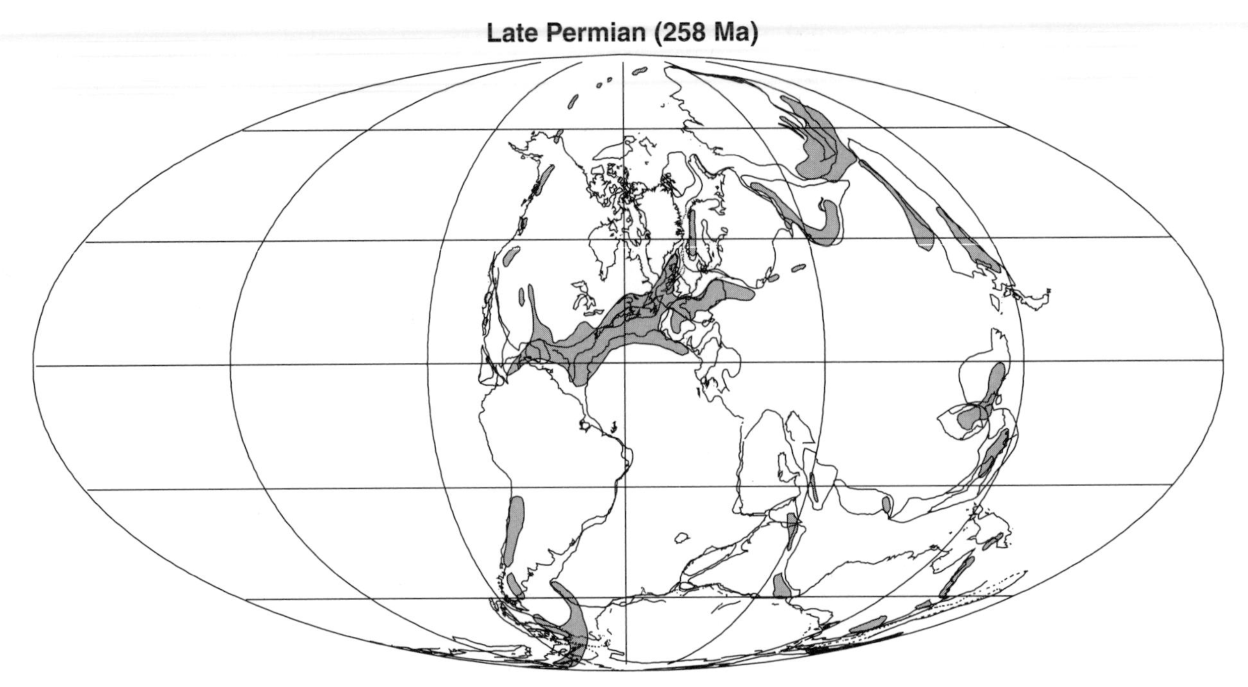

Figure 5.2 Approximate continental positions by the late Permian. Also indicated are the approximate distribution of mountain and highland areas (Scotese, 2001).

oceanic circulation would have resulted (as is the present-day situation in Asia) in a monsoonal climate, with warm and wet summers and dry winters. In addition, there would have been significant cross-equatorial winds (Erwin, 1993).

The pattern of increased warming and aridification of Permian climates occurred against a backdrop of increasing atmospheric CO_2 concentrations. Long-term carbon cycle models, for example, estimate that CO_2 concentrations rose from levels similar to those of the present day in the early Carboniferous (~300 Ma) to levels almost four times greater by the end of the Triassic (~240 Ma) (Berner, 1991, 1997) (Figure 5.3). These Permian CO_2 estimates are further supported by palaeobotanical evidence from fossil stomatal densities (McElwain and Chaloner, 1995), and palaeosols (Mora *et al.*, 1996). One of the most important geological processes thought to have contributed to increased atmospheric CO_2 during this period was enhanced tectonic activity.

5.2 Evolution of cycads, bennettites, ginkgos, and glossopterids

It was against a background of major environmental change, with the climate becoming warmer and drier, that evolution of the next groups of seed plants, namely cycads, bennettites and ginkgos occurred. In addition, in the southern

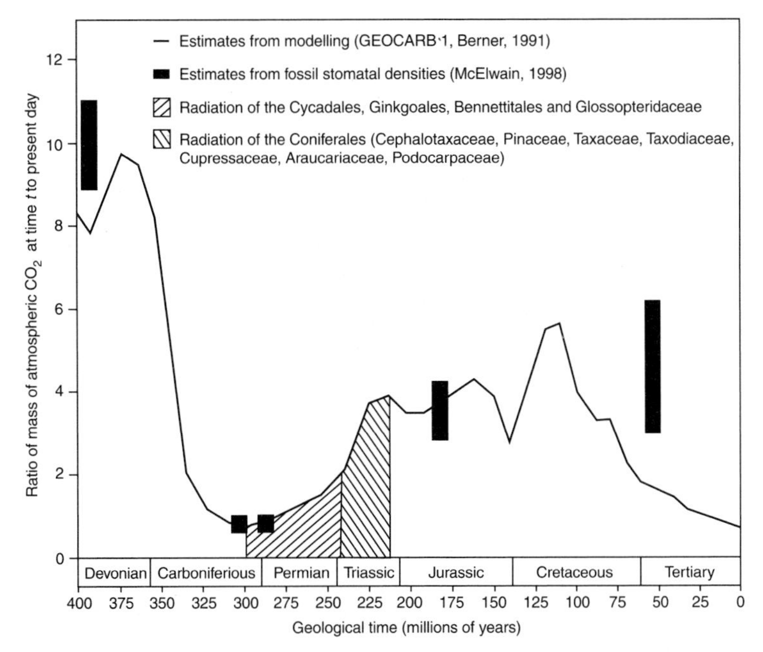

Figure 5.3 Estimated variations in atmospheric CO_2 during the past 400 million years (redrawn from Berner, 1991; GEOCARB I). The general direction of the trends are supported by independently derived results from carbon isotope analysis of palaeosols (e.g. Mora *et al.*, 1996) and stomatal densities on fossil leaves (e.g. McElwain, 1998). For comparison, the latter are indicated on the graph. Both are plotted against ratio of mass of atmospheric CO_2 at time t to present day (where present is taken at the pre-industrial value of 300 p.p.m.v.). The mid-Permian to late Triassic rise in atmospheric CO_2 is highlighted, along with the major changes apparent in the plant fossil record during this time. Note that the Quaternary (last 1.8 Ma) is not marked on the geological timescale.

hemisphere there was the major expansion of the seed fern group, the glossopterids. The cycads and ginkgos contain living and fossil representatives, although in Mesozoic time they had a much greater global distribution. Fossil evidence suggests, for example, that by the mid-Triassic (~240 Ma) their distribution extended as far north as Siberia and Greenland. The bennettites and glossopterids are now extinct.

Cycadales

Vegetative and reproductive features

There are approximately 10 genera (and 100 species) of Cycads living in tropical and warm temperate regions of the world today (Mexico and the West Indies, Australia, and South Africa). All types are dioecious (i.e. they have male and female plants, rather than both sexual organs on one plant) and there is no fossil evidence to suggest that there were ever monoecious plants in this group (Thomas and Spicer, 1987). The first fossil evidence for the cycads is from approximately 280 million years ago (early Permian) and indicates that, like their present-day counterparts, some species grew up to 15 m in height, although some of the earliest plants had short (<3 m), stout, and unbranched trunks with a crown of leaves.

In extant groups of cycads, there are two types of leaves, pinnate leaves that look similar to some simple fern fronds and are borne usually as an apical rosette (i.e. growing out of the top of the trunk), and scale leaves which cover the upper

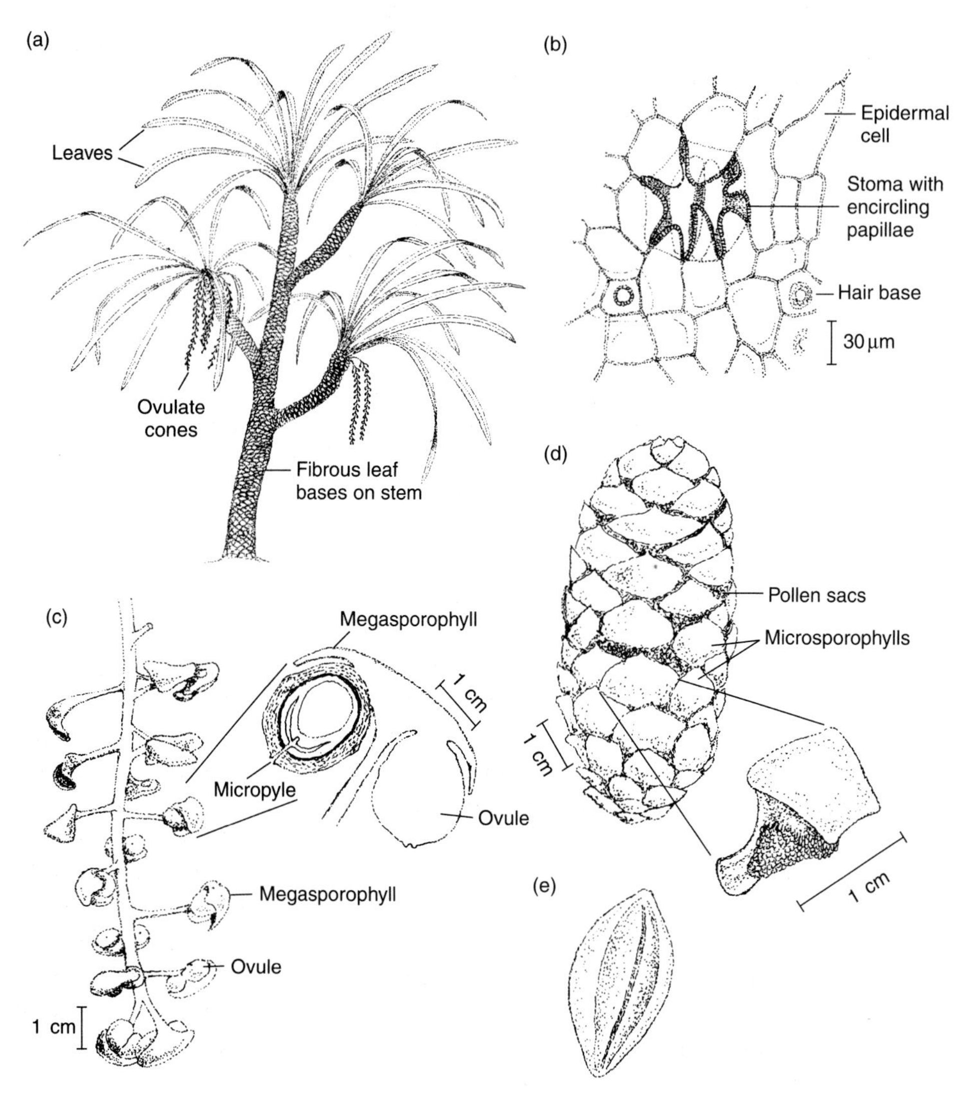

Figure 5.4 Early cycad tree: (a) habit; (b) detail of leaf cuticle indicating epidermal and stomatal features; (c) female reproductive structure (ovulate cone) with detail of megasporophyll and ovules (enlarged); (d) male reproductive structure (pollen cone) with detail of microsporophyll and pollen sacs (enlarged); (e) monolete pollen (~60 μm in diameter) (redrawn from Harris, 1964).

part of the stem (Figure 5.4a). The scale leaves often disintegrate to form a fibrous sheath around the stem (Bell, 1992), whereas the pinnate leaves leave behind distinctive fibrous leaf bases. All the leaves are evergreen. Cycad foliage is extremely common in the fossil record (especially as compression/impression fossils). Usually the apical 'fronds' are preserved and these demonstrate a similar growth position to present-day cycads (i.e. at the top of the trunk). However, fossil leaflet morphology is highly varied and suggests that there were some early species of cycad that had entire leaves whereas others had divided, pinnately compound leaves. All these fossil leaves can be identified as cycad, however, by the similarity of their epidermal and stomatal features and also, in most cases, by simple or open venation patterns (Brashier, 1968) (Figure 5.4b).

There is much evidence in the fossil record for nearly complete cycad stems with morphological and anatomical features showing remarkable similarity to present-day taxa (Stewart and Rothwell, 1993; Taylor and Taylor, 1993). These features include a thick, often unbranched stem, which in cross-section is a eustele. This is composed of a massive central core of pith surrounded by vascular tissue and then bark. The primary vascular tissue surrounding the pith is made up of vascular bundles containing xylem, and by a thin layer of secondary xylem surrounding the primary xylem. Additional support to the stem is also provided by the leaf bases, which provide a firm outer coating to the trunk. Short, squat cycads that are typical of those in the extant flora did not appear, however, until the Tertiary, and the majority of Mesozoic (~248–65 Ma) cycads were characterized by a more slender and branching habit (Delevoryas and Hope, 1976). This slender branching habit is generally not seen in extant cycads but has been observed as a response to physical trauma (Gifford and Foster, 1989).

The reproductive organs of the cycads are well documented in the fossil record. The female plants possessed clusters of ovules situated on modified leaves called megasporophylls (Figure 5.4c). The male reproductive structure also consisted of modified leaves (microsporophylls), but each of the leaves had small, compact pollen sacs attached to their lower surface (Figure 5.4d). In extant groups, each pollen sac produces numerous pollen grains, which are boat-shaped and thin-walled, with one broad germination furrow (monocolpate) (Figure 5.4e). Similar examples have been found in the plant fossil record dating back as far as the early Permian (270 Ma) (Traverse, 1988). It is suggested that germination in the early cycads was similar to that seen in extant groups, whereby the production of a sugary drop from the micropyle trapped the wind-distributed microspores, drawing them into the pollen chamber. The pollen then germinated to grow a pollen-tube like structure out of its distal face, from which male gametes were then released to swim to the archegonia, entering the egg to effect fertilization.

Origins and relationships of the Cycadales

Recent cladistic analyses show that extant cycads are most closely related (i.e. a sister group) to a group including extant ginkgos, conifers, and Gnetales

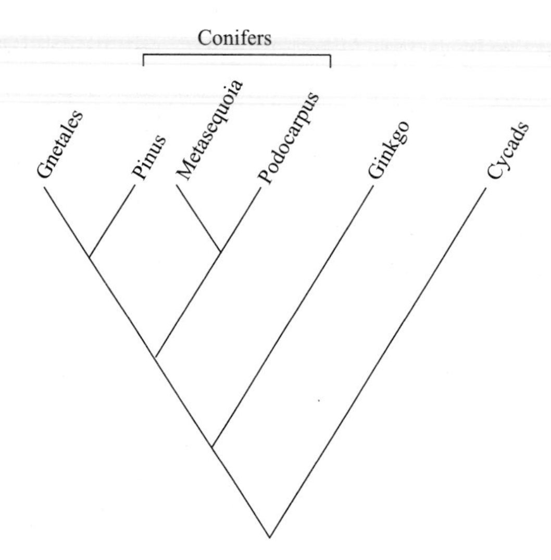

Figure 5.5 A simplified phylogenetic tree to indicate the possible evolutionary relationship of the cycads to other seed plants (redrawn from Qui *et al.*, 1999; Pryer *et al.*, 2001).

(Qui *et al.*, 1999; Pryer *et al.*, 2001) (Figure 5.5). Which of these three orders is the most closely related to the cycads, however, has yet to be resolved. Morphological analysis of fossil plant vegetative and reproductive structures suggests that the cycads are most probably derived from the Medulosaceae (seed ferns) (Stewart and Rothwell, 1993), which dominated the tropical and warm temperate biomes of the late Carboniferous (~300 Ma). This is supported by some cladistic analyses (Crane, 1985), although not unanimously (Doyle and Donoghue, 1986; Rothwell and Serbet, 1994).

Bennettitales

Vegetative and reproductive features

The bennettites have a fossil record indicating widespread global distribution and high diversity from the early Triassic to late Cretaceous (~248–140 Ma). This group bears many morphological similarities to both extant and extinct cycads, and for many years they were classified as such. However, a number of features have been found to suggest that in fact they were a discrete taxonomic group.

In the literature, one of the most commonly cited examples of a genus within the bennettites is *Williamsonia*, since it was originally proposed that these plants had reproductive structures with a superficial resemblance to angiosperm flowers. It has since been demonstrated, however, that an evolutionary link between members of the bennettites and the earliest flowering plants is unlikely (Thomas and Spicer, 1987; Stewart and Rothwell, 1993), although some phylogenetic studies still support this interpretation (Crane *et al.*, 1995).

The shape of the bennettite leaves were similar to those of cycads, in that they were frond-like and composed of long, slender rachises with numerous pinnae (although some species had entire leaves) (Figure 5.6b). The fronds grew from an apical crown at the top of the trunk in a spiral arrangement, and fossil evidence suggests that they were periodically deciduous, leaving behind a distinctive pattern of leaf bases on the trunk (Figure 5.6a). On first examination, therefore, there is little difference between the cycads and bennettites. However, the epidermal structure of the leaf indicates significant differences. For example, whereas the walls of the epidermal cells in cycads are relatively straight (Figure 5.7), they were highly sinuous in bennettites (Bell, 1992). In addition, bennettites had lateral subsidiary cells (on the outside edge of the guard cells) that were derived from the same mother cells as the guard cell. This type of stomatal arrangement is referred to as syntedocheilic (Stewart and Rothwell, 1993). In cycads the subsidiary cells are not derived from the same mother cell as guard cells but from adjacent epidermal cells. This type of stomatal arrangement is referred to as haplocheilic (Figure 5.7). There are also major differences in their reproductive structures (compare Figures 5.4 and 5.6).

The stems of bennettites were composed of a large pith surrounded by a ring of vascular tissue and the overall structure was eustelic. Numerous examples of fossil bennittite cones indicate that with one or two exceptions, they were unisexual early in their radiation, but later became predominantly bisexual (Stewart and Rothwell, 1993). The female cones consisted of ovules, which were covered with sterile scales and surrounded by a long integument, extended as a micropylar tube (Bell, 1992) (Figure 5.6d). The male reproductive structure was composed of a whorl of leaves bearing small pollen sac-like structures which consisted of fused sporangia, known as a synagium (Figure 5.6e, f). The pollen grains produced by these sporangia were similar to those of cycads in that they had a single furrow (monocolpate) (Figure 5.6f). Finally, the whole reproductive structure was enclosed in spirally arranged bracts which were originally thought to have opened out in much the same way as a flower bud does, when conditions are suitable for pollination (Wieland, 1915). Reinterpretation by Crepet (1996), however, suggests that opening of the whole structure was unlikely, due to structural constraints. Little is known about the actual mechanism by which the pollen reached the micropyle. Although both wind pollination and selfing have been suggested, evidence for male and female development which is out of phase suggests that some form of animal pollination may have been utilized.

Bennettite origins and relationships

As with the cycads, a medullosan seed fern origin for the bennettites has been suggested on the basis of shared pollen organ characteristics (Figure 5.6f). Perhaps the most intriguing relationship, however, is the suggestion that bennettites were the precursor to the angiosperms. Many bennettite reproductive structures have been referred to as 'flowers' including, for example, two taxa discovered in rocks of middle Jurassic age (~176–169 Ma) from Yorkshire in England, namely *Williamsoniella coronata* and *Williamsonia gigas*. Although these pre-angiosperm fossil 'flowers' demonstrate many shared characteristics with true

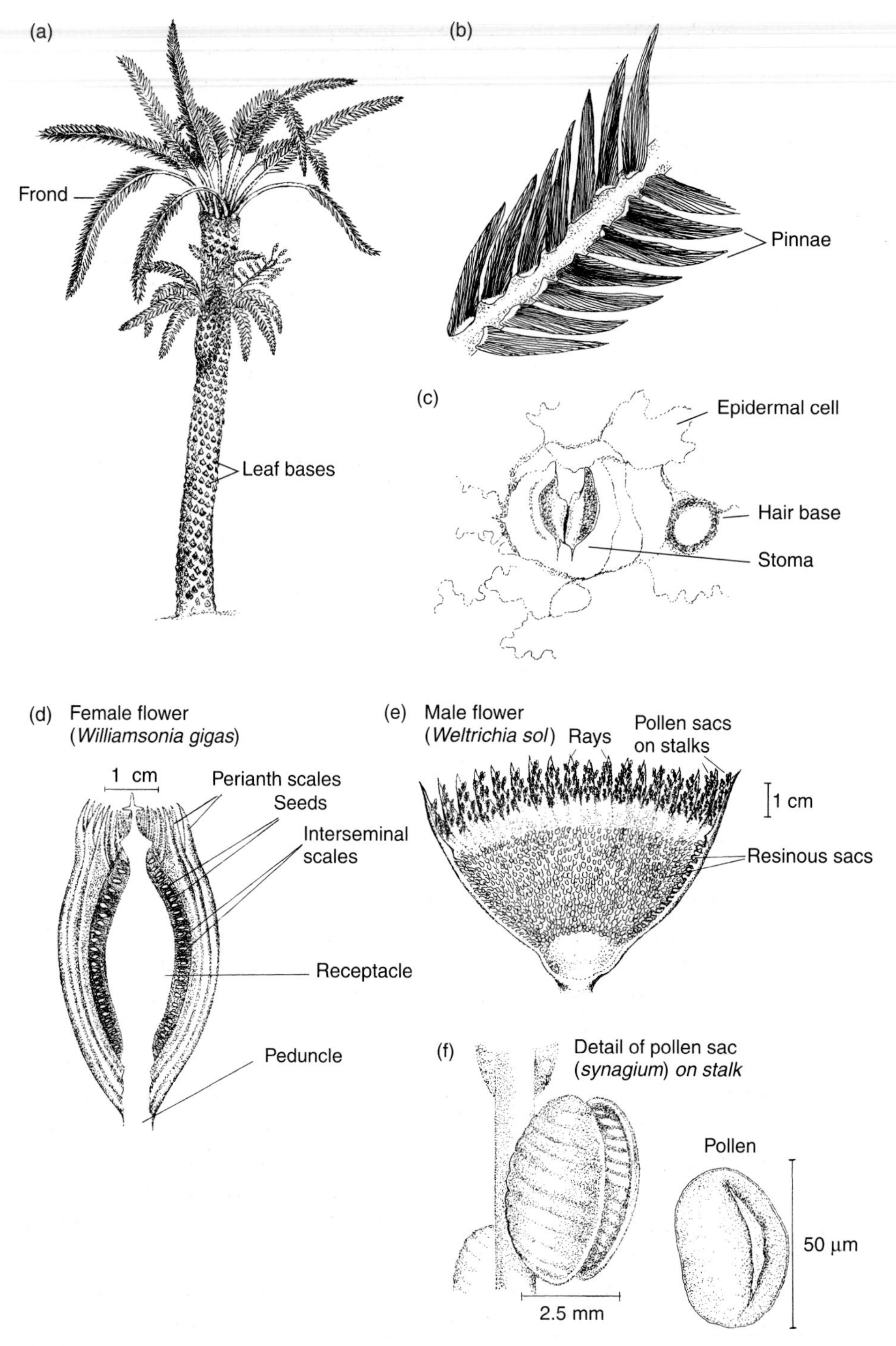

Figure 5.6 Early Bennettitalean tree, *Williamsonia*: (a) habit; (b) leaves (*Zamites gigas*) and their position on the frond; (c) epidermal structure of leaves (from Bell, 1992); (d) female reproductive structure; (e) male reproductive structure ('flower'); (f) pollen sac and pollen. (Redrawn from Crane (1985) (d,e,f), Harris (1969) (b,c) and Bell (1992) (a).)

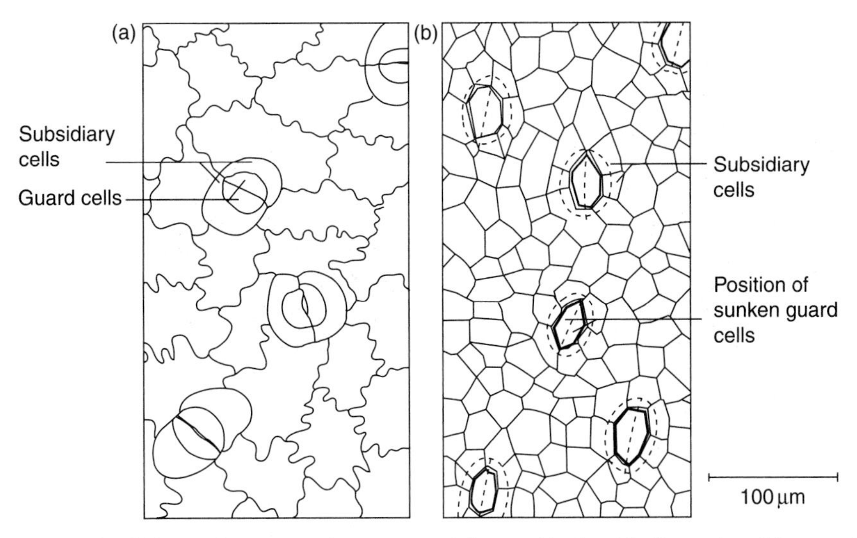

Figure 5.7 The difference between (a) bennettite and (b) cycad leaf cuticle illustrating differences in stomatal and epidermal cell arrangement, known as syndetocheilic (a) and haplocheilic (b) respectively (redrawn from Bell, 1992).

angiosperm flowers, and phylogentic studies suggest a close affinity (e.g. Crane *et al.*, 1995), the presence of only one rather than two integuments surrounding the ovule, and the lack of any convincing morphological intermediates, has led some to question this interpretation (Hughes, 1994).

Ginkgoales

Vegetative and reproductive structures

Ginkgos have a fossil record extending back to approximately the early Permian (~280 Ma). Today they are represented by a single species, *Ginkgo biloba* (the maidenhair tree), which grows wild in parts of China but is widely used as an ornamental tree in botanical gardens throughout the world. However, at the height of their global radiation, plant fossil evidence suggests that at least 16 genera formed a significant part of the world vegetation (Pearson, 1995). The remarkable morphological similarity between extinct fossil specimens and extant *Ginkgo biloba* (Figure 5.8) has often led to the description of *Ginkgo biloba* as a 'living fossil' (Stewart and Rothwell, 1993; Taylor and Taylor, 1993).

The plant fossil record indicates that ginkgo leaves were borne on both long and short shoots on lateral branches of the main stem (Figure 5.8a). There was great variation in leaf morphology between species (Figure 5.8b), ranging from wedge-shaped lamina with highly dissected margins, to lanceolate leaves with an entire margin (Taylor and Taylor, 1993). All ginkgo leaves (fossil and extant), however, are readily recognized by their distinctive leaf shape, open dichotomous venation pattern, and in some cases by their stomatal characteristics

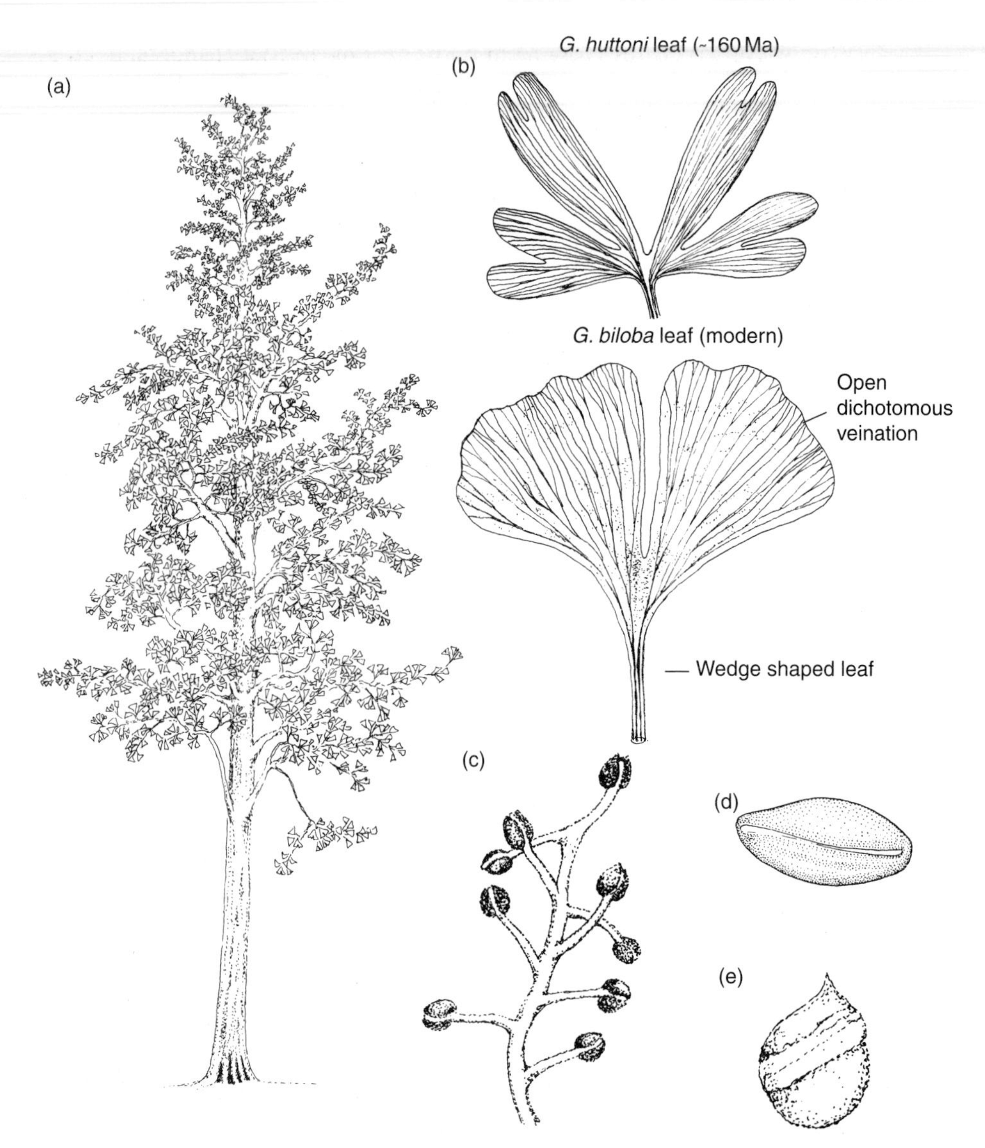

Figure 5.8 *Ginkgo*: (a) habit of *G. biloba*; (b) fossil (upper) and modern (lower) leaves; (c) fossil male reproductive structure; (d) fossil pollen; (e) female reproductive structure (fossil seed). (b) and (e) redrawn from Harris (1974), (d) and (c) redrawn from van Konijnenburg-van Cittert (1971).

(Bell, 1992). Ginkgo stomata are often sunken and surrounded by epidermal cells with overarching papillae (hairs) (Figure 1.3b). The modern *Ginkgo biloba* is deciduous and the large number of leaves found in the fossil record has led to the suggestion that this was also the case for its ancestors (Thomas and Spicer, 1987).

Presently *Ginkgo biloba* may grow up to 30 m in height with a laterally branching stem bearing both short and long shoots (Figure 5.8a). Leaves and reproductive structures grow from both the long and short shoots. In cross-section, the *Ginkgo biloba* stem is composed of a eustele containing large amounts of secondary xylem, and demonstrates characteristics that are difficult to separate from some coniferous wood such as *Pinus*. Fossil evidence for stems of *Ginkgo* is scarce but material that is available suggests a similar stellar arrangement.

Ginkgo biloba is dioecious and therefore has both male and female plants. The male plants bear catkin-like structures, which arise in pairs in the axil of the leaves (Figure 5.8c). Each fertile structure bears 2–4 pollen sacs at the tip, which contain monocolpate pollen grains that are wind dispersed. Female plants possess ovules, which are usually borne in a similar manner to the male reproductive structures, in the leaf axils on erect stalks, known as peduncles (Figure 5.8e). At maturity the ovules are 1.5–2.0 mm in diameter and are therefore much larger in size than most conifer seeds but are comparable to those of the cycads and medullosan seed ferns (Stewart and Rothwell, 1993). Each ovule has an integument and associated micropyle, which produces a pollination drop. When the pollen lands on the ovule, a pollination tube germinates out from the grain but, similarly to cycads, each grain produces two motile sperm. These are released into the liquid in the archegonian chamber above the female gametophyte, and one swims into the archegonium.

There is evidence in the fossil record to suggest that there was much variation in reproductive structures between species. Examples include species that had over 100 small ovules attached singly on short stalks to a central axis (Thomas and Spicer, 1987) and dichotomous branching stems bearing ovules (Archangelsky and Cuneo, 1990).

Ginkgo origins and relationships

Despite major advances in both molecular and morphological phylogenetic studies, the evolutionary history of the Ginkgoales remains unresolved. This is mainly due to the fact that the ginkgos, both fossil and the only living species *Ginkgo biloba*, share many vegetative characteristics with cordaites and conifers but their reproductive structures are more similar to those of cycads. These morphological complexities are also reflected in the molecular data, which show a variety of possible relationships between ginkgos and cycads and ginkgos and conifers. Furthermore, although there is a rich fossil record of ginkgo and ginkgo-like foliage, fossil reproductive structures are relatively rare and our ability to trace the evolutionary history of Ginkgoales and their possible ancestral group or groups is therefore made difficult. On the basis of the fossil leaves alone, it has been suggested that *Dicranophyllum moorei*, a late Carboniferous fossil plant with long slender forking leaves up to 10 cm long, is a possible Ginkgoalean ancestor. Without co-occurring ginkgo-like reproductive structures however this suggestion remain controversial. To date, two lower Permian fossil plant genera, *Trichopitys* and *Polyspermophyllum* are the oldest genera, which share both vegetative and reproductive affinities with Gingkoales and are thus believed to be possible ancestral forms (Andrews, 1961, Archangelsky and

Cúneo, 1990). Discovering more Ginkoalean and ginkgo-like reproductive structures in the fossil record will certainly be key to unraveling their evolutionary relationships in the future.

Glossopteridaceae

Vegetative and reproductive structures

At least six groups of seed ferns have been recognized in the fossil record (see Taylor and Taylor, 1993). Although their general characteristics and early fossil appearance are described in Chapter 4, the fossil distribution of the seed fern group Glossopteridaceae needs to be mentioned here. This group dominated the flora of the southern hemisphere continent of Gondwana during the Permian (~290–248 Ma) and, like other groups, has also been suggested as a possible angiosperm ancestor. Plants in this group are thought to have been predominantly arborescent and deciduous, with reproductive organs attached to highly modified leaves. Over 200 species have been described, with fossil examples found in present-day localities of Australia, Africa, South America, Antarctica, and the Indian peninsula (Stewart and Rothwell, 1993; Taylor and Taylor, 1993).

Evidence from fossil wood suggests that glossopterids grew to approximately 10 m in height, with thick trunks of eustelic structure indicating a close similarity to the wood of modern *Araucaria* (White, 1990) (Figure 5.9a). The trees were deciduous (as indicated by the presence of thick leaf mats of glossopteris leaves in Permian-aged rocks) and had leaves that were arranged in whorls (Figure 5.9a). The deciduous nature of glossopterids, taken together with evidence of clear growth rings, suggests that these trees flourished in a climate with a temperature and/or precipitation regime typical of our current temperate biome (Ziegler, 1990).

The fossil leaves are distinctive by their cross-connections between veins (anastomosing venation) (Alvin and Chaloner, 1970), resulting in a net-venation pattern remarkably similar to that typical of many angiosperm dicotyledon leaves (see Chapter 6; Figure 5.9b). The leaves typically have stomata only on the lower leaf surfaces. The stomata of glossopterids are very similar to those of the cycads, as both possess a ring of subsidiary cells surrounding the guard cells, derived from the adjacent epidermal cells (haplocheilic) (Figure 5.7b).

The female reproductive structures of glossopterids are extremely diverse, including both uni-ovulate and multi-ovulate structures borne in both solitary and compound leaf and leaf-like structures (Taylor and Taylor, 1993). These have been grouped these into two broad categories, namely the megafructi which, as the name suggests, had large fructifications (seeds) located on unaltered foliage leaves, and the microfructi in which smaller seeds were attached to modified scale leaves (White, 1990). The male reproductive structures were composed of fertile scales, sometimes aggregated into cones (Figure 5.9d) and had distinctive sporangia containing bisaccate, striated, pollen grains (Figure 5.9e).

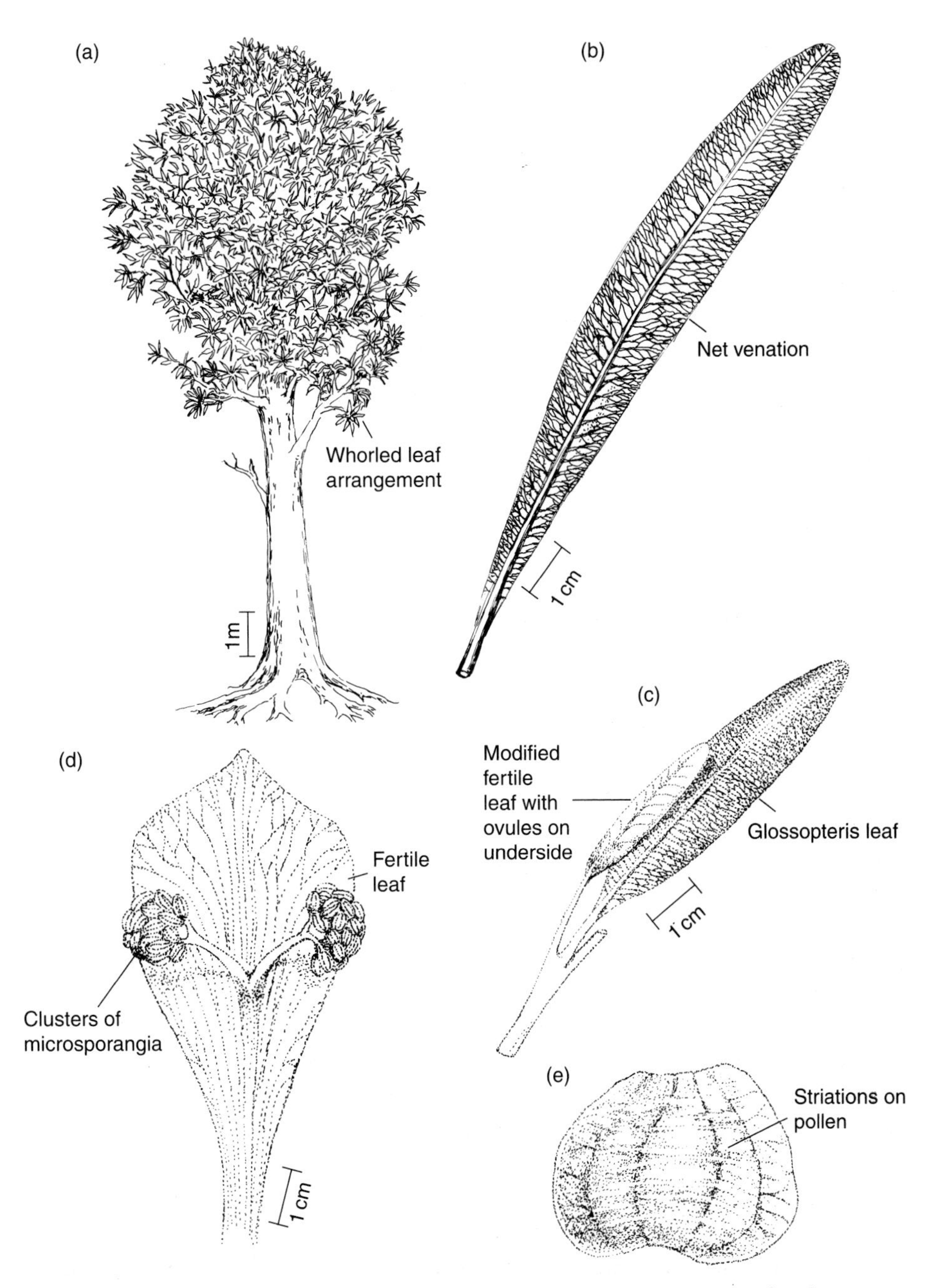

Figure 5.9 Glossopteris tree: (a) habit; (b) leaf indicating net-venation pattern; (c) female reproductive structure; (d) male reproductive structures; (e) pollen (a to d redrawn from Stewart and Rothwell).

Glossopteridales: origins and relationships

The Glossopteridales appear to combine the characteristics of many other seed plant groups and have, yet again, been proposed as a plausible ancestor to the flowering plants (angiosperms), on the basis of similarity in both venation pattern (Melville, 1983; Srivastava, 1991) and female reproductive structures (Retallack and Dilcher, 1981). However, this is now thought to be unlikely (Hughes, 1994). Based on a detailed morphological comparison, Schopf (1976) hypothesized that the glossopterids have affinities with the cordaites and may even have given rise to the Gnetales. Indeed, Taylor and Taylor (1993) have noted that close relationships with other pteridosperms, cycads, conifers, cordaites, Gnetales, and angiosperms have been variously suggested, but that glossopterids have the closest affinity with the pteridosperms. The evolutionary position of the glossopterids within the seed plant as a whole therefore remains ambiguous; however, detailed investigation of anatomically preserved reproductive and vegetative structures will certainly help to resolve this further (Pigg and Trivett, 1994).

5.3 Biogeographical distribution of global vegetation during the middle Permian (267–264 Ma)

The Permian (~290–248) was characterized by major global climate change, from a glaciated (icehouse) to completely ice-free (hothouse) state (see Section 5.1). Not surprisingly, this dramatic climate change is reflected by marked changes in composition, dominance patterns, and biogeographical distribution of world vegetation. One of the most striking features of these vegetation changes was the relatively increased proportion of seed plants in global floras by the middle Permian and reduction of the swamp-dwelling lycopsids and sphenopsids (Ziegler, 1990) (Figure 5.10). These floristic changes, together with evidence for increased provinciality, provide a strong signature of increased aridity and greater pole to equator temperature gradients. Biome-level biogeographical analyses of the middle Permian, based on a sophisticated multivariate statistical approach, have identified seven distinct biomes (Ziegler, 1990; Rees *et al.*, 1999), which will be discussed below.

Cold temperate biome

The cold temperate biome in the middle Permian (267–264 Ma) occupied high palaeolatitudes between 60° and 90° in both hemispheres. In the northern hemisphere this biome incorporated present-day eastern Siberia, and in the southern hemisphere, present-day Antarctica and southern Australia. In the northern hemisphere (also known as the Angaran realm) the vegetation was dominated by cordiates and sphenopsids, whereas in the southern hemisphere (the Gondwanan realm) glossopterids (with at least six different species) and sphenopsids were predominant. Lithological data and climate model projections, together with the relative low-diversity fossil floras in these areas, are indicative of cool climates typical of the cool temperate biome (Rees *et al.*, 1999). Seasonality in temperature, another typical climatic feature of this biome, is supported by evidence for growth rings in fossil wood and extensive leaf mats of both glossopterids and cordaites, indicating deciduousness.

Cool temperate biome

The cool temperate biome, which covered a significant proportion of the high latitudes in both hemispheres, included present-day South Africa, India, Madagascar, and Australia in the southern hemisphere, and Siberia in the northern hemisphere. Fossil floras from these regions show relatively greater diversity than in the higher-latitude cold temperate biome. Palaeobotanical evidence such as growth rings and the presence of deciduousness indicates that despite relatively high species diversity, climates were distinctly seasonal. In the northern hemisphere the vegetation was composed mainly of cordaites and sphenopsids with some ferns, but in the southern hemisphere floras contained a high proportion of glossopterids. Intriguingly, in the northern hemisphere, the cool temperate biome, together with the tropical everwet biome, is characterized by the presence of the last remnants of the water-loving lycopsid flora that were so extensive during the late Carboniferous.

Warm temperate biome

The vegetation spanning northernmost North America, Greenland, and Scandinavia in the northern hemisphere, and northern India, western Australia, and Chile in the southern hemisphere has been classified as a warm temperate biome. In the northern hemisphere this was characterized by high diversity and composed of cordaites, peltasperms (a pteridosperm group), ginkgos, sphenopsids, and ferns; whereas in the southern hemisphere glossopterids were predominant.

Mid-latitude desert biome

The mid-latitude desert is marked in the southern hemisphere by an abundance of evaporite deposits indicating aridity. There are no fossil floras present in this biome during the middle Permian.

Subtropical desert biome

The predominance of evaporites in what is now northern Africa, northern South America, and much of Europe and North America, is used to designate these low

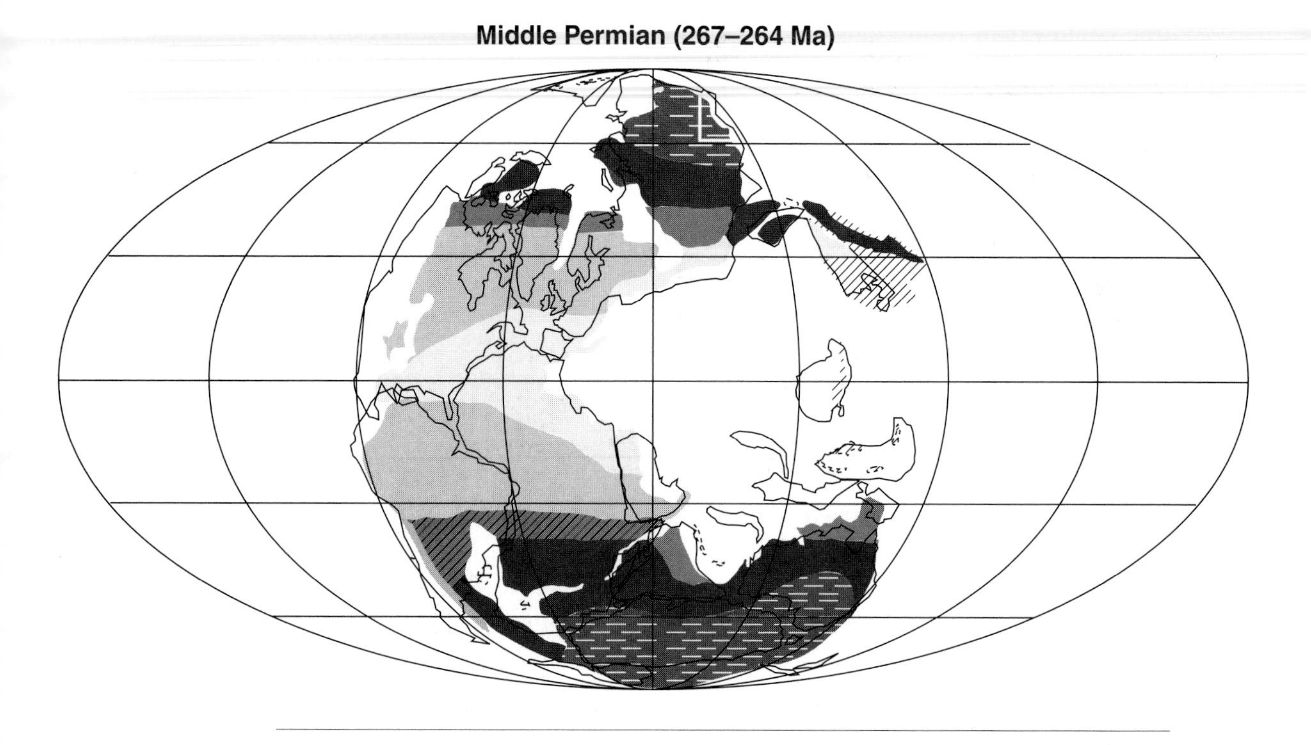

palaeotropical latitudes as the subtropical desert biome (Rees *et al.*, 1999) (Figure 5.10). Palaeobotanical evidence indicates that these great Permian deserts were not totally devoid of vegetation, but on the coast supported cycads, ginkgos, and conifers (Volziales) at some localities (Rees *et al.*, 1999).

Summerwet (tropical)

The low equatorial latitudes were characterized by a more depauperate flora than the temperate belts, composed mainly of Ginkgoales and conifers (Volziales) with some cordaites. The xerophytic nature of the vegetation, including small, scale-like leaves of the Volziales in particular, and, to a certain extent, the finely divided leaves of the early ginkgos, are suggestive of a climate prone to seasonality in precipitation regimes.

Tropical everwet biome

The increased aridity and seasonality of the global climate between 267 and 264 Ma is also strikingly demonstrated by the displacement of the tropical everwet biome from the continental interior of Pangea, where it straddled the equatorial latitudes in the upper Carboniferous, to China (referred to as the Cathasian Realm). This biome, which occupied what is now northern China, was composed of medulosan pteridosperms, sphenopsids, ferns, and

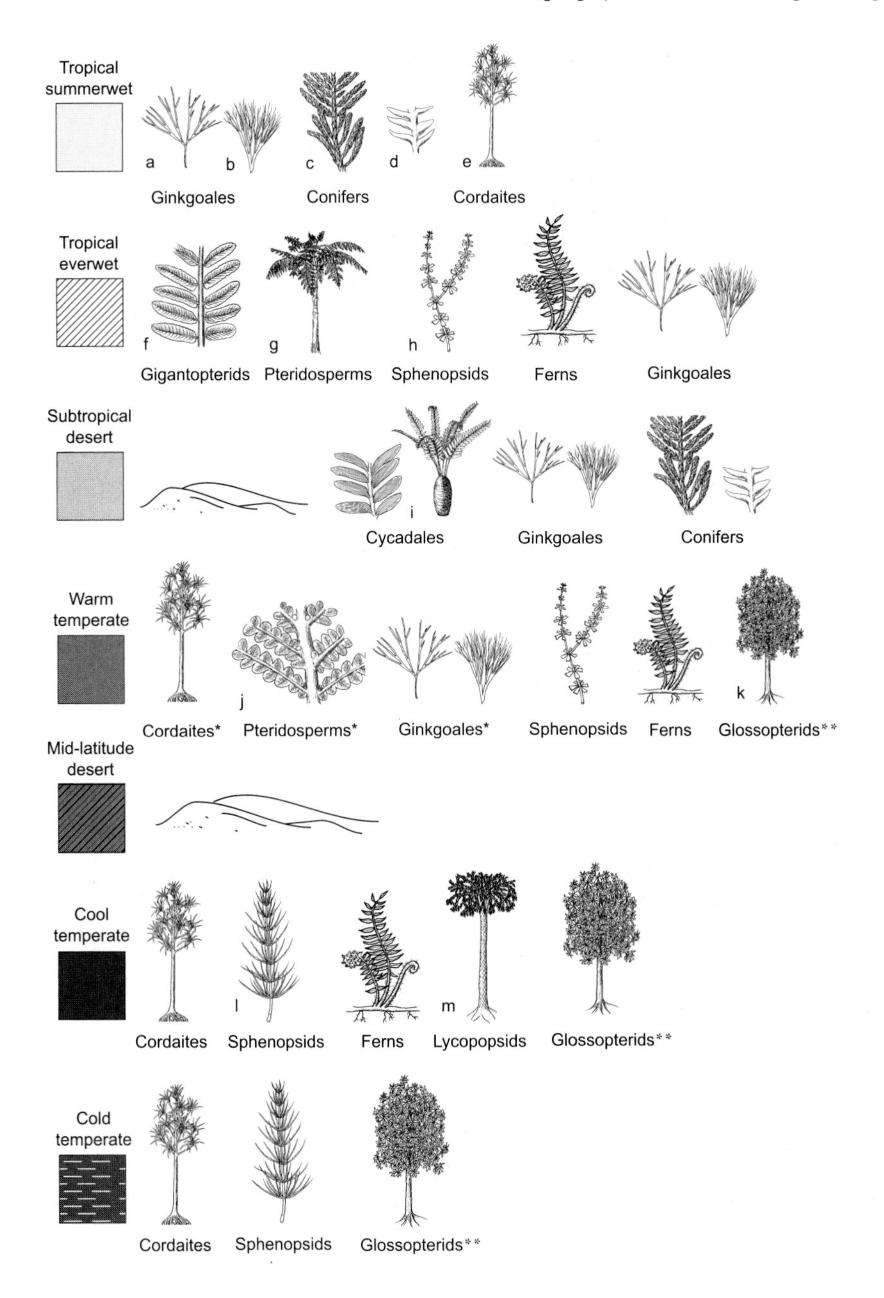

Figure 5.10 Suggested biomes for the Middle Permian (267–64 Ma) with representatives of the most abundant and/or dominant fossil plant taxa shown (modified from Rees *et al.*, 1999). The biomes are superimposed on a global palaeogeographic reconstruction for the Wordian (~267 Ma) (courtesy of A. M. Ziegler, PaleoAtlas Project). (a) *Baiera* leaf; (b) *Sphenobaiera* leaf; (c) fertile branch of *Lebachia piniformis*; (d) *Ullmannia frumentaria* leaves; (e) cordaites tree; (f) *Gigantoclea guizhouensis*; (g) *Medullosa noei*; (h) *Sphenophyllum*; (i) stylized cycad with details of *Plagiozamites oblongifolius* leaves; (j) *Callipteris conferata* frond (peltasperm); (k) *Glossopteris* tree; (l) stylized equisetites; (m) *Lepidodendron* tree. See Appendix 3 for sources of plant reconstructions and line drawings.** Dominant or abundant in the southern hemisphere.

Ginkgoales, and dominated by large-leaved gigantopterids (another group of seed ferns or pteridosperms) (Ziegler, 1990; Rees *et al.*, 1999).

5.4 Major radiation of the conifers

The earliest conifer in the fossil record, *Swillingtonia denticulata*, was discovered from an upper Carboniferous locality (Westphalian B, ~310 Ma) in Yorkshire. Although conifers were not a major component of lowland global vegetation at this time, it is suggested that they most probably inhabited the drier environments of upland areas (Scott and Chaloner, 1983), from which they subsequently radiated. The presence of conifers in the plant fossil record increased throughout the Permian (Miller, 1988). However, it was not until the Triassic period

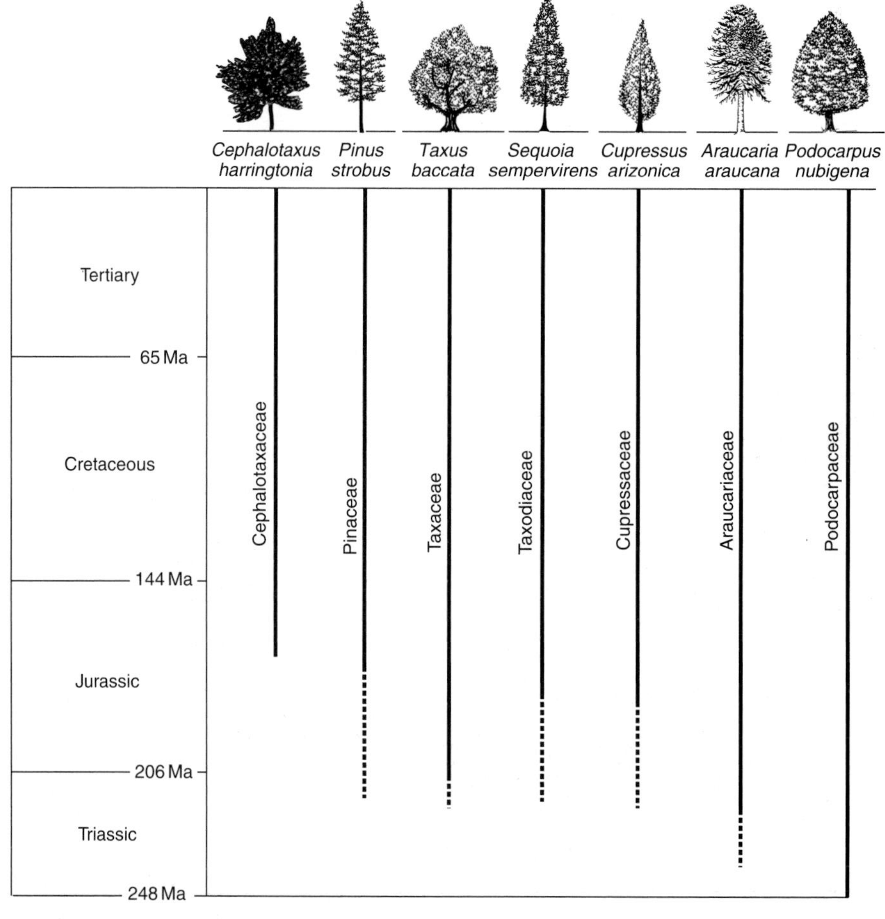

Figure 5.11 First appearance of the extant conifer families in the geological record (redrawn from Miller, 1982). Also indicated are silhouette drawings of extant species in each of the families (redrawn from Riou-Nivert, 1996).

(~245–208 Ma) that this group underwent a major radiation (Figure 5.1) and there was the first occurrence and radiation of the eight conifer families (Podocarpaceae, Taxaceae, Araucariaceae, Cupressaceae, Taxodiaceae, Cephalotaxaceae, and Pinaceae) (Figure 5.11) that have a widespread distribution today (Archibold, 1995). This radiation occurred against a climatic background of increasingly warm and seasonally wet/dry conditions, with further increases in atmospheric CO_2 (Figure 5.3) and the suggestion that a 'megamonsoonal' system had developed over the interior of Pangea (Parrish, 1993).

Characteristics that distinguish extant Coniferales include a pyramidal arborescent growth form (although some types, such as *Juniperus*, are shrubby), small simple leaves that are often needle-like in appearance, mature stems that are mostly composed of secondary wood, and water-conducting elements (tracheids) that are frequently (although not universally) arranged into distinct annual rings with resin canals. The rooting systems of conifers are simple in structure and consist of a branching tap-root system. Most conifers are monoecious but bear the male and female reproductive structures in separate cones on different parts of the tree.

Conifers are more advanced than cycads and ginkgos in that they produce pollen which delivers sperm nuclei to the egg by means of a pollen tube, and pollen grains that often have two asymmetrically placed air bladders (Figure 5.12) and germinate distally (Bell, 1992).

The earliest fossil evidence for trees bearing some or all of these characteristics spans a period from the late Carboniferous to early Permian (~310–260 Ma).

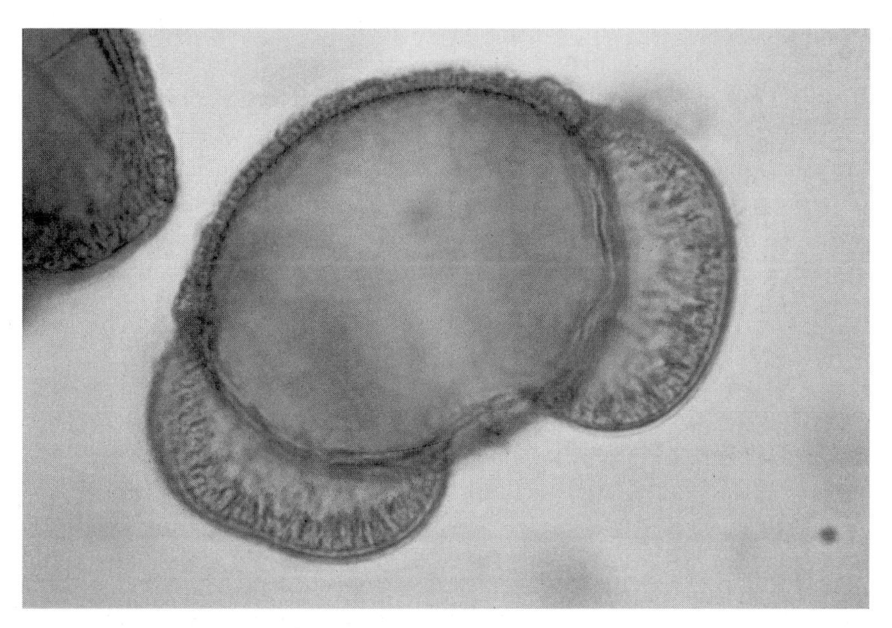

Figure 5.12 Extant *Pinus sylvestris* pollen. The grain is approximately 100 μm in diameter and has two large, asymmetrically placed air bladders (photograph K. D. Bennett).

The latter part of this range is comparable in time to the first appearance in the fossil record of ginkgos and cycads. These early conifers have been placed in a group called the Utrechtiaceae (Taylor and Taylor, 1993), formerly known as the Lebachiaceae (Florin, 1938) and Walchiaceae (Clement-Westerhof, 1988). There are no extant members of this group.

Coniferales

Vegetative and reproductive structures

One of the most commonly cited fossil examples of Permian conifers is *Utrechia* (Figure 5.13). Fossil evidence indicates that *Utrechia* was a tree or shrub that reached heights of 5 m and had a stem composed of a eustele with a relatively large pith and tracheids showing distinct annual rings. These conifers bear a close morphological similarity to many extant conifers (Bell, 1992).

The leaves of *Utrechia* were microphyllous and needle-like in appearance and spirally arranged about twigs. In turn, the twigs were attached to branches that were arranged in whorls (usually 5–6 branches in a whorl) about a slender monopodial stem (Figure 5.13a). In general, this simple branching pattern closely resembles the extant Norfolk Island pine *Araucaria heterophylla*.

The male and female reproductive structures of *Utrechia* were probably borne on separate shoots (Taylor and Taylor, 1993). Fossil evidence for cone-like reproductive structures borne on leafy shoots are thought to represent the female structure (Figure 5.13b), whereas evidence for the male structure producing pollen cones suggests that it was very similar to those of extant conifers (Thomas and Spicer, 1987) (Figure 5.13c). The female cones consisted of an axis, approximately 8 cm long that was borne on a reduced leaf with a reproductive structure (a fertile scale) in its axil (Figure 5.13b). This fertile, short shoot seen in the fossil *Utrechia* is regarded by many palaeobotanists as the morphological intermediate between the female reproductive structure of cordaites and the ovuliferous scale of a modern conifer (Bell, 1992). In contrast, the male reproductive pollen-producing cones were composed of helically arranged flattened scales, each producing a stalk bearing two pollen sacs. The pollen of *Utrechia* was monosaccate (with a single sac enclosing the grain: Figure 5.13d), although nothing is known about the process of fertilization in these early conifers.

Coniferales: origins and relationships

A detailed investigation on the origin of conifers led to the hypothesis that all modern conifers (excluding the Taxaceae) are derived from the Utrechtiaceae (Lebachiaceae) via the Voltziales (Florin, 1951). This interpretation is supported by early cladistic analysis (Miller, 1982). Furthermore, the Voltziales are often regarded as having shared a common ancestor with the cordaites. However, the extent to which all extant and extinct conifers can be regarded as being derived from a single common ancestor (i.e. a monophyletic group or clade) remains

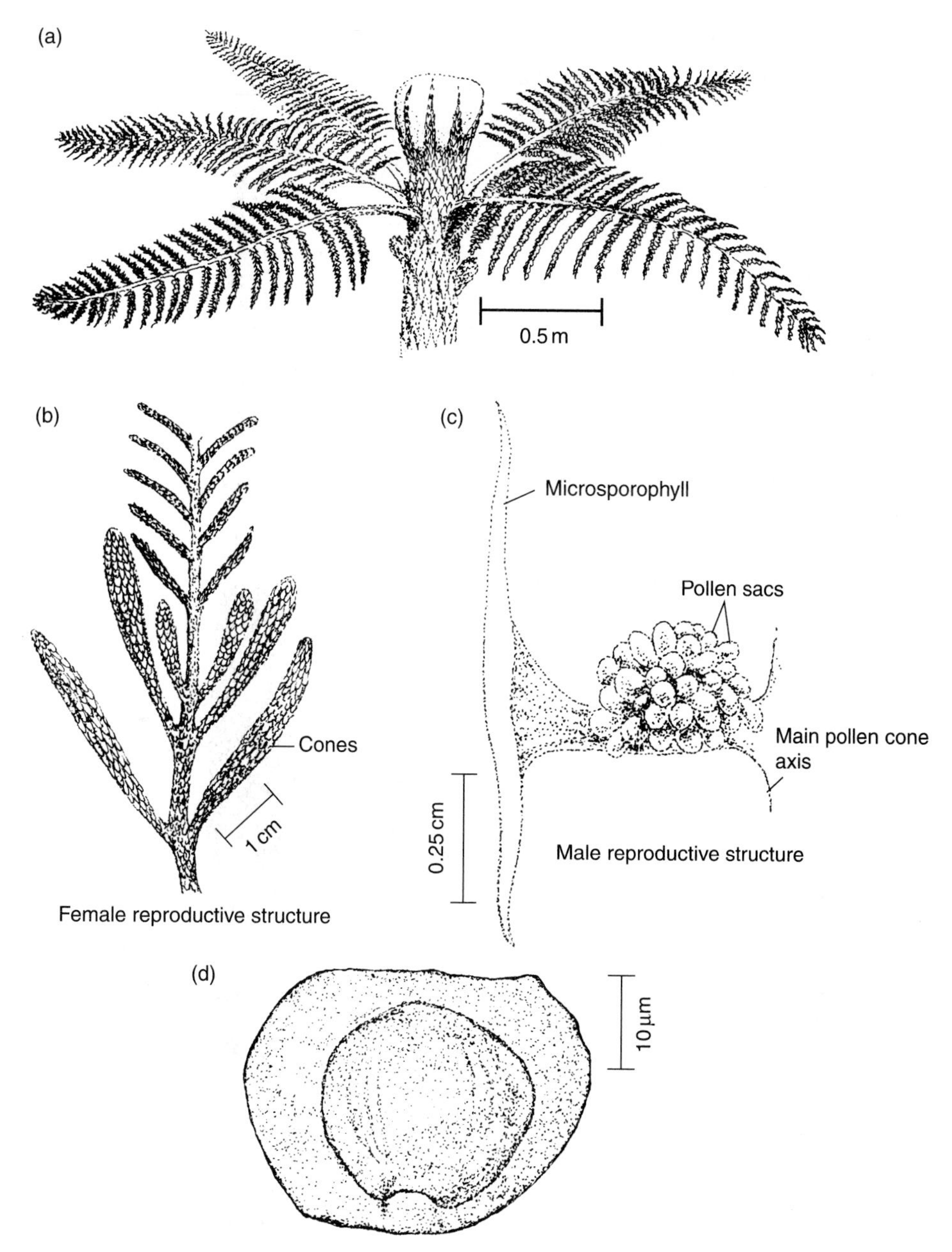

Figure 5.13 Coniferales, *Utrechia*: (a) habit; (b) female reproductive structure; (c) male reproductive structure (enlarged portion of male cone); (d) pollen (redrawn from Meyen, 1987).

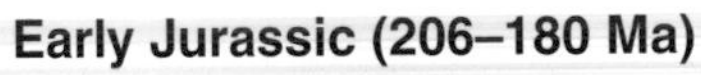

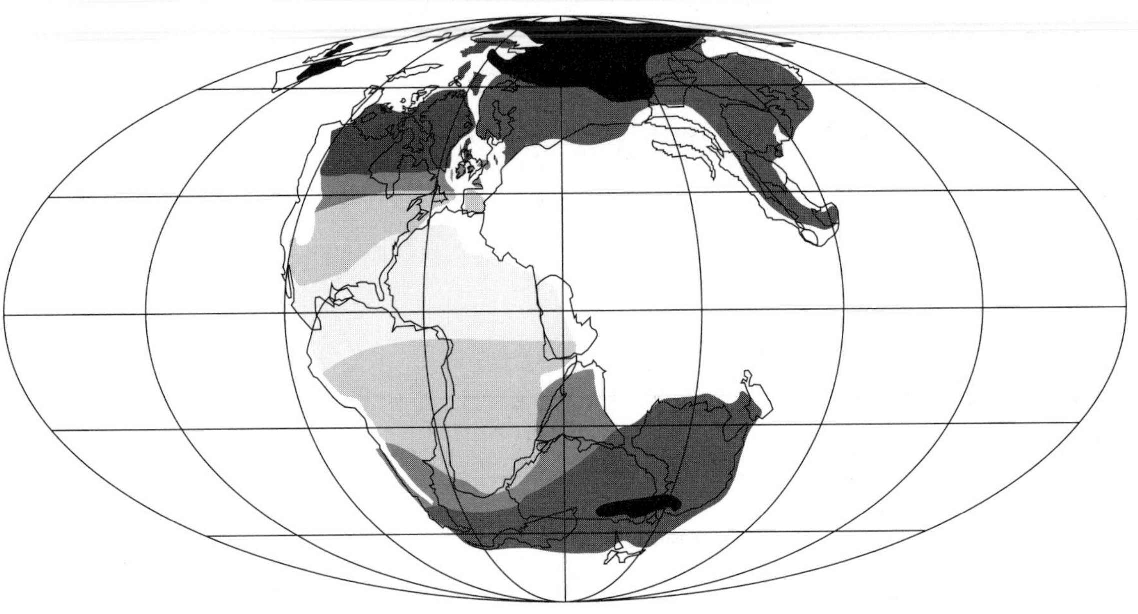

ambiguous, since evidence for both monophyly (Crane, 1985; Doyle and Donoghue, 1986) and polyphyly (Miller, 1988; Rothwell and Serbet, 1994) has been demonstrated (Figure 5.14).

5.5 Biogeographical distribution of global vegetation during the early Jurassic (206–180 Ma)

By the early Jurassic, both the composition and distribution of global vegetation had altered dramatically. Cordaites, gigantopterids, and glossopterids no longer dominated world floras. Instead they were replaced by cycads, bennettites, ginkgos, and conifers, and for the first time global floras contained a significant component of vegetation with recognizable forms in the present day. Five distinct vegetation biomes have been identified for the early Jurassic (Pliensbachian, ~195–190 Ma) (Rees *et al.*, 1999) (Figure 5.14). The methods used in the recognition of these biomes build on earlier statistical work (Ziegler, 1990, 1993) and utilize correspondence analysis to arrange the occurrences of fossil plants according to their degree of association (Rees *et al.*, 1999).

Cool temperate biome

In the high latitudes (>60°) of both hemispheres a cool temperate biome has been recognized (Figure 5.14; Rees *et al.*, 1999, 2000). The fossil floras which

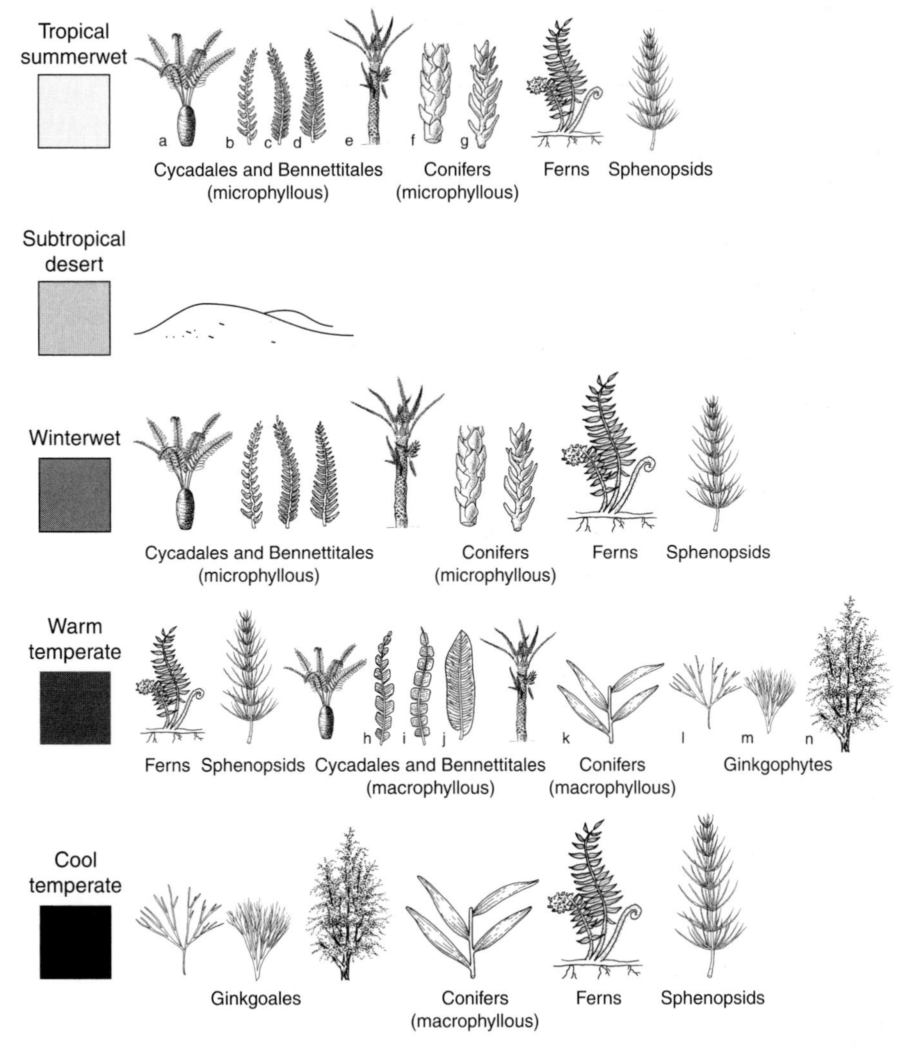

Figure 5.14 Suggested biomes for the Early Jurassic (206–180 Ma) with representatives of the most abundant and/or dominant fossil plant taxa shown (from Rees *et al.*, 2000). The biomes are superimposed on a global palaeogeographic reconstruction for the Pliensbachian (~196 Ma) (courtesy of A. M. Ziegler, PaleoAtlas Project). (a) Stylized cycad; (b) *Otozamites*; (c) *Ptilophyllum*; (d) *Zamites*; (e) *Williamsonia sewardiana*; (f) *Brachyphyllum*; (g) *Pagiophyllum*; (h) *Pseudocycas*; (i) *Anomozamites*; (j) *Nilssoniopteris*; (k) *Lindleycladus*; (l) *Baiera*; (m) *Sphenobaiera*; (n) *Ginkgo* tree. See Appendix 4 for sources of plant reconstructions and line drawings.

define this biome are characterized by relatively low species diversity with a high proportion of deciduous species. Fossil woods from these floras display well-defined growth rings, indicating seasonality (Vakhrameev, 1991). This biome is further characterized by the dominance of ginkgos and macrophyllous (large-leaved) conifers, together with ferns and sphenopsids. The dominant conifers included woody and herbaceous members of the Pinaceae and Voltziales (Meyen, 1987). In total it has been estimated that there was a minimum of 70 genera and over 120 species established in the northern hemisphere cool temperate biomes (or Siberian region) by the early Jurassic (~200 Ma) (Meyen, 1987).

Warm temperate biome

The cool temperate biome grades into a warm temperate biome at palaeolatitudes below 60° and centred on 40°. The fossil floras of this biome were of exceptionally high diversity, composed of ferns, sphenopsids, macrophyllous cycads and conifers and, to a lesser extent, ginkgos. Conifer families present in this biome included Pinaceae, Taxodiaceae, and Podocarpaceae (Vakhrameev, 1991). Macrophyllous cycads and bennettites were also extremely abundant. The high diversity of the warm temperate biome floras relative to those of the cool temperate biome is striking, and it is estimated that by 200 million years ago there were at least 200 genera and over 500 species established in this biome (Meyen, 1987). It is interesting to note that in the greenhouse world of the early Jurassic, the warm temperate biome of the mid-temperate latitudes were the most productive environments (Rees *et al.*, 2000).

Winterwet biome

A winterwet biome, which may have possessed a climate similar to that of the Mediterranean region today, occupied a narrow band of North America in the northern hemisphere, and western Australia, India, and the southernmost tip of South America in the southern hemisphere. This biome was dominated by microphyllous cycads, bennettites, and conifers, with some ferns and sphenopsids (Rees *et al.*, 2000). The relatively increased proportion of plants with small leaves and other xerophytic features, together with a decline in the proportion of vegetation with water-loving characteristics, clearly indicates seasonal water deficits. Such features are typical of the winterwet biome.

Subtropical desert biome

A subtropical desert biome occupied present-day western and southern North America, and southern Africa and South America in the northern hemisphere (Rees *et al.*, 2000) (Figure 5.14). The absence of fossil floras, together with an abundance of evaporites and wind-blown (aeolian) sediments, supports the presence and extent of this biome.

Summerwet (tropical) biome

A summerwet biome has been recognized in a broad equatorial band incorporating early Jurassic fossil localities in present-day southern Mexico, Cuba,

Columbia, northern Brazil, northern Africa, and Israel. This biome was dominated by microphyllous bennettites, numerous ferns, including *Weichselia* (which is thought to have possessed xerophilic characters) (Vakhrameev, 1991), and microphyllous conifers, including members of the Cupressaceae and Podocarpaceae. Cycads were rare, and groups such as ginkgos and conifer families (including Pinaceae) were also absent from these floras.

Summary

1. During the period spanning the Permian (290–248 Ma) a major transformation occurred in global vegetation, with the emergence and widespread radiation of the seed plants.

2. Although two seed plant groups evolved earlier, during the middle Devonian (cordaites and early pteridosperms), the new seed plant groups to emerge between 280 and 260 Ma included the cycads, gingkos, bennettites, and glossopterids (pteridosperms).

3. This time of major floral evolution occurred against a backdrop of increasing global warmth and aridity—often described as a change from a global 'icehouse' to a global 'greenhouse'.

4. The formation of the supercontinent Pangea by 300 Ma is thought to be one of the main reasons accounting for this global climate change. The new supercontinent would have altered global climate patterns through greatly increased continentality and the creation of extensive uplands.

5. Analyses of the biogeographical distribution of global vegetation between 267 and 264 Ma (middle Permian) indicates seven distinct biomes: cold temperate; cool temperate; warm temperate; mid-latitude desert; subtropical desert; summerwet; and tropical everwet.

6. Evolution and radiation of the conifers occurred between ~248–206 Ma (Triassic). Nine conifer families that radiated at this time still have widespread global coverage today.

7. Biogeographical distribution of the early Jurassic global vegetation (206–180 Ma) indicates that global vegetation by this time was dominated by seed plants and that five distinctive vegetation biomes can be recognized: cool temperate; warm temperate; winterwet; subtropical desert; and tropical summerwet.

8. By the early Jurassic, global floras, for the first time, contained a significant component of vegetation with recognizable present-day forms.

6 Flowering plant origins

Flowering plants (angiosperms) are the dominant plants in the world today, accounting for between 300 and 400 families and between 250 000 and 300 000 species (compared to approximately 10 000 species of pteridophytes and 750 species of gymnosperms). However, in evolutionary terms, flowering plants are relatively recent, with fossil evidence indicating their first appearance at around 140 million years ago in the early Cretaceous, followed by rapid diversification and radiation in the mid-Cretaceous. By the early Tertiary (~65 Ma), only 60 to 70 million years after their first appearance, angiosperms had attained ecological dominance in a majority of habitats and over a wide geographical area. As a group, therefore, even though they evolved some 300 million years later than the first vascular plants and 220 million years later than the first seed plants, they are of profound evolutionary interest. This chapter examines the evidence for the first flowering plants, considers various theories as to why their appearance was so late in the geological record and discusses the proposed evolutionary pathways leading to angiosperm evolution.

6.1 Evidence for the first angiosperms

Features used to separate extant angiosperms from other seed plants include the enclosed nature of the ovary (the carpel or carpels), the presence of flowers, specialized conducting cells in the xylem and phloem, ovules that have a double-layered seed coat (two integuments), and pollen with a distinctive grain wall made up of columellae. A number of characteristics of the angiosperm life cycle are also distinctive, including a process of double fertilization, whereby two sperms are released from the pollen tube into the ovary (which has one or several ovules inside it). One sperm fuses with the egg to form the zygote (which divides immediately after fertilization to form the embryo), and the other sperm fuses with the embryo sac to produce the primary endosperm (the storage tissue in the seed) (Figure 6.1). The angiosperms were traditionally divided into monocotyledons and dicotyledons; however, this division is no longer supported by recent systematic studies. Instead, two major monophyletic groups are now recognized, the monocots and the eudicots (which contain most, but not all, of the dicotyledons and are characterized by a pollen type with three apertures)

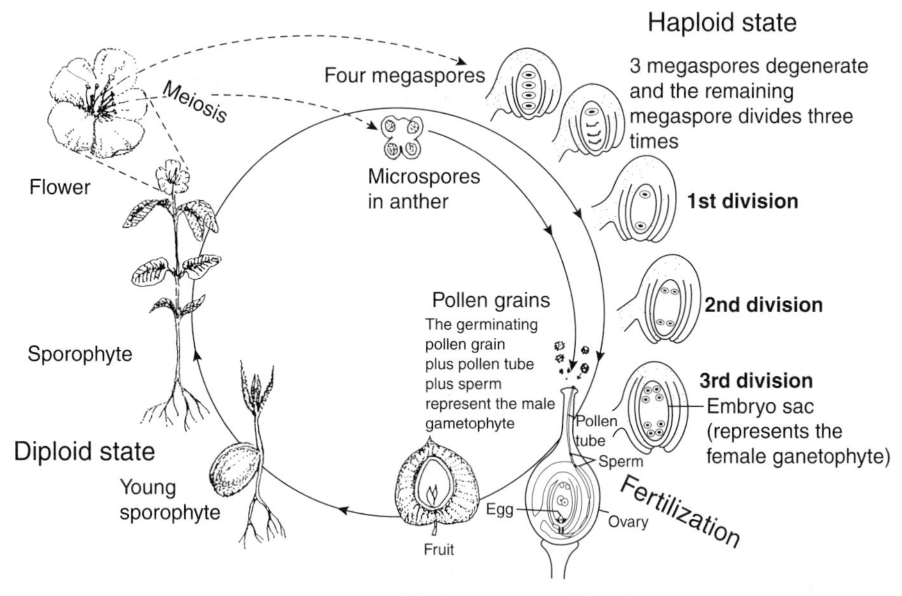

Figure 6.1 Simplified angiosperm life cycle.

Table 6.1 Distinguishing characteristics between monocotyledons and dicotyledons (redrawn from Raven, 1992; Magallon *et al.*, 1999)

Characteristic	Dicotyledons	Monocotyledons
Flower parts	In fours or fives (usually)	In threes (usually)
Pollen	Usually with three pores or furrows	Usually having one pore or furrow
Cotyledons	Two	One
Leaf venation	Usually netlike	Usually parallel
Primary vascular bundles in stem	In a ring	Complex arrangement
True secondary growth with vascular cambium	Commonly present	Commonly absent

(Crane *et al.*, 1995; Magallon *et al.*, 1999). The remaining dicotyledons constitute a third, smaller group, the magnoliids, which have pollen with a single aperture. Although the similarities between the two major groups are far greater than the dissimilarities, the features that group angiosperms as a division and separate monocots from eudicots (Table 6.1) appear to have persisted from early stages of angiosperm evolution (see Section 6.3). Evidence from the geological record suggests that angiosperms first appeared approximately 140 million years ago in the Valanginian, with their major radiation, leading to a global

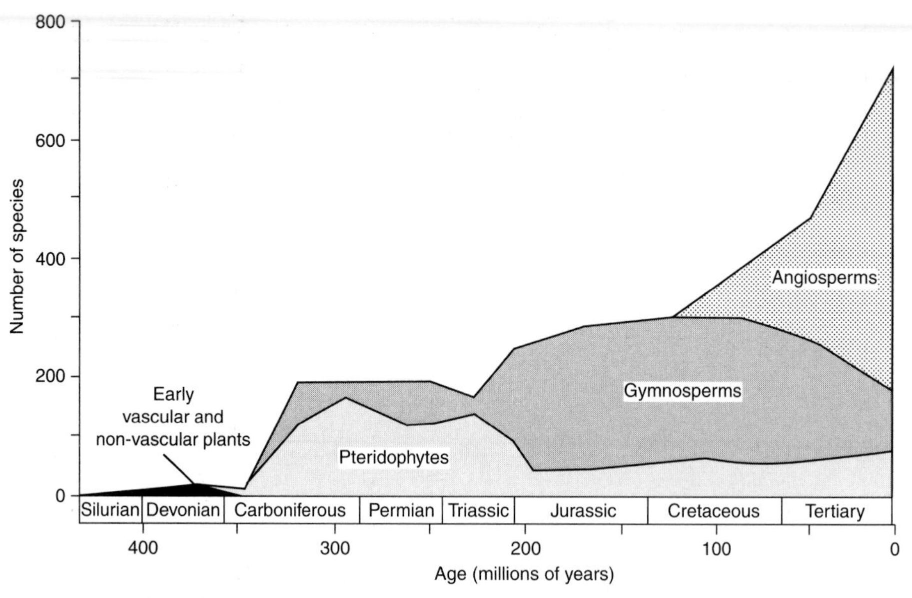

Figure 6.2 Evidence for the appearance and major expansion of the angiosperms from ~140 Ma and a dramatic increase in the number of angiosperms through the mid-Cretaceous (redrawn from Niklas *et al.*, 1983).

distribution, during the Albian–Cenomanian (~100–90 Ma) (Figure 6.2). The earliest fossil evidence for angiosperms is represented in the record as flower parts, fruits, leaves, wood, and pollen (Table 6.2).

Flower parts

Some of the earliest known fossil flowers, dating back to approximately 127–120 Ma, have been found in deposits from Portugal (Barremian or Aptian) (Friis *et al.*, 1999) and near Melbourne in Australia (Aptian) (Taylor and Hickey, 1990). The fossil flowers from Portugal show character combinations that indicate magnoliid or monocotyledonous affinity. Those from Melbourne appear to be similar in floral arrangement to extant perennial herbs in the Piperaceae family, and probably grew in marshy habitats. Although such early finds are rare (the majority of evidence for fossil angiosperm flowers dates back only to approximately 100 million years ago), they show the greatest potential for further insights on early angiosperm biology, ecology, and evolution (Endress, 1987; Crane *et al.*, 1995).

Most early fossil angiosperm flowers can be subdivided into two broad groups (Endress, 1987). Group (i) includes features such as relatively few flower parts (stamens, carpels, number of ovule/seeds per carpel, number of perianth members) and small (< 1 mm diameter), possibly unisexual flowers. Such flowers are compared to those of existing angiosperm families such as Chloranthaceae,

Table 6.2 Some of the earliest angiosperms in the fossil record

Order/family/species	Fossil evidence (leaves, pollen, flowers, wood)	Tree, shrub or herb	Fossil localities	Age (Ma)	Reference
Dicotyledons					
Magnoliales					
Lesqueria	Fruits, flowers	? Herbaceous	Kansas, USA	112(99 (AL)	Crane and Dilcher (1984)
Archaeanthus	Fruit, leaves	? Herbaceous	Kansas, USA	112–99 (AL)	Dilcher and Crane (1984)
Protomonimia	Reproductive organ, wood	Woody shrub or tree	Japan	112–99 (AL)	Nishida and Nishida (1988)
Prisca	Fruit, leaves, stem	Woody tree or shrub	Kansas, USA	99–93 (CE)	Retallack and Dilcher (1981)
Winteraceae					
Walkeripollis	Pollen, wood	Woody tree or shrub	Israel, California	121–99 (AP/AL)	Walker *et al.* (1983)
Laurales					
Amborellaceae	Fruit	? Shrub		121–112 (AP)	Friis *et al.* (1995)
Crassidenticulum	Leaves	? Woody tree/shrub	Nebraska, USA	99–93 (CE)	Upchurch and Dilcher (1990)
Mauldinia	Flowers	? Woody tree/shrub	North America	99–93 (CE)	Drinnan *et al.* (1990)
Chloranthaceae					
Chloranthus	Flowers, pollen, seeds	? Herbaceous	North America	112–99 (AL)	Friis *et al.* (1986)

Table 6.2 (*Continued*)

Order/family/species	Fossil evidence (leaves, pollen, flowers, wood)	Tree, shrub or herb	Fossil localities	Age (Ma)	Reference
Hedyosmum-like	Flowers, pollen	Herbaceous	Portugal	127–112 (BA/AP)	Friis *et al.* (1999)
Piperales	Flowers, pollen	Herbaceous	Portugal	127–112 (BA/AP)	Friis *et al.* (1999)
Platanaceae					
Platanus potomacensis	Flowers, pollen	? Woody tree/shrub	North America, Sweden	112–99 (AL)	Friis *et al.* (1988)
Hamamelidales					
Hamamelidae	Leaves	? Herbaceous	Patagonia, Argentina	127–112 (BA/AP)	Romero and Archangelsky (1986)
Monocotyledons					
Pandanaceae	Leaves	?	?	71–65 (MA)	Jarzen (1978)
Palmae	Leaves, pollen, and stem	Tree/shrub	New Jersey	89–85 (CO)	Christophee (1979)

BA, Barremian; AP, Aptian; AL, Albian; CE, Cenomanian; TU, Turonian; CO, Coniacian; SA, Satonian; CA, Campanian; MA, Maastrichtian.

Figure 6.3 Fossil *Archaeanthus linnenbergeri* ($\sim$100 Ma), indicating many similarities with members of extant Magnoliaceae (redrawn from Dilcher and Crane, 1984).

Piperaceae (the pepper family), and Platanaceae (the plane-tree family). In comparison, group (ii) has numerous flower parts and large, bisexual flowers (up to 65 mm in diameter). These flowers are compared with those of extant angiosperm families such as Magnoliaceae, Degeneriaceae, and Winteraceae (Friis and Crepet, 1987). An example of a fossil flower classified in this second group is *Archaeanthus* (Figure 6.3). This has a fossil record dating back to the Albian ($\sim$100 Ma) (Dilcher and Crane, 1984), and demonstrates many features in common with extant species of *Magnolia*, such as numerous stamens (between 50 and 60) and free carpels (between 100 and 130).

It was originally suggested that the first flowers to evolve were similar to extant *Magnolia*, with numerous bisexual flowers (therefore those classified above in group (ii)), and that group (i) originated either by extreme reduction in the floral parts from group (ii) or from two different sources in the gymnosperms (Endress, 1987). Fossil evidence, however, indicates that both types of flowers were present at around the same time (Friis and Crepet, 1987). Also, in extant orders of angiosperms, extreme variations in the number of floral organs are not restricted to comparatively unrelated orders or families, but can occur within a genus, and even a single species (Endress, 1987), thus indicating that floral organization is very plastic. It is highly probable therefore that the number and arrangement of floral organs changed many times during evolution, and that extremes in these features in the earliest angiosperms was not necessarily an expression of distant relationship.

Fruits

Earliest evidence for fossil angiosperm fruits dates back to the Aptian and Albian (~121 Ma), with examples from localities in Asia and North America, including fruits of ceratophyllales (Dilcher, 1989), juglandales (Krassilov and Dobruskina, 1995), and ranunculids (Friis *et al.*, 1995). Many of these early seeds were small (1–40 mm in length) in relation to later groups in the fossil record—a feature that is thought to be indicative of the 'weedy' stature of these early flowering plants (Tiffney, 1984). Comparison of seed sizes in extant floral groups, for example, broadly demonstrates that small propagules with thin seed walls and little storage material are associated with early successional plants that can be classified as 'r' strategists (weedy generalists) (Wing and Boucher, 1998).

Pollen

Angiosperm pollen is non-saccate (without bladders), and in the eudicots has numerous symmetrically arranged pores and furrows. In all angiosperms the pollen wall is divided into an outer layer (the tectum) supported on numerous short, radial structures (looking like columns and often referred to as columellae), which provide an extensive chamber system for the deposition of biologically active substances that act as 'recognition substances' on reaching the stigma (Traverse, 1988b; Hughes, 1994) (Figure 6.4).

The earliest unequivocal fossil angiosperm pollen has been found in late Valanginian (~130 million years ago) deposits of Israel (Brenner, 1996) and Morocco (Gubeli *et al.*, 1984). There are also reported occurrences of angiosperm pollen from older sediments in Libya (Berriasian, ~140 Ma) (Thusu *et al.*, 1988) and China (Hauterivian) (Li and Liu, 1994), but the ages of these sediments are not well constrained (Barrett and Willis, 2001).

All the earliest angiosperm pollen grains are small, between 10 and 50 μm in diameter, and distinguishable by their wall construction and the number and types of germination-furrows. Four morphological groups have been identified in the earliest angiosperm pollen, namely *Clavatipollenites*, Pre-*Afropollis*, *Spinatus*, and *Liliacidites* (Brenner, 1996). Those classified in the *Clavatipollenites* group have a characteristic columellate wall with usually one germination-furrow (Figure 6.5a). Those in the Spinatus group can be distinguished, along with other features, by short spines on the margin of the grain (Figure 6.5c). The pre-*Afropollis* group contains grains that are inaperturate (i.e. no furrows) but have a grain wall pattern that is described as wedge-shaped, with fluted walls (Figure 6.5b) (Brenner, 1996). Finally, those in the *Liliacidites* groups are distinguished by their larger size (up to 50 μm in diameter), their single germination-furrow, and a cell wall composed of very high columellae (Figure 6.5d).

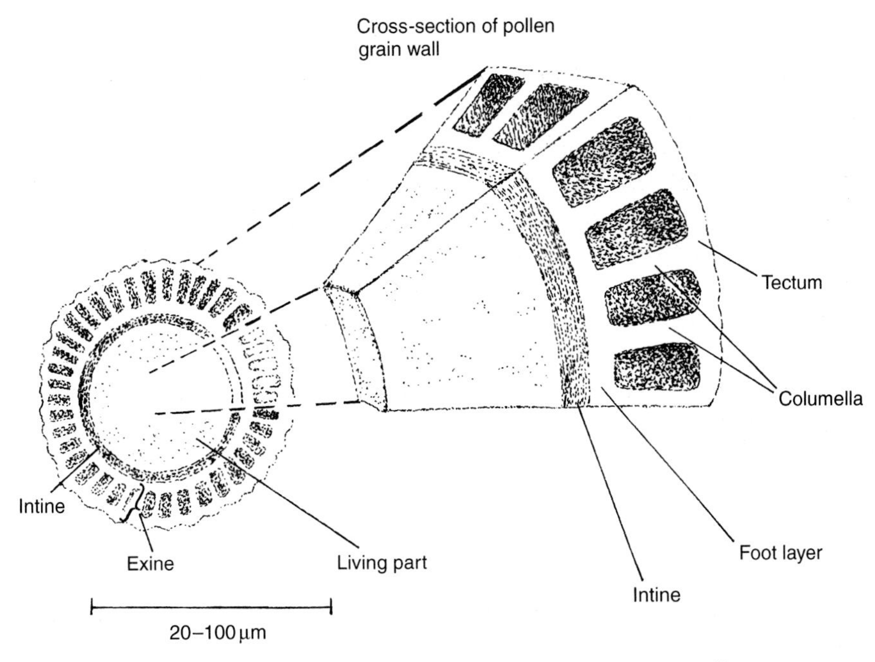

Figure 6.4 Distinguishing characteristics of fossil angiosperm pollen.

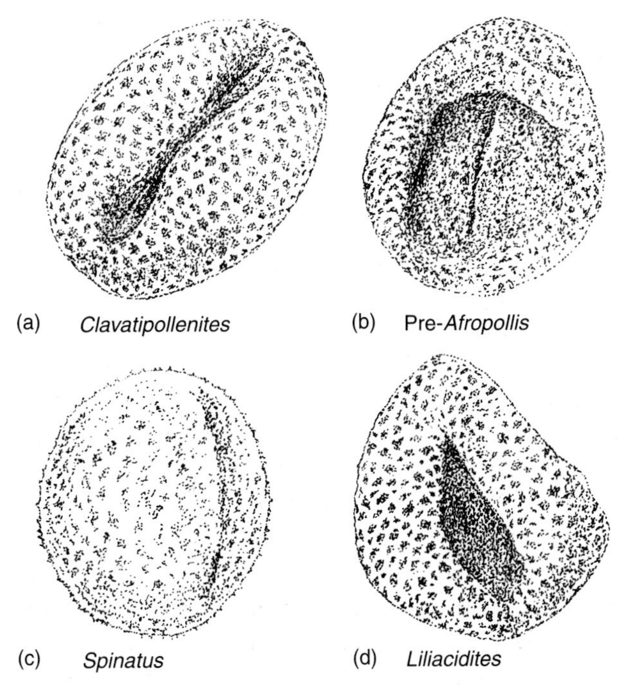

Figure 6.5 Four morphological types identified in the earliest fossil angiosperm pollen: (a) *Clavatipollenites*; (b) Pre-*Afropollis*; (c) *Spinatus*; and (d) *Liliacidites* (redrawn from Brenner, 1996). All are between 10 and 50 µm in diameter.

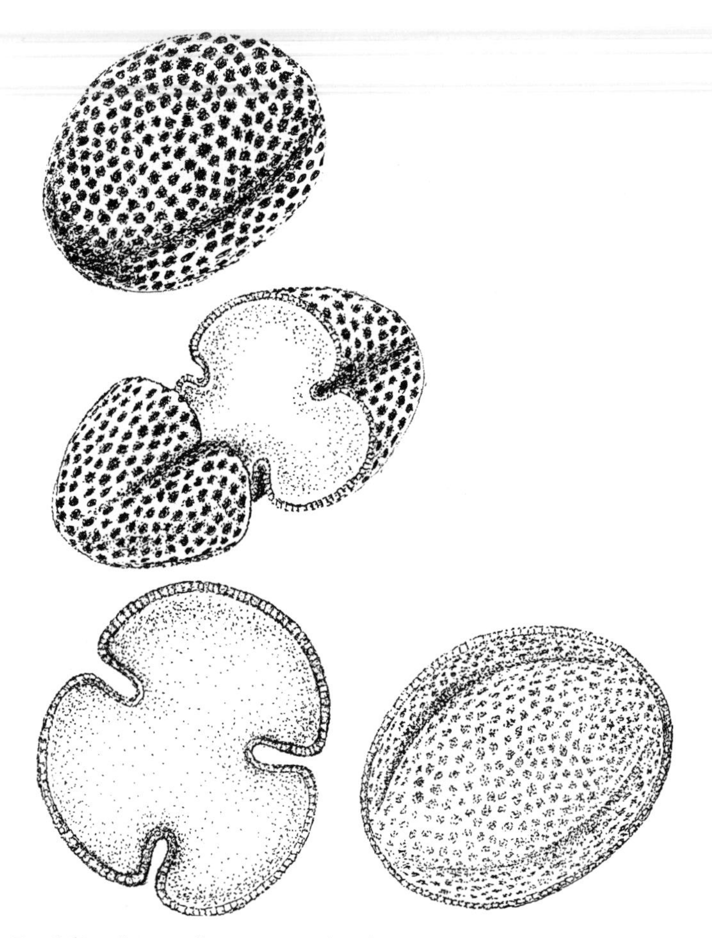

Figure 6.6 Fossil *Tricolpites* pollen, equatorial and transverse view (redrawn from Stewart and Rothwell, 1993). These grains were approximately 50 μm in diameter.

A number of associations have been made between the early fossil angiosperm pollen and extant types. The fossil grains of *Clavatipollenites*, for example, apparently bear a close morphological resemblance to pollen of the extant family Chloranthaceae (Traverse, 1988b). *Clavatipollenites* has also been found within fossil flowers identified as members of the Chloranthaceae family, supporting the suggestion that it is associated with this family (Crane *et al.*, 1995). Those in the *Liliacidites* group, in comparison, have an outer layer (the sexine) that is similar in structure to that of extant monocotyledons such as Liliaceae (Brenner, 1996).

In increasingly younger sediments (Barremian/Aptian boundary), there is the appearance of more complex fossil pollen grains with various arrangements of pores and furrows. One type, named *Tricolpites*, for example, had three symmetrically arranged furrows (often referred to as tricolpate; Figure 6.6).

Evolution of three furrows is thought to be an adaptation that facilitated germination on a stigmatic surface (Endress, 1987), as any orientation of a tricolpate grain would result in at least one germination furrow being positioned near or on the surface of the stigma. *Tricolpites* pollen grains have been found inside the anthers of fossil Platanaceae flowers (Friis and Crepet, 1987), indicating a probable association with this family. Another type of pollen grain to appear from the Cenomanian (~95 Ma) was spherical with numerous apertures resembling pores. These grains, known as polyporate, had pores either arranged around the equatorial margins of the grain or all over the surface of the grain, which presumably would also have facilitated germination (Muller, 1981, 1984).

Leaves

Angiosperm leaves are thought to be megaphyllous in origin. Distinguishing characteristics include reticulate venation, forming areoles on dicotyledon leaves, together with veins that end blind within the areoles, and parallel major veins arranged in sets of various sizes and interconnected by smaller veins on the lamina of monocotyledons (Stewart and Rothwell, 1993). In monocotyledons, the leaf is usually differentiated into a blade and sheath, whereas in dicotyledons the leaf is usually differentiated into a blade and petiole (Figure 6.7a).

Angiosperm leaves are first recorded in the fossil record from approximately 120 million years ago (Barremian). Fossil localities range from central Asia, the Russian far-east, Portugal, and the eastern United States (Hughes, 1994), and indicate broadly similar assemblages containing small leaves (approximately 2–4 cm in diameter) with expanded laminae and reticulate venation patterns (Figure 6.7b).

Attempts to characterize these earliest leaf types and compare them with extant leaves and their associated environmental conditions (based on features such as presence/absence of drip-tips, entire or dissected margins, size of leaf blade) has led to the suggestion that most were similar to plants that grow in streamside situations, semiaquatic habitats, or as an understorey (Upchurch and Wolfe, 1987; Wolfe and Upchurch, 1987). In comparison, fossil leaves present in younger assemblages of Albian age (~110–100 Ma) possessed features characteristic of extant early successional plants, namely pinnately compound leaves, palmately lobed leaves, and shallow cordate leaves with serrated margins and palmate venation (Figure 6.8) (Hickey and Doyle, 1977; Taylor and Hickey, 1990). Leaves present in deposits of Cenomanian age (~100–90 Ma) included leaf physiognomies (shape and size) that in extant species are typical of late successional plants. These marked changes in angiosperm leaf physiognomy spanning the latter part of

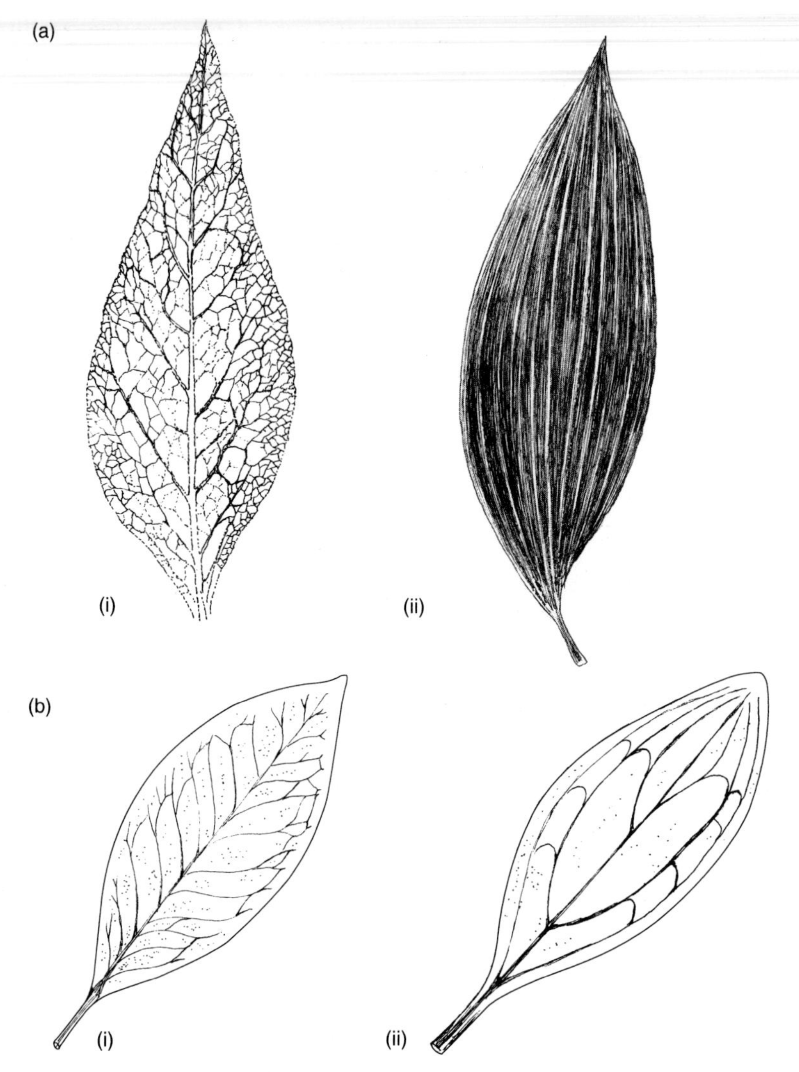

Figure 6.7 (a) Extant angiosperm leaves: distinguishing characteristics include (i) reticulate venation forming areoles on dicotyledon leaves, together with veins that end blind within the areoles, and (ii) parallel major veins arranged in sets of various sizes and interconnected by smaller veins on the lamina of monocotyledons (Friis *et al.*, 1987; Stewart and Rothwell, 1993). (b) Primitive angiosperm leaf types from the early Cretaceous: (i) dicotyledon, showing pinnate venation and entire leaf margin; (ii) monocotyledon, showing convergence of main veins at top of leaf and interconnecting veins.

the early Cretaceous has led to the suggestion that although angiosperms were initially early successional plants, within a matter of 20 million years they had formed the canopy of late successional forests (Doyle and Hickey, 1976).

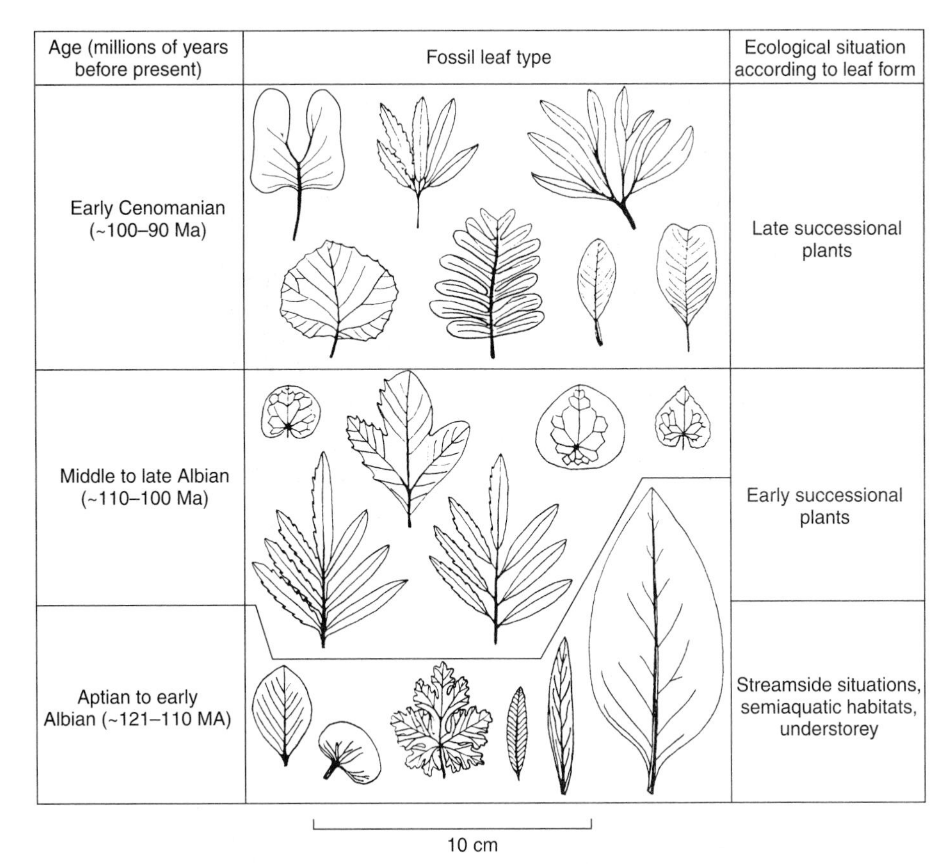

Figure 6.8 Trends in the earliest fossil angiosperm leaves through the Cretaceous and classification of their probably ecological situation according to leaf form (after Upchurch and Wolfe, 1987).

6.2 Nature and distribution of the earliest Angiosperms

Trees, shrubs, or herbs?

Various lines of evidence from the fossil record, including fossil flowers, fruits, leaves, pollen, and also wood, suggest that by 100 million years ago (Albian/ Cenomanian) there was an increasing diversity of angiosperms in the global flora (Table 6.3). There is some difficulty, however, based on the available fossil evidence, in determining whether the earliest angiosperms were trees, shrubs, or herbs. Many of the earliest fossil families (i.e. Chloranthaceae, Piperaceae, Platanaceae, Magnoliaceae, Degeneriaceae, and Winteraceae) have both arborescent and non-arborescent forms. Hence, many competing hypotheses on the likely vegetative morphology of ancestral angiosperms have been proposed. These fall broadly into three schools of thought. According to the first, the

Table 6.3 Diversification and radiation of aborescent angiosperms between 100 and 70 Ma (adapted from Wing and Boucher, 1998)

Family/species	Dicot./ monocot. (tree, shrub, herb, vine)	Present-day distribution	Earliest fossil evidence
Hamamelidaceae	Dicot. (T and S)	Chiefly subtropical, especially E. Asia	~93–89 Ma (Turonian)
Platanaceae	Dicot. (T)	N. hemisphere (7 species); S.E. Asia (1 speices); S.E. Europe (1 species); N.E. America (1 species)168	~112–99 Ma (Albian)
Myrothamnaceae	Dicot. (S)	S. Africa and Madagascar	~99–93 Ma (Cenomanian)
Ulmaceae			
Ulmus	Dicot. (T and S)	N. temperate to N. Mexico	~71–65 Ma (Maastrichtian)
Celtis	Dicot. (T)	Tropical (70 species); temperate (4 species)	~93–89 Ma (Turonian)
Fagaceae			
Castanea	Dicot. (T)	N. temperate	~85–83 Ma (Santonian)
Northofagus	Dicot. (T)	S. hemisphere	~85–83 Ma (Santonian)
Betulaceae			
Alnus	Dicot. (T)	N. temperate; S.E. Asia; Andes	~85–83 Ma (Santonian)
Betula	Dicot. (T)	N. hemisphere	~85–83 Ma (Santonian)
Myricales			
Myrica	Dicot. (S)	Cosmopolitan except Mediterranean and Australia	~85–83 Ma (Santonian)
Juglandaceae	Dicot. (T and S)	Temperate and warm N. hemisphere to S. America and Malaysia	~83–71 Ma (Campanian)
Theaceae	Dicot. (T, S, and V)	Tropical (520 species) with few warm temperate	~93–89 Ma (Turonian)
Moringaceae	Dicot. (T)	Semi-arid Africa to Asia	~93–89 Ma (Turonian)
Actinidiaceae	Dicot. (T, S, and V)	Tropical and warm Asia mountains	~83–71 Ma (Campanian)
Clethraceae	Dicot. (T and S)	Tropical America; Asia; N. America	~71–65 Ma (Maastrichtian)
Ericaceae	Dicot. (T and S)	Cosmopolitan except deserts (and scarce in Australasia)	~93–89 Ma (Turonian)

Diapensiaceae	Dicot. (S and H)	Arctic and N. temperate to Himalayas	~83–71 Ma (Campanian)
Symplocaceae	Dicot. (T)	Tropical and warm America	~71–65 Ma (Maastrichtian)
Sapotaceae	Dicot. (T and S)	Tropical	~71–65 Ma (Maastrichtian)
Bombaceaceae	Dicot. (T)	Old World Tropical	~71–65 Ma (Maastrichtian)
Buxaceae	Dicot. (T)	Nearly cosmopolitan	~112–99 (Albian)
Araliaceae	Dicot. (T)	Tropical; America, few temperate	~71–65 Ma (Maastrichtian)
Cornaceae	Dicot. (T)	N. temperate; rare in Tropics and S. temperate	~71–65 Ma (Maastrichtian)
Rhamnaceae	Dicot. (T)	Cosmopolitan especially tropical and warm N. temperate	~71–65 Ma (Maastrichtian)
Rosaceae	Dicot. (T, S, and H)	Subcosmopolitan but especially temperate and warm N.	~71–65 Ma (Maastrichtian)
Sabiaceae	Dicot. (T and S)	S.E. Asia; Malasia; tropical America	~71–65 Ma (Maastrichtian)
Caesalpinaceae	Dicot. (T and S)	Tropical; warm America	~93–89 (Turonian)
Myrtaceae	Dicot. (T and S)	Tropical, warm and temperate	~85–83 (Santonian)
Onagraceae	Dicot. (T and S)	Cosmopolitan, especially temperate and warm America	~71–65 Ma (Maastrichtian)
Combretaceae	Dicot. (T and S)	Tropical and warm, especially Africa	~83–71 (Campanian)
Gunneraceae	Dicot. (T and S)	S. tropical and S. hemisphere	~93(89 (Turonian)
Sapindaceae	Dicot. (T, S, and H)	Tropical and warm, few temperate	~89–85 (Coniacian)
Malpighiaceae	Dicot. (T, S, and V)	Tropical and warm, especially S. America	~71–65 Ma (Maastrichtian)
Aquifoliaceae (e.g. Ilex)	Dicot. (T and S)	Almost cosmopolitan	~93–89 (Turonian)
Olacaceae	Dicot. (T, S, and V)	Tropical; S. Africa	~71–65 Ma (Maastrichtian)
Santalaceae	Dicot. (T, S, and H)	Subcosmopolitan, especially tropical and warm dry	~71–65 Ma (Maastrichtian)
Proteaceae	Dicot. (T and S)	Tropical and subtropical, especially S. hemisphere	~83–65 (late Cretaceous)
Palmae	Monocot. (T)	Tropical and warm	~112–99 (Albian)
Pandanaceae	Monocot. (T and S)	Old world tropical to New Zealand	~71–65 Ma (Maastrichtian)

T, tree; S, shrub; V, vine; H, herb.

earliest angiosperms were arborescent shrubs or small trees (Arber and Parkin, 1907). The second suggests that they were herbaceous and rhizomatous in habit, such as in extant Chloranthaceae or Piperaceae (Taylor and Hickey, 1996). A third intermediate hypothesis suggests that they were most probably herbaceous, weedy, small shrubs (Stebbins, 1974; Crane, 1987; Wing and Tiffney, 1987).

Increasingly, evidence from the fossil and molecular record appears to support the third hypothesis. Angiosperm wood is rare in the early Cretaceous fossil record compared with that of gymnosperm wood, and it is not until the late Cretaceous (~70 Ma) that a diverse angiosperm wood flora is apparent (Wing and Tiffney, 1987; Wheeler and Baas, 1993; Wing *et al.*, 1993). Most of the specimens of early Cretaceous angiosperm wood are also extremely small (< 10 cm in diameter) (Herendeen, 1991). It is generally assumed, therefore, that this lack of angiosperm fossil wood is a reflection of the herbaceous nature of the earliest angiosperms. This assumption is further supported by the fact that, in the fossil record, the earliest angiosperm seeds tend to be small (1–40 mm in length) with thin seed walls, and the leaves, small (2–4 cm in diameter) with expanded laminae and reticulate venation. All of these characteristics in extant plants are usually indicative of small, weedy plants with a rapid life cycle (Taylor and Hickey, 1996; Wing and Boucher, 1998; Friis *et al.*, 1999). This evidence therefore appears to also support a herbaceous hypothesis. However, recent molecular phylogenetic analysis, based on evidence from mitochondrial, plastid, and nuclear DNA in extant plants, has identified *Amborella*, a small shrub, as the basal branch of angiosperm phylogeny (Qui *et al.*, 1999). This finding therefore supports the intermediate hypothesis, that the earliest angiosperms were herbaceous, weedy, small shrubs. It is interesting to note, however, that the second most basal group identified by the molecular study of Qui *et al.* (1999) are Nymphaeales, which are aquatic rhizomatous herbs. These results therefore also lend support to the herbaceous origin hypothesis.

Dicotyledons or monocotyledons?

In the fossil record, the earliest flowers and most of the leaves and pollen appear to be from dicotyledons and, although several lineages of monocotyledons are now known from both the palynofloras and megafloras of Barremian–Albian age (127–112 Ma) (e.g. Friis *et al.*, 1999; Pole, 1999), early monocotyledon fossils are rare (Table 6.3). Various suggestions have been made as to why there is this apparent paucity of early monocotyledons. First, it has been suggested that the majority of the monocotyledons were herbaceous (as is the case for extant monocotyledons) and therefore would not have been as well preserved as the woody dicotyledons (Taylor and Taylor, 1993). A second suggestion is that, similar to the present-day situation, there were many more genera of dicotyledons than monocotyledons (Stewart and Rothwell, 1993). Presently there are approximately six times as many dicotyledons as monocotyledons in the world's flora, and it is possible that this had a major influence on representation in the early fossil record of angiosperms (Daghlian, 1981). Thirdly, it is suggested that the monocotyledons evolved from the dicotyledons (Walker and Walker, 1984) and therefore were later in evolutionary terms. This last suggestion appears to be increasingly unlikely, however, in the light of a number of recent cladistic and phylogenetic studies (e.g. Davis, 1995; Stevenson and Loconte, 1995; Gandolfo *et al.*, 1998; Qui *et al.*, 1999). Recent estimates, based on a phylogeny of monocotyledonous angiosperms, suggest that the major radiation of monocotyledons occurred during the early Cretaceous (Bremer, 2000). Thus, their late appearance in the fossil record is most likely due to taphonomic bias rather than a later evolution.

Place of origin and radiation

There is much discussion in the literature on where exactly angiosperms originated. Currently, the most favoured hypothesis suggests that angiosperms originated in the palaeotropics (0–30°), radiating out to colonize higher-latitude environments some 20 to 30 million years later (Hickey and Doyle, 1977; Crane and Lidgard, 1989; Drinnan and Crane, 1990; Hughes, 1994; Lupia *et al.*, 1999; Barrett and Willis, 2001). This hypothesis is based primarily on the evidence in the fossil pollen record (Figure 6.9). The earliest well-dated angiosperm pollen has been found in late Valanginian (~135 Ma) fossil localities in Israel (Brenner, 1996) and Morocco (Gubeli *et al.*, 1984). During the early Cretaceous these regions lay between the palaeoequator and 25°N (Smith and Littlewood, 1994) (Figure 6.9a). Angiosperms then appear to have spread relatively rapidly into the higher latitudes, with evidence for slightly younger (Hauterivian, ~132 Ma) pollen grains from fossil localities in England (Hughes and McDougall, 1987, 1994) and China (Li and Liu, 1994). By the Barremian (~127 Ma), angiosperms appear to have been widespread, with fossil localities in central Africa, Australia, Europe, and China (Barrett and Willis, 2001 and references cited therein).

Although angiosperms appear to have spread relatively quickly, they did not become floristically prominent in the low palaeolatitudes (between ~30°N and 30°S) until the Aptian (~120 Ma), and in higher palaeolatitudes (between 40 and 65°S) until the Cenomanian (~ 100 Ma) (Crane and Lidgard, 1989; Drinnan and Crane, 1990; Lidgard and Crane, 1990) (Figure 6.9b). However, differences in both diversity and abundance between the low and high latitudes persisted for at least another 30 million years. In the Maastrichtian (~70 Ma), for example, angiosperms dominated low latitude pollen assemblages, accounting for 60–80% of pollen, whereas in the high latitudes, they accounted for only between 30 and 50% of the total pollen (Crane, 1987). The remaining percentage of pollen in high-latitude floras was made up of gymnosperms and pteridophytes. The apparent paucity of early angiosperms in these high-latitude environments is thought to be due, in part, to limitations of light and temperature for a substantial part of the year. Such environmental conditions would have offered few opportunities for the replacement of existing well-adapted vegetation,

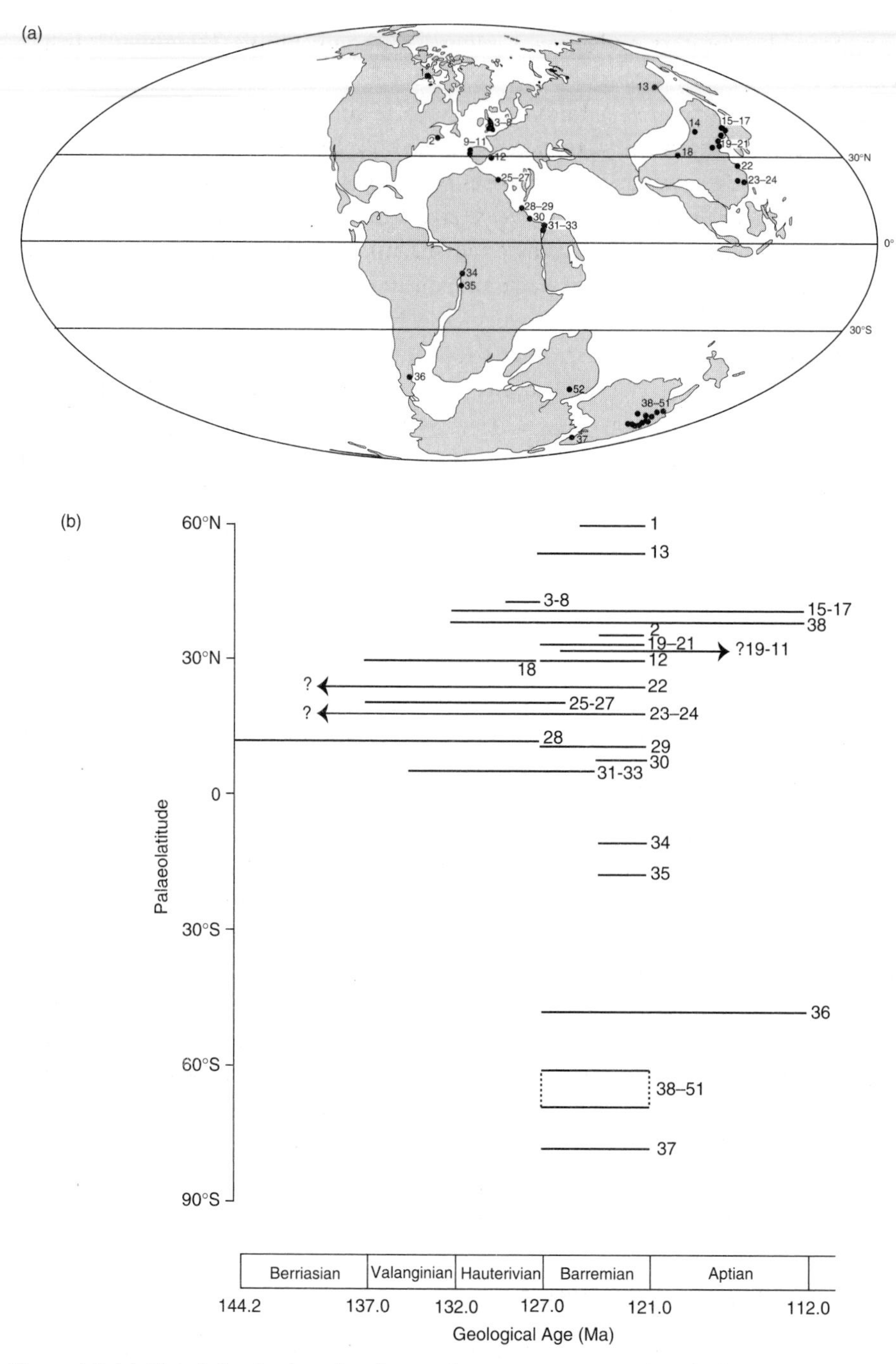

Figure 6.9 (a) Global distribution of earliest angiosperm occurrences (pre-Aptian, < 121 Ma) recorded in the fossil pollen record, plotted onto a palaeogeographical map of age 135 Ma (Valaginian). (b) Graphical representation of the data presented on the map above. Horizontal lines represent the temporal ranges suggested for the specific angiosperm taxa, the numbers correspond to the localities listed on the map (redrawn from Barrett and Willis, 2000 and references cited therein).

dominated by conifers and ferns, by 'weedy' angiosperms (Wing and Boucher, 1998).

Angiosperm diversity and abundance in the high latitudes increased slowly through the early and mid-Cretaceous, most probably as a function of their biology rather than due to slow migration rates (Wing and Boucher, 1998). Even today in the tundra and boreal biomes of the high-latitude regions of North America and Eurasia, for example, angiosperms are of relatively low abundance compared to gymnosperms, in particular the conifers (Nikolov and Helmisaari, 1992; Sirois, 1992).

Although the early angiosperms were not an invasive element in the high-latitude environments, for later angiosperm groups the high-latitudes appear to have been centres of origin. Southern Gondwana, in particular, appears to have been a centre of origin for *Nothofagus* (Dettmann and Jarzen, 1990; Dettmann *et al.*, 1990; Dettmann, 1992; Cantrill and Nichols, 1996). *Nothofagus* is presently an important element of the southern hemisphere flora, and according to the fossil record of this genus, evolved in southern high latitudes during the Maastrichtian (~70 Ma) (Hill and Scriven, 1995).

6.3 Why so late?

Even though angiosperms were present from as early 140 Ma (Berriasian–Valanginian), in plant evolutionary terms this is late. The first unequivocal evidence for angiosperms in the fossil record is up to 300 million years later than the first vascular land plants, and, as a group, they are therefore the most recent. The 'enigma of angiosperm origins' and their late arrival in the fossil record have long been areas of palaeobotanical interest and research (Hughes, 1994). There are a number of hypotheses are to why they appear so late in the fossil record. These include bias in the fossil record (therefore they evolved much earlier but went undetected), and the suggestion that their evolution was triggered by a particular set of environmental conditions, and/ or biotic interactions (such as co-evolution with faunal groups). A broad overview of these different hypotheses will be considered in the following section.

Nature of the fossil evidence

One commonly cited explanation for the late appearance of angiosperms is due to taphonomy. It is suggested that there is a bias in the fossil record against the preservation of the earliest angiosperm vegetative or reproductive parts. Thus it is argued that angiosperms may have been part of the global vegetation much earlier than the Cretaceous, but were situated in dry upland environments, where preservation potential would have been poor (Axelrod, 1952; Taylor and

Taylor, 1993). Both fossil and molecular evidence exists to support this hypothesis, although neither is particularly convincing.

Fossil evidence for a pre-Cretaceous origin of angiosperms is based on early examples of angiosperm-like pollen, fruiting axes, and leaves. Angiosperm-like pollen, recognized by a tectate pollen wall, has been found in deposits dating as far back as the Triassic (~ 220 Ma) and includes several species of grains with a single furrow that have been classified loosely in the *Crinopolles* group (Traverse, 1988). However, many discount this evidence since this early 'pollen' has never been found within angiosperm flowers. In addition, other morphologically similar types of early pollen, originally classified as angiosperm-like, have since been demonstrated to belong to earlier groups (i.e. gymnosperms or pteridophytes), or be contained in deposits that were stratigraphically misplaced and much younger than originally thought (Hughes, 1994).

Fossil evidence for angiosperm fruiting axes has been described recently from a late Jurassic (~145 Ma) fossil locality in China (Sun *et al.*, 1998). The angiosperm megafossil named *Archaefructus liaoningensis*, and placed within the Magnoliophyta division, displays a number of key angiosperm features, including, for example, ovules (seeds) that are completely enclosed in carpels, and leaf-like structures subtending each axis, which are probably flowers. However, there has been much debate about the age of the formation in which this 'early' angiosperm is preserved. The fossil was extracted from the lower part of the Yixian Formation of the Liaoning Province, China, which has been dated by biostratigraphical correlations to late Jurassic (~145 Ma) (Sun *et al.*, 1998). However, other lines of evidence, including radiometric dating (^{40}Ar–^{39}Ar) and palaeontological evidence, suggest that the lower part of the Yixian Formation is of late early Cretaceous age (~125 Ma) (Luo, 1999; Swisher *et al.*, 1999; Barrett, 2000), casting some doubts on the claim that this is the first unequivocal pre-Cretaceous angiosperm megafossil.

Angiosperm-like leaves have been found in deposits dating back to the late Triassic (~ 210 Ma), the most commonly cited being a leaf-type named *Furcula*. This leaf appears to have a typical angiosperm-like venation pattern (Stewart and Rothwell, 1993). However, *Furcula* leaves also display forking laminae (Figure 6.10), which are much more common in certain pteridosperms (Harris, 1932; Stewart and Rothwell, 1993). The classification of this leaf type as angiospermous therefore remains ambiguous, and from the fossil evidence alone (pollen, fruiting axes, and leaves) there is a justifiable scepticism concerning the pre-Cretaceous occurrence of angiosperms (Crane *et al.*, 1995).

However, evidence from molecular analyses has rekindled some support for a pre-Cretaceous origin for angiosperms. Construction of a molecular clock, based on the number of substitutions that have occurred since the divergence of monocotyledons from dicotyledons, suggests that this event may have occurred as early as 300 million years ago in the late Carboniferous (Martin *et al.*, 1989). Thus by inference, this study suggests that angiosperm origins must have occurred much earlier in plant evolutionary history. But a note of caution must

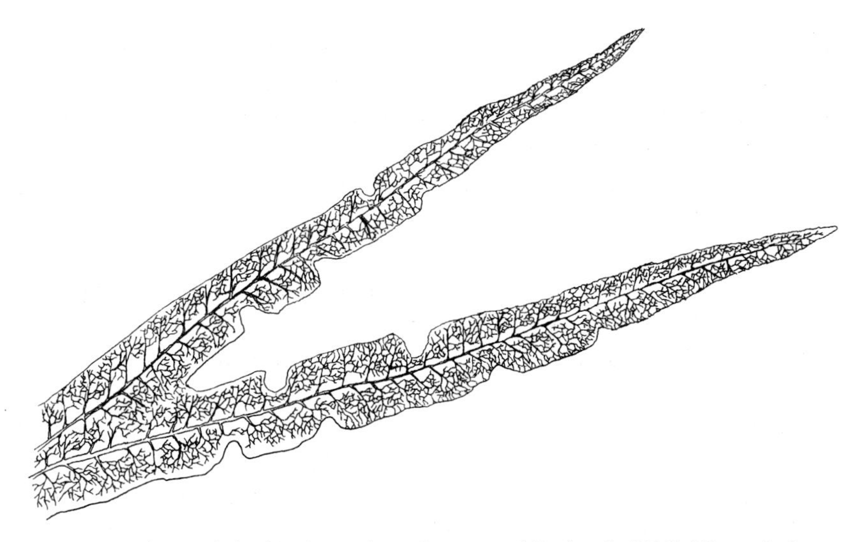

Figure 6.10 Fossil *Furcula* leaf (redrawn from Stewart and Rothwell, 1993). The typical angiosperm-like venation pattern seen in this fossil, which has been found in deposits dating back to the late Triassic (~210 Ma), led to the suggestion that this was an early angiosperm and therefore supports a pre-Cretaceous origin of this group.

be added since the method used in this study compared the number of substitutions that occurred in DNA of a slowly evolving glycolytic enzyme in nine extant species. When the same question (i.e. divergence of monocotyledons from dicotyledons) was asked using DNA from chloroplast sequences of extant species (Li *et al.*, 1989), the molecular clock suggested a date of between approximately 250 and 200 million years ago (Triassic). Both studies therefore suggest an earlier date for angiosperm origin than supported by the fossil evidence, but with up to a 100 million year discrepancy. However, the results of these molecular studies have not only been questioned on this discrepancy, but also on a number of other accounts. It is argued that their phylogenetic methodology is flawed (Crane and Lidgard, 1989), that the assumption that rates of molecular evolution in angiosperms has remained relatively constant through time is not adequately justified, and that the results are not supported by the angiosperm fossil record (Barrett and Willis, 2001).

Recent molecular phylogenetic studies of 106 extant angiosperms, however, have yet again reopened this whole debate (Qui *et al.*, 1999). This study analysed gene sequences for all three plant genomes (mitochondrial, nuclear, and chloroplast DNA), thereby avoiding the problems associated with single genome analysis (as above) (Kenrick, 1999a). Results from this study suggest that the split between gymnosperms and angiosperms may have occurred as early as the late Carboniferous (~290 Ma) (Qui *et al.*, 1999), therefore also supporting a pre-Cretaceous origin of the angiosperm lineage (Kenrick, 1999a).

Environmental considerations

Another hypothesis for the late appearance of angiosperms is related to the effects of major global environmental changes, including oceanic anoxia, increased tectonic activity, and sea-floor spreading, occurring in the mid-Cretaceous (~140–80 Ma). While there is no clear evidence (as yet) that the diversification and radiation of angiosperms was triggered by these environmental events, a number of floristic changes in mid-Cretaceous vegetation do correlate broadly with them (Lupia *et al.*, 1999). It is suggested that these major environmental changes may have conferred competitive advantages to the angiosperms at the expense of the previously dominant gymnosperms and pteridophytes.

In particular, the period between 124 and 83 million years ago (Aptian to Campanian) saw a dramatic change in continental configurations. Although the supercontinent Pangea had started to break up by the early Jurassic (~ 200 Ma) (Erwin, 1993), a period of rapid plate spreading was initiated in the Aptian (~124 Ma) until the Cenomanian (~83 Ma) (Sheridan, 1987, 1997). This resulted not only in changing continental configurations (the continents of Africa and South America were formed, and India, Australia, and Antarctica were distinguishable as attached plates), but also a 50–100% increase in the Earth's ocean crust production. Large-scale changes in global sea levels and atmospheric composition would have been associated with continental break-up. Geological evidence indicates that this entire period of major environmental perturbation is most probably attributable to an extraordinary upwelling of heat and material from the Earth's core–mantle boundary, known as a superplume episode (Larson, 1991a, b).

The effect of these geological events would have had a significant impact on global environments. It is estimated, for example, that a 100-m rise in long-term eustatic sea level occurred (Valentine and Moores, 1970; Haq *et al.*, 1987; Hallam, 1992), resulting in the formation of extensive epicontinental seas covering the interiors of North America, southern Europe, Australia, Africa, and South America (Briggs, 1995; Condie and Sloan, 1998). In addition, increased volcanism, which accompanied the plate movements, would have pumped greenhouse gases into the atmosphere, most notably CO_2 (Figure 6.11). Carbon cycle modelling indicates that CO_2 levels were 4–5 times greater than those of the present day (Tajika, 1999). This could have raised global temperatures by as much as 7.7 °C (Caldeira and Rampino, 1991). Evidence from various palaeoclimatic indicators also supports the suggestion that this was a time of global climate warming (Figure 6.12). Coral reefs, for which warm water is essential for survival and growth, increased their ranges as much as 1500 km closer to the poles, and isotopic analysis of deep ocean sediments ($\delta^{18}O$) suggests that deep ocean water, presently hovering near freezing, was 15 °C warmer. There is also no evidence for polar ice during this period of time (Barron and Washington, 1984; Condie and Sloan, 1998).

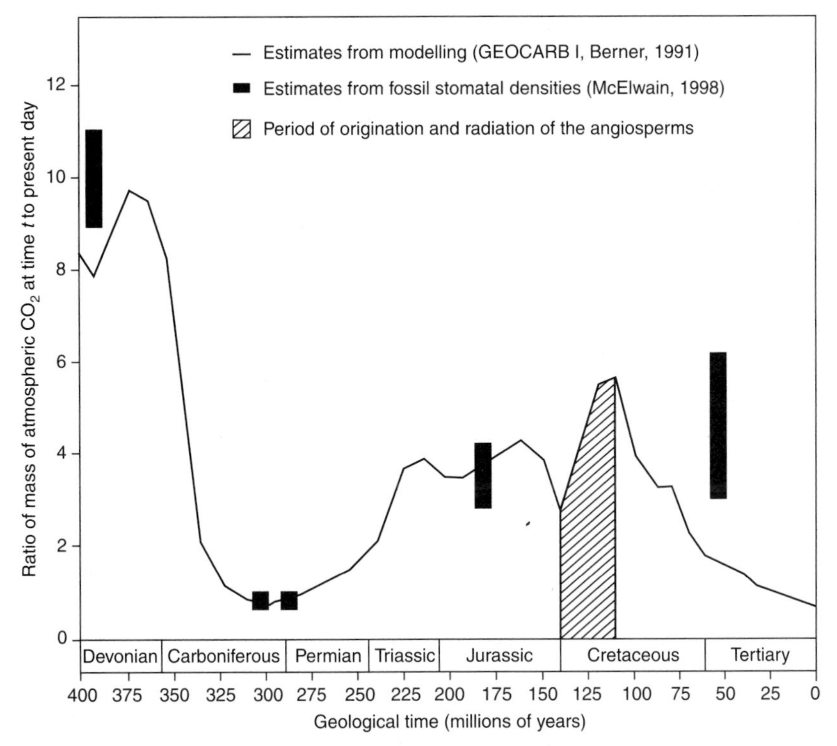

Figure 6.11 Estimated variations in atmospheric CO_2 during the past 400 million years (redrawn from Berner, 1991; GEOCARB I). The general direction of the trends are supported by independently derived results from carbon isotope analysis of palaeosols (e.g. Mora *et al.*, 1996) and stomatal densites on fossil leaves (e.g. McElwain, 1998). For comparison, the latter are indicated on the graph. Both are plotted against ratio of mass of atmospheric CO_2 at time t to present day (where present day is taken at the pre-industrial value of 300 p.p.m.v.). The early to mid-Cretaceous rise in atmospheric CO_2 is highlighted, along with the major change apparent in the plant fossil record during this time.

But why should increasing warmth promote angiosperm evolution or favour their diversification and radiation? A number of key innovations are recognizable among angiosperms which may have made them more drought resistant and therefore at a competitive advantage. These include tough, leathery leaves that were commonly reduced in size; a tough, resistant seed coat that protected the young embryos from drying out; vessel members providing much more efficient water-conducting cells than in previous groups; and a deciduous habit (Taylor and Taylor, 1993). The latter characteristic would have been critical in periods of drought but is not, however, unique to angiosperms. For instance, a number of gymnosperms (e.g. glossopterids, Ginkgoales, and Cycadales) are thought to have been deciduous, particularly at higher latitudes. The predominance of drought-resistant features, as well as a weedy life history and rapid reproduction, in the early angiosperms may have given them a competitive advantage in increasingly disturbed environmental conditions and warm

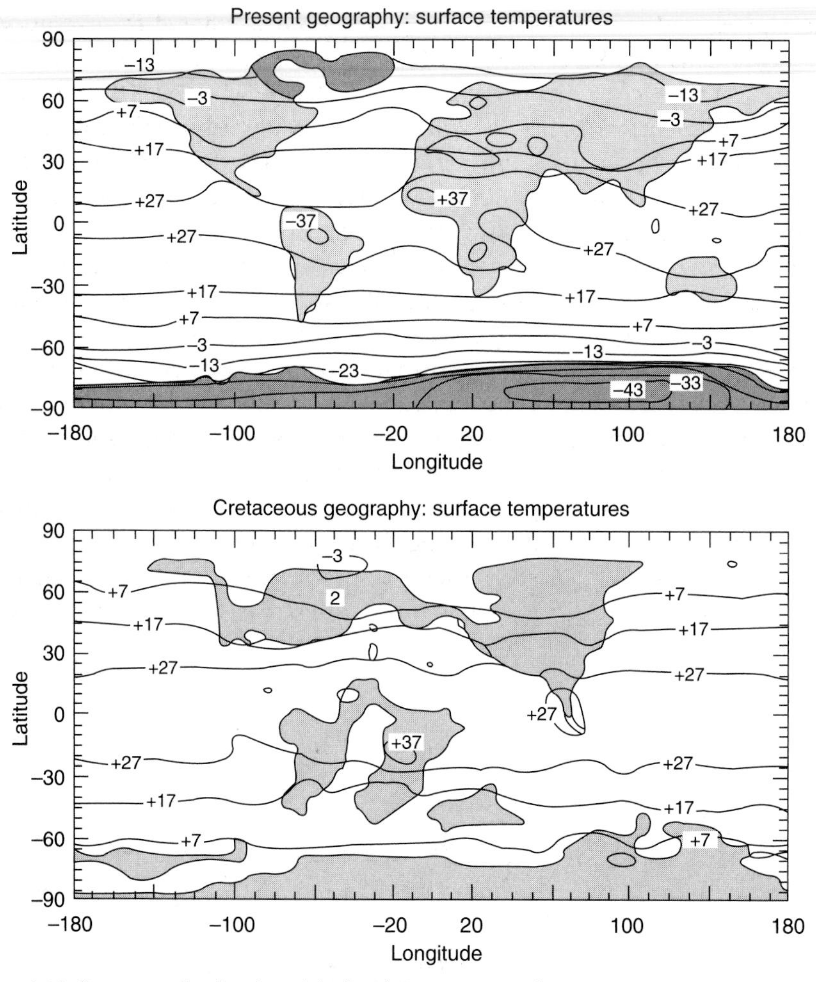

Figure 6.12 Computer-simulated model of mid-Cretaceous surface temperatures (~100 Ma) compared to present (redrawn from Barron and Washington, 1984). The model used predictions of Cretaceous topography, ice cover (stippled), continental positions, and sea-level in comparison with the present day in order to determine global surface temperatures. From this model it is predicted that the average global temperature during the Cretaceous was approximately 4.8 °C warmer than present. Temperatures are contoured in °C.

climate. An alternative explanation, is that dramatically accelerated speciation rates, which are characteristic of the angiosperms (Doyle and Donoghue, 1986), simply led to an overwhelming diversity of adaptive types (Lupia *et al.*, 1999).

Biotic interactions

Dinosaur–angiosperm coevolution

One hypothesis proposed for the late appearance of the angiosperms is that their evolution was closely associated with the large-scale radiation of certain groups of tetrapods, and that dinosaur feeding behaviour promoted the evolution of flowering plants (Bakker, 1978, 1986). This hypothesis is based on apparent evidence in the fossil record to suggest that a change in herbivore communities from high to low browsers occurred at approximately the same time as the initial evolution and radiation of angiosperms. Dinosaur fossil evidence suggests, for example, that approximately 160 million years ago, in the late Jurassic, 95% of the preserved biomass of dinosaurs was made up of sauropods and stegosaurs (Figure 6.13). These were high-browsing herbivores with a diversity of cranial and dental adaptations indicating a diet of a wide range of conifer tissue. It is argued that these high-browsing forms would have put intense pressure on the canopies of the mature trees, but permitted the development of gymnosperm saplings. However, from approximately 144 million years ago (Jurassic/Cretaceous boundary) the herbivore communities changed considerably and new groups of big, low-browsing ornithischian dinosaurs (Figure 6.13) appeared in the fossil record. It is suggested that these intense low-browsing dinosaurs would have increased mortality among the gymnosperm seedlings and thinned out the forest structure, thus creating gaps in the canopy and highly disturbed environments (Bakker, 1978, 1986). Early angiosperm traits, such as small

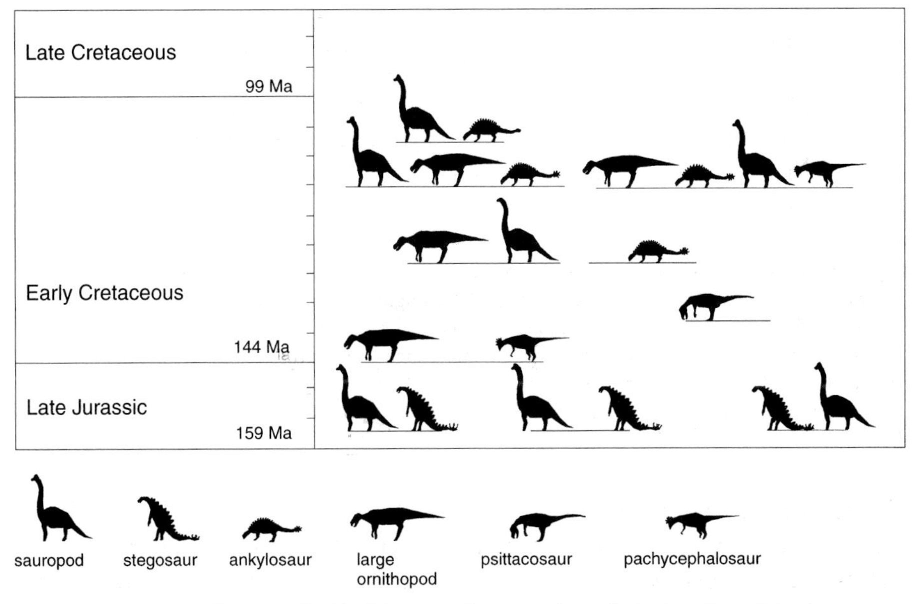

Figure 6.13 Stratigraphic record of herbivorous dinosaurs through the late Jurassic/early Cretaceous (redrawn from Bakker, 1978, 1986). It is proposed that the transition from high-browsing forms to low browsers in the early Cretaceous resulted in increased mortality among the gymnosperm seedlings and thinned out the forest structure, thus creating gaps in the canopy and highly disturbed environments (Bakker, 1978, 1986). Early angiosperm traits, such as small structure, rapid life cycle, and high colonizing ability, would have given them the competitive advantage on these disturbed substrates, thereby promoting their radiation.

structure, rapid life cycle, and high colonizing ability, would have given them the competitive advantage on these disturbed substrates, thereby promoting their radiation (Bakker, 1978; Wing and Tiffney, 1987).

Additional work on plant–herbivore interactions adds further detail to the above in suggesting that it was the mechanics of chewing among herbivores that aided angiosperm evolution during the Cretaceous. In particular, the radiation of ornithischian dinosaurs between 160–120 million years (mid-Jurassic–mid-Cretaceous) may have played an important role, since these were the first group of herbivores to develop a transverse chewing motion (which ground the food before passing it into the gut), thereby enabling these animals to exploit more fully the newly evolving plant resource (Weishampel and Norman, 1989).

However, recent work comparing the timing and location of the evolution of angiosperms with major events in the evolution of dinosaur herbivory has rejected this hypothesis, for a number of reasons (Barrett and Willis, 2001). First, the fact that angiosperms did not comprise a significant proportion of the global flora until the late Cretaceous suggests that it is highly unlikely that they formed a major constituent of dinosaur diets during the early Cretaceous. Secondly, detailed examination of the timing of the dinosaur fossil record suggests that no major event in the evolution of herbivorous dinosaurs can be correlated with angiosperm origins. Innovations in chewing and the onset of factors such as low browsing, for example, either precede or post-date the appearance of the first angiosperms in the fossil record. In addition, changes in dinosaur browsing behaviour at the Jurassic–Cretaceous boundary may not have been so marked as previously suggested. Thirdly, fossil evidence currently available suggests that there was no spatial overlap between the earliest angiosperms and the major clades of herbivorous dinosaurs. Most early Cretaceous dinosaur localities, for example, were situated at palaeolatitudes higher than 30°N and 30°S, with only around 10% of the total number within 20° of the palaeoequator, where the earliest angiosperms occurred (Barrett and Willis, 2001).

Insect–angiosperm coevolution

Another proposed biotic interaction of importance is that of angiosperm–insect coevolution. Early and mid-Cretaceous flowers contained many features to suggest that they were insect pollinated, including stamens with small anthers and low pollen production, and pollen grains often covered with pollenkitt-like material, and they were larger than the most effective size for wind dispersal (Crepet and Friis, 1987; Crane *et al.*, 1995).

Insect pollination would have been highly advantageous to the early angiosperms, enabling genetic exchange between widely spaced individuals or small populations. Furthermore, the suggestion that self-incompatibility mechanisms were present in the earliest angiosperms makes processes such as insect pollination even more critical for cross-pollination (Zavada, 1984). It is therefore suggested that the late evolution of angiosperms is closely related to that of insect evolution.

Fossil evidence for the coevolution of pollinating insects with angiosperms is ambiguous (Barrett and Willis, 2001). There is some fossil evidence for insect herbivory on Cretaceous angiosperms, such as leaf mines and other damage to leaves caused by feeding (Scott *et al.*, 1992; Labandeira *et al.*, 1994; Labandeira, 1998; Wilf *et al.*, 2000). However, comparison of the times of appearance of innovations in insect feeding systems with the timing of angiosperm radiation does not indicate a strong relationship (Labandeira, 1997). The advanced pollinator groups, including some *Hymenoptera* (certain wasps that are the sister group of the bees) and *Lepidoptera* (butterflies and moths), indicate a certain amount of synchronicity between their first fossil appearance (from approximately 140 million years ago) and that of angiosperms (Crepet and Friis, 1987; Labandeira and Sepkoski, 1993) (Figure 6.14). However, other early groups such as *Diptera* (e.g. crane flies and fungus-gnats), *Coleoptera* (beetles), and some *Hymenoptera* (e.g. saw-flies) have a fossil record indicating that their first appearance pre-dates that of the angiosperms, whereas other species of *Hymenoptera*, including the *Apoidea* (honeybees), have a fossil record that only extends as far back as the Albian (~100 Ma). Moreover, in analysis of insect familial-level diversity through time there is no marked increase in insect diversity with the time of angiosperm origin (Labandeira and Sepkoski, 1993) (Figure 6.15). There is increasing evidence to suggest, therefore, that advanced pollinators (wasps, bees, and moths) may have played an important role in the coevolution and major radiation of certain groups of flowering plants (Crepet, 1996; Grimaldi, 1999), but that the timing of angiosperm evolution as a whole cannot be explained by insect coevolution alone.

6.4 Evolutionary trends: gymnosperms to angiosperms?

Two questions that inevitably arise when discussing angiosperm origins are, from which lineage did they evolve, and when and how did divergence of the monocotyledons from the dicotyledons occur? Originally these questions were tackled by examination of only the fossil record, but more recent techniques, including morphological and molecular phylogenetic analyses of extinct and extant groups, have allowed the construction of detailed evolutionary relationships between the gymnosperms and angiosperms, and the monocotyledons and dicotyledons.

Despite early suggestions that angiosperms were of a polyphyletic origin (a number of difference ancestors), almost all recent evidence (morphological and molecular) suggests that angiosperms were derived from a single common ancestor (that is they had a monophyletic origin) (Doyle and Donoghue, 1986). Two of the earliest suggestions for possible evolutionary pathways between gymnosperms and angiosperms were via the Bennettitales (Arber and Parkin, 1907, 1908) and Gnetales (Von Wettstein, 1907).

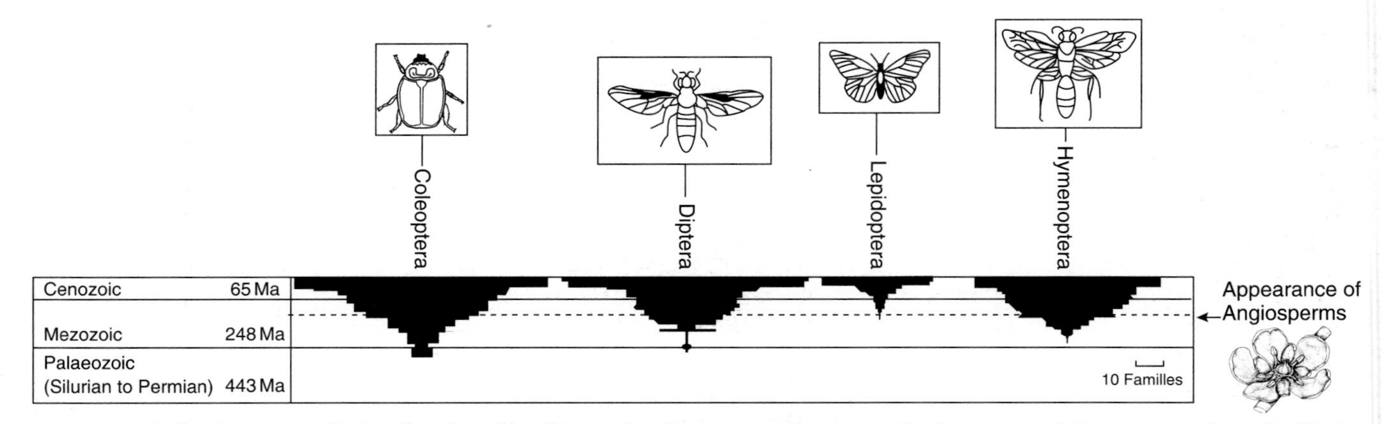

Figure 6.14 Spindle diagrams to display diversity of fossil insect families Diptera, Coleoptera, Lepidoptera, and Hymenoptera from the Silurian (~443 Ma) to present. The temporal position of the first appearance of angiosperms in the fossil record is also indicated (redrawn from Labandeira and Sepkoski, 1993).

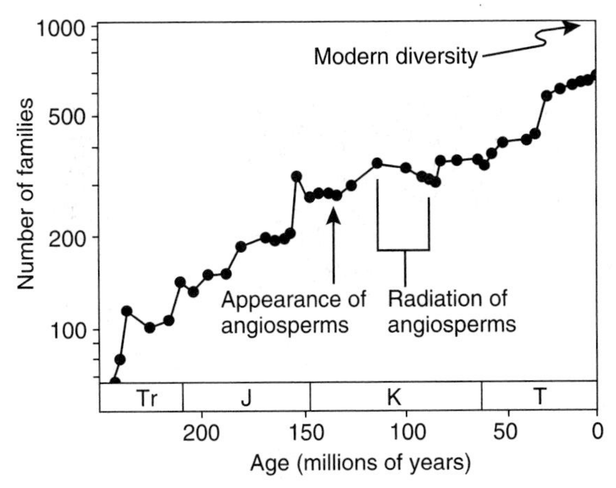

Figure 6.15 Insect familial diversity from the Triassic (~248 Ma) to present, plotted on a semi-logarithmic scale. The temporal position of the first appearance of angiosperms in the fossil record, and their subsequent radiation, are also indicated. (Redrawn from Labandeira and Sepkoski, 1993.)

Evidence for Bennettitales as the precursor to early angiosperms was based upon the fact that certain species in this fossil group had flower-like bisexual reproductive organs and similar wood anatomy. One species in particular that is often cited is *Williamsoniella*. This late Jurassic fossil plant had a reproductive organ that consisted of a bisexual reproductive axis bearing naked ovules above a series of pollen-bearing structures, the whole enclosed by large bracts (Stewart and Rothwell, 1993; Taylor and Taylor, 1993). These large bracts were thought to be the equivalent to petals in an angiosperm flower. Also, the position of the reproductive organ, erect and at the end of the branch, bore some similarities to extant insect-pollinated flower structures. *Coleoptera* and *Diptera* flies have been suggested as pollinators for this plant (Crepet and Friis, 1987).

Gnetales are present in the fossil record from the early Cretaceous (~140 Ma). The group has three extant genera in biogeographically distinct regions of the world: *Ephedra* (35 species), which exists in arid and semiarid regions, including parts of the Mediterranean, Asia, and the Americas (Bell, 1992); *Welwitschia mirabilis*, which is restricted to the Namibian desert (Figure 6.16); and *Gnetum* (30 species), which is exclusively tropical, occurring in Asia, Africa, and South America.

Morphological similarities between many species of Gnetales and angiosperms first led to the suggestion that they were probably close in evolutionary terms. These include, for example, reproductive organs that are bisexual (in some species), the presence of vessels, leaves with a venation pattern closely approximating that of dicotyledons, and a pollen wall (e.g. in certain species of

Figure 6.16 A model of extant *Welwitschia mirabilis* (courtesy of The Field Museum, neg. # B83079c).

Ephedra) that is tectate (Doyle and Donoghue, 1986). Some extant Gnetales are also insect pollinated (Friedman, 1996). Phylogenetic analysis based on these shared characteristics, in both extant and extinct species (Hill and Crane, 1982; Crane, 1985; Doyle and Donoghue, 1987), also confirmed a close relationship between angiosperms and Gnetales (Figure 6.17).

More recent cladistic analyses, which also include the molecular characteristics of extant species, are in agreement with the earlier morphological analysis and demonstrate a close relationship between Bennettitales, Gnetales, and the earliest angiosperms (Doyle *et al.*, 1994; Crane *et al.*, 1995). The phylogenetic trees obtained from these analyses (Figure 6.17) indicate that angiosperms from a clade with Bennettites and/or Gnetales but also sometimes with an extinct group, the Pentoxylales. However, the majority of studies suggest that the Gnetales are the most likely closest living relative of angiosperms.

There are a number of hypothesis as to which of the earliest angiosperms forms an evolutionary link with the Gnetales (Figure 6.18 a, b, c). Most suggest that early members of the Nymphaeales and Piperales were the evolutionary link between Gnetales and angiosperms (Taylor and Hickey, 1992; Doyle *et al.*, 1994); others have suggested a single genus, *Ceratophyllum*, within the Nymphaeales (Les *et al.*, 1991), and members of the Laurales (Loconte and Stevenson, 1991; Loconte, 1996). However, there is also strong support for woody Magnoliales (for a review, see Wing and Boucher, 1998).

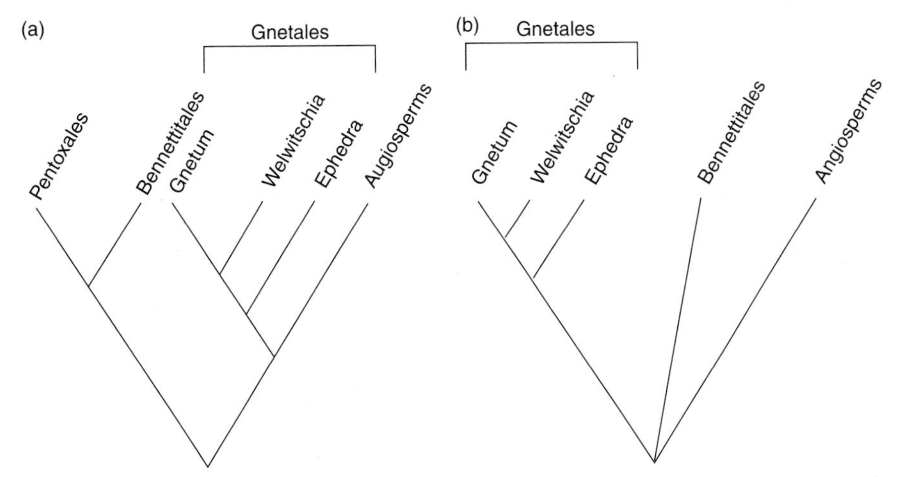

Figure 6.17 Phylogenetic trees showing two alternative relationships between angiosperms and related seed plants, (a) supports the hypothesis that Gnetales are the closest relative of the angiosperms and (b) supports the hypothesis than Bennettitales are the closest relative of the angiosperm (adapted from Crane (1985) and Doyle *et al.* (1994).

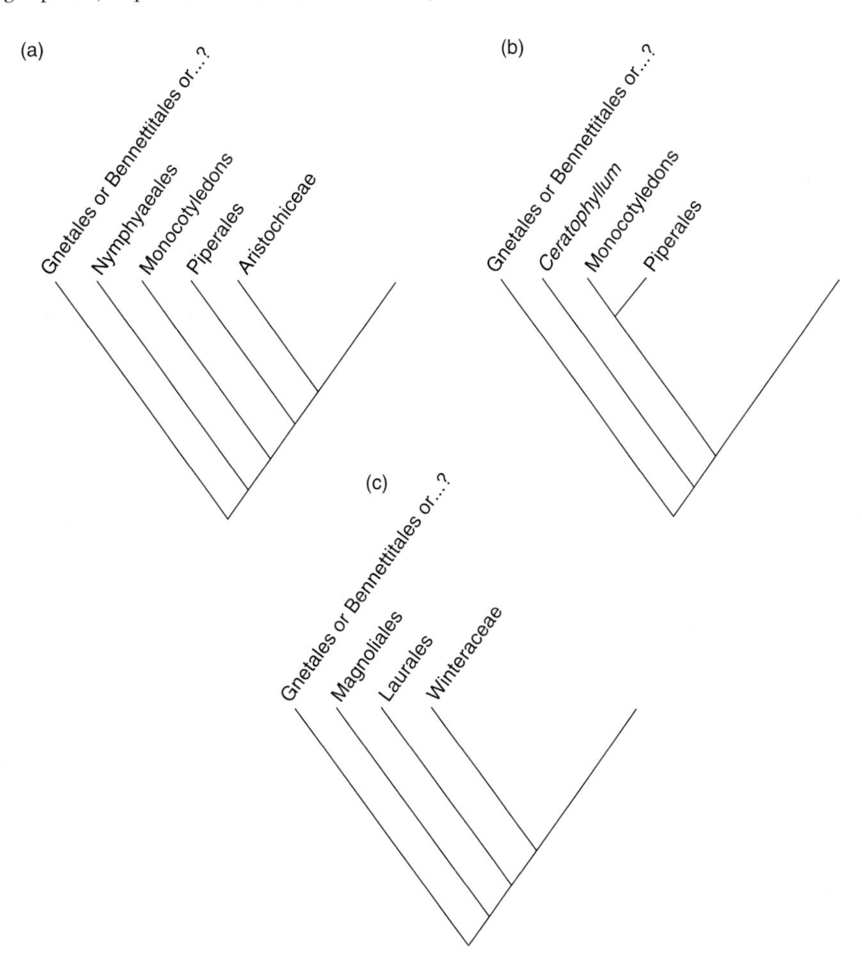

Figure 6.18 (a–c) Phylogenetic trees (from Wing and Boucher, 1998) to indicate alternative relationships for major basal clades within the angiosperms.

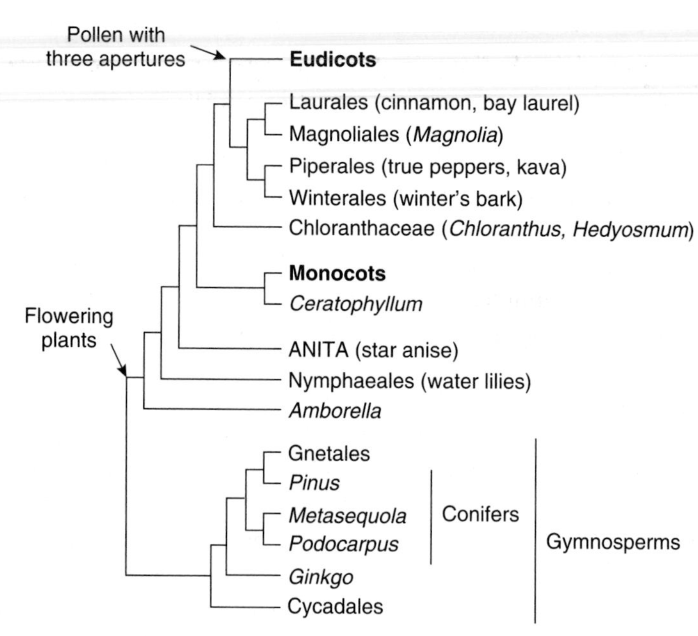

Figure 6.19 Phylogenetic tree based on recent molecular studies of angiosperms (redrawn from Qui *et al.*, 1999).

More recent and extensive molecular studies of angiosperm phylogenetics have, however, turned a number of these relationships on their head (Qui *et al.*, 1999; Soltis *et al.*, 1999). In the first instance, the results of both studies do not support the hypothesis that the Gnetales are the closest living relatives of the angiosperms. Instead, these analyses group the Gnelates with the conifers (Figure 6.19). Secondly, the results do not agree with any of the existing ideas about which angiosperms are most primitive. Instead, the analyses indicate that *Amborella trichopoda*, the only extant species of Amborellaceae, represents the most primitive (basal group) of all flowering plants (Qui *et al.*, 1999; Soltis *et al.*, 1999). This is followed by the Nymphaeales, and then a group including Illiciaceae, Trimereniaceae, *Austrobaileyaceae*, and Schisandraceae. Together they have been referred to as ANITA.

Evidence from the various cladistic analyses described above have also indicated the relationship between early monocotyledons and dicotyledons. Again, the analyses support the fossil evidence in suggesting that monocotyledons were an early branch in angiosperm evolution (Crane *et al.*, 1995; Niklas, 1997; Soltis *et al.*, 1999). However, the species or group that forms the evolutionary link between the monocotyledons and dicotyledons is still under debate.

6.5 Biogeographical distribution of global vegetation during the late Cretaceous (~84–65 Ma)

During the late Cretaceous (~100–65 Ma) angiosperms increased in both species number and diversity (Figure 6.20). Angiosperm trees and shrubs evolving during this time (Table 6.3) included a number of families that constitute a significant part of the present-day global flora. Evidence from the fossil record suggests, for example, that a number of extant northern and southern hemisphere families appeared for the first time. These include Ulmaceae (including evidence for *Ulmus*), Betulaceae, Juglandaceae, Fagaceae (especially *Nothofagus*), and Gunneraceae (Table 6.3). The striking number of angiosperm fossils present by the late Cretaceous with close affinities to extant families has led to the suggestion that eventually it will be discovered that all angiosperm families originated during this remarkable period (Wing and Boucher, 1998).

The majority of trees to appear in the late Cretaceous (~100–65 Ma) have a present-day distribution that is mainly tropical or subtropical (Table 6.3). It is interesting to note that most present-day angiosperm families are basically tropical in their requirements, with over one-half of angiosperm families confined to tropical regions and over three-quarters of all angiosperm families attaining optimum development and diversity in a tropical environment (Axelrod, 1966). Therefore, although many may now be classified as northern/southern temperate species by their distribution, it is probable that they still possess the traits that would enable them to survive in conditions similar to those characteristic of the early environments where they originated.

Detailed 'biome level-analyses' of global plant biogeography and palaeoclimate have been carried out for the late Cretaceous (Maastrichtian, ~71–65 Ma)

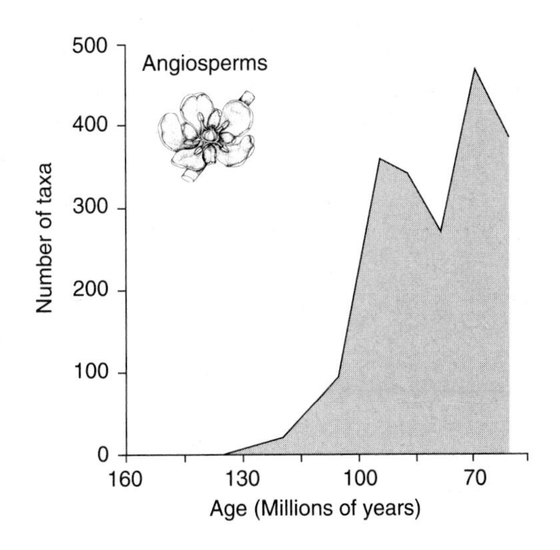

Figure 6.20 Evidence from fossil leaf assemblages (summed genus and species diversity) indicating the major expansion of the angiosperms from ~140 Ma and a dramatic increase in the absolute number (summed diversity) of angiosperms through the mid-Cretaceous (~10 Ma). Data taken from 147 late Jurrasic to Palaeocene macrofossil floras (redrawn from Lidgard and Crane, 1988; Lupia *et al.*, 2000).

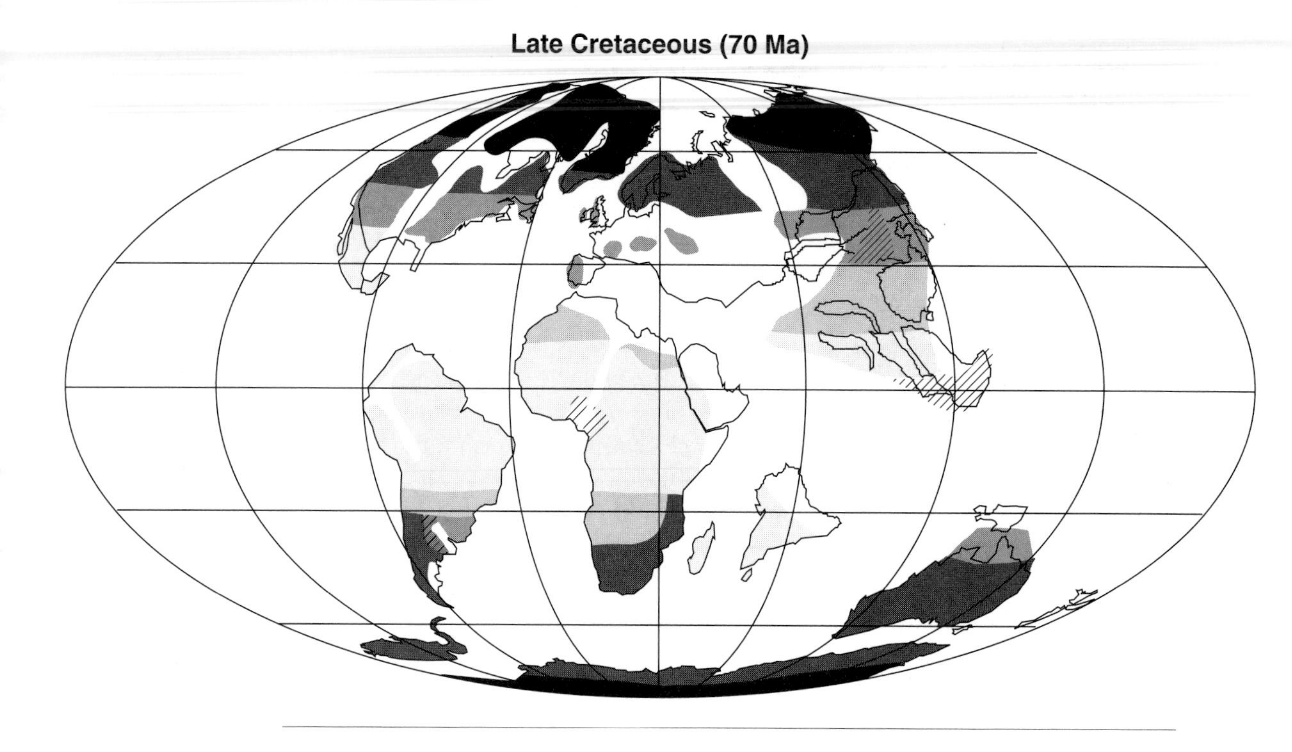

(Horrell, 1991; Upchurch *et al.*, 1999). Six global biomes are recognized, as follows (Figure 6.21).

Cool temperate biome

The cool temperate biome in the Maastrichtian coincides roughly with the present-day Arctic circle, comprising Canada, Greenland, and Siberia in the northern hemisphere and Antarctica in the southern hemisphere (Horrell, 1991). The vegetation of the cool temperate biome has also been referred to as polar deciduous forest (Upchurch *et al.*, 1999) and was clearly one of three remaining biomes in the late Cretaceous not dominated by angiosperms. Instead, vegetation in this region was dominated by deciduous and evergreen conifers with ferns and ginkgos. Angiosperms are believed to have been present as understorey and included members of the Betulaceae and Juglandaceae. The prevalence of leaf-shedding among the angiosperm species (Vakhrameev, 1991), relative low-diversity floras, evidence for growth rings (Spicer and Parrish, 1990), and the presence of some, but not abundant, coals, supports the suggestion of seasonality, typical of a cool temperate climate (Horrell, 1991). Members of the Pinaceae and Taxodiaceae (*Sequoia* and *Taxodium*) were common conifers, and the fossil record also includes evidence for 'modern' genera such as *Pinus* and *Abies*. Other conifers, widely spread but less significant in terms of their representation, included Cupressaceae and Araucariaceae. In the southern

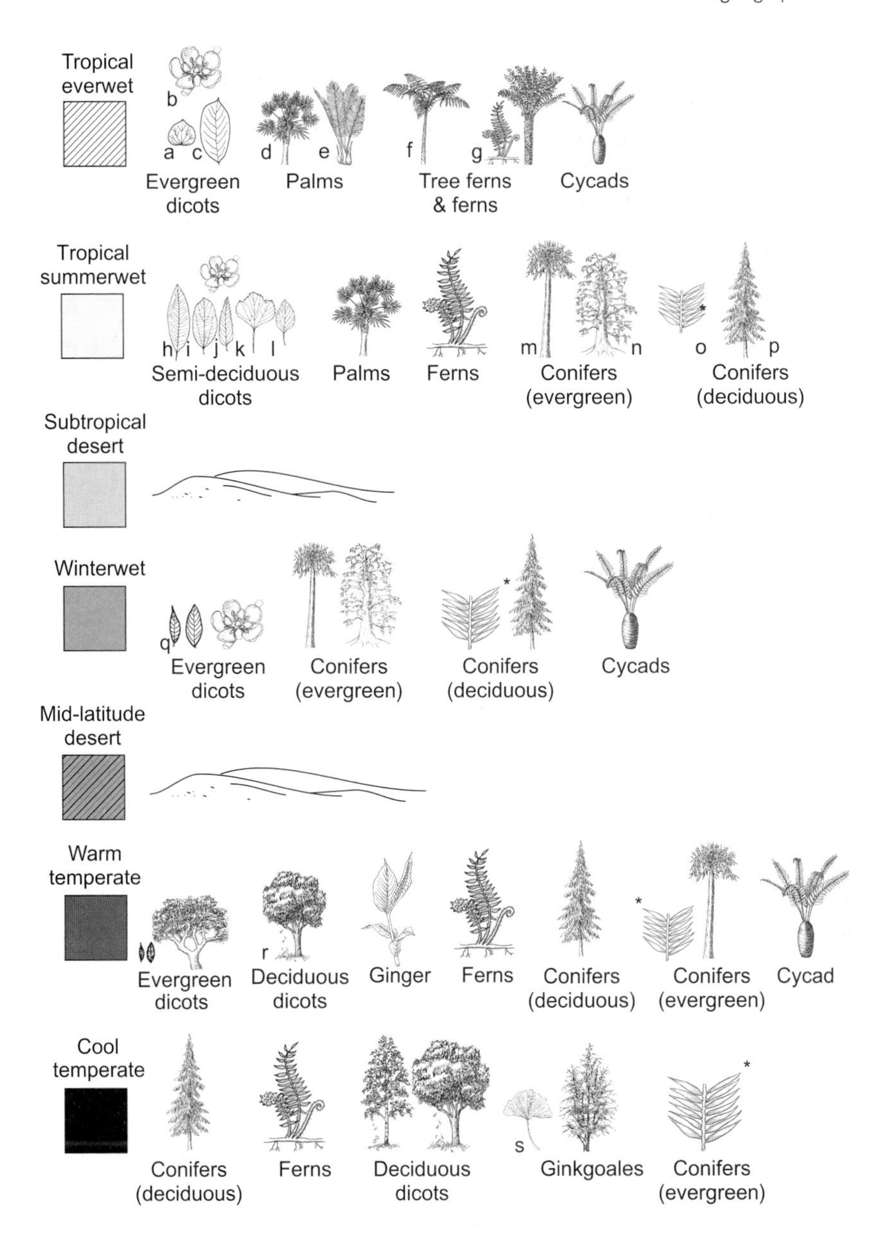

Figure 6.21 Suggested biomes for the late Cretaceous (84–65 Ma) (modified from Horrell, 1991; Upchurch *et al.*, 1999), with representatives of the most abundant and/or dominant fossil plant taxa shown. The biomes are superimposed on a global palaeogeographic reconstruction for the Maastrichtian (~70 Ma) (courtesy of A. M. Ziegler, PaleoAtlas Project). (a) and (c) Stylized megaphyllous leaves; (b) generalized Cretaceous flower; (d) *Sabalites* palm; (e) *Nypa* palm; (f) *Cyathea*; (g) *Dicksonia*; (h) *Myrica* leaf; (i) *Ficus* leaf; (j) *Dryophyllum* leaf; (k) *Viburnum* leaf; (l) *Grewiopsis* leaf; (m) *Araucaria*; (n) Cheirolepidaceous tree; (o) *Podocarpus* leaves; (p) Taxodiaceae conifer; (q) stylized microphyllous leaves; (r) generalized deciduous tree; (s) *Ginkgo* leaf. See Appendix 5 for sources of plant reconstructions and line drawings. * Dominant or abundant in the southern hemisphere.

hemisphere, the cool temperate biome was characterized by an abundance of podocarpacean and araucarian conifers and the angiosperm dicot genus *Nothofagus* (southern beech). The abundance of *Nothofagus* pollen has led to the suggestion that the climates of Antarctica and southern Australia were cooler and drier than those of the southern hemisphere lower latitudes (Herngreen and Chlonovo, 1981).

Warm temperate biome

The warm temperate biome between 45° and 65° palaeolatitude, encompassed present-day northern North America, southern Greenland, parts of western Europe, Russia, and northern China in the northern hemisphere, and Australia and coastal Antarctica in the southern hemisphere (Horrell, 1991). The vegetation characteristic of this biome included abundant dicotyledonous and monocotyledonous angiosperms, evergreen and deciduous conifers, ferns, and cycads. Upchurch *et al.* (1999) have referred to the vegetation of this region as 'sub-tropical broad-leaved evergreen forest and woodland' as many of the fossil leaf shapes are common among modern subtropical plants. However, due to the palaeolatitudinal position of this region, well outside the subtropics, Horrell's climatic assessment and interpretation of this region as warm temperate rather than subtropical is used here (Figure 6.21). Common angiosperms included members of the Fagaceae (e.g. *Castanea*), Betulaceae (e.g. *Betula*), Juglandaceae (e.g. *Juglans*), Ulmaceae (e.g. *Ulmus, Zelkova*), Proteaceae (southern hemisphere only), and Winteraceae among the dicotyledons and corphoid palms were abundant monocotyledons. Conifers included Araucariaceae and Taxodiaceae.

The high diversity of angiosperm dicotyledons in fossil floras from this region was thought to indicate their dominance. However, detailed analysis of a fossil flora from Wyoming, preserved *in situ* by an ash fall, has shown that although dicotyledons constituted 61% of the total floral diversity, they accounted for a mere 12% of cover, compared with 49% by ferns (Wing *et al.*, 1993). It is apparent, therefore, that although angiosperms had reached a dominant position in stream-side vegetation of the warm temperate biome by the Maastrichtian (~70 Ma), ferns still maintained dominance in certain habitats, even after 30 million years of angiosperm diversification and radiation (Wing *et al.*, 1993). This study also demonstrates the complexity and mosaic of different habitats that exist within each individual biome, and highlights the importance of taking into account preservational biases when undertaking biogeographical analyses.

Winterwet biome

The vegetation between palaeolatitudes of approximately 30° and 45° was markedly less diverse than that of the warm temperate biome, and is characterized mainly by the occurrence of evergreen dicotyledons together with both evergreen and deciduous conifers with some cycads. A notable feature of the vegetation of this region was the relatively lower abundance of monocotyledons. Floras from the winterwet biome indicate the presence of abundant *Araucaria* and cheirolepidaceous conifers, and evidence from fossil wood suggests that angiosperms may have formed part of an understorey of both shrubs and small trees rather than large canopy trees (Upchurch and Wolfe, 1987).

In the southern hemisphere, Patagonian floras contained abundant ferns and angiosperm families such as Lauraceae. The presence of huntite, a mineral which today is only formed in warm environments with high rates of evaporation, typical of Mediterranean climates, together with an absence of clear growth rings in fossil woods, provides good support for the designation of a winterwet biome to this region (Horrell, 1991). However, this is not supported unanimously and Upchurch *et al.* (1999) suggest that the absence of fossil leaves with spinose margins, which are typical of the modern winterwet biome vegetation, questions whether this biome really existed during the late Cretaceous. It is worth noting, however, that despite the absence of one vegetative characteristic, a number of model simulations of the late Cretaceous climate indicate that a Mediterranean (i.e. 'winterwet') climate may indeed have prevailed in certain areas during this time.

The northern limit of the winterwet biome in the northern hemisphere is marked by coals, whereas to the south it is marked by the presence of evaporites, indicating humid and arid conditions, respectively. It is thought therefore that this biome represents a transition between the arid desert biomes of the lower latitudes and the humid biomes of the higher-latitude temperate belts (Horrell, 1991).

Subtropical desert biome

Evidence for extensive evaporite deposits during the late Cretaceous (~70 Ma), indicating high rates of evaporation over precipitation, mark the subtropical desert biome (Horrell, 1991). These deposits occur in a northern hemisphere belt which incorporates present-day north Africa, China, and the Yukatan peninsula, and in the southern hemisphere include present-day south-western Africa and southern South America. No fossil floras have yet been found in these biomes, most probably reflecting both the low diversity and productivity which would have been apparent, but also the low preservation potential of fossil plants in arid environments.

Tropical summerwet biome

The tropical summerwet biome incorporates the majority of present-day Africa, South America, and India, from palaeolatitudes 0° to 25° (Horrell, 1991; Upchurch *et al.*, 1999) and is characterized by a 'tropical semi-deciduous forest' type vegetation. Common elements in the vegetation of this region include dicotyledons and monocotyledons, ferns, conifers, and cycads. Evidence from fossil pollen and wood suggest that common conifers included members of the Araucariaceae, Cheirolepidaceae, and Podocarpaceae. India contained floral elements typical of both the southern hemisphere (e.g. *Notofagus*, Proteaceae, and Podocarpaceae) and also elements of the northern hemisphere vegetation, leading to the suggestion that by the late Cretaceous, the Indian plate was already part of the equatorial biogeographical region, even though in terms of palaeogeography it was still 30°S (Briggs, 1995).

The combined evidence from fossil floral composition, leaf physiognomy, and sedimentological indicators suggests that this region was characterized by a hot, subhumid to semiarid climate typical of present-day summerwet or savannah regions (Horrell, 1991).

Tropical everwet biome

The tropical everwet biome of the late Cretaceous was much reduced in comparison with the present-day extent of tropical rain forest, and was restricted to an area including present-day subequatorial west Africa and Malaysia (Horrell, 1991) and possibly Somalia, in east Africa and Colombia in South America (Upchurch *et al.*, 1999). Palynological assemblages recovered from these areas suggest that during the late Cretaceous the equatorial region was dominated by species of Arecaceae (palms) (Vakhrameev, 1991), including the extant genus *Nypa* (Figure 6.21). Other angiosperms present included Proteaceae and many other dicotyledon groups, while ferns and tree ferns were also abundant (Meyen, 1987). A striking feature of this biome is the almost complete absence of evidence for either evergreen or deciduous conifers, with the exception of Araucariaceae in Malaysia.

Comparison of the biogeographical patterns present in the late Cretaceous (~70 Ma) with those from the early Carboniferous (~360 Ma), middle Permian (~267 Ma), and early Jurassic (~196 Ma) reveals a consistent pattern, whereby the tropical everwet biome is either absent or severely restricted in extent. This phenomenon has also been observed for the early Cretaceous (~137 Ma) (Ziegler *et al.*, 1987). In contrast, however, the extent of the tropical everwet biome during the late Carboniferous (~300 Ma) spans from the equator to approximately 25°, similar to the observed extent of this biome today. It is noteworthy that the features common to both the upper Carboniferous and present day are the presence of polar ice caps, high latitudinal temperature gradients and comparatively low atmospheric CO_2 concentration. It can therefore be suggested that these characteristics may be important global environmental prerequisites to the development of an extensive 'tropical rainforest' biome.

Summary

1. The first flowering plants (angiosperms) appeared in the fossil record from approximately 140 Ma (early Cretaceous). Rapid diversification from $\sim$100 Ma led to their global dominance by the early Tertiary ($\sim$65 Ma).

2. Evidence for the first angiosperms includes fossil flowers, fruits, pollen, and leaves.

3. The earliest angiosperms originated in the palaeotropics (0–30°), radiating out to colonize higher latitudes some 20–30 million years later.

4. In evolutionary terms, the angiosperms are the most recent group to appear in the fossil record, approximately 300 million years later than the first vascular plants, and 220 million years later than the first seed plants.
5. Various hypotheses have been suggested to account for their relatively late appearance in the fossil record, including a bias in the fossil evidence, a particular suite of environmental conditions and/or biotic interactions that led to their later evolution.

6. Two biotic interactions that are of particular interest are the coevolution of flowering plants with low-browsing dinosaurs and with pollinating insect groups.

7. There is increasing evidence to support the suggestion that the earliest angiosperms were herbs or small shrubs.

8. Molecular phylogenies indicate that monocotyledons probably evolved from dicotyledons early in angiosperm evolution.

9. Examination of cladistic analyses based on morphological and molecular traits of both extant and fossil material indicates that angiosperms were probably derived from a single common ancestor.

10. Traditionally the suggested precursor to the angiosperms included members of the Bennettitales and/or Gnetales. However, recent phylogenetic analyses have indicated that *Amborella trichopoda* (which is a small shrub and the only extant species of Amborellaceae) is the most primative of all angiosperms.

11. Reconstruction of the biogeographical distribution of vegetation during the Maastrichtian ($\sim$71–65 Ma) indicates that six global biomes can be recognized during this period, with all but the highest latitudes dominated by angiosperms.

7 The past 65 million years

A consideration of the past 65 million years (Tertiary and Quaternary) are vital to understanding the evolution and distribution of the present-day flora. During this time continental plates moved into their present arrangement, prominent mountain ranges, such as the Himalayas, Alps, Carpathians, Caucasus, Zagros, and Rockies, were formed, and the global climate, which was initially very warm, became increasingly cool, culminating in a build-up of ice at the poles. Against this backdrop of changing environmental conditions, major changes occurred in the overall composition and distribution of global vegetation. Angiosperm groups continued to diversify and modernize throughout the Tertiary, grasses evolved in the Palaeocene (~60 Ma), and a long-term decline in the aerial extent of forests and expansion of grasslands occurred (~35 Ma). The past 65 million years is therefore regarded as 'the period of time that we must look to for the precursors of most modern taxa and plant communities' (Tallis, 1991).

7.1 Environmental changes over the past 65 million years (Tertiary and Quaternary)

Early Palaeocene to middle Eocene (65–45 Ma)

Evidence from the geological record indicates that the interval spanning the early Palaeocene to middle Eocene (65–45 Ma) was one of the warmest periods in Earth history. Reconstructed global temperatures (from both terrestrial and oceanic sources) indicate that a general pattern of warming occurred through the Palaeocene, and that maximum Tertiary global temperatures were reached between the early and middle Eocene (~55–45 Ma). For example, deep ocean temperature estimates of between 9 and 12 °C higher than present, and mean annual sea surface temperatures in the Antarctic ocean exceeding 15–17 °C have been recorded (Rea *et al.*, 1990; Kennett and Stott, 1991). Reasons suggested for high global temperatures during this period include changing continental configurations, increased mantle degassing, volcanic-induced greenhouse warming, and methane-induced polar warming (Sloan and Barron, 1992).

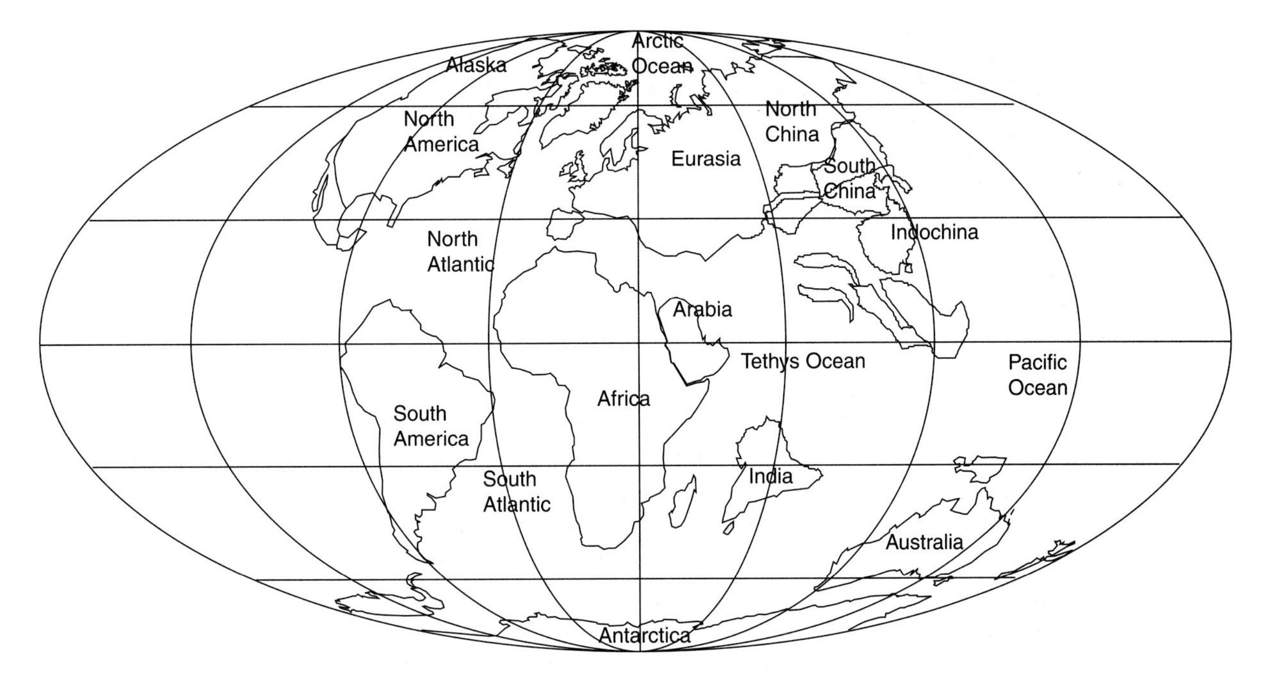

Figure 7.1 Continental configurations at 70 Ma (map courtesy of A. M. Zieglar, PaleoAtlas Project, University of Chicago).

In the early Palaeocene (~65–55 Ma) the global position of continental plates (with the exception of Australia and New Zealand) had moved close to their present-day position (Figure 7.1). However, the configuration of the continents was still unfamiliar, as Greenland was attached to North America and Europe was intermittently connected to North America via Greenland. Australia was attached to Antarctica, and there was probably an archipelago connection between South America and the Antarctic Peninsula. In addition, the Mediterranean Sea was unformed and instead, the Turgai Sea separated Europe from Asia, and Africa was separated from Europe by the Tethys Sea (Figure 7.1). North America was connected to Asia via Berringia (Europe).

These continental configurations had a significant impact on ocean currents and consequently on global climate. The greatly expanded Pacific Ocean, with vast ocean gyres, carried heated tropical waters to high latitudes in both hemispheres. Also, the close proximity of South American and Australian land masses to the Antarctic prevented the development of a cold circumpolar ocean current around Antarctica (as is the present-day situation) (Lawver and Gahagan, 1998) and warm equatorial ocean currents were deflected southward to Antarctica (Condie and Sloan, 1998). As a consequence, permanent ice cover at the poles was absent, and the prevailing low relief of the continents, coupled with high seas, resulted in rain-bearing winds penetrating far into the interior of all the main land masses. The temperature gradient between the equator and poles was therefore considerably lower during the Palaeocene and early Eocene (65–45 Ma) than exists in the present day.

Geological evidence also suggests that the Palaeocene and early Eocene was a period of global carbon imbalance, where the output of carbon dioxide into the atmosphere exceeded the uptake by carbon burial (Compton and Mallinson, 1996). This is believed to have resulted in higher atmospheric CO_2 concentrations, thereby further increasing global temperatures through the greenhouse effect (Miller *et al.*, 1987; McElwain, 1998).

Middle Eocene to end Miocene (~45–5 Ma)

Over the subsequent 40 million years, however, dramatic changes occurred in the configuration of the continents, both in terms of the relative positions of land to sea and also through changing topography of the continents through mountain building activity (Hallam, 1994). These changes had a dramatic effect upon global climates, by decreasing temperatures and increasing aridity.

In the southern hemisphere, Australia, New Guinea, Africa, and India continued to move northwards toward their present positions and Antarctica became separated from Australia and the South American archipelago. This separation resulted in the development of a seaway between Australia, Antarctica, and South America, known as the Drake Passage, in the early Oligocene (~30 Ma), which enabled a circum-Antarctic ocean circulation system to develop for the first time (Figure 7.2). The development of this cold circumpolar ocean current around Antarctica would have prevented warm equatorial currents from penetrating into the southern polar regions. It is suggested that thermal isolation of Antarctica, resulting from this change in ocean currents, was one of the main processes responsible for climatic cooling at the South Pole, the initiation of Antarctic ice sheets, and an associated fall of eustatic sea level (Lawver and Gahagan, 1998). The splitting of Antarctica from Australia and South America was effectively the end of Gondwana and initiated a trend toward cooler glacial climates (Kennett, 1977, 1995).

In the northern hemisphere, by the late Oligocene (~30 Ma) sea-floor spreading between Greenland and Norway had created marine connections between the Arctic Ocean and the North Atlantic. By the Miocene (~21.5 Ma), Africa had become connected to Asia via the Arabian peninsula, eliminating the Tethys sea and creating the Mediterranean (Figure 7.3). Opening of the Labrador Sea and the Norwegian Sea enabled the establishment of ocean current systems bringing Arctic sea-waters southwards for the first time (Williams, 1986; Wright, 1998). This would have resulted in the development of cooler and drier climates in the mid to high latitudes of the northern hemisphere (Briggs, 1995).

During the Tertiary a number of major mountain-building episodes are also thought to have had a significant impact on the global climate (Partridge *et al.*, 1995). Between ~55 and 40 Ma, the uplift of the Himalayas, Tibetan Plateau, and Cordilleras (which include the Rocky Mountains) occurred (Chase *et al.*,

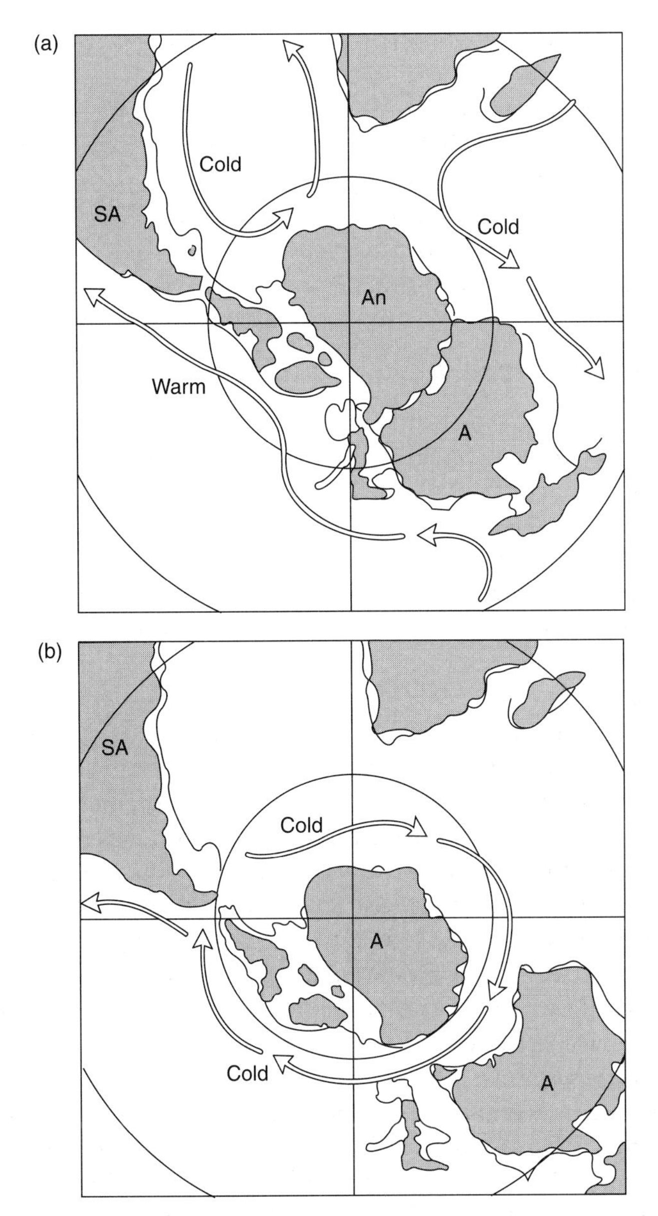

Figure 7.2 The separation of Australia from Antarctica during the late Eocene (~37 Ma), resulting in the development of a seaway between Australia (A), Antarctica (An), and South America (SA), known as the Drake Passage, which enabled a circum-Antarctic ocean circulation system to develop for the first time (from Condie and Sloan, 1998).

1998; Harrison *et al.*, 1998) followed by the Pyrenees, Carpathians, Caucasus, and Zagros mountains between ~35 and 25 Ma (Condie and Sloan, 1998). It is suggested that the changing topography resulting from these mountain-building episodes would have contributed to a general climatic trend of increasing aridity and the formation of desert and arid biomes. Increased aridity was driven by

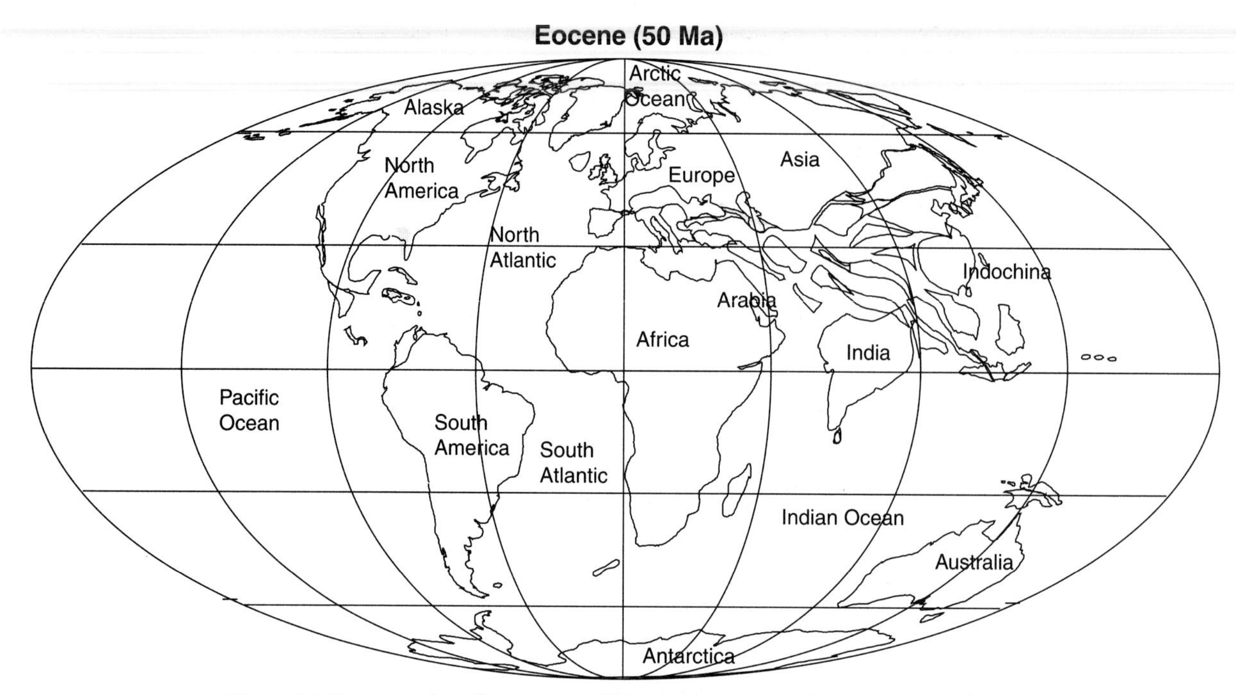

Figure 7.3 Continental configurations at 50 Ma (map courtesy of C. R. Scotese, PaleoMap Project, University of Texas).

both the physical mass of emerging mountain ranges restricting the movement of moisture-bearing winds into continental interiors, and an increased proportion of water becoming land-locked in polar ice-caps, thus decreasing moisture availability in the global water budget (Ruddiman and Raymo, 1988; Ruddiman and Kutzbach, 1989; Patridge, 1997).

It is also suggested that these mountain-building episodes would have had a significant draw-down effect on atmospheric CO_2 , contributing to a reduction in global temperatures through a reduced greenhouse effect (Berner, 1995). It is estimated, for example, that increased rates of uplift and erosion associated with the formation of the Himalayan mountains and Tibetan Plateau would have provided a large, new source of calcium silicate minerals for chemical weathering. Weathering of the silicate minerals would have transferred carbon from the atmosphere to carbonate minerals ($CO_2 + CaSiO_3 \leftrightarrow CaCO_3 + SiO_2$), which ultimately were buried as ocean sediments. This process is therefore thought to have resulted in the long-term removal of CO_2 from the atmosphere during the Eocene and Miocene (Raymo *et al.*, 1988; Raymo and Ruddiman, 1992; Ruddiman *et al.*, 1997). Estimates suggest, for example, that atmospheric CO_2 concentration declined from levels approximately four times higher than present in the late Cretaceous ($\sim$100 Ma) to levels equivalent to those of the present day by the Miocene (Figure 7.4) (Berner, 1991, 1994; Pearson and Palmer, 1999).

The overall effect of these changing continental configurations, ocean circulation patterns, and a reduction in atmospheric CO_2 concentrations, was a

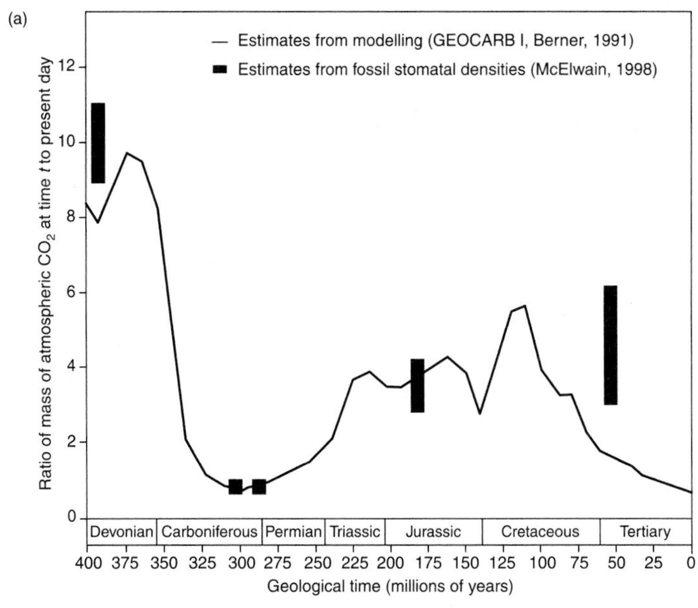

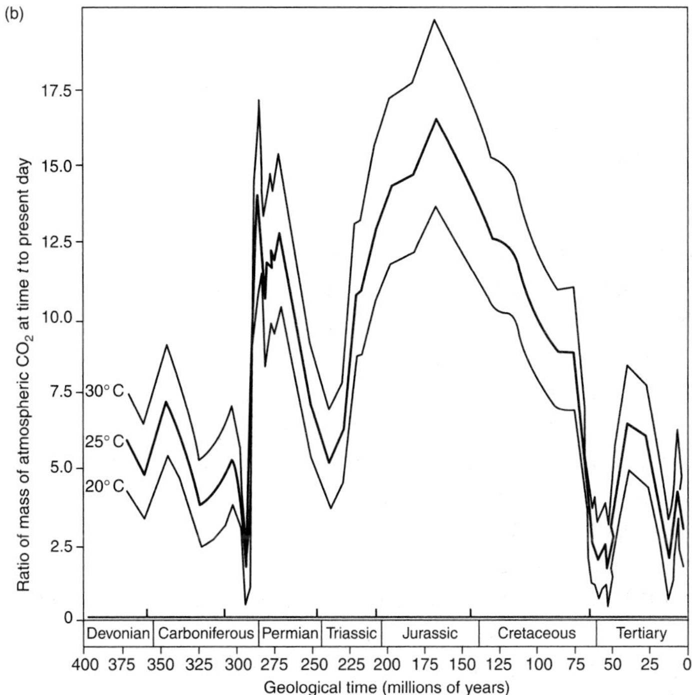

Figure 7.4 Levels of atmospheric CO_2 over the past 400 million years to present: (a) results from the GEOCARB I model (Berner, 1991), and fossil stomatal densities (McElwain, 1998); (b) estimates from fossil soil carbonates (Ekart *et al.*, 1999). Both are plotted against ratio of mass of atmospheric CO_2 at time *t* to present day (where present day is taken at pre-industrial levels of ~300 p.p.m.v.). In (b) three predictions are given according to soil temperature, since isotopic fractionation factors for reactions vary with temperature, giving slightly different results. Although the general trends are similar, it is interesting to note that there are considerable differences in the predicted atmospheric CO_2, most notably over the past 100 million years. In particular, evidence both from the fossil stomatal densites (on graph a) and fossil soils (on graph b) indicate a peak in atmospheric CO_2 at approximately 60–55 Ma. (Note that the Quaternary (the past 1.8 Ma) is not indicated on the geological timescale.)

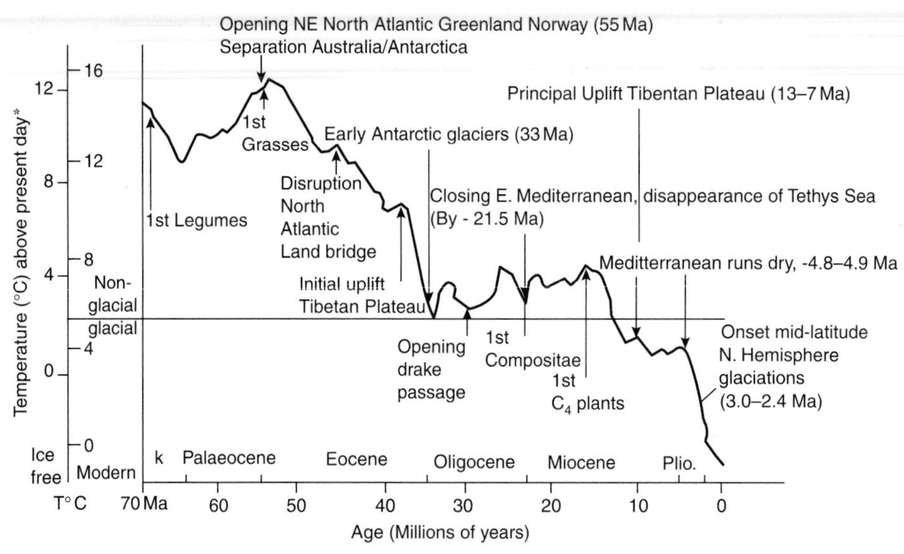

Figure 7.5 Temperature trends in the Tertiary, provided from oxygen isotope analysis of composite benthic foraminiferal records from Atlanctic DSDP sites (based on Miller *et al.*, 1987), where temperature is °C above present-day values. *Note that the temperate scale determined from oxygen isotope analysis varies for an ice-free/modern world, and that the transition between the two scales should be applied at the Eocene–Oligocene boundary (~35 Ma). Against this temperature record are indicated major geological and botanical events during the Tertiary (based on Graham, 1999).

long-term decline in global temperatures through the Tertiary (Figure 7.5). Overall declining temperatures and build-up of ice at the poles also increased equator to polar temperature gradients. Glaciological evidence suggests that major ice-sheet formation began at the South Pole from as early as 35–30 Ma. In the North Pole, ice-sheet formation was initiated somewhat later, in the late Pliocene (3–2.5 Ma) (Kennett, 1995).

Although a relatively warm period occurred during the earliest middle Miocene (~19.5–15 Ma) possibly due to volcanic-induced greenhouse warming (Graham, 1999), this episode was short lived and immediately followed by a major decline in global temperatures, between 15 and 12.5 Ma. This decline marked the major expansion and permanent growth of the East Antarctica ice sheet (Flower and Kennett, 1994), a trend toward a lower global sea level (Haq *et al.*, 1987) and the initiation of glacial climates in the Arctic.

A long-term pattern of global climatic cooling therefore emerges for the Tertiary, on which shorter-term fluctuations of both warming and cooling can be superimposed (Figure 7.5).

The Quaternary (1.8 Ma to present)

The end of the Tertiary (from ~2.75 Ma) was heralded by further decreases in global temperatures resulting from a combination of environmental factors, including a long-term decline in atmospheric CO_2 (e.g. Maslin *et al.*, 1996). This resulted in the build-up of large ice-sheets in the northern hemisphere, and from this point onward the Earth was locked into recurrent glacial–interglacial cycles. During the Quaternary (1.8 Ma to present) there have been at least 10 of these cycles, with glacial conditions accounting for almost 80% of the time (Shackleton *et al.*, 1995).

The timing of the glacial–interglacial cycles has been closely linked to variations in incoming solar radiation, resulting from variations in the Earth's orbit around the sun and the tilt and wobble of the Earth on its axis (Imbrie and Imbrie, 1979) (termed Milankovitch cycles). During the glacial periods, continental ice-caps covered large areas of land in mid to high latitudes in both hemispheres, with low global temperatures and precipitation creating harsh environments. In the low latitudes, aridity and slightly cooler temperatures prevailed. In contrast, during the interglacials conditions similar to, or warmer than, present occurred (Williams *et al.*, 1998). Concentrations of atmospheric CO_2 also became greatly reduced during the glacial periods, with estimates suggesting concentrations on average around 200 p.p.m.v., contrasting with an average value of 300 p.p.m.v. during interglacial periods (Petit *et al.*, 1999).

Viewed over the past 410 million years, however, the environmental changes that have occurred over the past 1.8 Ma (Quaternary), appear minor. No major continental plant movements have occurred, there have been no major tectonic events (comparable, for example, to the superplume episodes in the Cretaceous; see Chapter 6) (Larson, 1991a,b), and levels of atmospheric CO_2 have been at their lowest in Earth's history. The only thing of real significance, therefore, during the Quaternary are the extended periods of time when Earth has been locked in a glacial phase. The impact of these glacial phases upon long-term patterns of plant evolution will be discussed in Chapter 10.

7.2 Biogeographical distribution of global vegetation between ~60 and 50 Ma (late Palaeocene to early Eocene)

Evidence from diverse fossil fruits, seeds, leaves, pollen, and wood all indicate that the vegetation of the late Palaeocene and early Eocene (~60–50 Ma) was adapted to a warm and moist global climate, with remarkably low equator to polar temperature gradients. Forests and woodlands of angiosperm trees and shrubs, conifers, and ferns ranged from pole to pole. The vegetation of this interval was markedly more diverse, and possessed more tropical elements than that of the warm, but slightly cooler, preceding early Palaeocene (Wing *et al.*, 1992). Six biomes have been recognized for this interval, based on the original biogeographical analysis of Wolfe (1985), Meyen (1987), Tallis (1991), Janis (1993), and reviews by Collinson (1990) and Manchester (1999), in conjunction

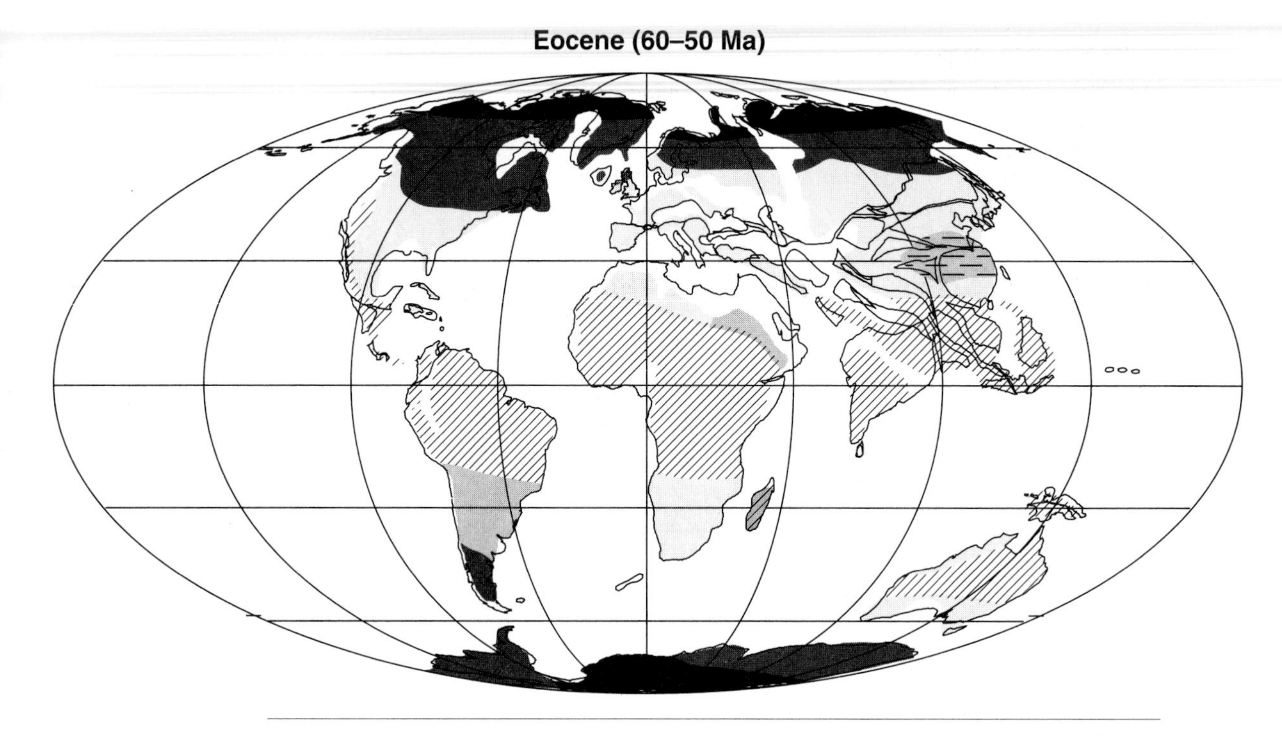

with evidence from climatically sensitive sediments (Boucot *et al.*, 2001) (Figure 7.6). Seasonality and daylength are thought to have been primarily responsible for the distribution of major vegetation types during this interval (Wolfe, 1985).

Tropical everwet biome

In terms of area covered, the tropical everwet biome of the late Palaeocene to middle Eocene provided the most extensive humid tropical climate conditions since the appearance of angiosperms (Figure 7.6), or indeed since the expansive tropical swamps of the late Carboniferous. In this vast region, spanning the majority of present-day South America, Africa, South-East Asia, and southern and western North America, a closed multistratal tropical rainforest developed. Angiosperms underwent rapid diversification and modernization in this biome, and many of the features so characteristic of tropical rainforest vegetation today emerged. These included closed multistratal canopies, the presence of tall trees, lianas, epiphytes, and shade-tolerant trees (Upchurch and Wolfe, 1987; Kellman and Tackaberry, 1997; Morley, 2000). Vegetation of the tropical everwet biome was diverse and dominated by evergreen angiosperms with large, entire-margined leaves greater than 12 cm in length (megaphyllous) often with drip-tips, indicating humid and warm climatic conditions without cold or dry seasons (Wolfe, 1985). Many angiosperm families found in present-day tropical environments, such as Tiliaceae, Elaeocarpaceae, Simaroubaceae, Sapindaceae, Araliaceae, Proteaceae, Dipterocarpaceae, and Olacaceae, were first recorded in

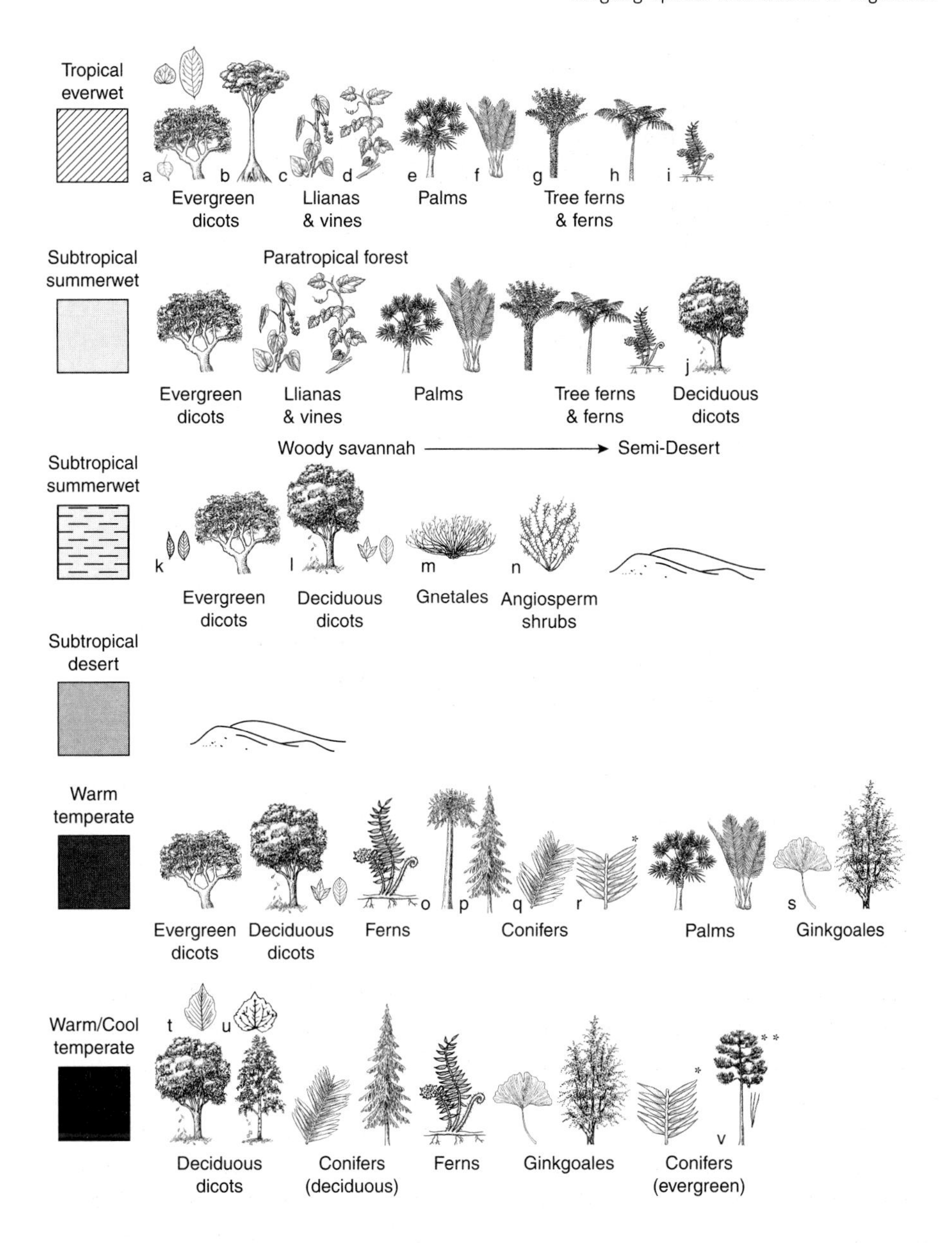

Figure 7.6 Suggested biomes for the late Palaeocene and early Eocene (60–50 Ma) (modified from Wolfe, 1985; Janis, 1993) with representatives of the most abundant and/or dominant fossil plant taxa shown. The biomes are superimposed on a global palaeogeographic reconstruction for the Eocene (50 Ma) (courtesy of C. R. Scotese, PaleoMap Project). (a) Generalized evergreen tree; (b) generalized tropical rainforest tree with buttress roots; (c) generalized Menispermaceae climber; (d) generalized vine (Vitaceae); (e) *Sabalites* palm; (f) *Nypa* palm; (g) *Dicksonia* tree fern; (h) *Cyathea* tree fern; (i) generalized fern; (j) generalized deciduous tree; (k) microphyllous leaves; (l) typical deciduous leaves; (m) *Ephedra*; (n) generalized sclerophyllous bush; (o) *Araucaria*; (p) generalized Taxodiaceae conifer; (q) *Metasequoia* leaves; (r) *Podocarpus*; (s) *Ginkgo* leaf and tree; (t) *Corylites*; (u) *Trochodendroides*; (v) generalized Pinaceae conifer. See Appendix 6 for sources of plant reconstructions and line drawings. * Dominant or abundant in the southern hemisphere; ** mainly in upland fossil floras.

this tropical zone (Lakhanpal, 1970; Morley, 2000). Palms were diverse and abundant (Mai, 2000). In contrast, conifers, including the families Araucariaceae and Podocarpaceae, and ginkgos were present but rare.

The exact low-latitude extent of this biome is difficult to gauge, as fossil floras from the equatorial latitudes are relatively rare. However, the poleward extent is marked by narrow bands of evaporites, indicating seasonally arid (summerwet) or perhaps even desert conditions, in northernmost Africa (with the exception of the coasts), central Asia, and southern South America and Africa. Nevertheless, the presence of abundant floras of tropical rainforest aspect along the western coast of North America up to 50°N clearly indicates that this biome extended much further poleward along coastal regions than it does today (Wolfe, 1985).

Summerwet biome

Paratropical forest

The vegetation spanning much of northern Europe, North America, and Russia in the northern hemisphere, and eastern Australia and parts of Argentina in the southern hemisphere, between 60 and 50 Ma was truly unique, in terms of composition and poleward extent. This vegetation, which has been referred to as paratropical rainforest (Wolfe, 1985), contained an unusual mixture of what are now either tropical or temperate elements, and was fringed along the coasts by mangrove swamps (Collinson, 2000). Angiosperm families present in this biome included members of Anacardiaceae, Anonaceae, Burseraceae, Cornaceae, Lauraceae, Sapindaceae, and Sabiaceae, all of which are almost entirely tropical to subtropical in their present distribution (Collinson, 2000). Lianas and climbers belonging to Vitaceae, Menispermaceae, and Icacinaceae were also common elements of the fossil floras making up this biome, as were palms, which were diverse.

Mangrove swamps fringed the paratropical forests along the coast lines of present-day England, France, and Belgium in the northern hemisphere, and along the coastline of Tasmania at 65°S in the southern hemisphere. These coastal swamps were dominated by the palm *Nypa*, and members of the Rhizophoraceae, including *Ceriops*, were also common (Pole and Macphail, 1996; Collinson, 2000) (*Nypa* is now restricted to mangrove swamps of South-East Asia).

The poleward extent of the paratropical forest was approximately 50° in both hemispheres and up to 65° along the coasts. The presence of mid- to high-latitude bauxites, in association with coals (Boucot *et al.*, 2001) help to define the poleward extent of the summerwet biome at this time. Such a high poleward extent of vegetation characterized by a high proportion of tropical elements clearly represents a peak in 'tropicality' for the entire Cenozoic (65–0 Ma) (Graham, 1999). Paratropical forest occurs today in slightly seasonal, humid climates of southern China and adjacent highlands of South-East Asia (Olson, 1985).

Subtropical woodlands and woody savannah

More arid areas within the summerwet biome are demarcated by the presence of evaporites (Boucot *et al.*, 2001) and a decrease in overall leaf size of the vegetation. These areas included northern Tibet (Li *et al.*, 1984), southern Australia (Greenwood, 2000), and parts of Argentina (Romero, 1986).

In southern Australia, for example, the vegetation was dominated by evergreen dicotyledon angiosperm trees with leaf sizes that were typically notophyllous to mesophyllous, similar to those of subtropical forests of this region today (Greenwood, 2000). 'Notophyllous' and 'microphyllous' refer to the leaf lengths, which were typically between 8 and 12 cm and less than 8 cm, respectively. In contrast, leaves of the tropical and paratropical rainforests were typically greater than 12 cm in length, and are referred to as 'megaphyllous'. Members of Lauraceae were dominant among the angiosperms in the more arid regions of Asia in the northern hemisphere, and these were intermixed with more xeromorphic groups such as Casuarinaceae, and groups with microphyllous leaves such as Cunoniaceae, Myrtaceae, and Elaeocarpaceae, indicating more seasonal subtropical climates (Li *et al.*, 1984). Evergreen conifers, including Araucariaceae and Cupressaceae (*Libocedrus*), were also common.

Summerwet/semi-desert biome

The absence of coals, together with the presence of evaporites and calcretes, in parts of present-day North Africa, central Asia, and parts of South America, mark the most arid regions, which were probably transitional between summerwet and semi-desert conditions during the early Eocene. In addition, fossil floras from north, west, and central China contain significant percentages (as much as 80%) of dry-adapted (xerophytic) shrubs, such as *Nitraria* and *Ephedra*, which are indicative of semi-desert or desert conditions (Li *et al.*, 1984). However, these shrubs were apparently intermixed with deciduous forest taxa, such as *Alnus*, *Betula*, *Enhelhardia*, *Juglans*, and *Liquidamber* (Li *et al.*, 1984), indicating that full desert conditions had not yet developed in these regions. Today both *Ephedra* and *Nitraria* are extremely xerophytic desert dwellers. The areas which were probably full subtropical desert are indicated in Figure 7.6.

Warm temperate biome

The dominant vegetation of this biome tended to be evergreen with entire-margined leaves but without drip-tips and has been termed 'notophyllous broad leaved evergreen forest' (Wolfe, 1985). The warm temperate biome encompassed an area including present-day Canada, southern Greenland, and much of north-eastern Asia in the northern hemisphere, and Argentina and coastal Antarctica in the southern hemisphere. Fossil evidence suggests that the floras of this biome can be classified as an 'oak–laurel' tree group, namely Fagaceae,

Lauraceae, Theaceae, and Magnoliaceae (Olson, 1985). A number of fossil plant taxa in this biome also appear to occur in both the subtropical and warm temperate biomes of the northern hemisphere. However, the warm temperate floras differ in that they tend to lack a number of the more thermophylic families, such as Icacinaceae and Menispermaceae, and generally tend to be of lower diversity (Manchester, 1999; Collinson, 2000). Members of Icacinaceae are exclusively tropical today, while those of Menispermaceae are mainly, but not exclusively, tropical to subtropical.

The conditions thought to have prevailed in the warm temperate biome were of a humid climate, with a contrast of mean monthly temperature spanning from 13 to 20 °C (Wolfe, 1985).

Warm cool temperate biome

The cool temperate biome of late Palaeocene and early Eocene times is distinguished by a unique vegetation type which has been referred to as a 'polar broad-leaved deciduous forest' (Wolfe, 1985). It is remarkable that this polar deciduous forest, which stretched poleward of approximately 70° in both hemispheres, occupied an area of the globe that today is either treeless polar desert or under ice, including northernmost Greenland, Siberia, Canada, and, in the southern hemisphere, Antarctica. These polar forests have no modern analogue. In the northern hemisphere, angiosperm leaf sizes were large but mainly deciduous, including *Alnus*, *Betula*, *Quercus*, *Juglans*, *Populus*, and *Acer*. Many of the conifers present in the northern hemisphere high latitudes were also deciduous (e.g. *Larix*, *Metasequoia*, *Pseudolarix*, and *Taxodium*, as well as *Ginkgo*). The presence of these mixed deciduous forests in the high latitudes is attributed to much higher polar temperatures and the specific light conditions (no daylight in winter but 24 hours' daylight in summer) that would have been of primary significance during growth periods (Axelrod, 1966; Meyen, 1987). In addition, the large leaf size probably represents an adaptation to low incidence of solar radiation during the summer and long growing season days, combined with high temperatures (allowing long photosynthetic periods) (Wolfe, 1985).

In the southern hemisphere, the composition of these high-latitude forests was somewhat different in that deciduous angiosperms and needle-leaved gymnosperms were rare and the predominant trees were typically evergreen. Rather, trees present included the conifers *Araucaria*, *Podocarpus*, and *Dacrydium*, and evergreen southern beeches (*Nothofagus*). Other angiosperms present, but rarer in their occurrence, included members of the families Loranthaceae, Myrtaceae, Casuarinaceae, Ericales, Liliaceae, and Cunoniaceae (Truswell, 1990).

It is suggested that the relative rarity of deciduous trees in these southern polar latitude forests was probably a result of a climate characterized by lower seasonal ranges of temperature but higher levels of precipitation (in comparison to the northern hemisphere high latitudes), which were well distributed throughout the year (Axelrod, 1966).

7.3 The evolution of grasses

During the Palaeocene and early Eocene (65–50 Ma) the appearance of grasses in the fossil record marks one of the most important evolutionary events for humankind. All grasses belong to the angiosperm family Poaceae. Presently there are more than 10 000 species of grasses on the planet, and grass-dominated ecosystems, including tropical and subtropical savannah, temperate grasslands, and steppes, cover more than 30% of Earth's land surface (Archibold, 1995). Archaeological evidence suggests that grasses (in the form of wild cereals) were some of the earliest crops to be cultivated, and it is estimated that at present they provide up to 52% of the carbohydrates in human diets worldwide (Raven *et al.*, 1992).

The first grasses

Evidence from fossil pollen and macrofossils indicate that grasses evolved sometime in the latest Cretaceous and early Tertiary, between 65 and 55 Ma (Jacobs *et al.*, 1999) (Figure 7.7). However, although Poaceae pollen has been recorded as far back as the late Cretaceous, this evidence remains ambiguous (Daghlian, 1981; Collinson *et al.*, 1993) and the earliest unequivocal macrofossil evidence is from early Eocene (~55 Ma) fossil deposits in North America (Crepet and Feldman, 1991) and southern England (Chandler, 1964; Thomasson, 1987). Palaeobotanical evidence from these deposits includes whole plants with spikelets and inflorescences. The appearance of grasses during the late Cretaceous/early Tertiary also represents the earliest fossil evidence for wind-pollinated herbaceous monocotyledons.

The subsequent global expansion of grasses and their increasing relative abundance in global floras proceeded relatively slowly throughout the Tertiary and it was not until the early to middle Miocene (~20–10 Ma) that there were widespread grass-dominated ecosystems (Janis, 1993; Webb *et al.*, 1995; Jacobs *et al.*, 1999). Fossil grass pollen, for instance, indicates that grasses accounted for less than 1% of total pollen abundance between 55 and 40 Ma (Figure 7.7) (Chandler, 1964; Litke, 1968). Grass pollen is also absent from tropical sediments before the Palaeocene (~55 Ma) (Germeraas *et al.*, 1968; Daghlian, 1981; Jacobs *et al.*, 1999). By the early Oligocene (~25 Ma), however, there is evidence from plant macrofossils and pollen, palaeosols (caliche horizons), phytoliths, and grass-like root traces (Jacobs *et al.*, 1999), and the dentition and skeletal structure of fossil vertebrates (Janis *et al.*, 2000) to indicate that grasses had

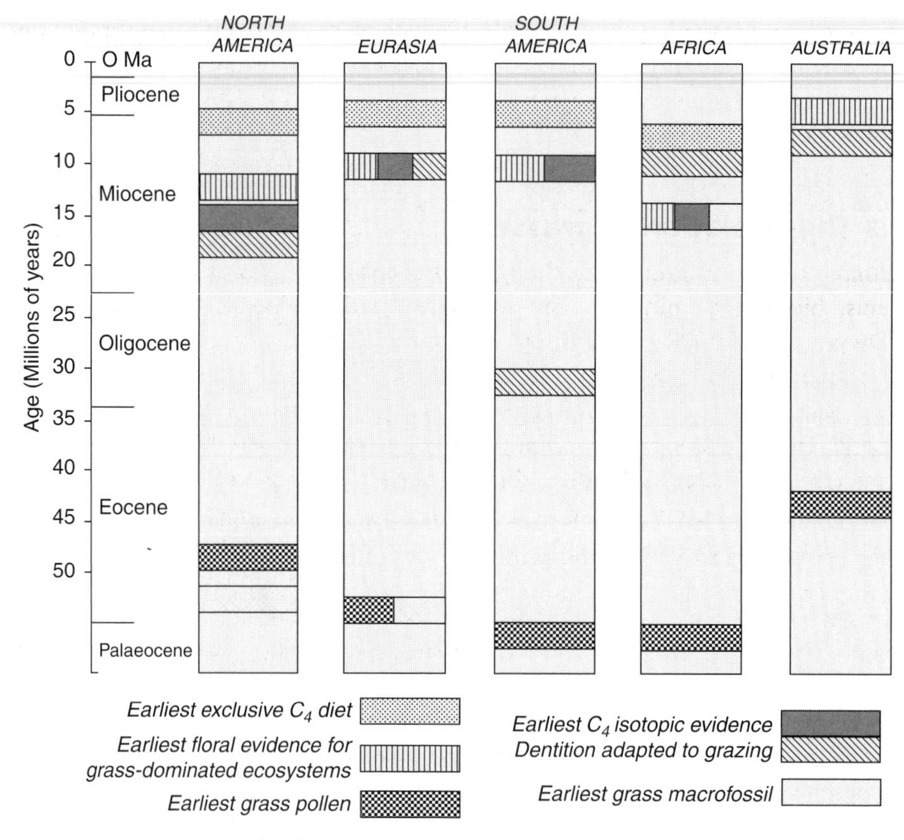

Figure 7.7 Evidence in the fossil record for the first appearance of grass pollen, macrofossils, faunal dentition adapted to grazing, and grass-dominated ecosystems. Also indicated is the earliest evidence for the emergence of C_4 plants and faunal assemblages that had a diet that was exclusively C_4 plants. (Redrawn from Jacobs *et al.*, 1999.)

started to form a significant component of global vegetation. All five subfamilies of the Poaceae had evolved by the early Miocene.

Why so late?

A number of hypotheses have been proposed for the apparently relatively late evolution of the grasses compared with other angiosperm families. These include changing climatic conditions, increased frequency of fires, and changing patterns of herbivory among global faunas.

One of the most compelling of these hypotheses suggests that increasing higher latitude aridity, especially in the continental interiors, coupled with decreasing temperatures during the Tertiary promoted the evolution and expansion of the grasses (Leopold and Denton, 1987; Wing, 1998). Many species of extant grasses are particularly well adapted to aridity, with drought-resistant features such as increased root and decreased shoot growth, a general reduction in physiological activity in periods of drought, sunken stomata, and thick dense cuticles (Archibold, 1995). The combined characteristics are suggested to have conferred a competitive advantage to grasses in conditions of increasing global aridity.

However, many of the features that enable grasses to withstand periods of drought also provide adaptations to the stresses associated with frequent fires and grazing. Both of these factors may have been equally important as global climate change in promoting the global expansion of grasses. In extant ecosystems, biomass burning is often associated with periods of increased aridity (Dwyer *et al.*, 2000), and in late Miocene pollen sequences from Africa, for example, grass pollen is often accompanied by charred Poaceae cuticle (Morley and Richard, 1993; Morley, 2000). It is highly probable, therefore, that the frequency of fires, especially in the continental interiors, would have increased in concert with increased aridity, thus favouring those plants better adapted to these frequent burning regimes. In grasses, the apical meristem (the main area from which growth occurs) is in the soil or at ground level. Therefore burning of the landscape would not damage the part of the plant responsible for growth. This is not the case for other plants. In addition, clearance of vast swathes of forest through burning would have created new niches for colonization and expansion of grasses.

Another hypothesis for the comparative late evolution of grasses proposes that a major shift in the balance of global faunas may have played an important role (Janis, 1993). It is suggested that an overall decline in numbers of reptiles (and the final decline of dinosaurs) and a major radiation of mammals during the late Cretaceous and early Tertiary (Janis, 1993) may have promoted the coevolution of Poaceae. Although the first mammals were initially generalized omnivores or herbivores, palaeozoological evidence suggests that they gradually became larger and their diets more specialized. By the middle Eocene (~55 Ma), for example, evidence for hoofed mammals (i.e. early horses, camels, and antelopes) with limbs adapted for high-speed running are well established in the fossil record (e.g. Honey *et al.*, 1998; MacFadden, 1998). Furthermore, the dentitions of these animals were clearly adapted to a diet high in cellulose and the silica-rich leaves of grasses. As these anatomical features are characteristic of species which today inhabit open grasslands and exist predominantly on a diet of grasses, a coevolutionary link between the expansion of grass-dominated ecosystems and hoofed mammals has thus been hypothesized (Stebbins, 1981). Again, however, the fact that grass leaves have intercalary growth (i.e. the top of the leaf can be eaten (or mowed!) and the leaf goes on growing at the base) was probably an important factor in promoting grasses in regions with high levels of grazing (Janis, 1984; Vaughan, 1986; Wing, 1998).

In addition to the grasses, a number of other angiosperm families which today contain members that are predominantly xerophytic shrubs and herbs adapted to arid ecosytems also made their first appearance in the late Palaeocene (~55 Ma) (Singh, 1988; Herendeen and Crane, 1995). These include Asteraceae and Agavaceae. Members of the Asteraceae (Compositae) are particularly well represented in semiarid regions of the tropics and subtropics today, such as in the tropical and subtropical summerwet biomes, but are also well represented in arctic, arctic–alpine, temperate, and montaine floras. Members of the Agavaceae, on the other hand, do not have such a cosmopolitan distribution but grow particularly well in arid and semiarid ecosystems of the tropics.

7.4 Decline of the forests and spread of aridland vegetation

At least 32 extant families which today possess one or more taxa adapted to climates and habitats of extreme or seasonal aridity evolved between the Eocene and Pliocene (~55–5 Ma) (Table 7.1). Most of these 'arid' types evolved within highly mixed communities containing both tropical and temperate species. Furthermore, a number of distinctive arid-adapted vegetation formations, which today occur exclusively in the arid climatic regions of the world, gradually became more recognizable in the fossil record from the late Eocene onwards (~47 Ma) (Singh, 1988; Leopold *et al.*, 1992; Morley, 2000). These included tropical and subtropical savannahs, temperate grasslands and steppes, deserts (including tropical and polar), tundra, and Mediterranean-type sclerophyllous vegetation.

Table 7.1 Date of appearance in the fossil record of angiosperm families now having one or more aridland members each (data from Singh, 1988)

Number of 'arid' angiosperm families to appear in interval	Age (Ma)	Geological interval	Cumulative number of 'arid' Angiosperm families
1	1.8–5.3	Pliocene	43
5	5.3–11.2	Late Miocene	42
2	16.4–23.8	Early Miocene	37
11	23.8–33.7	Oligocene	35
4	33.7–37.0	Late Eocene	24
4	41.3–49.0	Middle Eocene	20
4	49.0–54.8	Early Eocene	6
6	54.8–65.0	Palaeocene	12
4	65.0–71.3	Maastrichtian	8
2	83.5–85.8	Santonian	4
1	85.8–89.0	Coniacian	2
1	89.0–93.5	Turonian	1

Tropical and subtropical woody savannahs

Presently, the term 'savannah' is used to describe a broadly defined vegetation formation of continuous grass with trees and/or shrubs exhibiting aridland structural and functional characteristics. As such, the term savannah includes pure grasslands, parklands, and low tree and shrub savannahs, to open deciduous woodlands, thicket, and scrubs (Cole, 1986). Tropical savannahs now cover over 23 km^2 between the equatorial rainforests and the mid-latitude deserts and semi-deserts. This vegetation formation is characteristic of the modern summerwet biome (Walter, 1986), which incorporates all regions with a strongly seasonal summer rainfall regime and a dry period lasting for four to seven or eight months in the cooler season (Cole, 1986). The severity or length of the dry season in the savannah creates a graded series of vegetation types; the shorter the dry season, the greater the proportion of trees present; and the longer the dry season, the greater the proportion of grasses.

The first unequivocal evidence for the establishment of woody savannah vegetation of more modern aspect comes from middle Eocene (~40 Ma) aged fossil soils in the Great Plains of North America (Retallack, 1992). However, it has also been suggested that subtropical savannahs may have existed within the arid continental interiors from as early 50 Ma, in the early Eocene (Janis, 1993). Savannah vegetation developed relatively early in South America, Africa, India, Eurasia, and Australia, and many angiosperm trees that are presently used as indicators of savannah vegetation, such as *Eucalyptus* and *Acacia*, first became apparent in the late Eocene and Oligocene (~38–24 Ma) (Lange, 1982; Caccavari, 1996; Morley, 2000). Savannah vegetation containing a high proportion of grasses and herbs does not, however, appear until the mid-Miocene (~16–11 Ma) (Wolfe, 1985; Leopold *et al.*, 1992).

Temperate grasslands

Presently, temperate grassland occupies the continental interiors of the mid-latitudes, where seasonally dry climates favour the dominance of perennial grasses. The terminology used to describe these temperate grasslands varies considerably, including, for example, the prairies of North America, the steppes of Eurasia, the pampas of Argentina, and the grassveld of southern Africa. The climatic conditions which are associated with modern temperate grassland vegetation have been broadly classified as regions where there is recurring drought (Walter, 1986; Archibold, 1995).

Grass-dominated ecosytems did not appear in the fossil record until the mid-Miocene (~15–10 Ma) (Wolfe, 1987; Webb *et al.*, 1995; Wing, 1998; Jacobs *et al.*, 1999). Fossil evidence from a multitude of both direct and indirect sources suggests that these grassland formations occupied vast areas of the continental interiors of North America, Eurasia, and parts of Africa and south-west America. How similar these grass-dominated ecosystems were in composition to present-day temperate grasslands is still under debate, and some classify them as 'open-savannah', suggesting that true temperate grasslands did not evolve until the Pleistocene (~2 Ma) (Janis, 1993).

Mediterranean-type sclerophyllous vegetation

There are presently five regions of the world classified as having Mediterranean-type sclerophyllous vegetation. These include the Mediterranean basin, California, Chile, the western cape of South Africa, and southern Australia. All of these regions are part of the winterwet biome *sensu* Walter (1986) and are characterized by a Mediterranean climate with cool, wet winters and hot, dry summers (Groves and DiCastri, 1991). The dominant plant types are shrubs and low trees with small xerophytic, evergreen, and usually leathery leaves (Larcher, 1995). Typical Mediterranean plants of the northern hemisphere include evergreen species of *Quercus* (e.g. *Quercus ilex*, *Quercus coccifera*), *Laurus*, *Arbutus*, *Olea*, *Myrtus*, and *Ceratonia*. In the southern hemisphere, *Banksia* and members of Casuarinaceae (desert oaks and she oaks), Restionaceae, and Chenopodiaceae are common (White, 1990).

Although winterwet (Mediterranean) climatic conditions are thought to have existed in the Permian (Ziegler, 1990; Rees *et al.*, 1999), Jurassic (Rees *et al.*, 2000), and late Cretaceous (Horrell, 1991), dry Mediterranean-type sclerophyllous vegetation of truly modern aspect did not become fully established until the Miocene (~10 Ma) (White, 1988; Janis, 1993). The presence of a number of evergreen sclerophyllous taxa, including *Laurus*, *Arbutus*, *Cercis*, *Rhus*, *Cupressus*, *Juniperus*, and *Quercus* in middle Eocene fossil floras from North America led Axelrod (1975) to hypothesize that Mediterranean-type vegetation originated from a subhumid vegetation belt which occupied much of North America and Eurasia in the early Tertiary (Axelrod, 1975). Alternatively, it has also been suggested that this vegetation type originated from the older 'oak–laurel' tree group, which comprised the notophyllous evergreen forest formation of the warm temperate biome in the late Palaeocene and early Eocene (Wolfe, 1975). Although the exact origin of Mediterranean-type dry sclerophyllous vegetation is still not fully resolved, it is clear that this vegetation evolved as an adaptation to climates characterized by increased aridity and seasonality, sometime from the late Eocene (~37 Ma).

Tundra

Present-day tundra is found in polar latitudes and high mountain regions, where the climatic conditions for most of the year are extremely cold and dry with low summer temperatures (Archibold, 1995). Tundra regions are typified by treeless vegetation dominated by shrubs and herbs. In the northern hemisphere, characteristic taxa include sedges (*Carex*) cotton-grass (*Eriophorum*), various species of Ericaceae, and dwarf trees and shrubs, including species of

Salix, *Betula*, *Alnus*, and *Dryas*, such as *Betula nana* and *Salix herbacea* (Chernov, 1985). In the southern hemisphere, grasses, sedges, and mosses are most common (Archibold, 1995).

Janis (1993) speculated that tundra vegetation could have been present in the southern hemisphere as early as the middle Eocene (~40 Ma) when ice sheets were initiated on Antarctica (Janis, 1993). Similarly, in the northern hemisphere a late Miocene (~10 Ma) origin of tundra vegetation has been speculated on the basis of climatic changes associated with the initiation of arctic climates (Denton and Armstrong, 1969; Wolfe, 1985). There is, as yet, little fossil evidence to support either of these suggested origins of tundra vegetation (Truswell, 1990). The first unequivocal fossil evidence for tundra vegetation of modern aspect in the northern hemisphere is from a Plio-Pleistocene (~2–3 Ma) aged locality at 70°N (Kolyma lowland) (Wolfe, 1985 and references therein).

Deserts

Much of the land between 15° and 30° latitude is presently classified as subtropical desert, including, for example, over 40% of Australia, the Sahara desert, and the deserts of South Africa (Nambid, Kalahari, and Karoo). Polar deserts are present in high polar latitudes in both the northern and southern hemispheres. Desert climates are characterized by the scarce and variable nature of their precipitation. Although vegetation in deserts is scarce and conditioned by available moisture, they possess a remarkably diverse flora. Typical desert plants include small shrubs with narrow, scale-like and/or water-storing leaves, geophytes, succulents, prostrate cushions and subshrubs, tussock grasses, and ephemeral therophytes (Archibold, 1995; Larcher, 1995).

Many of these plants now found in desert floras were present in the fossil record from as early as the Eocene (~50 Ma) (Table 7.1), although evidence for the first desert vegetation formation of modern aspect is not apparent until 2 Ma (Wolfe, 1985). The apparent lack of evidence for desert vegetation formations is thought to be due to a variety of factors. First, by their very nature, deserts provide poor preservation potential for plant macrofossils. Secondly, the majority of desert plants are insect pollinated and therefore produce little pollen. Thirdly, desert vegetation is scarce for much of the year and therefore has less chance of being incorporated into the fossil record. Taking these factors into consideration, it is not surprising that deserts appear to have been anomalously absent for most of the Tertiary. However, detailed pollen and macrofloral analysis (Leopold *et al.*, 1992; Guo, 1993) and mapping of desert indicator sediments, such as evaporites, point to the presence of deserts and desert vegetation in the Eocene. These were, however, much restricted compared to those of the present day and Mesozoic desert belts, and will be discussed further in the following sections on Oligocene and Miocene biogeography.

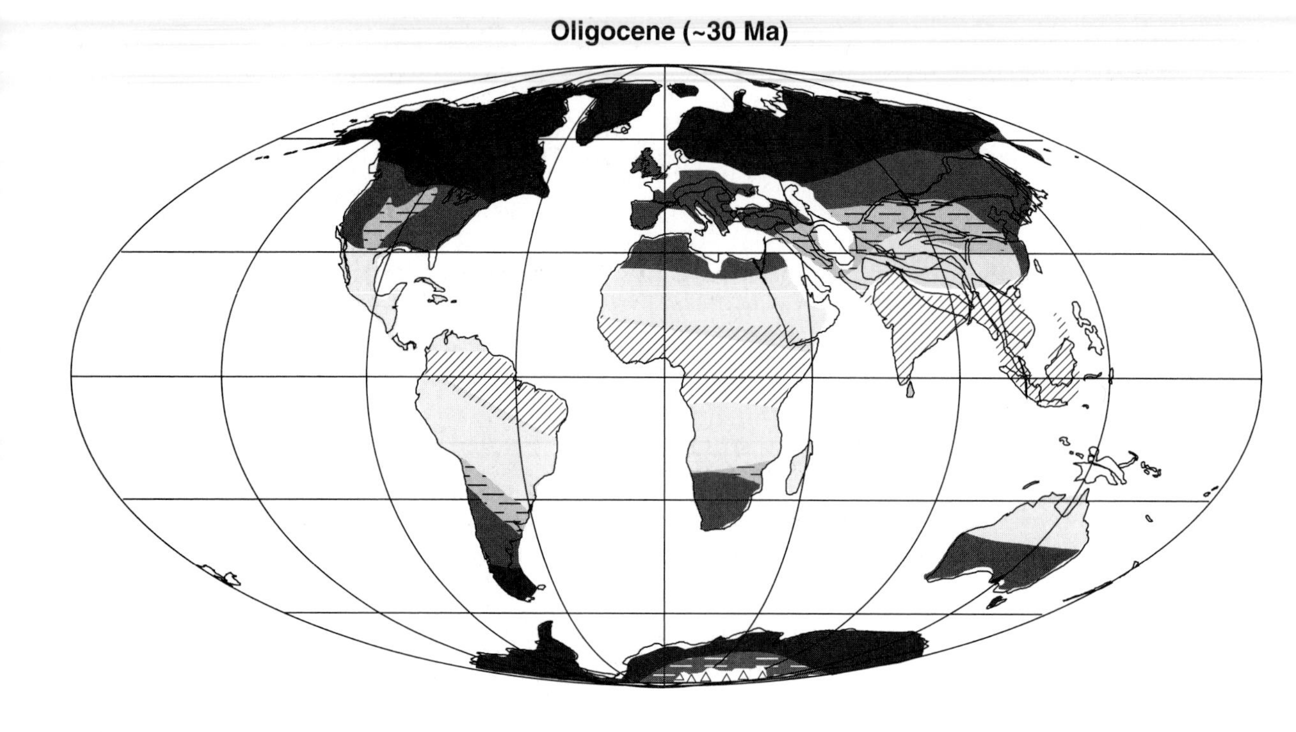

7.5 Biogeographical distribution of global vegetation between 34–25 Ma (Oligocene)

The transition from the Eocene to the Oligocene (at ~34 Ma) was a period of significant global climatic cooling and increased aridity, major ocean circulation changes, and the initiation of ice on Antarctica (see Section 7.1). Not surprisingly, therefore, major reorganization and redistribution of global vegetation mirrored these climatic trends, with evidence to suggest that the poleward extent of tropical and paratropical forests became severely restricted and the equatorward extent of temperate vegetation expanded (Figure 7.8). Seven distinctive biomes have been recognized in the Oligocene (38–24 Ma), based on the original biogeographical analyses and reviews of Wolfe (1985, 1992), Collinson (1992), Leopold *et al.* (1992), and Janis (1993), and in association with climatically sensitive sedimentary data (Boucot *et al.*, 2001).

Tropical everwet biome

There is little fossil evidence to indicate the exact composition or location of tropical forest during the Oligocene, but it is envisaged that these were similar in composition to those of the late Palaeocene and early Eocene (i.e. dominated

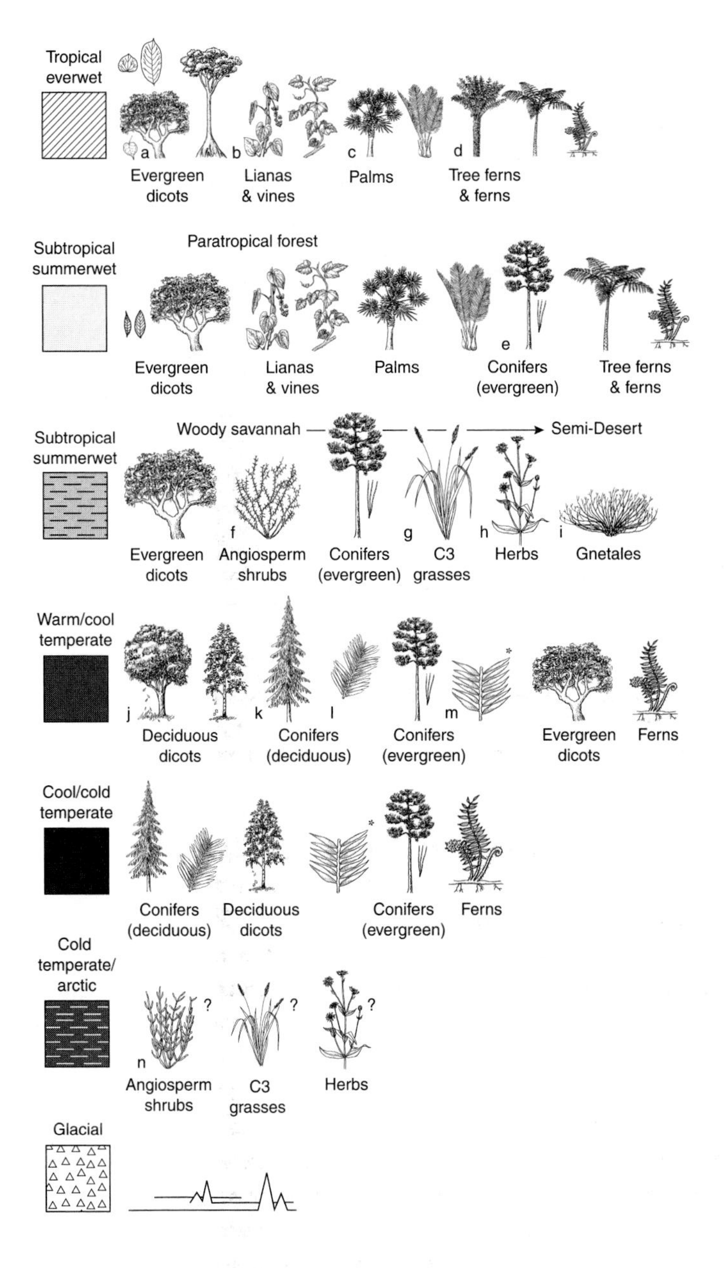

Figure 7.8 Suggested biomes for the Oligocene (34–25 Ma) (modified from Wolfe, 1985; Janis, 1993), with representatives of the most abundant and/or dominant fossil plant taxa shown. The biomes are superimposed on a global palaeogeographical reconstruction for the Oligocene (30 Ma) (Scotese, 2001). (a) Generalized evergreen trees; (b) generalized Menispermaceae and vine (Vitaceae) climbers; (c) *Sabalites* and *Nypa* palm; (d) *Dicksonia* and *Cyathea* tree ferns and generalized fern; (e) generalized Pinaceae tree; (f) generalized sclerophyllous bush; (g) generalized grass (Poaceae); (h) generalized herbs (Compositae), (i) *Ephedra*; (j) generalized deciduous trees; (k) generalized Taxodiaceae conifer; (l) *Metasequoia occidentalis*; (m) *Podocarpus* sp.; (n) salt bush (Chenopodiaceae). (?) Denotes uncertainty in the fossil record; *dominant or abundant in the southern hemisphere. See Appendix 7 for sources of line drawings.

by evergreen angiosperms with large, entire-margined leaves, often with drip-tips). The tropical everwet biome, however, was greatly reduced in latitudinal extent (Figure 7.8). The available fossil floras from the tropical latitudes provide evidence for numerous tropical angiosperm trees, including Elaeocarpaceae, Burseraceae, Sapindaceae, Euphorbiaceae, and Fabaceae. Mangrove vegetation also remained abundant along the coast, but was restricted to the tropical and subtropical latitudes, unlike the case in the late Palaeocene and early Eocene, when mangrove swamps extended high into the temperate latitudes of both hemispheres. It has been suggested that during the Oligocene, the range of tropical and paratropical forests may have been even more restricted than at present (pre-20th century) (Janis, 1993); however, this suggestion is difficult to evaluate.

Summerwet biome

The summerwet biome of the tropics and subtropics encompassed areas of North and South America, Africa, Asia, and Australia between palaeolatitudes of approximately 10–30° in both hemispheres. The presence of evaporite deposits indicating climatic aridity (Boucot *et al.*, 2001) help to define the more arid poleward limits of this biome, whereas the presence of bauxites and coals define the equatorward limit. By the Oligocene, although relatively rare, fossil pollen and macrofloral evidence suggests that this biome possessed two distinct plant formations, discussed below.

Paratropical forest

Broadleaved evergreen forest vegetation is believed to have occupied the less arid areas of the summerwet biome. These most probably included the areas immediately north and south of the everwet biome in all of the major continents. Common angiosperm families in fossil floras from Africa included Fabaceae, Anonaceae, Ebenaceae, and Sterculiaceae. It is suggested that the paratropical vegetation formation was similar in composition to the notophyllous broadleaved evergreen forest (Figure 7.6) of the warm temperate biome of the early Eocene (Wolfe, 1985). There are, however, few fossil records to confirm this suggestion. One of the few examples of this vegetation type comes from a late Oligocene fossil flora in southern Yunnan, China (Guo, 1993). The vegetation consisted of numerous and abundant evergreen trees of Fagaceae and Lauraceae with entire-margined leaves. The majority of fossil leaves were microphyllous, which is indicative of seasonal aridity, typical of monsoon climates of the summerwet biome. Although angiosperm trees and shrubs were dominant, two taxa of gymnosperms were also present: *Cephalotaxus* and *Calocedrus*.

Summerwet/semi-desert biome

This biome is demarcated by an increased abundance of evaporite deposits and was dominated by a vegetation classified as woody savannah. These regions included Mongolia, Kazakhstan, north-western China, and central North America in the northern hemisphere, and parts of South America in the southern hemisphere. Evidence from the fossil pollen and leaf floral record from the Great Plains of America indicates that many drought-adapted shrubs, such as *Celtis*, *Ephedra*, *Mahonia* and *Astronium*, became established from the middle Eocene (~40 Ma) (Leopold *et al.*, 1992). Additional evidence for increasing aridity in this region and a consequent opening up of the vegetation has been elucidated from an examination of fossil soils (palaeosols). The presence of caliche deposits and evidence for increasing distance between roots, coupled with a trend towards smaller root sizes (Retallack, 1992), in these fossil soils are thought to be indicative of a woody savannah-type vegetation.

In north-western China, arid-adapted shrubs such as *Ephedra* and *Nitraria* (which is presently found in salt deserts such as the Sahara), together with increased abundance of members of the salt bush family (Chenopodiaceae), became established. Fossil pollen records from south central Kazakhstan indicate a broadleaved woodland, composed of a mixture of subtropical angiosperm trees, shrubby taxa such as *Rhus*, needle-leafed conifers, herbs, and grasses. The vegetation from this region is therefore thought to indicate a trend towards open woody savannah (Leopold *et al.*, 1992).

Warm/cool temperate biome

A wide band of broadleaved evergreen and deciduous woodland became established throughout central Eurasia and North America and in northernmost Africa in the Oligocene. These woodlands and forests replaced the dominantly evergreen paratropical rainforest (see Section 7.2) (Figure 7.6) of the late Palaeocene and much of the Eocene (Wolfe, 1992; Meyer and Manchester, 1997; Collinson, 2000). The warm temperate biome vegetation was composed of cold temperate hardwoods, such as *Alnus*, *Betula*, *Corylus*, *Nyssa*, *Quercus*, and *Ulmus*, warm temperate trees such as *Carya*, *Liquidambar*, *Cercidiphyllum*, *Glyptostrobus*, and *Sequoia*, and subtropical taxa such as *Engelhardia* (Tallis, 1991). Changes in floristic composition between the Eocene and Oligocene in these regions were characterized by a notable decline in the abundance of plant taxa with tropical or subtropical nearest living relatives (Collinson, 1992). The percentages of entire-margined leaves also decreased considerably, and the relative percentage of non-entire-margined leaves increased. These floristic and leaf physiognomy changes reflect climatic cooling during the Oligocene.

Cool/cold temperate biome

The cool temperate biome incorporated Canada, Greenland, and most of Russia and Siberia in the northern hemisphere and encircled part of the Antarctic continent and the southern tip of South America in the southern hemisphere by the early Oligocene. The vegetation of this biome has been referred to as a mixed coniferous and deciduous woodland, similar in general composition to that of the highest latitudes in the early Eocene. The climatic cooling at the Eocene–Oligocene boundary appears to have altered the aerial extent of this biome by extending the low latitude limits further equatorward in both hemispheres. The dominant vegetation types in the northern hemisphere forests were *Metasequoia* and *Alnus* (Wolfe, 1992) and in southern hemisphere forests, *Nothofagus* and *Podocarpus* (Truswell, 1990). Polar broadleaved deciduous forest had disappeared from the high latitudes (see p. 206) by ~35 Ma; however, some of the important elements of these early Eocene forests (such as *Alnus* and *Metasequoia*) persisted in these high latitudes into the Oligocene. The abundance of Pinaceae conifers in high-latitude fossil floras of the northern hemispheres increased from the mid-Eocene (McIver and Basinger, 1994). This trend may have continued through the Oligocene; however, due to the paucity of high-latitude Oligocene-aged floras, this is difficult to evaluate.

Cold temperate or arctic biome?

It has been proposed that from approximately 38 million years ago, tundra communities became established on the land immediately surrounding the ice cap on Antarctica (Janis, 1993) (Figure 7.8). This is based on evidence from the fossil pollen record, which suggests that with increasing glaciation, the vegetation of Antarctica became less diverse and more open, with a trend towards vegetation dominated by grasses and Chenopodiaceae. As yet, however, there is little evidence to confirm a tundra or scrubland phase in the Oligocene (Truswell, 1990) and its presence therefore remains purely speculative.

7.6 The evolution of plants using the C_4 and CAM photosynthetic pathways

At present, all higher plants can be divided into three categories according to the method by which CO_2 is fixed in the photosynthetic pathway, namely C_3, C_4, or crassulacean acid metabolism (CAM) (Table 7.2). The appearance of plants possessing C_4 and CAM photosynthetic pathways represents one of the most recent evolutionary events to have occurred in the plant fossil record.

The majority of extant angiosperms, gymnosperms, and pteridophytes possess C_3 photosynthetic pathways, whereas C_4 plants are almost exclusively angiosperms. C_4 plants are most common amongst the Poaceae (especially many savannah grasses), summer annuals, and various members of the Cyperaceae, Portulacaceae, Amaranthaceae, Chenopodiaceae, and Euphorbiaceae. CAM plants include all of the cacti, most of the Asclepiadaceae, Bromeliaceae, and Orchidaceae, plus some Euphorbiaceae among the angiosperms (Ehleringer and Monson, 1993), and several species of polypodiaceous ferns and *Welwitschia mirabilis* (Gnetales; Figure 6.16) (Raven and Spicer, 1996). Experimental investigations of extant plant groups have demonstrated that the biochemical and physiological differences between C_3, C_4, and CAM plants (Table 7.2) directly influence their relative responses to differing conditions of drought, temperatures, and CO_2 concentration (Ehleringer and Monson, 1993). Conditions of high temperature and low precipitation favour plants with C_4 and CAM photosynthesis, as they show much more efficient CO_2 uptake and reduced water loss than C_3 plants. C_4 plants, however, are less photosynthetically efficient than C_3 plants in cold temperatures (below approximately 5–7 °C) (Larcher, 1995). The present-day global distribution of plants reflects these distinctions as CAM and C_4 plants tend to be found in hot and arid regions of the world, including north Africa, Asia Minor, and Australia, whereas C_3 plants have a predominantly warm temperate to arctic distribution.

A combination of anatomical, systematic, and chemical evidence has been utilized to track the fossil record of each photosynthetic group, in order to determine the timing of evolution of these three pathways. Based on these analyses, C_3 photosynthesis is known to have evolved not later than the late Silurian (~420 Ma), as the main photosynthetic pathway of the earliest terrestrial plants (Edwards and Selden, 1991; Raven, 1993). In contrast, evidence from the carbon isotopic composition of palaeosols (C_3 terrestrial plants have lower $\delta^{13}C$ values than C_4 plants) (Cerling and Quade, 1993; Cerling *et al.*, 1997) and fossil tooth enamel (which reflects the C_3/C_4 composition of the mammalian diet) MacFadden and Cerling, 1994), have indicated that plants with C_4 photosynthesis evolved in the middle Miocene (~16 Ma) and was followed by a rapid global expansion of C_4 plants in the late Miocene (~7–5 Ma) (Cerling *et al.*, 1993). The earliest unequivocal fossil evidence for CAM photosynthesis dates back to approximately 40 000 years ago (Troughton *et al.*, 1974). It is generally assumed, however, based on both physiological considerations and the antiquity of extant plants possessing CAM, such as *Welwitschia*, that this photosynthetic pathway evolved sometime in the late Cretaceous (> 130 Ma) (Ehleringer and Monson, 1993; Raven and Spicer, 1996).

The relatively late evolution of the C_4 and possibly the CAM photosynthetic pathways compared with that of C_3 has been the subject of much debate. A possible causal link has often been made between the ability of C_4 and CAM plants to thrive in hot, dry climates and the climatic evidence for increasing aridity (and temperature in the low latitudes) through the Miocene (Cerling *et al.*, 1997). However, as has been shown in Chapters 4 and 5, there have been many times in Earth history when hot and arid climate regimes prevailed over much greater expanses than evident in the Miocene. As yet, however, there is no clear evidence supporting evolution of C_4 (and possibly CAM) photosynthesis during any of these pre-Miocene intervals, with the possible

Table 7.2 Characteristics of C_3, C_4, and CAM plants

Characteristics	C_3 (e.g. oak tree)	C_4 (e.g. maize)	CAM (e.g. cactus)
Evolved (unequivocal evidence)	Mid-Silurian (400 Ma)	Mid-Miocene (16 Ma)	Quaternary (40 000 years ago)
Evolved (equivocal evidence/ speculation)		Late Carboniferous (300 Ma) and/or Mid-Cretaceous (120 Ma)	Late Cretaceous (>130 Ma)
Primary CO_2 acceptor for photosynthesis	RuBP (ribulose-1,5-biophosphate	PEP (phosphenol pyruvate)	In light RuBP, in dark PEP
First product of photosynthesis	3-Carbon acids: phosphoglycerate (PGA)	4-Carbon acids: oxaloacetate, malate, aspartate	In light PGA, in dark malate
Photosynthesis depression by O_2?	Yes, CO_2 and O_2 compete at the enzyme site (RuBP) for photosynthesis	No. CO_2 is concentrated by plant, thereby alleviating competition with O_2 for the enzyme binding site	No. CO_2 is concentrated by plant, thereby alleviating competition with O_2 for the enzyme binding site
CO_2-concentrating mechanism to avoid/reduce photorespiration	None. Rubisco, which catalyses the carboxylation of RuBP during photosynthesis also catalyses the oxygenation of RuBP in the process of photorespiration, resulting in CO_2 release	10- to 20-fold elevation of CO_2 at Rubisco binding site from breakdown of C_4 acids (produced in the mesophyll cells from carboxylation of PEP) in spatially separated bundle sheath cells	Breakdown of malate in the day to elevate CO_2 at Rubisco binding site. Malate is produced at night and stored in vacuoles

CO_2 release in light (photorespiration)	Yes	No	No
Net photosynthetic capacity at current CO_2 levels	Slight to high	High to very high	In light: slight In dark medium
Light saturation of photosynthesis	At intermediate irradiance	No saturation even at highest irradiance	At intermediate to high irradiance
Water use efficiency	Low to moderate	High	Very high
Current distribution	Dominate most terrestrial ecosystems, especially cool climates	Dominate warm to hot open ecosystems. Tropical and temperate grasslands	Xeric ecosystems (deserts, epiphytic habit)
% Composition of global flora	*c.* 85%	<5%	*c.* 10%
Limits to distribution	High light and temperatures. Low water availability	Low temperatures, very low water availability	Competetiton from C_3 and C_4 plants
Speculated future effects of elevated CO_2?	C_3 plants will have a competitive advantage, leading to expansion of their ranges	C_4 plants will be less favoured as C_4 photosynthesis is saturated at current ambient CO_2 concentration	
Speculated future effects of elevated CO_2 plus global warming?	Contraction of C_3 range during drought due to competition from C_4 plants	Increased WUE conferring competitive advantage to C_4, leading to expansion of range during drought	

exception of the late Cretaceous (Bocherens *et al.*, 1994). The lack of any evidence for C_4 and CAM plants, particularly in the Mesozoic, when global temperatures and aridity reached a zenith, suggests that other environmental or biological factors must have played a more important role. One of the main difference between earlier periods of high temperature and high aridity, and the Miocene, is that levels of atmospheric CO_2 were significantly lower during the Miocene (Figure 7.4). In competition with C_3 plants, C_4 plants are favoured under conditions of low atmospheric CO_2 when temperatures are high, due to their ability to concentrate CO_2 within the leaf, thereby reducing CO_2 loss by photorespiration (Table 7.2). It has therefore been suggested that a combination of high temperatures in the low latitudes and low levels of atmospheric CO_2 would have increasingly 'starved' C_3 plants and provided C_4 plants with a competitive advantage, leading to their rapid global expansion (Cerling *et al.*, 1997).

7.7 Biogeographical distribution of global vegetation by 11.2–5.3 Ma (late Miocene)

Following a brief period of climatic warming in part of the early and middle Miocene (~18–13 Ma), there was a steady decline in global temperatures and a continuation of the drying trend in the high latitudes (Janis, 1993) (see Section 7.1). In contrast, palaeoclimatic reconstruction suggests that temperatures increased once again in the equatorial regions, resulting in steep equator to polar thermal gradients (Figure 7.9). Major ice caps were already established on Antarctica and the initiation of ice rafting and build-up of terrestrial ice sheets was taking place in the northern polar regions (Rea and Schrader, 1985; Tallis, 1991). Continental ice caps at the poles caused sea levels to drop and decreased moisture availability of the global water budget. In consequence, continental interiors became increasingly arid, and large areas of shorelines were exposed due to falling sea level. In the northern hemisphere, for example, sea levels were so low that with the development of a land bridge between north Africa and southern Europe, the Mediterranean Sea dried out (Figure 7.10) (Hsu *et al.*, 1977; Hsu, 1983).

Eight major biomes are recognizable by the late Miocene (~10 Ma) (based on the biogeographical analyses and reviews of Wolfe (1985); Meyen (1987); Leopold *et al.* (1992); Collinson (2000)) in association with climatically sensitive sediments from Boucot *et al.* (2001). The vegetation of this period, which has been described as the 'age of the herbs' (Stanley, 1989; Briggs, 1995), clearly reflects the steep equator to polar thermal gradients that were established.

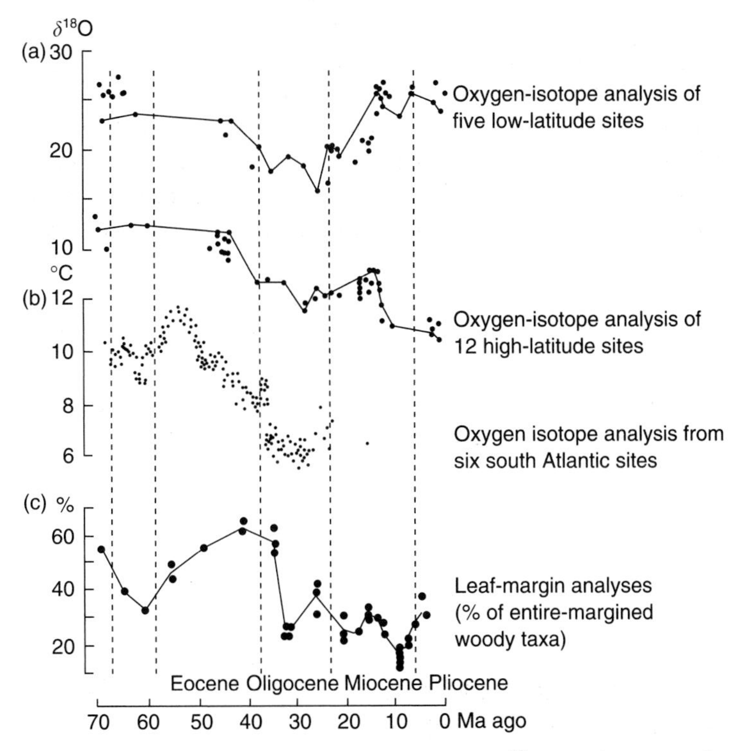

Figure 7.9 Temperature trends in the Tertiary, recorded in the δ^{18}O record in marine foraminifera from ocean cores (redrawn from Tallis, 1992, using data from (a) Savin *et al.*, 1975; (b) Shackleton, 1986). (c) Leaf margin analyses from North America – indicating a linear relationship between mean annual temperature and the proportion of woody taxa with entire-margined leaves (Axelrod and Bailey, 1969).

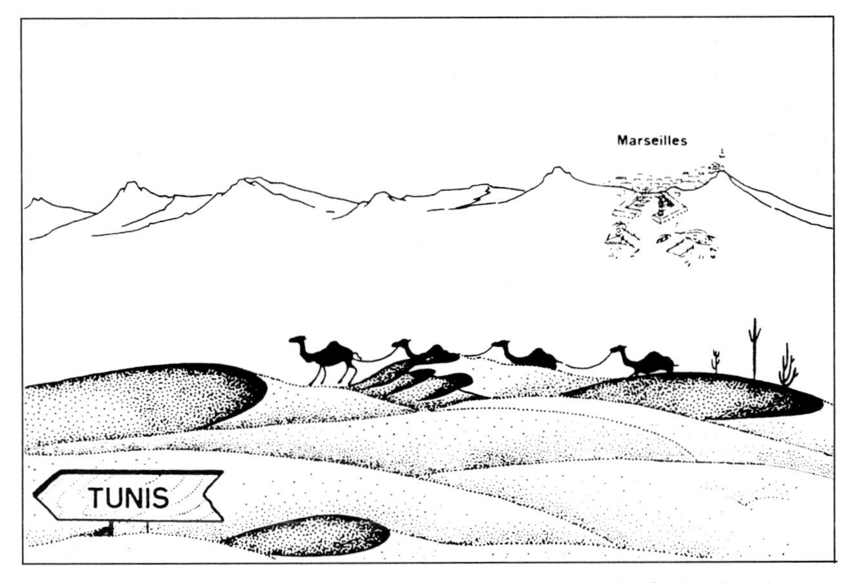

Figure 7.10 Cartoon from a North Borneo newspaper that reported the finding that the Mediterrenean Sea was a desert at approximately 5 million years ago (from Hsu, 1983).

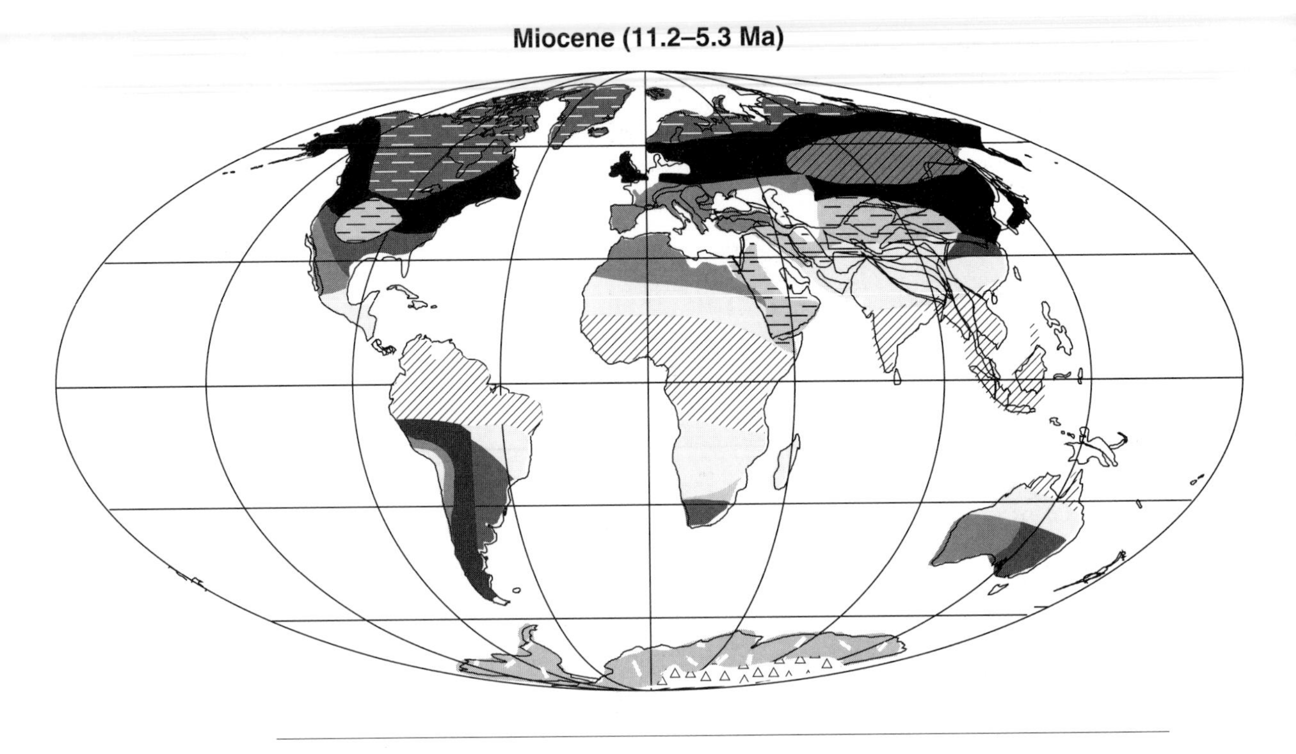

Tropical everwet biome

By 10 million years ago, a band of tropical rain forest extended across central Africa, northern South America, southern Asia, and the northern edge of Australia (Figure 7.11). Although the overall range covered today by tropical forest is much reduced (due primarily to anthropogenic activity), the latitudinal distribution that was in place by approximately 10 Ma has remained much the same to the present day. The dominant vegetation was extremely diverse, with abundant evergreen angiosperm trees, palms, lianas, vines, and some conifers, including Araucariaceae and Podocarpaceae. Nearly all of the tropical families of rainforest vegetation today were in place by the Miocene. The Dipterocarpaceae, which is one of the most important families of the Old World tropics today, appears in the fossil record of India for the first time in the Miocene (Meyen, 1987).

Summerwet biome

Paratropical forest

Tropical forest graded into paratropical forest poleward of approximately 25° in both hemispheres. Fossil floral evidence for paratropical vegetation is

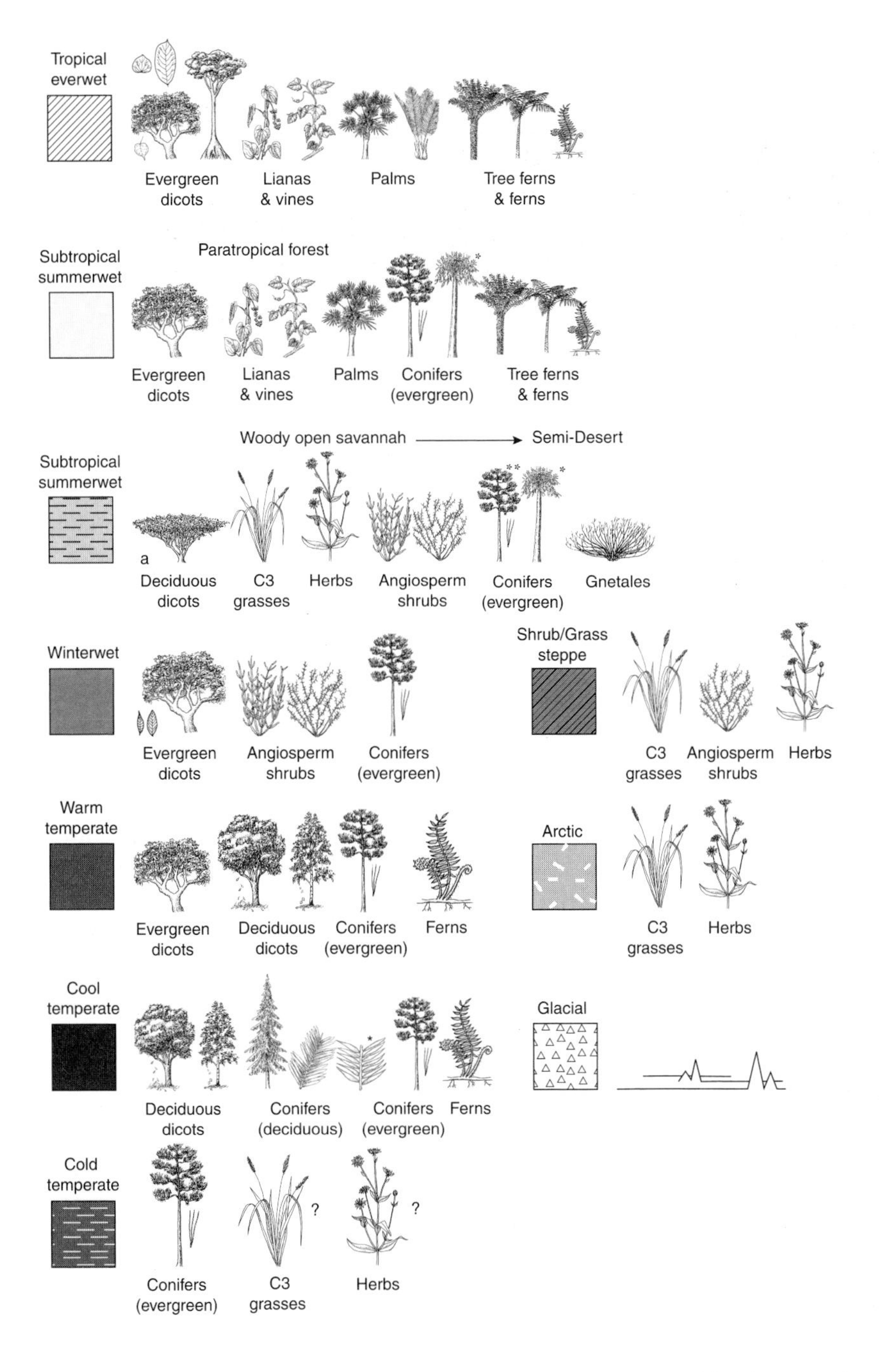

Figure 7.11 Suggested biomes for the late Miocene (11.2–5.3 Ma) (modified from Wolfe, 1985; Janis, 1993) with representatives of the most abundant and/or dominant fossil plant taxa shown. The biomes are superimposed on a global palaeogeographic reconstruction for the Miocene (10 Ma) (Scotese, 2001). (a) Acacia tree. For details on other plant drawings see Figures 7.6 and 7.8. * Dominant or abundant in the southern hemisphere. (?) Denotes uncertainty in the fossil record. **Present at higher elevations.

particularly abundant in Yunnan, China, indicating a vegetation dominated by evergreen angiosperm trees with microphyllous entire-margined leaves with some conifers (Guo, 1993). The dominant angiosperm families in this forest included Fagaceae and Lauraceae, and species of *Pinus*, *Picea*, and *Glyptostrobus* were the most abundant conifers. Guo (1993) has interpreted these floras as mixed broad-leafed evergreen and deciduous forests growing under a monsoon climate generated by the continued uplift of the Himalayas and Tibetan plateau (Ruddiman *et al.*, 1989; Raymo and Ruddiman, 1992; Partridge *et al.*, 1995) (Figure 7.11).

Summerwet/subtropical desert biome

Woody and grassy savannah

Paratropical forest graded into more impoverished deciduous forest dominated by Fagaceae in the more arid regions of the summerwet biome, such as northwest India (Meyen, 1987) and eventually into open woody savannah in areas of extremely high evaporite occurrence (Boucot *et al.*, 2001), such as Arabia and central and western Asia. In South America, for example, savannahs with a dominantly grassy ground cover became widespread (Van der Hammen, 1983). Indeed, many regions previously covered by broadleaved evergreen forests became areas of open woody savannah, with increasing proportions of grasses, by the late Miocene (~10 Ma) (Figure 7.11). It is also probable that woody savannah occupied a continuous belt from eastern Africa to western India (Wolfe, 1985) and many low latitude regions presently dry, between the equator and 35°, also developed savannah vegetation from this time, possibly heralding the beginning of desert vegetation formations of more modern aspect. The vegetation of these regions consisted of angiosperm shrubs and bushes with xerophyllous leaves and spinose margins. Among the conifers, *Pinus* and *Juniperus* increased in abundance and members of the salt bush family (Chenopodiaceae) and other herbs and grasses all show increased abundance in pollen assemblages from the Miocene (Leopold *et al.*, 1992). The most significant changes in the vegetation of the summerwet biome appears to be a further opening up of woodland vegetation and increasing abundance of grasses and herbs at the expense of the xerophytic bushes of the Oligocene.

Desert vegetation

Demarcation of a true subtropical desert biome or vegetation in the late Miocene is difficult. Similar to the Oligocene and previous stages, there is a paucity of good fossil plant evidence in the most likely arid areas, although an abundance of evaporite sediments at this time is taken to indicate extreme aridity in western Africa, Arabia and central Asia (Boucot *et al.*, 2001). Fossil pollen data from Kazakhstan and the Gobi desert region of China also show high percentages of grasses and herbs, which may indicate grassy savannah or even desert conditions. Further investigation is, however, required to establish whether these regions were part of the subtropical desert or summerwet biome, or a transitional type of biome which has no modern analogue as indicated here.

Winterwet biome

The winterwet biome, which is suggested to have been present in the late Cretaceous (~65 Ma) (Horrell, 1991), is detectable again in the Miocene (~23 Ma) after a long absence. By 10 Ma this biome extended along the western coasts of both North and South America, much of southern Europe and northern Africa, and south-western Australia, covering a more extensive area than the present-day Mediterranean region. The vegetation of this biome was typically composed of pine–oak woodlands with microphyllous shrubs and trees, including *Arbutus*, *Rhus*, and *Ceanothus*. The only region that is presently classified as having a Mediterranean-type sclerophyllous vegetation that was not part of this formation 10 million years ago is the South African Cape. It is often suggested that Mediterranean environments are a relatively recent phenomenon, strongly shaped by anthropogenic influence (Naveh and Vernet, 1991; Groves and Rackham, 2001). Evidence from the palaeobotanical record and climatically sensitive sediments suggests, however, that winterwet climates were probably in place by the late Cretaceous (Horrell, 1991), and potentially the Permian (Rees *et al.*, 1999).

An important factor in shaping the formation of Mediterranean-type sclerophyll of a more modern aspect in north Africa and southern Europe is thought to have been the desiccation of the Mediterranean Sea. It is estimated that this would have created cooler and drier local climates and therefore the formation of sclerophyll rather than paratropical forest in north Africa (Hsu *et al.*, 1977; Wright and Cita, 1979; Janis, 1993).

Warm temperate biome

During the Miocene, the warm temperate biome consisted of lowland *Taxodium* swamps and deciduous forests with *Acer*, *Populus*, *Salix*, and *Quercus*, to name a few. Broadleaved evergreen components of the vegetation were not as abundant and the whole biome was very much restricted in extent compared with that of the late Palaeocene and early Eocene. For instance, fossil floras from regions of Japan, which today possess notophyllous broadleaved evergreen vegetation, indicate the presence of broadleaved deciduous forests (Tanai, 1961, 1972) in the early Miocene. The evidence from these fossil leaves therefore suggests that the maximum extent of the warm temperate biome in Asia was south of Japan in the Miocene (Wolfe, 1985). The warm temperate biome incorporated parts of western North and South America, the southern cape of Africa, south-western Australia, and parts of western China. In North America there are no fossil floras from the south-eastern states that conform to this biome-type; however, Graham (1999) suggests that it is most likely that the vegetation of this region underwent rapid modernization during middle and late Miocene cooling. As a result of cooling, exotic Asian and Neotropical elements were ultimately lost from south-eastern USA, and today warm temperate vegetation is restricted to parts of Florida.

Cool temperate biome

The range of broadleaved deciduous woodland had changed significantly by the late Miocene. The equatorward limits of this biome were extended from approximately 45° in the Oligocene to 35° in the lowlands and 20° in the highlands by the late Miocene. The poleward extent of this biome, on the other hand, became restricted from above 80° in both hemispheres to below 70°, particularly in the continental interiors and in the southern hemisphere, probably reflecting cooler global temperatures and increased seasonality. Furthermore, by the late Miocene (10 Ma) Antarctica had become deforested and tundra vegetation developed in its place. The composition of vegetation in the cool temperate biome consisted of a mosaic of different plant communities, mainly dominated by diverse angiosperm trees such as *Alnus*, *Acer*, *Ulmus*, *Quercus*, *Betula*, *Platanus*, and *Tilia*, to name a few, with mixed conifers, including both deciduous and evergreen Pinaceae taxa. *Taxodium* swamps, which were present in the Oligocene, also persisted in both the cool and the warm temperate biomes.

Shrub/grass steppe. Cold temperate biome?

As the arid continental interiors of Eurasia and North America became progressively cooler and more seasonal, the vegetation characteristic of these regions in the Oligocene (34–24 Ma) became increasingly open. By the late Miocene (10 Ma) the composition of vegetation of these regions became closer to present-day grasslands, prairies, and steppes (Figure 7.11). Steppes which developed in north-west China were probably related to the orographic effect of the rising Himalayas (Wolfe, 1985). New groups such as the aster family (Asteraceae) begin to appear consistently in the Miocene sediments of China (Leopold *et al.*, 1992), indicating an increasing importance of herbs. Similarly, the continued uplift of the Rocky Mountains through the Miocene prevented moisture-laden winds from reaching the continental interior of North America. The resultant vegetation in the late Miocene appears to have been a mosaic of open deciduous forest woodland and widespread grasslands, with prairie-like assemblages including Juniper and *Celtis* (Graham, 1999). More xeric near-desert type formations containing *Ephedra* and *Artemisia* are also evident, for example in western Wyoming (Barnosky, 1984). However, grasslands of modern aspect did not appear in North America until ~1.6 Ma and later in the Quaternary (Wolfe, 1985; Graham, 1999).

Cold temperate biome

The appearance of taiga-like vegetation in the late Miocene heralded another vegetation formation of modern aspect but also indicated further climatic cooling, particularly at the poles, since the Oligocene. This cold temperate biome vegetation spanned the northern hemisphere continents from palaeolatitudes of approximately 50° in the continental interiors to the pole. Presently, taiga is located in the high latitudes of the northern hemisphere (the northern part of the boreal forest) but does not extend all the way to the pole, suggesting that, although Miocene polar climates had cooled significantly since the Oligocene, they remained warmer than those of the present day. Modern taiga vegetation is dominated by coniferous trees forming an open-crown lichen woodland (Sirois, 1992) and exists in areas that have low (< 0 °C) annual temperatures but summer temperatures and humidity that provide sufficient warmth and moisture for tree growth. Fossil evidence from the Canadian arctic (75–80°N) indicates that by 10 million years ago vegetation dominated by narrow-leaved coniferous trees, including *Pinus*, *Picea*, *Tsuga*, and some broadleaved deciduous angiosperm trees, such as *Juglans*, had developed (Hills *et al.*, 1974). This combination of predominantly coniferous forest also containing a few deciduous species, such as *Juglans*, *Betula*, *Alnus*, and *Acer*, is found in many regions of the present-day global boreal forest and classified as taiga (Nikolov and Helmisaari, 1992). Fossil evidence also suggests that taiga developed in the coastal area of central Siberia from approximately the same time, although it is estimated that it was not until approximately 2 million years ago that taiga attained its present aerial extent (Wolfe, 1985).

Arctic biome

It is speculated that all regions of the Antarctic continent which remained free of ice in the late Miocene, were covered by tundra vegetation (Singh, 1988). As was the case for the Oligocene (38–24 Ma) unequivocal fossil evidence for tundra is not apparent until 3Ma. However, the fossil pollen record does suggest a trend towards depauperization of high-latitude Antarctic floras and deforestation from approximately 10 Ma (Truswell, 1990).

Thus by the late Miocene (~10 Ma) not only had all major groups of plants evolved, but also most of the major vegetation formations that are presently recognizable (e.g. savannah, tropical rainforest, tundra, taiga, etc.) had also become established. Plant formations continued to modernize through the following 10 million years, but the broadscale biogeographical distribution of vegetation from this point in geological time was more a case of 'rearrangement' of composition and distribution of existing vegetation formations, rather than the establishment of entirely new groups. This rearrangement of the biomes has been particularly pronounced over the past 1.8 Ma (Quaternary) in response to the climatic/environmental changes associated with the glacial–interglacial cycles.

Summary

1. The climates of the Tertiary (65–1.8 Ma) can be broadly defined into two periods. The first, spanning the early Palaeocene to middle Eocene (65–45 Ma), was one of the warmest periods in Earth history. Deep ocean temperatures were between 9 and 12 °C higher than present, and sea surface temperatures in the Antarctic exceeded 15–17 °C. In contrast, during the second period, from the middle Eocene to the end of Pliocene (45–1.8 Ma), global climates became increasingly cool and arid, and the temperature differential between the poles and equator became much more pronounced.

2. Major ice-sheet formation began in the South Pole from as early as 50 Ma and the North Pole during the late Pliocene (3 Ma).

3. Analysis of the biogeographical global distribution of vegetation between 60 and 50 Ma indicates that forests and woodlands of angiosperm trees and shrubs and conifers ranged from pole to pole. Five biomes are clearly recognized at this time and possibly six.

4. The evolution of grasses occurred around 60 Ma, with a slow but substantial expansion, such that by the early to middle Miocene (~20–10 Ma) there were widespread grass-dominated ecosystems.

5. There are a number of hypotheses accounting for the appearance of grasses in the fossil record at this time, including that their evolution was closely linked with increasing aridity in the higher latitudes, increasing wildfires, and coevolution with hoofed mammals.

6. At least 32 families which today possess one or more taxa adapted to climates of extreme and seasonal aridity evolved between the Eocene and Pliocene (~55–5 Ma).

7. Increased cooling and aridity from the Eocene onwards (34 Ma) led to the restriction of the tropical and paratropical forests, and the equatorward extent of temperate vegetation expanded. Biogeographical analysis of the vegetation for this period indicates that seven distinctive biomes can be recognized in the Oligocene (34–24 Ma).

8. Plants with C_4 photosynthetic pathways evolved in the middle Miocene. A combination of high aridity and low CO_2 are thought to have been the main contributing factors accounting for the evolution of this additional photosynthetic pathway.

9. By the late Miocene ($\sim$10 Ma) steep pole to equator gradients had formed. Ice sheets covered the South Pole and were starting to build up in the high latitudes of the northern hemisphere.

10. Ten major biomes are recognized by the late Miocene ($\sim$10 Ma), clearly reflecting the steep equator to pole gradients. These biomes are still recognizable today.

8 Mass extinctions and persistent populations

Mass extinctions in the faunal fossil record are a well-versed and much debated subject. Unfortunately, the same is not true of the plant fossil record, and relatively little consideration or emphasis has been given to the question of whether plants have undergone mass extinctions and, if so, whether these were coincident with faunal extinction events and/or share the same triggering mechanisms. Similarly, the persistence of certain groups of plants in the fossil record is rarely considered, yet there are just as many 'living fossil' plants as there are animals, with plenty of examples showing little, if any, morphological change over the past 200 million years. This chapter describes briefly the 'big five' mass extinction events viewed in the marine faunal record. The nature and magnitude of these events are then compared to changes apparent in the plant fossil record at the same time interval. The apparent lack of evidence for mass extinction in the plant fossil record is then discussed with particular reference to the physiological characteristics that might enable plants to be more resilient to the types of environmental change associated with mass extinction events. Finally, examples of the long-term persistence of various families in the plant fossil record are described, and the implications of this in terms of understanding the driving mechanisms behind plant evolution are reviewed.

8.1 Definition of mass extinction

Mass extinction is a term that has been traditionally used in geology and palaeontology to describe relatively short intervals of geological time when a high proportion of diverse and geographically widespread taxa underwent extinction. Mass extinction differs from normal 'background' species-level extinction in that it is characterized by the relatively rapid extinction of groups of organisms, usually at a higher taxonomic level, that is families and even orders. Thus each mass extinction is a real catastrophe brought about by extraordinary environmental factors. It involves substantial biodiversity losses that are global in extent, taxonomically broad, and rapid relative to the average duration of the taxa involved (Jablonski, 1986a, b, 1995; Sepkoski, 1986).

There are a myriad of viewpoints as to how and why mass extinctions occur. The underlying causes can, however, be divided into those that are Earth bound (physical and biological) and those that are extraterrestrial (Chaloner and Hallam, 1994; Hallam, 1994; Huggett, 1997). Earth-bound causes (Table 8.1) include the environmental impacts associated with global cooling and glaciation (Stanley, 1988), massive volcanism (Loper *et al.*, 1988; Rampino and Stothers, 1988; Courtillot, 1990; McElwain *et al.*, 1999), a reduction of salinity and/or oxygen in the oceans (Stevens, 1977; Kennett and Stott, 1995; Wignall and Twitchett, 1996), and changing sea levels (Hallam, 1989). In comparison, extraterrestrial causes include single large impacts from meteorites, comet storms, radiation from supernovae, and large solar flares (Alvarez *et al.*, 1980) (Table 8.1). Suggested environmental changes associated with these extraterrestrial impacts include shock-waves, heat-waves, impact winters (long periods of darkness shutting down photosynthesis), super-acid rain, toxic oceans, superwaves, and superfloods (Schindewolf, 1963; Terry and Tucker, 1968; Reid *et al.*, 1978; Huggett, 1997).

8.2 Evidence in the geological record: plants versus animals

Five mass extinction events have been recognized in the marine faunal record over the past 600 million years. These are often referred to as 'the big five' (Figure 8.1). The first occurred approximately 443 million years ago at the end of the Ordovician, when 12% of marine families (Raup and Sepkoski, 1982), encompassing up to 96% of marine species, became extinct (Brenchley, 1990). The second mass extinction event occurred approximately 364 million years ago at the Frasnian/Famennian boundary in the Devonian. This event is marked by a 14% reduction in marine family diversity, 60% reduction in marine deep-water species, and 95% reduction in shallow-water species (McGhee, 1990, 1996). The next mass extinction event, at the Permian–Triassic boundary (248 Ma) resulted in the extinction of up to 90% of all durable skeletonized marine invertebrates and 54% of all marine families (Erwin, 1990, 1993). The fourth mass extinction occurred at approximately 206 million years ago at the Triassic–Jurassic boundary. This boundary is characterized by the virtual disappearance of ammonoids, a loss of over 20% (approximately 300 families) of marine invertebrates and vertebrates, and major extinctions among insects in the terrestrial realm (Benton, 1990). Finally, the fifth, and perhaps most famous, mass extinction event occurred at the Cretaceous–Tertiary boundary (~65 Ma). An 80% reduction in marine invertebrates (Surlyk, 1990), total extinction of dinosaurs (with the exception of birds) and drastic reduction in the number of mammal species (Halstead, 1990) has been recorded at the Cretaceous–Tertiary boundary.

Table 8.1 Physical and biological causes of mass extinctions (adapted from Huggett, 1997)

Ultimate cause	Proximate cause	Effects	Examples of mass extinctions
Extraterrestrial			
Single large impacts	Shock-waves, heat-waves, wildfires, impact winters (shutting down photosynthesis), super-acid rain, toxic oceans, superwaves and superfloods (from an oceanic impact)	A grand global dying	Cretaceous–Tertiary event (Alvarez *et al.*, 1980; Smit, 1994)
Comet storms	Shock-waves, heat-waves, wildfires, impact winters, super-acid rain, toxic oceans, superwaves, and superfloods	Stepwise extinction events	Cenomanian–Turonian, Cretaceous–Tertiary, Eocene–Oligocene (Donovan, 1987)
Radiation from supernovae	Direct exposure to cosmic rays and X-rays. Ozone destruction and exposure to excessive amounts of ultraviolet solar radiation	Sterilizes and kills organisms, causes mutations. Selective mass extinctions (exposed animals, including shallow-water aquatic forms, but not plant life)	Possibly any event (Schindewolf, 1963; Terry and Tucker, 1968)
Large solar flares	Exposure to large doses of ultraviolet radiation, X-rays, and photons. Ozone depletion	Mass extinctions	Events during magnetic reversals (Reid *et al.*, 1978). Sporadic faunal extinctions (Hauglustaine and Gérard, 1990)
Terrestrial–physical			
Geomagnetic reversals	Increased flux of cosmic rays	Mass extinctions	Late Ordovician, late Permian, late Devonian, and late Cretaceous events (Whyte, 1977)

Continental drift	Climatic change: glaciations when continents encroach upon the Poles. Aridity increase on moving into low latitudes	Global cooling. Extinctions because species find themselves in inhospitable climatic zones	Late Ordovician, late Devonian, and late Permian marine events associated with encroachment of land masses on Poles (Stanley, 1988a, b)
Volcanism	Cold conditions, acid rain, and reduced alkalinity of oceans resulting from release of sulphate volatiles. Toxic trace elements. Climatic change from release of ash and carbon dioxide	Stepwise mass extinctions	Terminal Cretaceous flood—basalt eruptions (McLean, 1981, 1985; Officer *et al.*, 1987) Triassic–Jurassic mass extinction event
Sea-level change	Loss of habitat	Mass extinctions of susceptible species (e.g. marine reptiles)	Cretaceous–Tertiary event (Bardet, 1994)
Salinity changes	Reduced salinity	Mass extinction in marine realm	Permian extinctions (Fischer, 1964; Stevens, 1977) Triassic extinctions (Holser, 1977, 1984)
Anoxia	Shortage of oxygen	Mass extinctions in oceans	Frasnian–Famennian event (Geldsetzer *et al.*, 1987; Buggisch, 1991). Late Palaeocene (Kennett and Stott 1995)
Earth-bound biological			
Spread of disease and predators	Direct effects (made possible by changes in geography)	Mass extinctions	Late Cretaceous mass extinction (Bakker, 1986)
Evolution of new plant types	Changed biogeochemical cycles reducing the ocean nutrient supply	Gradual extinctions of marine biota	Late Permian mass extinction (Tappan, 1982, 1986)

Table 8.2 Magnitude of the major mass extinctions in the Phanerozoic (from Erwin, 1993; data from Sepkoski, 1989, 1990)

Mass extinction	% Family extinction	% Generic extinction
End-Cretaceous	17	50
End-Triassic	23	48
End-Permian	57	83
Late Devonian	19	50
End-Ordovician	27	57

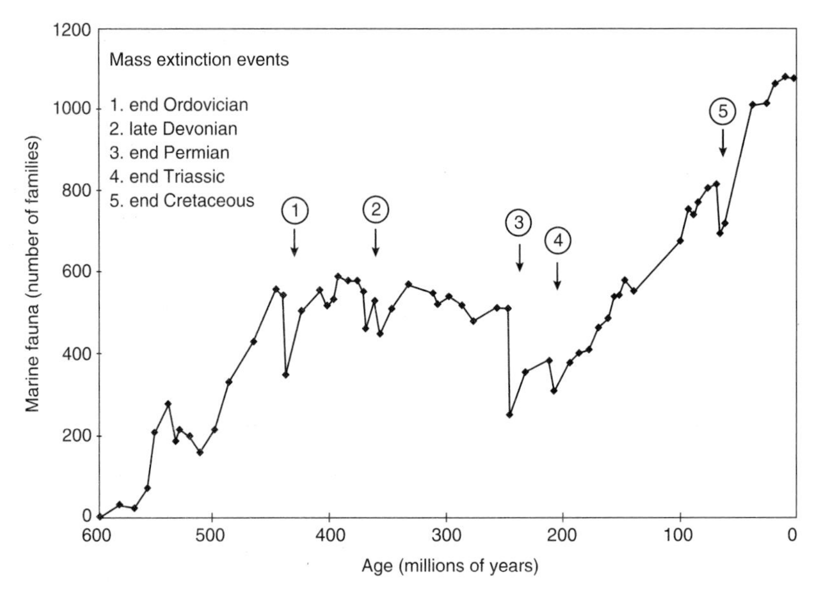

Figure 8.1 Five big mass extinctions in the fossil marine record (after Raup and Sepkoski, 1982).

Although the pattern and timing of mass extinction events in Earth history are most sharply defined by family-level diversity losses in the marine fossil record, similar trends are also observed, to a certain extent, among terrestrial faunas for at least three of the 'big five' mass extinctions. These include the Permian–Triassic (reviewed by Erwin, 1993), Triassic–Jurassic (Benton, 1986), and Cretaceous–Tertiary (Archibald and Bryant, 1990) boundaries (Figure 8.2). The trends in plant extinctions, and overall diversity changes, on the other hand, are more ambiguous. There are no major peaks of extinction in the plant fossil record comparable to those of the faunal mass extinction events (Niklas *et al.*, 1983; Knoll, 1984; Niklas, 1997), with the exception, perhaps, of the Permian–Triassic boundary (Figure 8.3).

After examining briefly the main trends in the faunal fossil record, plant responses across all four mass extinction events of the Phanerozoic (that is those

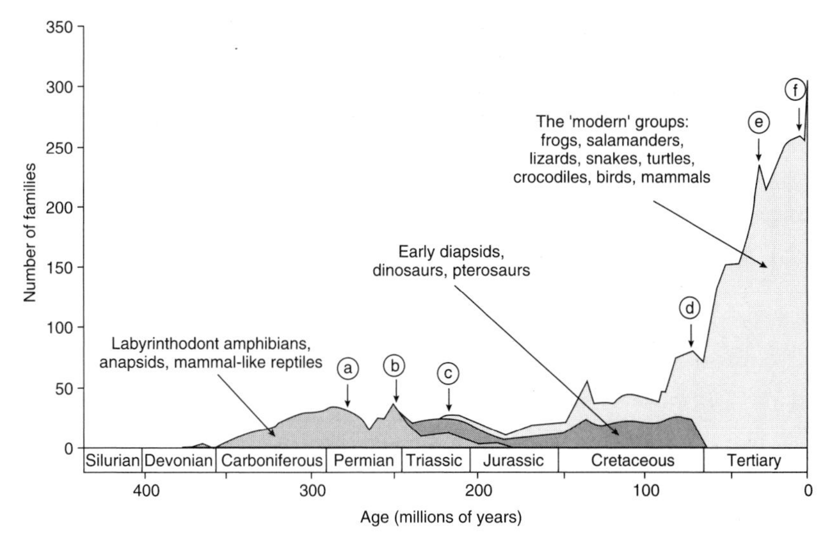

Figure 8.2 Mass extinction events in the fossil terrestrial record. Diversity curve for terrestrial tetrapod families (after Benton, 1989, redrawn from Huggett, 1997). Six mass extinction events are recognized (letters a–f).

which occurred over the past 400 million years) will be discussed in detail in the following sections.

Frasnian–Famennian boundary (late Devonian) (364 Ma)

Over half of the marine genera present on Earth, accounting for more than a fifth of all families, did not survive the Frasnian–Famennian mass extinction (McGhee, 1996). This event, which is often referred to as the late Devonian mass extinction, lasted for an estimated 3 million years, and resulted in the decimation of most major marine taxa, including cnidarians, stromatoporoids, brachiopods, foraminiferids, phyllocarids, echinoderms, benthic algae, cephalopods, conodonts, fishes, phytoplankton, and zooplankton. A number of causes have been suggested, including an asteroid or multiple asteroid impacts (McLaren, 1983), oceanic anoxia (Buggisch, 1972, 1991), global climatic cooling (Copper, 1986), and even global climatic warming (Brand, 1989).

There is much debate as to whether similar levels of extinction occurred among terrestrial plants at the Frasnian-Famennian boundary. The late Devonian was a time of marked speciation and radiation of many different plant groups, including lycopsids, sphenopsids, filicopsids, and progymnosperms, and is characterized by the establishment of the first extensive forests (Scheckler, 1986) (see Chapter 4). A number of studies have demonstrated that this trend of increased diversity and radiation through the Devonian period (417–434 Ma) was punctuated by a number of marked plant extinction events. The suggestion has therefore been made that these extinction events were coeval with those

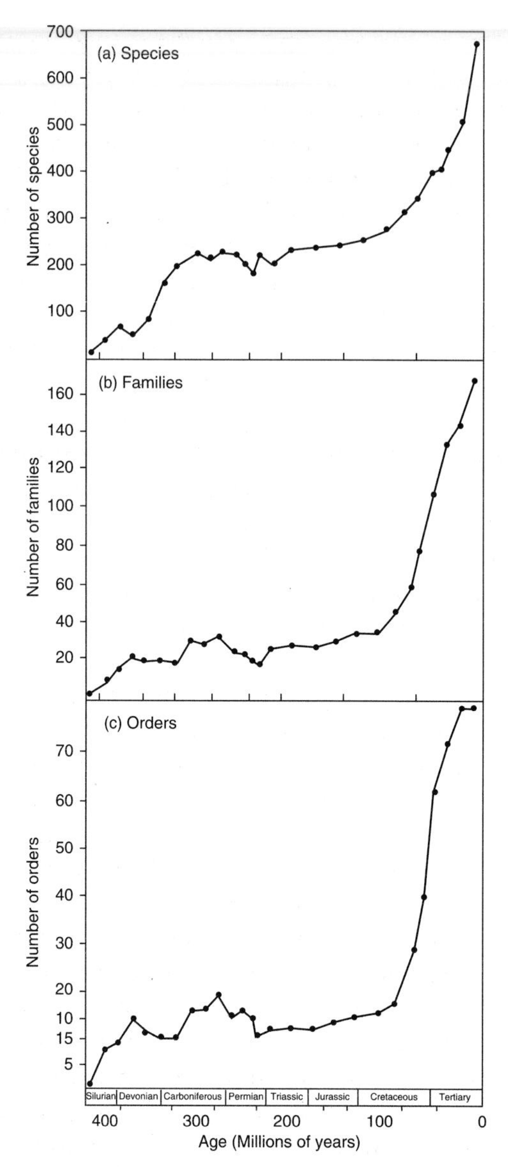

Figure 8.3 Taxonomic richness of vascular plants during the past 420 million years. (a) Species richness (data from Niklas *et al.*, 1983); (b) family richness (from Knoll, 1984); (c) ordinal richness (from Knoll, 1984). There appear to be no major episodes of mass extinction comparable to the faunal record at any level (species, family, order), with perhaps the exception of the Permian–Triassic boundary.

occurring at the Frasnian–Famennian boundary amongst diverse marine faunal groups (Banks, 1980; Richardson and McGregor, 1986; Boulter *et al.*, 1988). Evidence for extinction in late Devonian fossil plant assemblages, for example, indicates that a reduction in overall species number occurred at the Givetian–Frasnian boundary (~370 Ma) (Figure 8.4) (Banks, 1980). Similarly, a decrease in overall plant species number at the Frasnian–Famennian boundary

(~367 Ma) was observed in a sequence from New York State (Richardson and McGregor, 1986). Using the evidence from these studies, the suggestion was therefore made that a mass extinction of a comparable nature to that seen in the marine faunal record occurred among terrestrial plants during the late Devonian. However, this interpretation has been criticized on a number of counts (Traverse, 1988; McGhee, 1995). First, it was suggested that it was impossible to draw this conclusion from two studies, both with very small sample sizes. Secondly, it was suggested that a biased selection of taxa was chosen in the analyses and that this gave a false impression of the scale of the decline (Figure 8.3). Thirdly, it was argued that since these two episodes of extinction apparent in the plant fossil record were asynchronous (approximately 3 million years apart), they did not suggest a single large event. Fourth, and perhaps the most significant, was the observation that neither macrofossil nor microfossil records from these two sequences demonstrated extinction among high taxonomic levels, such as families or orders, thus in the strict sense of the definition, neither represents 'mass extinction'.

A more recent study (Raymond and Metz, 1995) has sought to address a number of the limitations associated with previous studies through the detailed analyses of 338 assemblages of compression/impression fossils compiled from the literature, including assemblages from North America, Europe, the Russian platform, Spitzbergen, and Bear Island. Of the five different measures of diversity that were examined, all demonstrated declines in diversity of fossil plants at the Frasnian-Famennian boundary (Figure 8.4). It remains somewhat questionable, however, whether these results demonstrate 'mass' extinction (Willis and Bennett, 1995) since a maximum reduction in generic diversity of between 30 and 40% is recorded, which is much smaller than the reduction observed in the faunal record. In addition, Raymond and Metz (1995) suggest that low diversity at this boundary may reflect sampling bias or may be the result of low origination rates during this interval rather than high extinction rates. Despite these potential biases in the fossil plant diversity record, it seems apparent from the multiple measures employed in this detailed study that they do indeed identify a 'plant diversity minimum' during Frasnian–Famennian time. What they do not support, however, is a plant mass extinction event, i.e. that there was widespread extinction among many and diverse higher taxa of terrestrial plants.

End-Permian (248 Ma)

The end-Permian marks an interval in Earth history when between 90 and 96% of all durable skeletonized marine invertebrate species (Raup, 1979) and 54% of all marine families (Sepkoski, 1986) became extinct. As such, this mass extinction event was the greatest in the history of life, and has been referred to as 'the mother of mass extinctions' (Erwin, 1993), 'Palaeozoic nemesis' (Hallam and Wignall, 1997), and other equally dramatic terms. As in the case of the

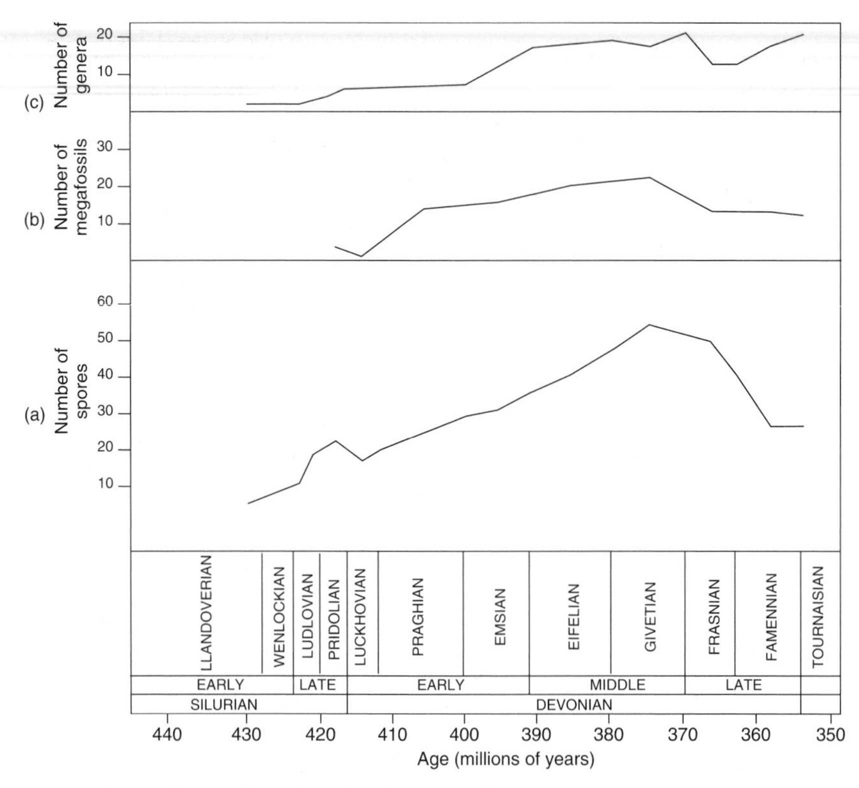

Figure 8.4 Three palaeoecological studies thought to demonstrate mass extinction: (a) total species curve based on 171 fossil spore taxa (Richardson and McGregor, 1986); (b) total species curve based on 64 megafossil taxa (Banks, 1980) (both redrawn from Boulter *et al.*, 1988); and (c) standing diversity at interval boundaries based on 338 fossil assemblages (Raymond and Metz, 1995).

Frasnian–Famennian extinctions, many causal mechanisms have been suggested for those of the end-Permian, and most, if not all of them, are associated in some way with changing continental configurations, tectonic instability (see Chapter 6), and/or global climatic change. More specifically, suggested causal mechanisms include an intense period of volcanism (Renne *et al.*, 1995), a major drop in sea level due to a decline in sea-floor spreading (Holser and Magaritz, 1987; Erwin, 1993), a bolide impact (Rampino *et al.*, 1984), global cooling (Stanley, 1988), global warming (Retallack, 1999), and oceanic anoxia (Wignall and Hallam, 1992; Wignall and Twitchett, 1996) (Figure 8.5). Of these many potential mechanisms (reviewed in detail in Erwin, 1993; Hallam and Wignall, 1997), it is most likely that the greatest biotic crisis in Earth history was driven by a multitude of interacting factors rather than just one individual mechanism (Hoffman, 1989a, b; Erwin, 1993; Martin, 1998).

How did plants respond to the multitude of environmental changes of the end-Permian? Although there has been much discussion in the literature that a 'mass' extinction event occurred among plants, the general picture remains equivocal

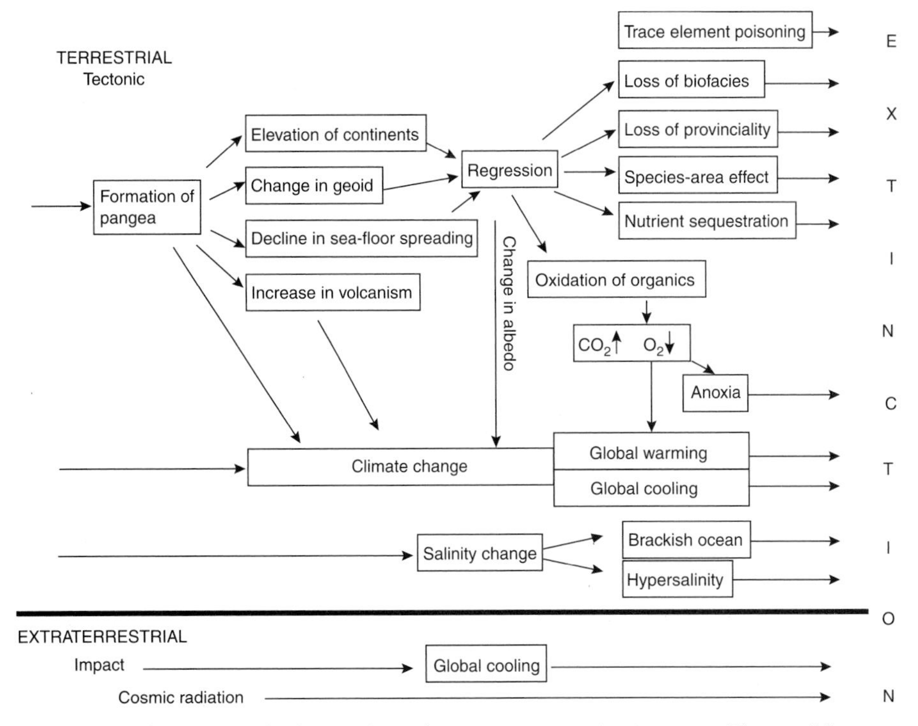

Figure 8.5 Multiple causes leading to the end-Permian mass extinction event. The possible direct causes of the end-Permain mass extinction are to the right, with the more indirect causes progressively to the left. Some events produce a number of secondary effects that may have contributed to the extinction (redrawn from Erwin, 1993).

(Erwin, 1993). Prior to the Permian–Triassic boundary, world vegetation was dominated by lycopsids, sphenopsids, filicopsids, cordaites, and pteridosperms (see Chapter 5). During the late Permian and early Triassic (256–242 Ma) a gradual transition occurred with the reduction, and eventual extinction, of Palaeozoic groups, and the emergence and major radiation of cycads, ginkgos, bennettites, and conifers (see Chapter 6) (Figure 8.6). It seems apparent from the available evidence, therefore, that the predominant longer-term trends over the Permian–Triassic boundary involved major reorganization of plant communities and evolution of new species (Knoll, 1984). The question is, therefore, whether there is evidence for a more rapid and catastrophic reduction in fossil plant diversity at the Permian–Triassic boundary, that can be superimposed on, and could account for, this longer-term trend. A number of studies have addressed this question, with mixed results.

For instance, fossil floras from Siberia indicate the replacement of a high-diversity vegetation dominated by cordaites by a low-diversity, fern-dominated flora at the Permian–Triassic boundary (Dobruskina, 1987). Retallack (1995) has suggested that the environmental perturbations associated with the end-Permian extinction resulted in the proliferation of weedy opportunist taxa such

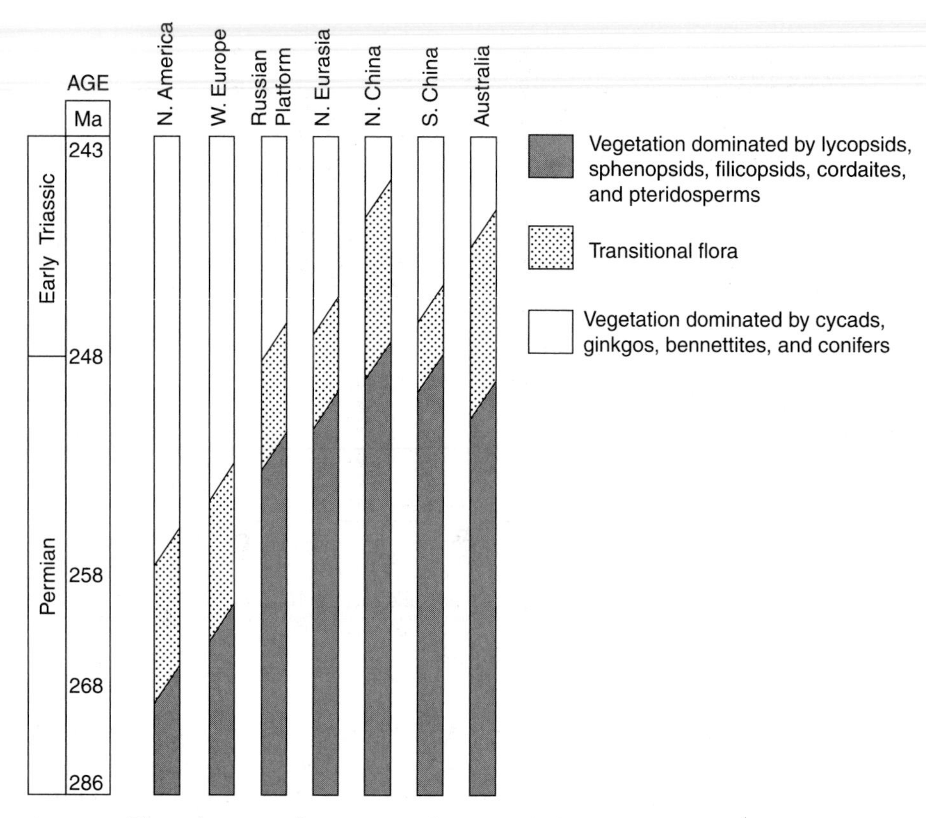

Figure 8.6 The replacement of a vegetation dominated by lycopsids, sphenopsids, filicopsids, cordaites, and pteridosperms with cycads, ginkgos, bennettites, and conifers during the late Permian to early Triassic (256–242 Ma). A transition phase, lasting approximately 20 million years, is recognized, but no evidence for a mass exinction event (redrawn from Knoll, 1984).

as *Dicroidium* and isoetalean relatives. In Gondwana, the dominant *Glossopteris* flora (see Chapter 5) was replaced by this *Dicroidium* flora (Briggs, 1995). In Europe, a massive dieback of equatorial conifer forests and their replacement by a low-diversity lycopsid-dominated flora has been inferred from detailed fossil pollen analysis (Looy *et al.*, 1999) (Figure 8.7). Furthermore, an abundance of fungal spores has been observed at the Permian–Triassic boundary, again in support of widespread and excessive dieback of terrestrial vegetation (Visscher and Brugman, 1981; Visscher *et al.*, 1996).

However, despite much evidence for ecological trauma at the Permian–Triassic boundary, analysis of changes at the higher taxonomic level indicates that total family diversity only dropped by approximately 50% (from 30 to 15 families), and that this occurred over a 25 million year period. This is not consistent with a sudden mass extinction event (Knoll, 1984) (Figure 8.6). In addition, it has been noted that although the environmental changes at the Permian–Triassic boundary had a devastating impact on regional ecosystems, such as in Europe, these were re-vegetated again by related conifer-dominated

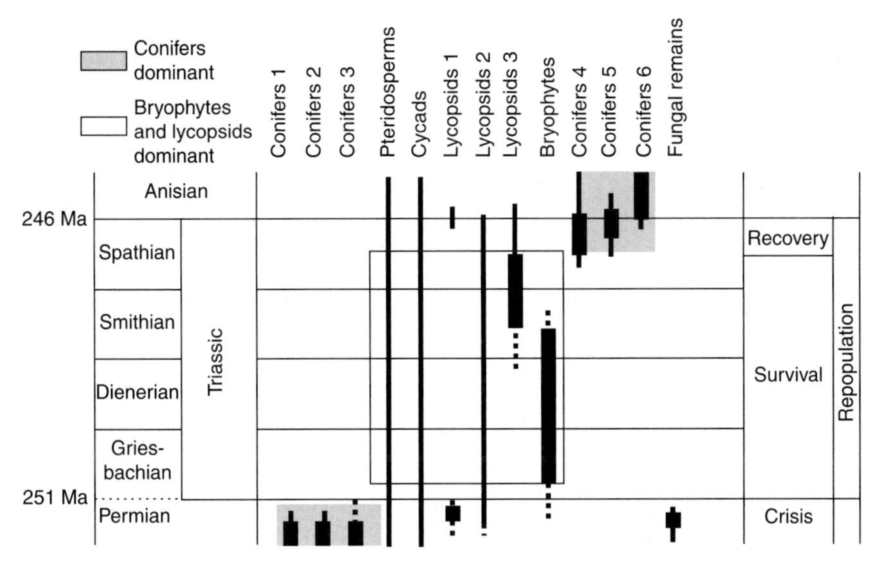

Figure 8.7 Distribution of selected fossils, spores, and pollen across the Permian–Triassic boundary (~251 Ma) (adapted from Looy *et al.*, 1999). Note the long gap (~5 million years) in conifer dominance after the Permian–Trassic extinction event.

forests (Looy *et al.*, 1999). This period of recovery and repopulation took approximately 4–5 million years (Looy *et al.*, 1999). Therefore, it seems apparent from the available evidence that the predominant longer-term trends over the Permian–Triassic boundary involved major reorganization of plant communities and evolution of new species (Knoll, 1984) (Figure 8.7). There is some evidence for major ecological upheaval (Visscher *et al.*, 1996; Looy *et al.*, 1999), but sudden extinction among diverse globally distributed groups of plants did not occur, and there is no evidence for a 'mass' extinction event comparable to that seen in the faunal record.

Triassic–Jurassic boundary (206 Ma)

The Triassic–Jurassic boundary marks the third greatest faunal extinction event in Earth history, and until very recently was the least well known of the 'big five' extinction events. However, a wealth of recent literature is beginning to shed light on this once enigmatic interval. This boundary is characterized by the almost complete extinction of ammonites, with the exception of one genus, and a dramatic disappearance of reef ecosystems. In the terrestrial realm, vertebrates underwent one, if not two, extinction peaks in the stages before or at the Triassic–Jurassic boundary (Benton, 1986, 1989; Olson *et al.*, 1987). Evidence from fossil pollen and macrofloras (fossil leaves, wood, and reproductive structures) indicates a 95% turnover of species at the boundary in northern Europe and North America (Harris, 1937; Visscher and Brugman, 1981; Fowell *et al.*, 1994). However, no significant change in the overall composition of macrofossils (fossil leaves and wood) or fossil pollen have been observed in more southerly latitudes in North America (Ash, 1986). The extinction pattern of fossil plants across the Triassic–Jurassic boundary, similar to the previous two extinction boundaries discussed, therefore appears equivocal. Does the 95% turnover of species in northern Europe and North America represent mass extinction or more local ecological upheaval?

Recent investigations of anatomical and morphological changes of fossil leaves from Greenland and Sweden have shown that major climatic warming occurred at the Triassic–Jurassic boundary due to massive increases in the concentration of atmospheric carbon dioxide (McElwain *et al.*, 1999). Fossil pollen data also indicate an overwhelming increase in the abundance of *Classopollis–Corallina* pollen at the Triassic–Jurassic boundary (Traverse, 1988a). This pollen type is known to belong to a particular arid/hot climate adapted extinct family of conifers, the Cheirolepidaceae (Vakhrameev, 1991), which supports the suggestion that climates became hotter and/or more arid at this time.

Both global warming and increased levels of atmospheric CO_2 are hypothesized to have contributed to the 95% turnover of fossil plant species in Greenland and Sweden, by causing high-temperature injury to many of the large-leaved plants within the vegetation at that time (McElwain *et al.*, 1999). However, because the large-leaved fossil plant taxa, which were present before the Triassic–Jurassic boundary, were replaced by related taxa with smaller or more dissected leaves (which were able to cool more efficiently and avoid high-temperature injury) after the boundary, this pattern merely represent 'pseudo-extinction', because the lineage has been preserved, albeit in a modified form (Figure 8.8). Again, therefore, evidence is apparent for ecological upheaval and adaptation by plants to that upheaval, but there is no clear-cut evidence for 'mass' extinction of terrestrial plants at the Triassic–Jurassic boundary.

Cretaceous–Tertiary boundary (65 Ma)

The Cretaceous–Tertiary (K/T) extinction event is the most widely known and studied of the 'big five' mass extinctions, due to public and scientific interest in the extinction of dinosaurs. The boundary is also marked by up to 80% extinction of marine invertebrates species and a drastic reduction in the number of mammalian species. Among the marsupials, for example, only opossums persisted beyond this boundary. Some estimates suggest that between 60 and 80% of all living animal species became extinct (Raup, 1988). Suggestions to account for this mass extinction include an intense period of volcanism, global sea-level changes (Officer and Drake, 1985; Courtillot, 1990), and/or a catastrophic asteroid impact (Alvarez *et al.*, 1980), among many others.

Evidence in favour of a bolide impact hitting Earth approximately 65 Ma (Alvarez *et al.*, 1980) comes from both marine and terrestrial sources (Pickering, 2000). Evidence of glassy droplets and concentrations of elements scarce in the Earth's crust but abundant in asteroids (e.g. iridium) have been found, for

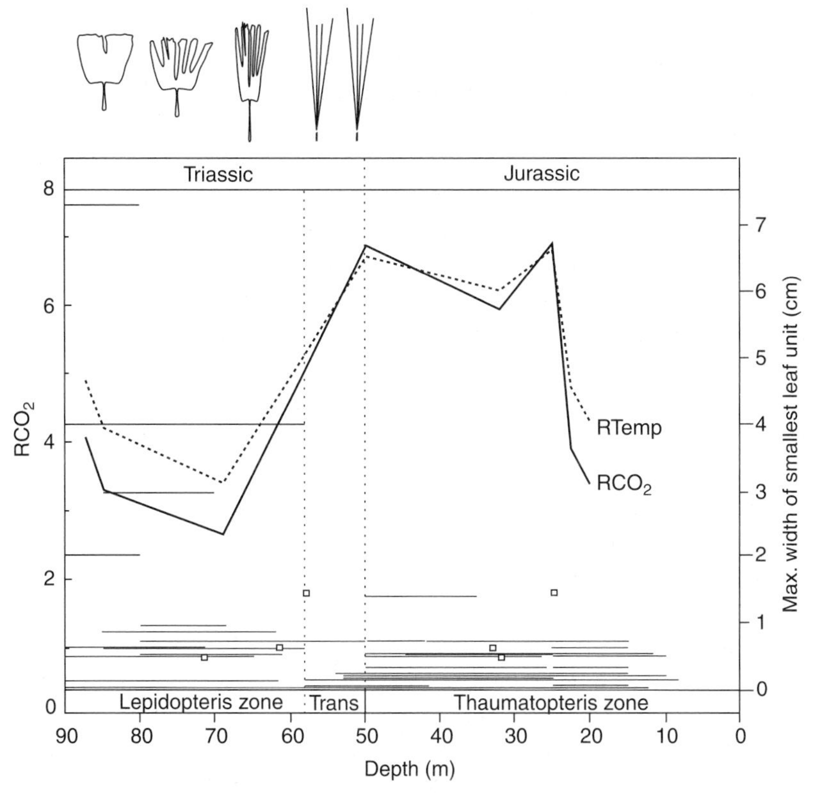

Figure 8.8 Changes in fossil leaf shape across the Triassic–Jurassic boundary (modified from McElwain *et al.*, 1999). Horizontal bars represent the stratigraphic ranges of individual fossil leaf species. The schematic leaf drawings illustrate the general changes in leaf shape, which occurred among the Ginkgoales. Highly dissected Ginkgoales belonging to the genera *Baiera* and *Sphenobaiera* were more common than the less dissected *Ginkgoites* when temperature and CO_2 levels increased. RCO_2 is a ratio of predicted CO_2 concentration to a pre-industrial level of 300 p.p.m. RTemp is the difference in mean global temperature in °C between that estimated during the Triassic–Jurassic and today. For instance at a depth of 50 m, which represents the Triassic–Jurassic boundary, estimated mean global temperature was more than 6°C higher than today.

example, in over 75 marine sedimentary sequences and dating to the Cretaceous–Tertiary boundary (Smit, 1990). There is also evidence of increased carbon aggregates from a number of terrestrial sequences of a similar age (graphitic carbon similar to the charcoal produced presently from forest fires), suggesting global wildfires and a massive global flux of carbon (Woolbach *et al.*, 1985). Suggested climatic conditions associated with such an impact include a prolonged period of darkness and intense cold, followed by global warming, and nitric acid rain. Such a suite of environmental changes would have had deleterious effects on many types of terrestrial and marine fauna.

The evidence for a magnitude of extinction in the plant fossil record similar to that seen in the faunal record is again ambiguous. Global vegetation at the

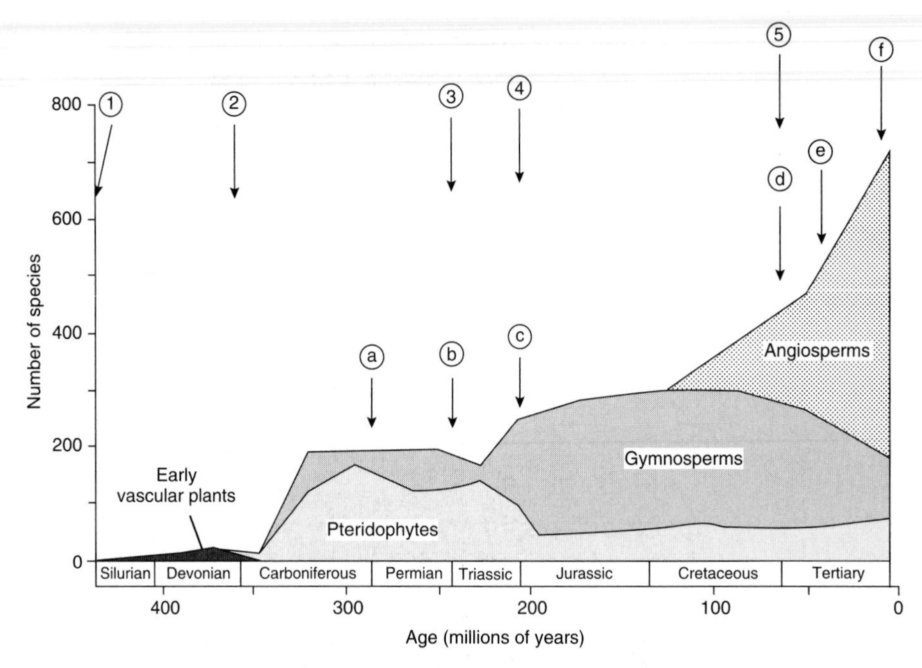

Figure 8.9 Diversity curves for vascular plant species over the past 440 years (redrawn from Niklas *et al.*, 1983; see also Figure 6.2). Also indicated are the big five mass extinction events in the marine record (numbers 1–5) and the six recognized in the terrestrial tetrapod record (letters a–f).

Cretaceous–Tertiary boundary was dominated by angiosperms in almost all of the major biomes, with the first unequivocal evidence for grasses also appearing in the fossil record around this time (between 65 and 55 Ma) (Jacobs *et al.*, 1999) (see Chapter 7). It was one of the most prolific times in terms of species diversity increases, with total numbers of plant species increasingly rapidly (Figure 8.9). It is therefore against this backdrop of increasing species number, that evidence for a mass extinction event must be sought.

Various sedimentary sequences have indicated floral changes associated with the Cretaceous–Tertiary (K/T) boundary. Analyses of fossil pollen sequences from the western interior of North America, for example, indicate that a sudden decrease in the abundance and diversity of fossil pollen grains occurred at the K/T boundary (Smit and Van der Kaars, 1984). This was followed by a sharp and rapid rise in fern spore abundance. Fossil pollen abundance then increased again after the 'fern-spike' but indicate a very different composition of fossil plants from that which existed prior to the K/T boundary (Figure 8.10). In addition, a series of palynological sequences in Japan indicate a sudden change in floristic composition at the K/T boundary. In this case, the pollen profiles show a transition from a high abundance of angiosperm and gymnosperm pollen, to a high proportion of fern spores, followed by an increase in gymnosperm (*Pinus*) pollen (Saito *et al.*, 1986). In the macrofossil record, studies of leaf characteristics across the K/T boundary also reveal some significant changes. Fossil floras from the Vermejo-Raton Basin

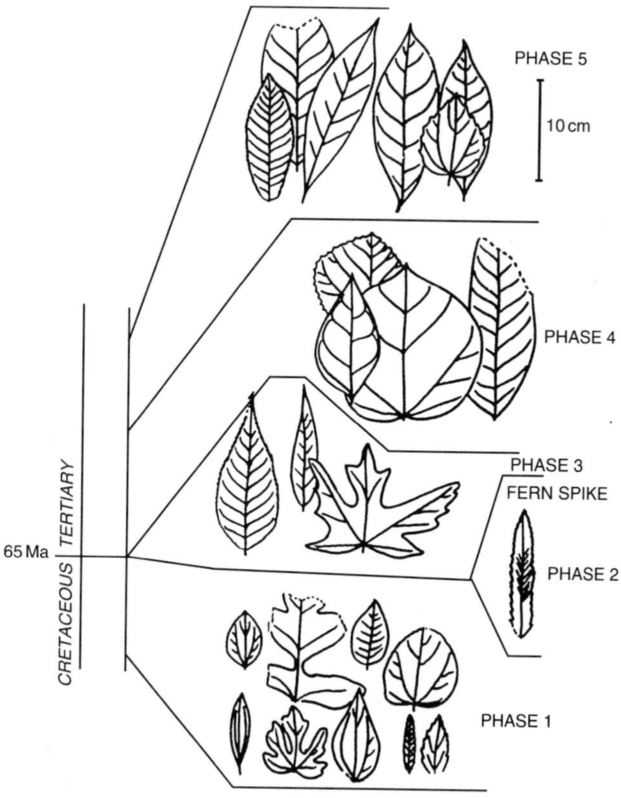

Figure 8.10 Changing leaf assemblages across the K/T boundary (redrawn from Boulter *et al.*, 1988, data from Upchurch and Wolfe, 1987). It has been suggested that the pattern of leaves seen across this boundary are typical of a floristic change with ecological trauma (Upchurch and Wolfe, 1987). Leaves in phase 1 are indicative of broadleaved evergreen vegetation with high diversity, probably representing subhumid conditions. Phase 2 consists of leaves, rhizomes, and cuticles, typical of ferns and herbs—taken to be indicative of wildfires. Phase 3 is indicative of a flora composed of plants with large leaves and with drip-tips and thick, smooth cuticles, typical of early successional vegetation in an environment of high precipitation. Phase 4 indicates a flora with increasing leaf diversity and physiognomically a warm, humid vegetation.

in North America (Wolfe and Upchurch, 1987) have been divided into five distinct floristic phases according to leaf size and shape (Figure 8.10). Phase 1 (pre-K/T boundary) indicates a vegetation consisting of a flora of both broadleaved and small-leafed evergreen angiosperms, with thick, hairy cuticles. Comparison with extant vegetation types indicates that this type of vegetation is found in dry, subhumid regions. The next phase (phase 2) (at the K/T boundary) indicates a fern spike, where the flora is dominated by leaves and rhizomes of ferns and other herbaceous vegetation types. Phase 3 (post-K/T boundary) indicates a flora composed of large leaves with drip-tips and thick, smooth cuticles (usually indicative of early successional vegetation in an environment of high precipitation), followed by phases 4 (post-K/T boundary) and 5 (post-K/T boundary), which indicate a more diverse assemblage typical of low-diversity rainforest.

It is suggested that these patterns, seen in both the microfossil and macrofossil records across the K/T boundary, are typical of floristic change associated with ecological trauma (Wolfe and Upchurch, 1986). In particular, the rise in abundance of fossil ferns at the K/T boundary is thought to indicate rapid re-vegetation after widespread wildfires, possibly caused by a massive asteroid impact (Saito *et al.*, 1986; Tschudy and Tschudy, 1986). Similarly, the marked changes in macrofloras (mostly fossil leaves) are thought to reflect the climatic changes associated with such an impact and, in particular, the effects of an 'impact winter' (Upchurch and Wolfe, 1987; Wolfe and Upchurch, 1987; Wolfe, 1991).

Unlike the animal record, however, the scale and extent of trauma suffered by vegetation was by no means global. Even within the American continent, fossil evidence illustrates great variation in vegetation changes between regions. Comparison of five Cretaceous–Tertiary assemblages located in a south to north transect (Figure 8.11) reveal, for example, that levels of extinction were high in the southern States but low in the northern States (Spicer, 1989). At a global scale, sites in higher latitudes and also the southern hemisphere appear to indicate no abrupt extinction events among plant species but rather a gradual reduction in diversity, probably related to climatic cooling (Spicer, 1989, 1990; Truswell, 1990; Briggs, 1995). This had led to the conclusion that the effect of the Cretaceous-Tertiary boundary event was only very weakly expressed in these regions (Spicer and Parrish, 1990). In support of this interpretation are experimental investigations of plant responses to a prolonged period of darkness and cooling, simulating the environmental conditions predicted to have followed an asteroid impact. These have shown that most plants present at the time of the K/T boundary would have been able to tolerate such adverse conditions (Read and Francis, 1992). Thus even in the regions that record dramatic floral

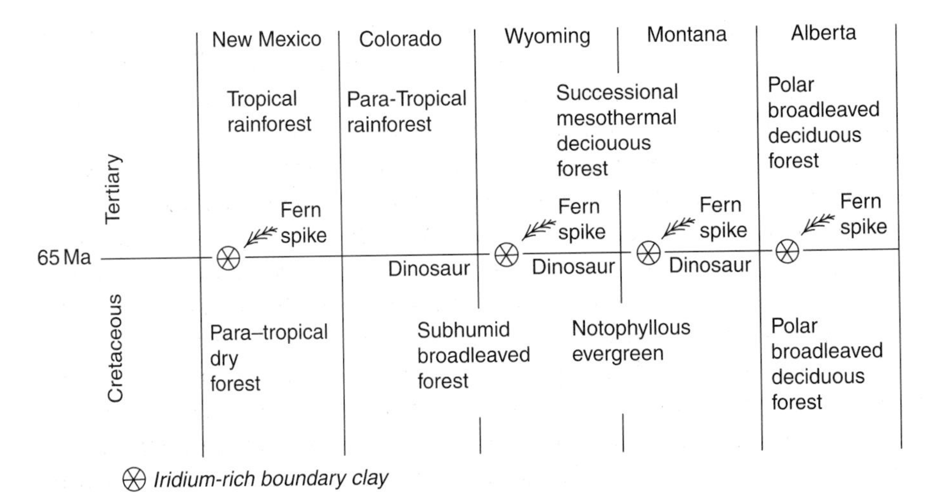

Figure 8.11 Assemblages across a south to north transect in North America, indicating the vegetation impact at the K/T boundary (redrawn from Halstead, 1990).

changes (e.g. North America), it remains debatable whether the event can be classified as a mass extinction (Halstead, 1990). Although a reduction in overall biomass occurred, there is currently no evidence supporting a major decline in species diversity ((Niklas, 1986; Lidgard and Crane, 1988, 1990) (Figures 8.9, 6.20). Furthermore, among angiosperms, which accounted for approximately 80% of global floral diversity at that time, a maximum species extinction of between 5 and 10% has been recorded. This has led Halstead (1990) to conclude that 'the pattern of change across the boundary does not seem to involve much in the way of extinction; rather it seems to indicate a climatic change with an increase in seasonality and change in pattern of distribution of vegetation zones'.

In summary, therefore, it would appear that at each of the major faunal mass extinction events (Figure 8.1), there is currently no strong evidence for a comparatively rapid blanket extinction in the fossil plant record. This is not to say that no changes are apparent in the plant fossil record during these intervals. There is evidence for major disruption of some ecosystems, usually involving changes in floral composition, but these tend to be local in distribution and did not involve mass reduction in overall species numbers or diversity.

8.3 Why no mass extinction in the plant fossil record?

Why are there differences in the levels of extinction between the animal and plant kingdoms? Furthermore, how do these differences comply with evolutionary models that propose mass extinction as the uppermost tier in evolutionary hierarchies, i.e. that 'mass extinction has the strongest impact on life's subsequent history' (Gould, 1995).

It has been suggested that the lack of mass extinction in the plant fossil record as opposed to the faunal record is due in part to the greater ability of plants to withstand major ecological trauma. As humorously noted by Knoll (1984), if you go into a zoo and hammer all the animals on the head, you effectively have the end of the zoo. However, if you go into a botanical garden and burn or scythe all the plants down, in 10 years time you will still have a botanic garden, with the majority of plants sprouting once again. Vascular plants, in comparison to animals, have a very limited set of basic needs, namely water, carbon dioxide (CO_2), nitrogen (N), magnesium (Mg), phosphorous (P), potassium (K), plus some trace elements (Knoll and Niklas, 1987; Niklas, 1997). This list is not only small in comparison to that required by animals, but also basically universal among all plant groups, and has barely changed throughout geological time. Thus it is assumed that this list was the same for the swamp-dwelling *Lepidodendron* trees of the Carboniferous as it is for modern orchids and daisies (Traverse, 1988). In terms of resources, therefore, plants have relatively basic energy requirements.

Plants also have a number of various in-built mechanisms for coping with environmental stress (Knoll, 1984; Traverse, 1988a). These include leaves or whole-branch systems that wilt or can be shed during periods of drought, the ability to die back to the ground and perennate as underground stems or rhizomes, lie dormant during periods of adverse weather, survive for long periods of time in seed banks, and reproduce vegetatively. These will be discussed in detail in the following sections. Most plants also have the potential to migrate away from unfavourable biological or environmental conditions through transportation of pollen and seeds via wind, animal, and water vectors. In addition, it has been demonstrated that plants are not nearly as sensitive to population size dynamics as animals. Thus a 'deme' of vascular plants, consisting of only a few individuals can persist far better than a small deme of tetrapods, for example. Moreover, if a population size is greatly reduced, plants within a small population are more likely to persist than animals (Traverse, 1988a). Plants can also undergo mosaic evolution, whereby different organs on the same plant can develop differently independently of each other. This is an extremely useful tool, especially during times of environmental stress, as it enables a plant to protect its total genetic make-up, by altering one organ (leaves for example) while retaining all others unchanged. A combination of these features has therefore enabled plants not only to cope with periods of intense ecological trauma, such as those associated with faunal mass extinction events, but also to persist for millions of years in the geological record.

8.4 Evidence for persistence in the plant fossil record

Evidence from extant taxa indicate that, in general terms, 'living fossils' are much more abundant among terrestrial plants than animals (Traverse, 1988a). Very few animals appear to be so closely related (morphologically) to their fossil relatives as plants, and from morphological evidence alone it would appear that a number of plant taxa have remained relatively unchanged for very long periods of time (Table 8.3). Evidence from the fossil record demonstrates that many families persist, even if only as one or two genera. Cycads, for example, still thrive presently in Australia, Africa, and Mexico, approximately 100 million years after they lost out competitively to angiosperms (Knoll, 1984). However, one of the most classic examples of plant persistence in the fossil record is that of the Ginkgoales (Figure 8.12).

Approximately 250 years ago *Ginkgo biloba* was introduced into Britain (1758), having been known in cultivation in Japan since 1690 (Gifford and Foster, 1989). Ginkgoales have a fossil record extending back to the early Permian (~280 Ma) (see p. 140) and diverse ginkgoalean species were abundant in Triassic and Jurassic vegetation, especially in the northern hemisphere (Figure 5.8). Estimates suggest that there were at least 16 different genera of

Table 8.3 Examples of persistent plant genera in the geological record (from Traverse, 1988a; Taylor and Taylor 1992; Stewart and Rothwell, 1993; Scott and Wing, 1998)

Group/family	*Genus*	Millions of years of existence of forms morphologically close to extant genus
Selaginellaceae	*Selaginella*	300
Lycopodiaceae	*Lycopodium*	325
Equisetaceae	*Equisetum*	280
Cycadales	*Cycas*	240
Ginkgoales	*Ginkgo biloba*	240
Gnetales	*Welwitschia*	200
	Ephedra	200
Araucariaceae	*Araucaria*	150
Taxodiaceae	*Cunninghamia*	160
	Metasequoia	140
	Sequoia	100
Cephalotaxaceae	*Cephalotaxus*	200
Pinaceae	*Pinus*	140
Taxaceae	*Taxus (?)*	214
Winteraceae	*Takhtajania*	120
Chloranthaceae	*Chloranthus*	120
Magnoliaceae	*Magnolia*	120
Platanaceae	*Platanus*	120
Ulmaceae	*Ulmus*	80
	Celtis	90
Fagaceae	*Castanea*	87
	Nothofagus	90
Betulaceae	*Alnus*	87
	Betula	87

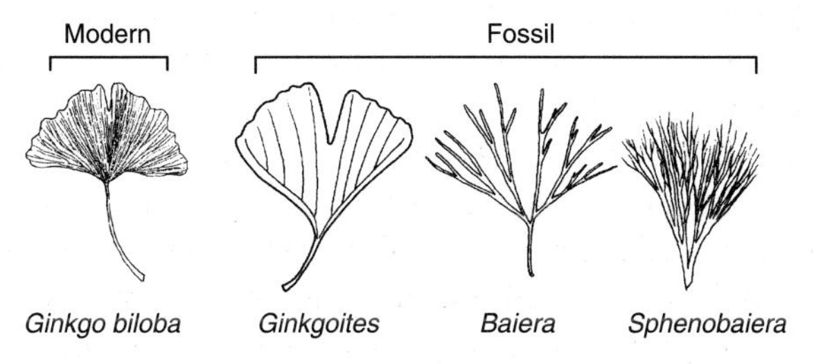

Figure 8.12 Extant and fossil Ginkgoales.

Mesozoic (248–65 Ma) Ginkgoales (Pearson, 1995). By the end of the early Cretaceous (~100 Ma), however, the global distribution of Ginkgoales became greatly reduced, probably due to radiation of the angiosperms. A rapid decline in the number of species and genera occurred, such that by the middle Eocene, only a few species, including *Ginkgo biloba*, survived. Ginkgoales have therefore persisted in the fossil record for at least 280 million years and even though other groups, especially the angiosperms, have drastically restricted their geographical range, *Ginkgo biloba*, the so-called 'living fossil' still remains viable as a taxon.

Certain angiosperms also demonstrate extreme persistency. Recently the discovery of a population of over 250 adults of *Takhtajania perrieri*, for example, the only extant African/Madagascan representative of the family Winteraceae (Schatz *et al.*, 1998), suggests that this family, and even possibly the species (from pollen evidence), have persisted for approximately 120 million years (Table 8.3) and are still going strong.

8.5 Adaptations of plants for persistence

Many of the characteristics that have enabled plants to persist relatively unchanged in the fossil record over long periods of time are the same as those that are thought to have conferred resistance to the environmental changes associated with faunal mass extinction events. These characteristics include hybridity, polyploidy, asexual reproduction, and persistence and dormancy of propagules.

Hybridity

Hybridization is the crossing of two individuals of dissimilar genetic constitution (i.e. the individual produced is from genetically different parents). The advantages of this process to both the continuation of a population and the evolution of new species are numerous. First, since hybrids recombine the characteristics of their parents, they contain new genetic combinations. In a changing environment such hybrids may potentially be better adapted to the new suite of environmental conditions than either of the parent plants. The characteristics of a population of hybrids may then become stabilized if they are better adapted to the new environmental conditions than either of the parents, and the continuation of the population, even if in a slightly altered genetic state, occurs.

In many plant species, hybridization is thus an important evolutionary mechanism, where the recombination of genetic material results in development of more resilient taxa that can grow in diverse habitats (Stebbins, 1959). This has been demonstrated in a number of extant species, including *Eucalyptus* (eucalypt), *Quercus* (oak), *Arctostaphylos* (bearberry), and *Ceanothus* (mountain lilac) (Raven *et al.*, 1992). Extant populations of *Ceanothus*, for example, grow in a number of different ecological niches on the west coast of

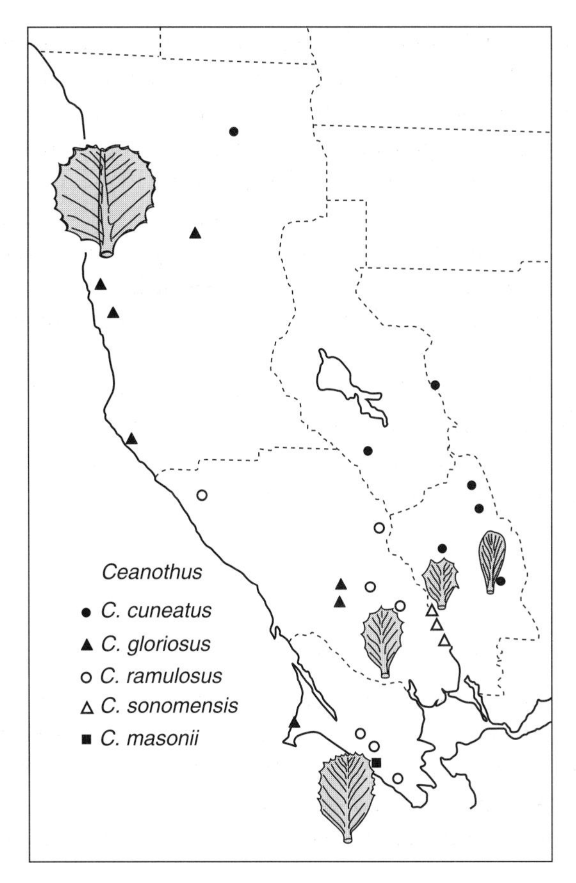

Figure 8.13 Distribution of extant populations of *Ceanothus* (mountain lilac) on the west coast of North America. Two widespread and distinct species of mountain lilac exist in this region, the coastal *Ceanothus gloriosus* and the interior *Ceanothus cuneatus*. However, there are also three hybrid populations from these two original species, each representing a stable hybrid population, that show improved growth in their own region than either parent from which they arose. (Redrawn from Raven *et al*., 1992.)

North America (Figure 8.13). Two widespread and distinct species of mountain lilac exist in this region, the coastal *Ceanothus gloriosus* and the interior *Ceanothus cuneatus*. However, there are also three hybrid populations from these two original species, each representing a stable hybrid population, that show improved growth in their own region compared to either parent from which they arose. There are also numerous examples of intergeneric hybridization, i.e. where hybridization has occurred between different genera. Although this is much rarer, since different genera are genetically and phenotypically more divergent than different species, extant examples include the shrubs *Purshia tridentata* and *Cowania stansburiana* (Strutz and Thomas, 1964; Niklas, 1997). These have combined to form a hybrid called *Purshia glandulosa*, which is genetically and phenotypically distinct.

One of the problems with hybridization, however, lies in the fact that many hybrids are sterile. This is because the chromosomes of the parent species of some hybrids are not homologous and are therefore unable to pair up during meiosis. An often-cited example from the animal kingdom is the cross between horse and donkey which has produced a mule. The mule, although renowned as a pack animal, is sterile, and cannot in effect form a stable hybrid population through sexual reproduction. Plants have a major advantage over animals in this respect, in that they have at least two mechanisms to override hybrid-induced sterility, namely polyploidy and vegetative (asexual) reproduction. Thus, even if the first generation is partially sterile, full fertility is often achieved in later hybrid generations (Rieseberg, 1995).

Polyploidy

Polyploidy is a term used to describe a condition where cells or individuals have two or more sets of chromosomes. These arise at a low frequency and are either the result of a 'mistake' in mitosis, where the chromosomes divide but the nucleus does not, thereby producing tetraploid cells (Figure 8.14a), or where two different species successfully mate to produce tetraploid offspring containing sets of chromosomes from both parents (Figure 8.14b). Polyploid plants occur in many natural populations and especially among flowering plant families (angiosperms). It is estimated, for instance, that over 40% of dicotyledons and 60% of monocotyledons are polyploid (Niklas, 1997). In contrast, polyploidy is extremely rare amongst animals. The advantages of polyploidy to the evolution and long-term persistence of a population are as follows:

1. If polyploids arise from chromosome doubling in sterile diploid hybrids, the doubling of chromosome number implies that 'spare' chromosomes become available for normal pairing and can restore fertility to the sterile hybrids (see above).

2. New species can evolve through polyploidy without the requirement of a geographic barrier and within the home range of the parent species. This is known as sympatric speciation (Brown and Lomolino, 1998). For example, when polyploid members of a population interbreed, they will produce tetraploid offspring. When these offspring reproduce, they will produce fertile tetraploid progeny that are genetically isolated from the rest of the population. Thus within a single generation there is a barrier to gene flow between the new species and its parent species without the involvement of any geographical isolation (Stebbins, 1971; Briggs and Walters, 1984; Niklas, 1997).

3. In polyploid plants, harmful recessive alleles are more likely to be masked by normal dominant alleles. As a result, mutations are less likely to be expressed, plants are more likely to be resistant, and the phenotypic variability in the populations is likely to be low (Price, 1996).

4. Polyploid plants have a greater store of genetic variability (because there are more alleles at any one locus), and thus adaptation to cope in environmental

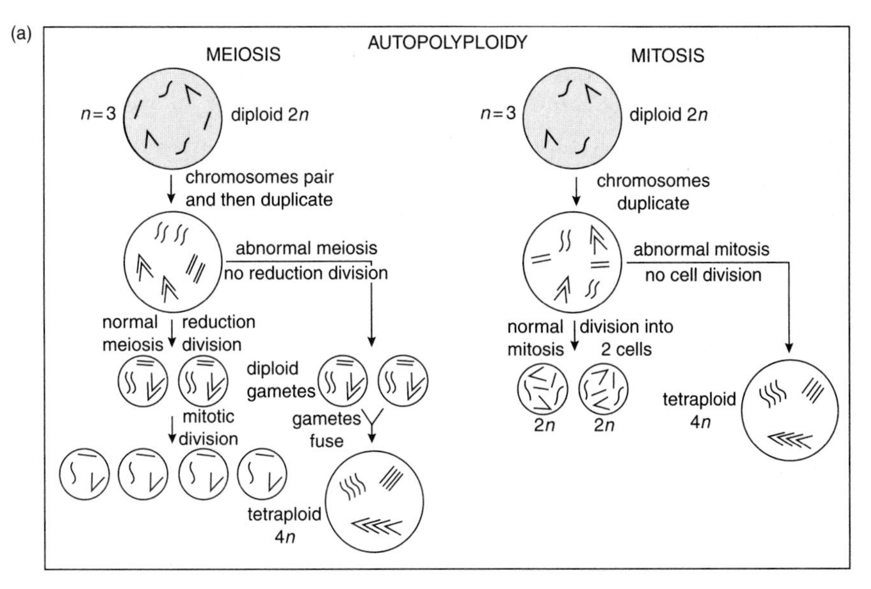

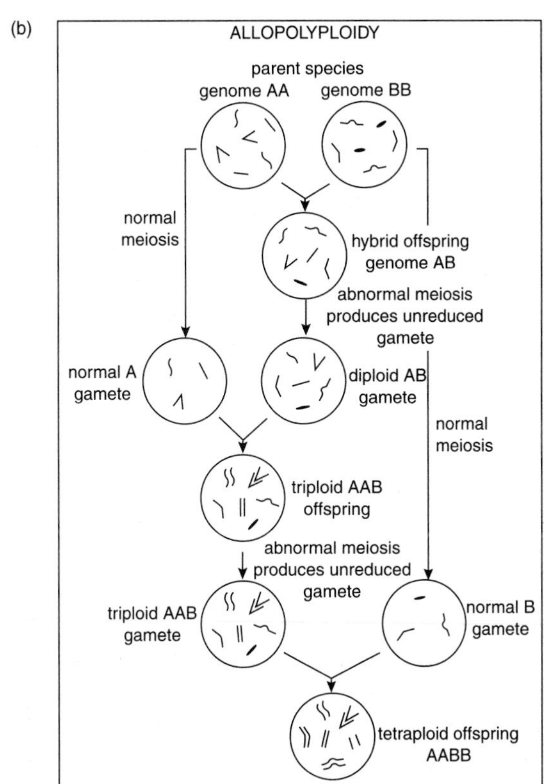

Figure 8.14 Diagrammatic representation of polyploidy in plants, where cells or individuals have two or more sets of chromosomes. These are either the result of a 'mistake' in mitosis, where the chromosomes divide but the nucleus does not, thereby producing tetraploid cells (a), or where two different species successfully mate to produce tetraploid offspring containing sets of chromosomes from both parents (b). (Redrawn from Skelton, 1993.)

extremes will be greater. Studies of extant populations, for example, have demonstrated that polyploid species are more widely distributed geographically and at the habitat margins than their non-polyploid parents (Grant, 1971).

Asexual reproduction

Asexual reproduction is a type of reproduction that occurs without the sexual process of haploid gamete formation. It is extremely common in plants and is thought to have been a crucial process in promoting the long-term persistence of vascular plants (Tiffney and Niklas, 1985; Traverse, 1988a; Niklas, 1997). Asexual reproduction in plants may involve the development of a new individual from an unfertilized egg, thus leading to seed formation without fertilization by pollen, or the production of individuals from somatic cells, i.e. cells not related to gametes. All individuals that are derived by uniparental or vegetative reproduction, known as clones, contain exactly the same genetic make-up as the parent plant (Price 1995). Clonal growth in plants may be divided into two types (*sensu* Tiffney and Niklas, 1985), namely non-linked or linked. Non-linked clones arise where the clones may be dispersed beyond the parent plant by biotic or abiotic vectors. These non-linked structures include bulbs, whole plants, and unfertilized seeds. Linked clones remain attached to the parent plant and are 'dispersed' by growth of parent root or branch. Linked structures include rhizomes, root and basal suckering, branch arching (stolons), and liana habits. Although these structures are initially attached to the parent plant, severance eventually leads to them becoming physiologically separate individuals.

There are at least two aspects of asexual reproduction that are thought to have been advantageous to the survival of plant species spanning intervals of geological time that resulted in the widespread extinction of many animal groups. First, it provides another means by which sterile plants (e.g. certain hybrids) may still become widespread and persist, both spatially and temporally. Secondly, vegetative reproduction in many plants is a mechanism to cope with stress. For instance, many plants will regenerate vegetatively after trauma, such as burning, cropping, and irradiation (Traverse, 1988a). This therefore enables plants to persist even during periods of adverse environmental conditions and even possibly disease.

Estimates for the presence of clonal versus aclonal plants in the fossil record (Tiffney and Niklas, 1985) has revealed some interesting trends (Figure 8.15). In general it would appear that when the environment has been favourable (here 'favourability' is determined by higher levels of atmospheric moisture), for example, during the Carboniferous (354–290 Ma), mid-Mesozoic (220–200 Ma), and early Tertiary (65–40 Ma), aclonal forms increased. In contrast, when the environment became more arid, for example in the late Permian and early Triassic (256–242 Ma) and from the Miocene through to the Quaternary (23.8–0 Ma), clonal forms increased. This apparent success of clonal forms during periods of increased aridity may indicate either the relative benefits of asexual reproduction and/or the benefits of a prostrate growth form (which is

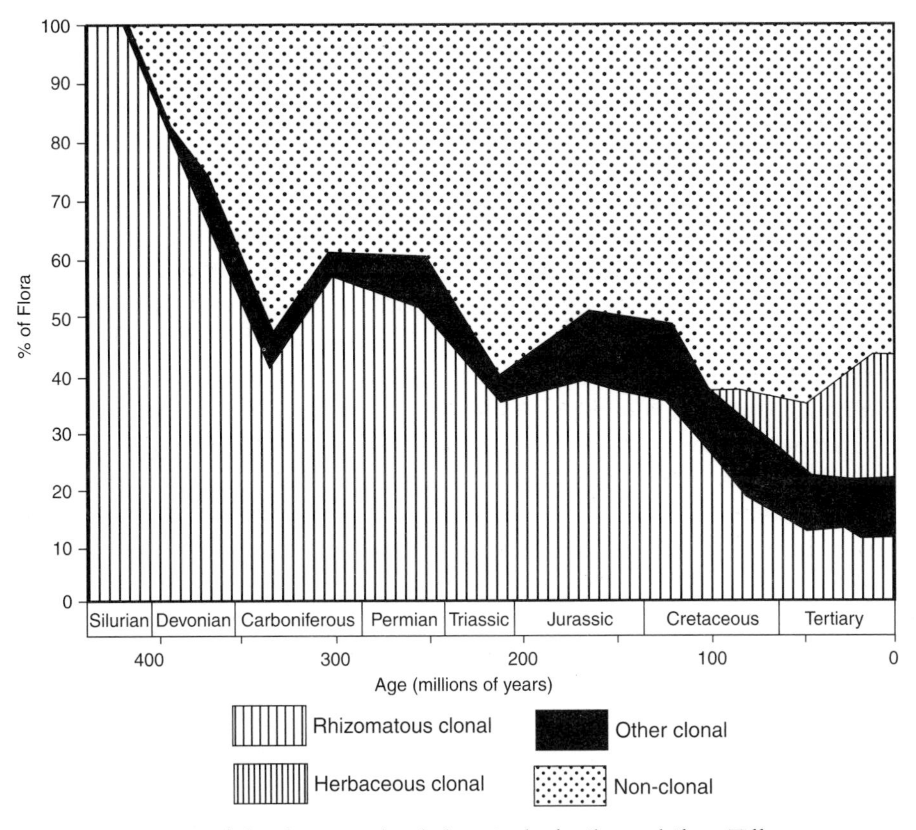

Figure 8.15 Presence of clonal versus aclonal plants in the fossil record (from Tiffney and Niklas, 1985).

usually associated with the clonal form) during times of environmental aridity. It is also interesting to note that presently within the angiosperms, monocotyledon families tend to be dominated by clonal growth (especially the grasses), whereas dicotyledons tend to be dominated by aclonal forms. With the prediction of increasing global aridity in the future it has been suggested that monocotyledons may become dominant in the global flora (Traverse, 1988).

Persistence and dormancy of propagules

A final mechanism by which plants are able to persist for a long period in the fossil record is through the dormancy of their propagules. Many seeds have the potential to go into long periods of dormancy in soil seed banks and still retain their viability for at least 100 years (Kivilaan and Bandurski, 1981; Bewley and Black, 1982). Such a long 'resting phase' would probably be of sufficient duration to survive the worst of the environmental catastrophes caused by, for example, a meteorite impact. There are also examples of seeds from the plant fossil record which have demonstrated that they can retain their viability for even longer than a century. For example, over two-thirds of seeds recovered from a lake bed in the Liaoning Province, China, and dated to between 1200 and 330 years old (Shen-Miller *et al.*, 1995) have been germinated, growing into robust living plants of sacred lotus (*Nelumbo nucifera*). It is suggested in this example that certain morphophysiological characteristics of fruits of the sacred lotus make them particularly adapted for longevity. These include impervious and mechanically strong fruit walls, the fact that they contain a protein-repair enzyme, high levels of reducing elements in embryonic axes, and the ability to maintain membrane fluidity (Shen-Miller *et al.*, 1995). The preservational environment, however, is also important, and it is suggested that the highly reducing anoxic environment of the sediments from which these fruits were obtained would have contributed to their exceptional longevity.

It would therefore appear that a large number of extant plant groups contain genera that have persisted for millions of years, as a result of one, or a combination, of these intriguing mechanisms. Such persistent mechanisms must be viewed as features that in many ways distinguish plants from other organisms and have guaranteed against the comparatively 'rapid blanket extinctions' (Leopold, 1967; Knoll, 1984; Traverse, 1988). If there is no mass extinction in the plant fossil record, a view which is consistent with the current evidence, and plants can persist for million of years, then what, if anything, is driving plant macroevolution? This question will be addressed in Chapter 10.

Summary

1. Mass extinctions are characterized by the relatively rapid extinction of groups of organisms, usually at a high taxonomic level, that is families and even orders.

2. At least five big mass extinction events have been recognized in the marine faunal record, with similar trends among terrestrial faunas for at least three of these.

3. The evidence for similar mass extinctions in the plant fossil record is ambiguous.

4. At the Frasnian–Famennian boundary (364 Ma) over half the marine genera present on Earth did not survive. The plant fossil record at this time indicates a reduction in plant diversity but there is no evidence for widespread extinction among many diverse higher taxa of terrestrial plants.

5. The mass extinction event at the end of the Permian (248 Ma) has been termed the 'mother of mass extinctions'. Up to 96% of all durable skeletonized marine invertebrate species went extinct. In the plant fossil record there is again a gradual reduction in plant diversity (over a 25 million year period). In addition, although in some regions there was a devastating impact on regional ecosystems (for example, in Europe), on recovery, these were re-vegetated by forests similar in composition to those growing previously. This evidence does not therefore suggest a mass extinction event among terrestrial plants.

6. The Triassic–Jurassic boundary (206 Ma) marks the third greatest faunal extinction event in Earth history. In the plant fossil record there is evidence of major ecological upheaval and that, in some regions, species adapted to hotter and/or more arid climates became dominant, but there is no clear-cut evidence for 'mass' extinction.

7. The Cretaceous–Tertiary boundary (65 Ma) is the most widely known of the 'big five' mass extinction events due to the extinction of dinosaurs, along with an 80% extinction of marine invertebrate species and a drastic reduction of mammalian species. In the fossil plant record, there is evidence of a massive 'wildfire event' which is consistent with the hypothesis that the cause of this extinction was related to meteoritic impact. However, although in some regions mass ecological upheaval and trauma occurred, in others areas there was relatively little disturbance. There is also no evidence to support a major decline in plant species diversity.

8. There does not (therefore) appear to be strong evidence for comparatively rapid blanket mass extinctions in the fossil plant record comparable to those seen in the fossil faunal records.

9. Reasons for the lack of mass extinction in the fossil plant record are probably closely related to the mechanisms that enable plants to withstand a certain amount of environmental stress. These include leaves or whole-branch systems that wilt or shed during periods of stress (e.g. aridity, high temperatures, low temperatures), the ability to die back to the ground and perennate as underground stems or rhizomes, and propagules that can survive for long periods of time in soil seed banks.

10. Many of these mechanisms also probably account for why there are so many 'living fossils' in our present-day flora. Other adaptations for persistence, however, include the ability to freely hybridize, polyploidy, and asexual reproduction.

9 Ancient DNA and the biomolecular record

In recent years there has been a surge of interest in the use of ancient DNA and other 'fossil' biomolecules to address the evolutionary record of fossil plants. There has, however, also been much controversy, particularly with regard to ancient DNA, surrounding the preservation and extraction of such material. This chapter considers the potential of ancient DNA in evolutionary research, with a particular focus on the plant fossil record. It examines conditions necessary for preservation and extraction of ancient DNA and considers the arguments for and against achieving reliable results. Specific case studies are also highlighted to examine how far current research in ancient DNA has contributed to further understanding the processes governing morphological and molecular plant evolution through geological time. The second part of the chapter examines the extraction of other ancient biomolecules and chemical constituents of fossil plants. These will be discussed in relation to their use in fossil plant systematics, long-term environmental and climatic reconstructions, and investigations of the plant fossilization process.

9.1 Potential of ancient DNA in evolutionary research

In recent years, as techniques to recover the molecular details of organisms have advanced, so too has the study of molecular evolution. This work has focused predominantly upon examination of extant organisms, with comparison between different individuals, species, and groups allowing the construction of molecular phylogenies, predictions of DNA diagenesis, and rates of sequence evolution (Lewin, 1997). With the advent of new extraction techniques, however, and in particular the ability to amplify a specific DNA fragment from a few intact DNA molecules (polymerase chain reaction), there has been increasing interest in reconstructing a 'fossil' molecular record through the amplification of ancient DNA. It is argued that there are three main areas where such a record would be of particular benefit to evolutionary research, namely systematics, DNA diagenesis and sequence evolution, and the calibration of molecular clocks (Golenberg, 1994; Herrman and Hummel, 1994).

Systematics

In the study of systematics, it is suggested that fossil DNA would be of particular use in the identification and classification of extinct species, the verification of previous fossil identifications, and in testing of classifications of well-studied fossil groups (Soltis *et al.*, 1992; DeSalle, 1994). At the simplest level this could include, for example, the identification of a disarticulated plant part in the fossil record, such as a branch, stem or leaf, as belonging to a particular species/genus. At a more complex level, however, ancient DNA could also provide an important test of phylogenetic trees where the evolutionary relationship between a group has previously been determined using molecular and morphological characteristics of extant taxa. One example where ancient DNA has been used in this way is in the evolutionary pathway proposed for termites (DeSalle *et al.*, 1992; DeSalle, 1994).

The classic phylogeny for termites (*Mastotermes*) based on extant morphological features suggested that they evolved as a grade from cockroaches (*Blaberus*). The morphological character differences used in this phylogenetic distinction were features such as the presence or absence of symbiotic gut flagellates, deciduous wings, and a mandibular tooth (Figure 9.1a). When a molecular phylogeny was constructed using extant material, however, results indicated that the evolutionary pathway for termites was ambiguous, and that they could have either evolved from cockroaches or mantids (*Mantis*) (Figure 9.1b). In order to resolve this ambiguity, fossil DNA was extracted and analysed from the extinct fossil termite *Mastotermes electrodominicus*, which was preserved in 25 million year old Dominican amber (DeSalle *et al.*, 1992). The resulting phylogeny indicated that the extinct *Mastotermes electrodominicus* was a sister to the other termites and that the genus *Mastotermes* (termites) was monophyletic (Figure 9.1c). The conclusion was therefore reached that termites are descendants of a single ancestral form, and neither direct descendants from cockroaches or mantids (DeSalle *et al.*, 1992; DeSalle, 1994). The inclusion of ancient DNA in this phylogenetic reconstruction thus radically altered the original interpretation.

DNA diagenesis and sequence evolution

A major area of interest in the study of molecular evolution concerns the rate of accumulation of nucleotide substitutions. The accumulation of nucleotide substitutions is monotonic (i.e. more substitutions occur over longer periods of time), but it is not clear whether the rate of accumulation is constant. Two opposing theories have developed to account for this rate of substitution through time (Lewin, 1997). The first is the so-called neutral theory, which states that the mutation rate itself should not be affected by selection, population size, or speciation events, and that extensive genetic variation occurs as the passive accumulation of chance mutations (Kimura, 1983, 1991). This theory therefore

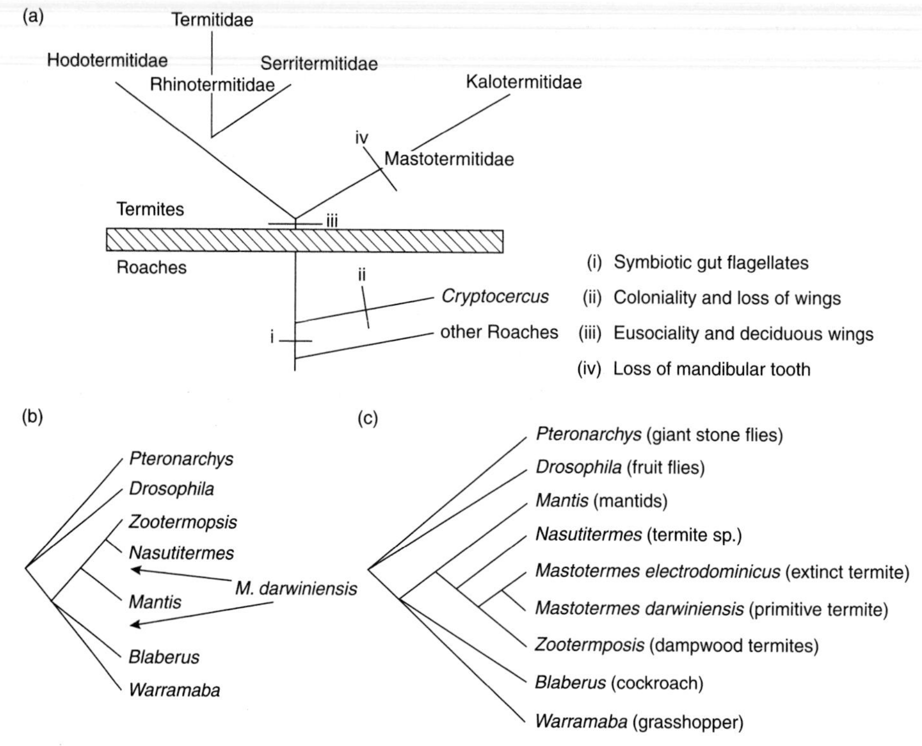

Figure 9.1 Phylogenetic relationship between termites (Termitidae) and roaches (Mastotermitidae). (a) Phylogeny based on extant morphological evidence, where characters used to define branches include (i) symbiotic gut flagellates; (ii) coloniality and loss of wings; (iii) eusociality and deciduous wings; and (iv) loss of mandibular tooth. (b) Phylogeny based on extant molecular evidence, where the position of Mastotermes is unclear. (c) Phloygeny based on extant and fossil molecular evidence which resolved the position of Mastotermes as a sister to the other termites and that the genus is monophyletic. (After DeSalle, 1994.)

predicts that rates of speciation should, on average, remain constant over time (Futuyma, 1998). The alternative view, the so-called selection theory, however, argues that molecular evolution is not a constant process because the rate of nucleotide substitution will be related to selection coefficients and population size. Thus if a mutation is disadvantageous it will be removed because the individual is less fit. Since selection coefficients and population size will vary over time, under a selection model, substitution rates should not remain constant (Gillespie, 1991). There is much discussion in the literature as to which theory is correct. Theoretically, ancient DNA has the potential to test these hypotheses (Golenberg, 1994). For example, if it were possible to extract DNA within the same genus from various datable horizons in the geological record, it should then be possible to measure the substitution rate. Such data would also provide a method to test predictions regarding the levels and structuring of genetic variation prior to and following speciation (see, for example, Mayr, 1963; Carson, 1982).

Calibration of molecular clocks

Another potential area of research for ancient DNA is in the calibration of molecular clocks. Molecular clocks take the neutral theory one step further, to suggest that if the rate of sequence evolution is sufficiently constant, then it should be possible not only to determine the rate of nucleotide substitution for a particular species (i.e. number of nucleotide substitutions/million years) but also, by comparing the genetic difference between two related species, calculate the length of time that has elapsed since they shared a common ancestor (Figure 9.2) (Zuckerkandl and Pauling, 1965). Thus the molecular clock adds a temporal perspective to phylogenetic trees in demonstrating relatedness amongst species (Lewin, 1997). Calibration of genetic distance as time, however, is difficult to determine. Usually, the number of substitutions that have occurred since a major recognizable event in the fossil record (such as the divergence/appearance of a specific group) is divided by the length of time that the group has been present in the fossil record, to calculate a substitution rate (Langley and Fitch, 1974).

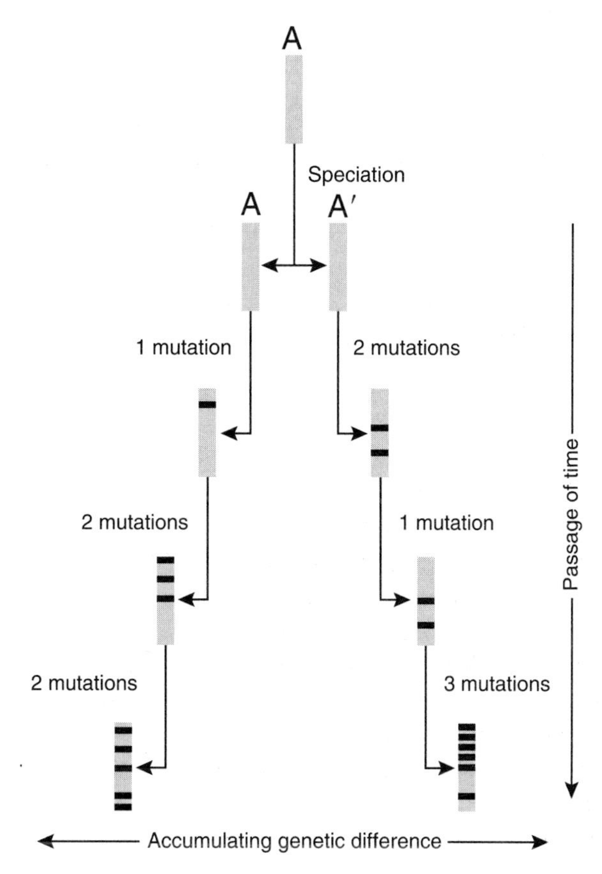

Figure 9.2 Accumulation of nucleotide substitutions over time, resulting in genetic differences between lineages (redrawn from Lewin, 1997).

However, this method of calibration is very approximate and appears to vary according to the DNA sequence measured (e.g. mitochondrial, nuclear, or chloroplast) (Wolfe *et al.*, 1987; Futuyma, 1998). In the plant record, for example, two molecular clocks have been constructed to calculate the length of time that has elapsed since the monocotyledons and dicotyledons shared a common ancestor. The first method, which looked at the number of substitutions in the DNA of the slowly evolving glycolytic enzyme, dated the origin of divergence to approximately 300 million years ago (note that this is far earlier than currently supported by the plant fossil record) (Martin *et al.*, 1989). However, the second method, which reconstructed the divergence time from chloroplast DNA (Wolfe *et al.*, 1989), suggested a date of between approximately 200 and 140 million years ago. The two methods therefore indicate a 100 million years discrepancy. Both are far older than supported by the fossil record, but the chloroplast data is much nearer to that estimated from the fossil record (see Chapter 6). Substitution rates therefore appear to vary greatly amongst different plant genomes, and a more direct method of calibration is needed if molecular clocks are to become more accurate. It has been proposed, therefore, that another huge area of potential for ancient DNA is in providing more direct calibration of genetic distance through time (Golenberg, 1994).

9.2 Deposition, preservation, and extraction of DNA

Plant remains containing ancient DNA have been obtained from herbarium specimens, charred seeds and cobs, mummified seeds and embryos, anoxic fossils, and amber (Brown and Brown, 1994; Austin *et al.*, 1997a). Depositional conditions are extremely important in the preservation of ancient DNA. The most successful depositional environments for the preservation of DNA are those where there is rapid and direct input of plant material from very local sources, thus reducing physical damage and minimizing externally and internally induced wound reactions. In addition, rapid burial is needed to reduce the exposure of tissue to biotic degradation, and if incorporation in a sedimentary sequence is to occur, a fine grain size and compaction is important to insulate the immediate environment of the fossil and prevent percolation (Golenberg, 1994). Plant fossils are more suitable than animals for the preservation of ancient DNA because they are less likely to be dead at time of deposition and, as a consequence, will have no significant internal bacterial fauna.

Extraction of ancient DNA involves a series of well-established techniques (see for example (Pääbo, 1989; Golenberg, 1999), all of which require sterile conditions free from any other possible sources of biological contamination (Figure 9.3). Once the fossil has been extracted from the deposits within which it is fossilized (i.e. sedimentary rocks, amber, etc.) the sample is immediately ground using a pestle and mortar, and the ground tissue transferred to tubes with

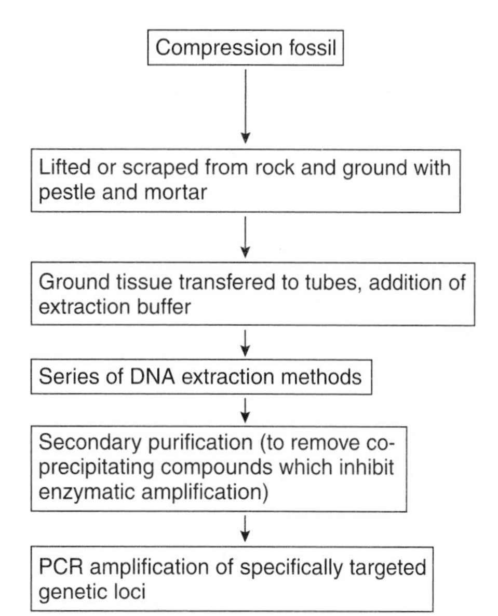

Figure 9.3 Simplified flow-chart of ancient DNA extraction methods.

the addition of extraction buffer. This part of the extraction procedure is usually carried out in the field. On returning to the laboratory, a series of DNA extraction methods are performed, including secondary purification to remove co-precipitating compounds which may inhibit enzymatic amplification. Amplification is then carried out on specifically targeted genetic loci using PCR (polymerase chain reaction).

The nature of the evolutionary questions being asked mainly determine which genomes to target. Different genomes in the same organism mutate at different rates, therefore if the evolutionary study involves the examination of genetic change across millions of years, then a gene with a slow mutation rate is appropriate, whereas a shorter timescale will require a gene with a rapid mutation rate. In plants there is the choice of three genomes to target, mitochondrial DNA (mtDNA), plastid DNA (cpDNA), and nuclear DNA (DNA) (Rollo *et al.*, 1994).

Mitochondrial DNA has the advantage of having several hundred copies per cell and also being inherited uniparentally, thus rendering relatively simple interpretation of population/evolutionary data. However, studies indicate that mtDNA evolves very slowly in sequence and, furthermore, that the mutation rate can be 100 times slower in plants than in animals (Palmer and Herbon, 1988). Thus mtDNA extracted from relatively young fossils (for example, seed remains from archaeological deposits dating back to no more than 2000 years) will not indicate any appreciable change when evolutionary related organisms are compared (Rollo *et al.*, 1994).

Plastid DNA consists of circular DNA molecules located in the chloroplast genome. Plastid DNA is again quite conservative in its mutation rate but, because it is located in the chloroplast genome, it is abundant in photosynthetic leaf tissue and therefore if the fossil being examined is a leaf, this may well be the

Table 9.1 Estimated rates of synonymous substitutions per 10^9 years in mitochondrial, chloroplast, and nuclear plant genomes (from Wolfe *et al.*, 1987)

Genome	Taxa compared	Rate
Mitochondrial	Maize/wheat	0.2–0.3
	Monocot/dicot	0.8–1.1
Chloroplast	Maize/wheat	1.1–1.6
	Monocot/dicot	2.1–2.9
	Angiosperm/bryophyte	1.4–1.6
Nuclear	Spinach/*Silene*	15.8–31.5
	Monocot/dicot	5.8–8.1

best genome to target. However, if another part of the plant, such as the seed, is being examined, the probability of extracting plastid DNA is small, since in seed tissue plastids account for only 1% of cellular DNA.

Nuclear DNA has a much higher evolution rate than the above two (at least twice that of chloroplast DNA) (Table 9.1) and is therefore very useful in discriminating between species even on the basis of very short tracks of DNA. However, there is an increased risk in false results with short tracks of DNA in the PCR amplification process and so there are obvious drawbacks in targeting nuclear DNA.

9.3 Examples of current research

To date there are relatively few instances in which evidence from ancient DNA has been used to address the potential research areas outlined above. In the early 1990s, several key papers were published describing the extraction and amplification of ancient DNA from the fossil record, but there has since been much controversy surrounding these results and many believe that, as yet, DNA of true antiquity has never been recovered (reviewed by Austin *et al.*, 1997b). However, before discussing the apparent limitations of this technique, three of these papers and their results will be described.

Ancient chloroplast DNA sequence from Magnolia leaves (17–20 million years old)

Some of the first results reporting successful extraction and amplification of ancient DNA from plants were from fossil deposits in northern Idaho, North America (Golenberg *et al.*, 1990; Golenberg, 1991). These deposits, known as

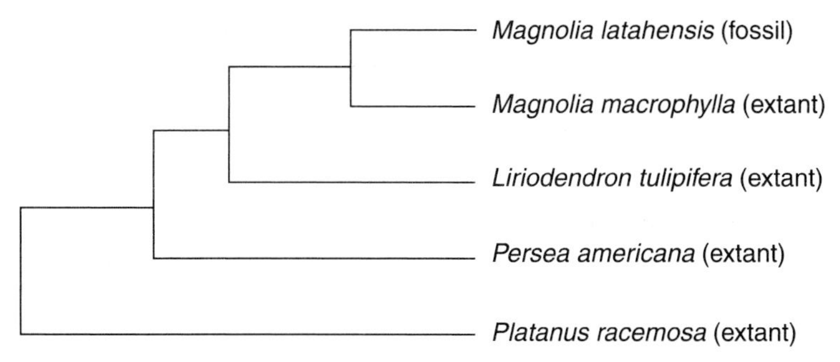

Figure 9.4 Phylogenetic relationship between extinct *Magnolia latahensis* with extant *Magnolia macrophylla, Liriodendron tulipifera, Persea americana*, and *Platanus racemosa*, based on comparison of sequences obtained from extant DNA and an 820 base-pair DNA fragment obtained from 17–20 million year old fossil leaves (from Golenberg *et al.*, 1990).

the Clarkia beds, consisted of an exposed sequence of lake sediments dated to between 17 and 20 Ma (early Miocene). Numerous leaves from these deposits, in an apparently excellent state of preservation, were extracted, including numerous *Magnolia latahensis* leaves. Ultrastructure studies of the leaves indicated that although they were chiefly compression fossils (see Chapter 1), they still contained a cellular structure and some secondary metabolites. In addition, the chloroplasts were extremely well preserved (Niklas *et al.*, 1985). Attempts to extract ancient cpDNA yielded an 820 base-pair fragment, with reproducible results obtained from both strands of DNA and from separate amplifications (Golenberg *et al.*, 1990). Comparison of this sequence with four other extant species, namely *Magnolia macrophylla, Liriodendron tulipifera, Persea americana*, and *Platanus racemosa* (Figure 9.4) revealed the least number of substitutions with *Magnolia macrophylla*, indicating that the fossil leaves were most similar to this species. The fossil DNA sequence showed an increasing number of substitutions compared with *Liriodendron tulipifera, Persea americana*, and *Platanus racemosa*, thus suggesting a possible evolutionary lineage.

Ancient chloroplast DNA sequence from *Taxodium* fossils (17–20 Ma)

Further work on the leaves contained in the Clarkia deposits also led to the apparently successful extraction of ancient cpDNA from fossil *Taxodium* leaves (Soltis *et al.*, 1992). In this case amplification of a cpDNA fragment of 1380 base pairs, representing nearly the entire 1431 *rbcl* chloroplast gene, was reported. Comparison of this fossil sequence to a sequence obtained from extant *Taxodium distichium* (swamp cypress) revealed that they differed by 11 base substitutions. It was therefore suggested that if the two *Taxodium* species shared a common ancestor 17–20 million years ago, then a minimum sequence

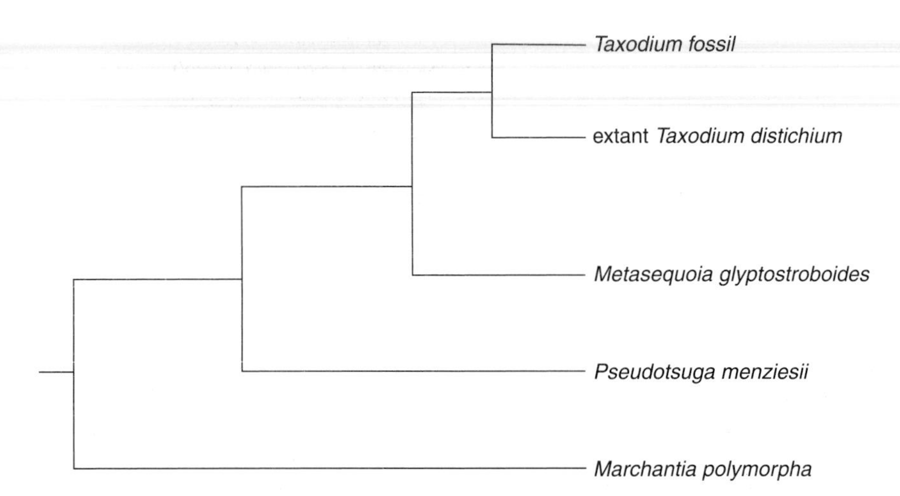

Figure 9.5 Phylogenetic relationship between extinct *Taxodium* fossil with extant *Taxodium distichium* (swamp cypress), *Metasequoia*, *Pseudotsuga*, and *Marchantia*, based on comparison of sequences obtained from extant DNA and a 1421 cp DNA fragment obtained from 17–20 million year old fossil leaves (from Soltis *et al.*, 1992).

divergence rate of 0.55–0.65 substitutions per million years could be calculated. In addition, the fossil and extant sequences obtained from the *Taxodium* species were also compared to a sequence obtained from extant *Metasequoia* (which is in the same family, i.e. Taxodiaceae). Results from these analyses indicated that the *Taxodium* sequences differed from the *Metasequoia* sequence by 38 substitutions, and were used to suggest an evolutionary lineage between these two groups (Figure 9.5).

DNA from the extinct tree *Hymenaea protera* (35 million years old)

Claims of the oldest DNA to be extracted from plant fossils comes from material extracted from Dominican amber deposits from the La Toca mine near Puerto Rica. These deposits are latest Eocene in age (dated to approximately 35 Ma) (Poinar *et al.*, 1993). Sufficient genetic material for the successful extraction, amplification, and sequencing of a 364 base pair fragment of the chloroplast gene *rbc*L was apparently obtained from a leaf of the extinct tree *Hymenaea protera* contained within the amber. Prior to the extraction of ancient cpDNA, the knowledge of past species diversity and distribution of *Hymenaea* was based on morphological characteristics. When a new phylogenetic tree was constructed using the sequence obtained from this ancient cpDNA, it confirmed the previously suggested phylogeny (based on floral and vegetative morphology), that there is a sister-group relationship between this extinct species *Hymenaea protera*, and the extant East African species *Hymenaea verrucosa* (Figure 9.6) (Poinar, 1994).

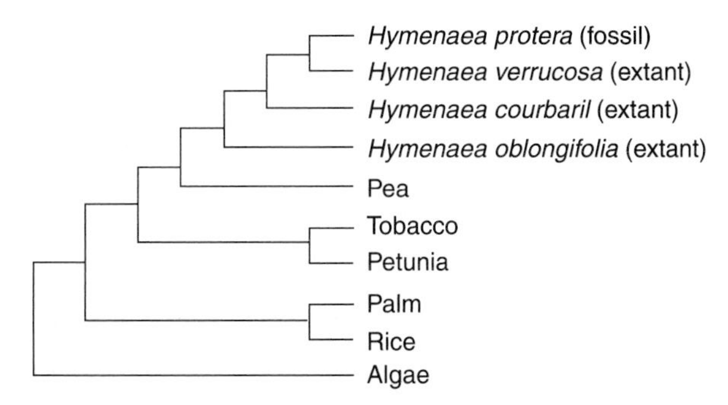

Figure 9.6 Phylogenetic relationship between the extinct tree *Hymenaea protera* and extant East African species *Hymenaea verrucosa*, based on comparison of a sequence of a 364 base pair fragment of the chloroplast gene *rbc*L from the 35 million year old leaf with sequences from extant species. These results confirmed the previously suggested phylogeny (based on floral and vegetative morphology) that there is a sister-group relationship between this extinct species *Hymenaea protera*, and the extant East African species *Hymenaea verrucosa* (redrawn from Poinar, 1994).

9.4 Limitations of the technique

Although the potential of ancient DNA in both micro- and macroevolutionary research has been recognized, there are remarkably few examples in the published literature, especially from the plant fossil record, of successful extraction, amplification, and use of ancient DNA to address evolutionary questions. In part, this is probably due to the exceptional environments required to preserve ancient DNA and thus the paucity of sites containing such fossils. However, it must also be due in part to the ongoing lively debate (and scepticism from some) about the preservation of DNA in the fossil record, and whether the material being amplified is 'ancient' or rather recent contamination. Main areas of concern with the amplification of ancient DNA are associated with predicted rates of DNA degradation, problems with PCR amplification, and the probability of contamination in short fragments of DNA (Austin *et al.*, 1997b).

DNA degradation

One of the main areas of scepticism lies in the ability of strands of ancient DNA to remain intact. It is argued that once an organism dies, the DNA strands are deprived of the repair mechanisms provided in living cells and therefore degradation will occur (Lindahl, 1993, 1997). Two types of degradation are thought to be of particular relevance to fossil deposits, namely the breaking down of the bonds linking the bases together by exposure to water (hydrolytic) and breaking down of the bonds due to exposure to air (oxidative) (Figure 9.7). For example, it has been demonstrated that under laboratory conditions hydrolytic and oxidative damage will degrade sequences to short fragments over several thousand years, and theoretically no fragments greater than 150 base

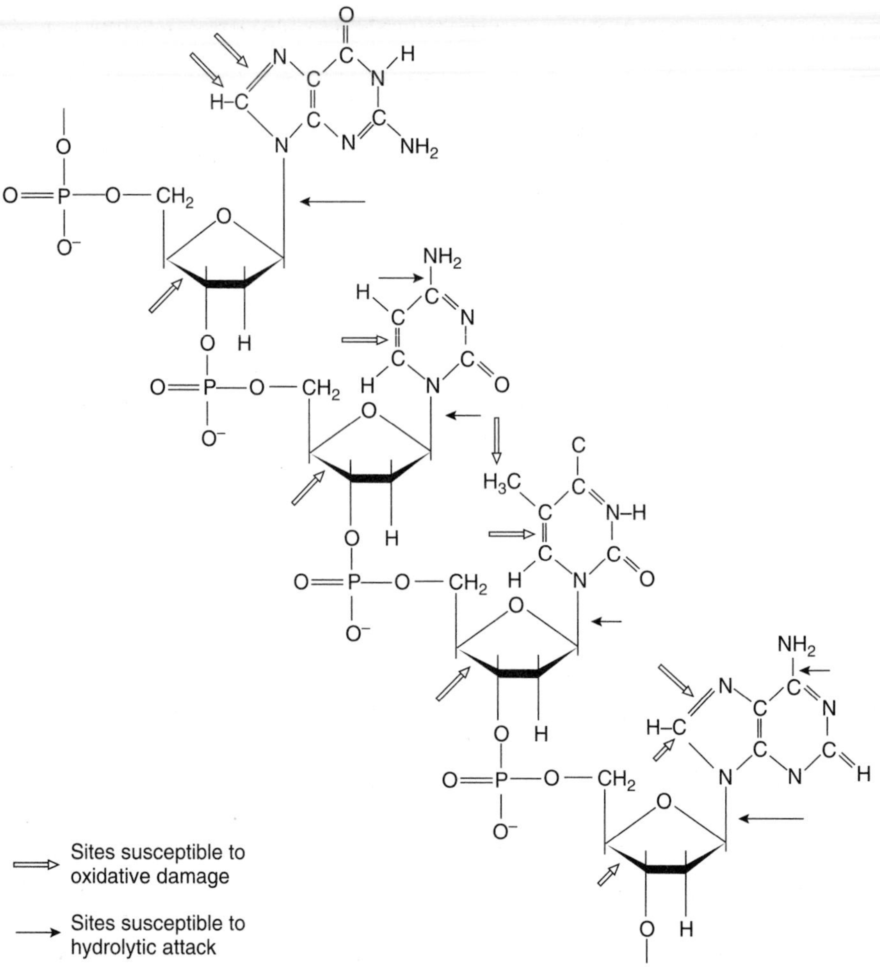

Sites susceptible to oxidative damage

Sites susceptible to hydrolytic attack

Figure 9.7 A short segment of one strand of the DNA double helix. Also indicated are the main sites for intracellular DNA damage (oxidative, hydrolytic) which ultimately break down the bonds linking the bases together, weakening the DNA strand, and causing it to fragment into smaller and smaller pieces (redrawn from Lindahl, 1993).

pairs should remain after approximately 2700 years, and no purines after 4 million years. It is suggested, therefore, that even if ancient DNA does survive, it will be highly fragmented and chemically modified (Austin *et al.*, 1997b).

PCR amplification

Another area of potential error is thought to be in the amplification process of ancient DNA. Ancient DNA is typically degraded and therefore the numbers of base pairs recovered tend to be small. Fragments of sufficient size, which span the region between designated primers, may be very few, and this will increase the likelihood of amplifying contaminant DNA (Pääbo *et al.*, 1988). There are inevitably some random errors introduced into the resulting sequencing in PCR amplification of modern samples, but studies have indicated that additional errors caused by the presence of damaged nucleotides are introduced in the amplification of ancient DNA (Brown and Brown, 1994). Thus minute amounts of modern or fossil contaminating DNA may be preferentially amplified when dealing with short fragments of ancient DNA.

Contamination

Another criticism is related to the increased probability of contamination from the micro-organisms (fungi and bacteria) that have been associated with the fossil over time. In addition, the many stages involved in getting the fossil from the field site to the laboratory will inevitably introduce further contamination from other specimens, micro-organisms, and humans (Lindahl, 1993, 1997; Brown and Brown, 1994; Austin *et al.*, 1997b).

With increasing research into the problems associated with preservation, extraction, and amplification of ancient DNA, it has become apparent that much of the earlier research, including the three studies outlined above, are highly problematical. There have been a number of attempts, for example, to reproduce the results from the Clarkia beds, and obtain ancient DNA from *Magnolia* and *Taxodium* leaves (e.g. Sidow *et al.*, 1991), with little success. In addition, examination of the preservation of other biomolecules in the Clarkia leaves, including polysaccharides, proteins, and amino acids, indicates that they are either not preserved or altered extensively (Logan *et al.*, 1993; Poinar *et al.*, 1996). The alteration of amino acids from a right-handed orientation (L-enatiomer) to a left-handed form (D-enatiomer) when isolated from active metabolic pathways is of particular interest, because this process, termed racemization, occurs at a similar rate to the depurination of DNA. Examination of the extent of racemization in the Clarkia beds has been taken to indicate that almost total depurination of the DNA in these deposits has occurred (Poinar *et al.*, 1996). Furthermore, even a number of the extremely resistant molecules, such as wax esters, certain sterols, and other biomolecules, have been shown to be extensively degraded in the Clarkia fossils (Logan *et al.*, 1995), providing further scepticism about the preservation of the much less resistant DNA (Briggs, 1999). Much of this work has therefore led to the conclusion that genetic sequences containing meaningful genetic information should not be preserved in most geological environments for more than 100 000-year-old sequences (Bada *et al.*, 1999).

There have, however, been a number of counter-arguments against these criticisms and the debate is ongoing (e.g. Golenberg, 1999). An early response to the criticism of DNA degradation, for example, argued that it is not possible to emulate, in the laboratory, the conditions that occur in the long-term fossilization process. For example, the work of Lindahl (1993, 1997) and others on the degradation of DNA was carried out under pH conditions ranging from

4.5 to 7.5 and at temperatures ranging from 80 to 45 °C. Results from this study indicated that under a constant pH of 5.0, the reactions decreased in a log linear fashion with a decrease in temperature. However, further work has indicated that rates of degradation with a change of pH under a constant temperature of 70 °C did not decrease in a log linear fashion. It is argued, therefore, that the breakdown of ancient DNA in fossil leaves over time may not be estimated by a simple linear extrapolation (Golenberg, 1994). In addition, it has been suggested that fossilization conditions peculiar to a site must also be taken into consideration. For example, the Miocene Clarkia fossil leaves are thought to have undergone a series of transformations leading to their extraordinary preservation, including natural dehydration before abscission. This would have resulted in rupturing of the tonoplasts (the membrane that separates the cell vacuole from the protoplasm) and the release of high concentrations of plant phenols (including tannin). It is suggested that the leaves were then rehydrated in tannin-rich waters, a process which may have been instrumental in their preservation, since plants with normally high tannin contents are often found in the best state of preservation (Niklas *et al.*, 1985b). Therefore it is argued that the conditions in which the DNA was preserved in the Clarkia fossil deposits are not similar to the normal physiological conditions encountered at fossil sites and therefore depurination may have been slower than estimated (Golenberg, 1994, 1999).

9.5 In defence of ancient DNA

There is some agreement that ancient DNA may be preserved for long periods of time in amber. This is a fossil resin formed by members of the coniferous family, Araucariaceae, and the angiosperm family, Leguminosae. Amber contains glucose, galactose, and arabinose sugars, as well as various acids, alcohols, and esters. When fauna or flora become trapped in the sticky resin, in the process of fossilization, the sugars in the resin withdraw the moisture from the original tissue and initiate a process of inert dehydration (Bada *et al.*, 1994; Poinar, 1994). Thus 'amber preservation appears to provide a uniquely advantageous way of retaining ancient DNA sequences because the DNA is largely dehydrated, partly protected from atmospheric oxygen and not exposed to microbial contamination' (Lindahl, 1993). Further support for the preservational potential of amber again comes from the study of racemization of amino acids. Results from insect tissue entombed in the 35 Ma Dominican amber, for example, indicate that little racemization has occurred (Poinar *et al.*, 1996). The conclusion is therefore reached that depurination of DNA would probably be similarly inhibited by the anhydrous conditions (Bada *et al.*, 1999). Again, however, there appear to be problems associated with reproducing the results of ancient DNA sequences from amber, and thus the preservational potential of this fossil medium is not without critics (Austin *et al.*, 1997a,b; Stankiewicz *et al.*, 1998a; Briggs, 1999).

With so many questions surrounding reproducibility of results, in conjunction with scepticism about the preservation, it is unsurprising that much of the research potential of ancient DNA has yet to be realized. One key area of research where ancient DNA is proving to be an invaluable tool, however, is in the study of the more 'recent' past. There have been numerous successful extractions of ancient DNA with meaningful and reproducible results from samples that are less than 100 000 years old, and these are starting to provide an important tool to archaeological research (e.g. Brown, 1999; Lister *et al.*, 1999; Stone and Stoneking, 1999; Jones, 2001). Although much of this research is beyond the scope of this book, there are two key areas of this 'archaeologically based' ancient DNA research, which broadly overlap with the type of evolutionary questions outlined in Section 9.1. First, ancient DNA is being used to provide a link between genetic information and archaeological chronologies. In particular, ancient DNA is starting to provide a direct calibration of genetic substitutions through time, which are more accurate, especially for shorter chronologies, than molecular clocks (Jones and Brown, 2000). Secondly, ancient DNA is making an important contribution to understanding rates and type of genetic change following speciation, and especially the appearance of domesticates.

One example demonstrating both of these contributions is in the analysis of the evolution rates of maize (*Zea mays*) (Goloubinoff *et al.*, 1993). It has long been suggested that domesticated maize (*Zea mays*), which has great morphological and genetic diversity, evolved from wild subspecies, either *Zea mays* subspecies *mexicana* or *Zea mays* subspecies *parviglumis*. The archaeological record indicates the sudden appearance in central America of domesticated maize (*Zea mays*) from approximately 5000 years ago, with the suggestion being made that the rate of molecular evolution proceeded much faster following domestication. Thus the hypothesis that the molecular clock might run faster following domestication was tested by examining an amplified part of a nuclear gene, *Adh2*, from ancient maize remains (~4500 years old) and modern domesticated maize and wild species of teosinte (Goloubinoff *et al.*, 1993).

On comparison of the sequence variation among ancient alleles with contemporary alleles, it was demonstrated that two 4500-year-old sequences were more related to some modern sequences (in numbers of base substitutions) than to each other (Figure 9.8). The use of ancient DNA in this study therefore demonstrated that ancient species of maize were as diverse as modern species, and that no accelerated molecular evolution has taken place since domestication. Furthermore, the conclusion was reached that there were a number of genetic lineages contributing to modern maize from throughout the wild *Zea mays* population, disproving the previously held view that maize domestication was a single isolated event (Goloubinoff *et al.*, 1993; Jones and Brown, 2000).

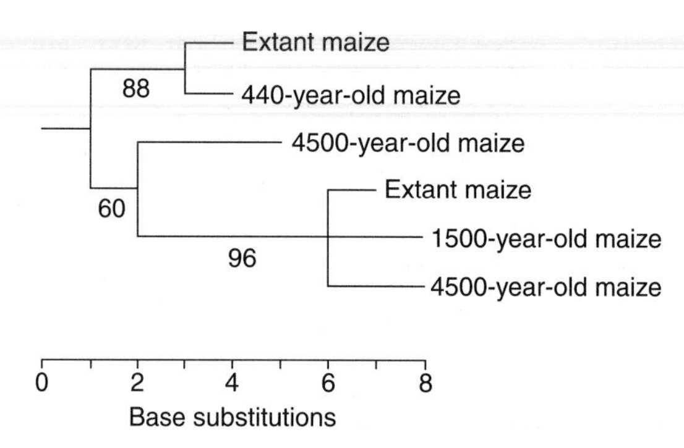

Figure 9.8 Comparison of an amplified part of a nuclear gene *Adh2* from ancient maize remains (~4500 years old), with modern domesticated maize and wild species of teosinte (Goloubinoff *et al.*, 1993). From this study it was demonstrated that the two 4500-year-old sequences were more related to some modern sequences (in numbers of base substitutions) than to each other. It was thus concluded that ancient species of maize were as diverse as modern species and that no accelerated molecular evolution has taken place since domestication.

9.6 Other fossil plant biomolecules, biomacromolecules, and chemical constituents

Although in recent years there has been much publicity surrounding the study of ancient DNA, it is becoming increasingly apparent that, in many ways, other biopolymers and chemical constituents of fossil plant tissues and organs hold the key to interpretation of the plant fossil record.

Plant cells and tissues are composed of a range of different organic materials, most of which are highly prone to decay (De Leeuw *et al.*, 1995). Some of the chemical constituents of these cells and tissues, however, and particularly those that provide the plant with structural support and protection, are highly resistant to decay, as they contain resistant biomacromolecules (Van Bergen *et al.*, 1995) (Figure 9.9). These include lignins, which are present in a range of plant cell types, such as xylem and cuticle; cutin and cutan, found in some leaf and stem cuticles, seed coats, and inner fruit walls; algaenans from algal cell walls; sporopollenin, which is present in spore walls and pollen walls; and suberan, found in periderm (Nip *et al.*, 1986; De Leeuw and Largeau, 1993; Hemsley *et al.*, 1994; Tegelaar *et al.*, 1995; Van Bergen *et al.*, 1995).

Investigation of the chemical composition of different fossil plant tissues has revealed many of these resistant biomacromolecules, or altered forms known as biomarkers (Van Bergen *et al.*, 1995). Cutan-like biomacromolecules, for example, have been isolated from fossil pteridosperm (seed fern) and ginkgo leaves of Carboniferous (~300 Ma) and Cretaceous (~70 Ma) age, respectively (Collinson *et al.*, 1994); fossil fruit and seed coats from the Eocene (~40 Ma) (Van Bergen *et al.*, 1993); suberan-like biomacromolecules have been found in fossil lycopsid

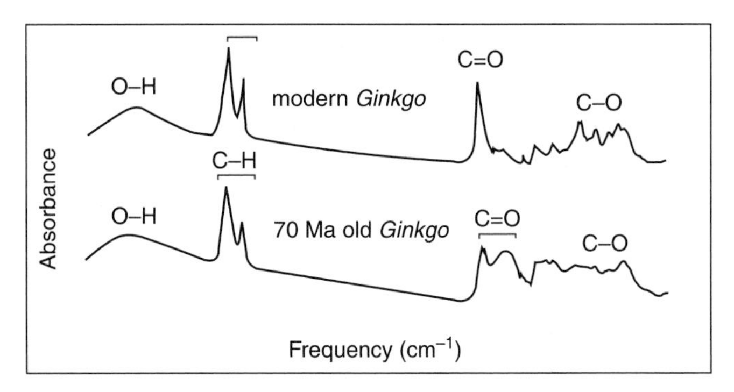

Figure 9.9 Infrared spectra of cuticle from extant and 70 Ma Ginkgo leaves. Results show clear similarities of the chemical bonds between the two samples, suggesting that they have remained fairly well intact in the fossil sample, but also that some bonds (e.g. the C = O bond) have undergone some alteration. Redrawn from Collisen *et al.*, NERC Ancient Biomolecular Initiative.

bark from the Carboniferous (Collinson *et al.*, 1994); and lignins have been isolated from a wide range of fossil plants of different ages (Van Bergen *et al.*, 1995).

Evidence from these studies therefore indicates that the chemical make-up of many of the more resistant fossil plant tissues has remained highly conserved through geological time. The ability to isolate recognizable chemical constituents of fossil plants (biomarkers) has far-reaching potential for the interpretation of the fossil record. To date, study of these ancient biomarkers has focused on three main areas of research (Farrimond and Eglinton, 1990; Eglinton and Logan, 1991; Van Bergen, 1999):

(1) the use of chemical biomarkers in the identification of morphologically indistinct organic remains (chemosytematics);
(2) as a tool with which to study the process of fossilization (molecular taphonomy); and
(3) the use of biomarkers in the reconstruction of long-term environmental and climatic change.

Identification of morphologically indistinct plant remains (chemosystematics)

The recognition that distinct differences exist in the chemical composition of different modern plant species has provided an opportunity to use chemical signals or 'biomarkers' from fossil plants as a tool in the identification of morphologically indistinct plant remains. This relatively new field of research, termed chemosystematics, can also be used to distinguish fossil plant organs, such as leaves and fruits, which are difficult to classify based on traditional morphological characters alone. Analyses of the chemical composition of waxes on the surface of modern plant cuticles (epicuticular waxes), for instance, have demonstrated that many genera, and even species, can be distinguished on the basis of the distribution of *n*-alkane and *n*-alkyl lipids in these waxes (Dyson and Herbin, 1968; Mimura *et al.*, 1998). This discovery has provided some important insights into the floristic composition of Tertiary fossil deposits, such as the famous Clarkia locality. The presence of different *n*-alkanes and *n*-alkyls in a number of different fossil leaf types from Clarkia, for example, has indicated that the waxes of these leaves had diagnostic chemical compositions before they were fossilized, thus confirming that these different leaf morphologies were indeed different species (Lockheart, 1997). The use of fossil wax *n*-alkanes has proved particularly helpful in identifying the species of oak in the Clarkia deposits as most probably very closely related to red oaks of the Appalachian region in North America rather than the white oak of northern Europe (Lockheart, 1997).

Chemotaxonomy has also been used widely in archaeological research to help identify the uses of archaeological artefacts. Lipid extracts from the inturned rim of a late Saxon bowl and jar, for example, indicated an extremely close correspondence to those of natural wax (Figure 9.10), leading to the conclusion that this bowl had been used in the storage of wax (Evershed, 1993). Similarly, desiccated seeds from a 6th century AD storage vessel from Egypt, which were unidentifiable by more traditional morphological means, were identified as ancient radish seeds based on the analysis of their preserved lipid profiles (O'Donoghue *et al.*, 1996). The presence or absence of particular lignins, in much older samples of indistinguishable organic matter in sedimentary rocks, can be used to infer the proportion of gymnosperms (i.e. conifers) versus angiosperm dicotyledons versus monocotyledons in the organic matter (Figure 9.11). This, in turn, can be used to provide a picture, albeit a crude one, of the composition of fossil ecosystems, because the lignin of each group (that is monocotyledons, dicotyledons, and gymnosperms) contains diagnostic units which can be identified readily by geochemical analysis (De Leeuw *et al.*, 1995). Indeed, much of the evidence for the first appearance of photosynthetic algae or the origin of life itself come from 'chemical fossils' or biomarkers of past metabolic activity from indistinguishable organic remains locked in the rock record.

Biomolecules and the plant fossilization process (molecular taphonomy)

Detailed studies of the chemical transformations and alterations that take place when biomolecules and resistant biomacromolecules are transferred from living plants to the fossil record are providing some very important insights into the underlying process of plant fossilization. The factors that are known to play an important role in the plant fossilization process (Eglinton and Logan, 1991) include:

(1) the original composition of the plant tissue or organ;

(2) the conditions prevalent at the time of death and burial of the plant material;

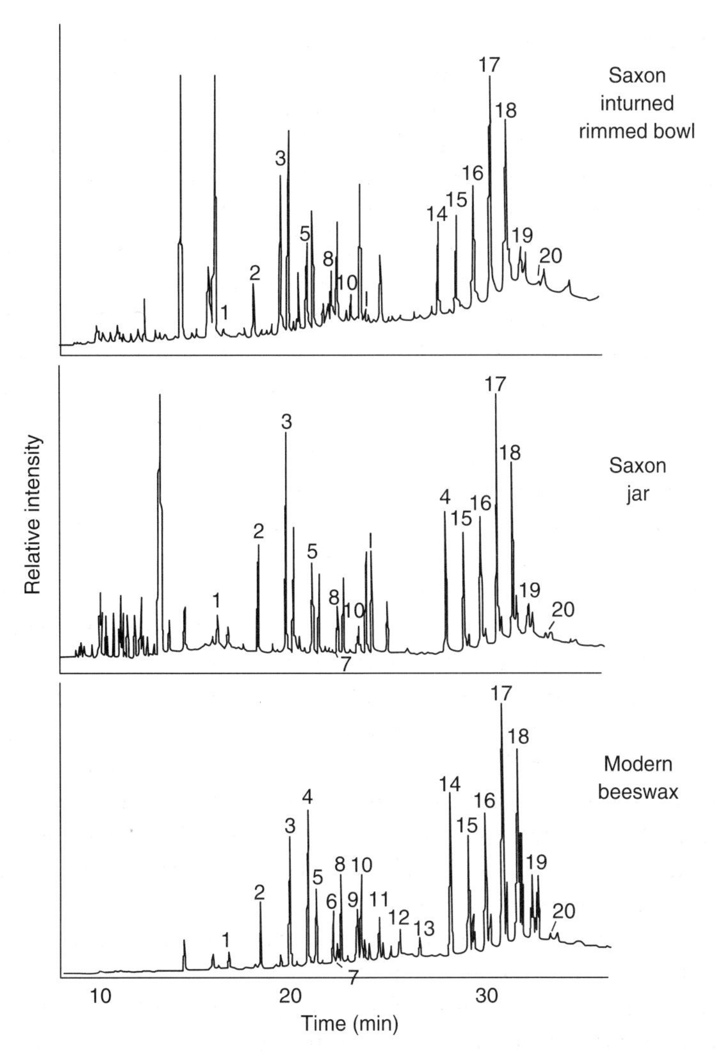

Figure 9.10 Gas chromatogram (GC) traces of lipid extracts from the inturned rim of a late Saxon bow and jar indicates an extremely close correspondence to that of natural wax, leading to the conclusion that this bowl was used in the storage of wax (Evershed, 1993).

(3) the process of rock formation (diagenesis) and fossilization of the plant material; and

(4) any further alteration to the rock through changes in temperature or pressure (i.e. the degree of metamorphism).

One of the most common forms of leaf fossil in the geological record is the coaly compression, which is made up essentially of a cuticular envelope surrounding a layer of coal which, before fossilization, consisted of mesophyll tissue. How can the cuticle, which is the transparent protective waxy layer that

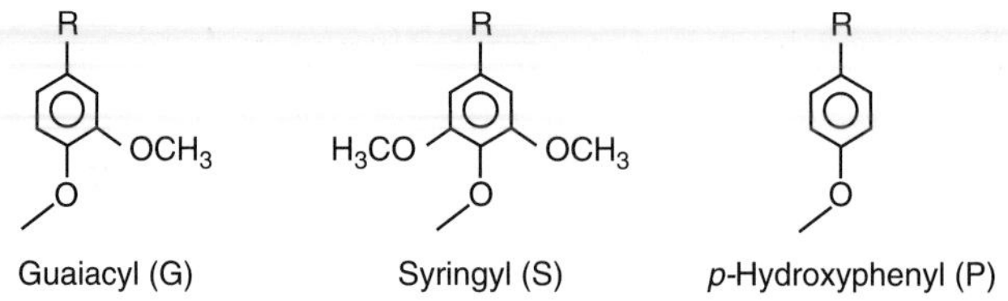

R R R

OCH_3 H_3CO OCH_3

O O O

Guaiacyl (G) Syringyl (S) *p*-Hydroxyphenyl (P)

Figure 9.11 Chemical composition of some of the lignin units in vascular higher plants: guaiacyl (G), syringyl (S), and *p*-hydroxyphenyl (P). Gymnosperms contain only G units, some angiosperms (dicoytledons) contain G and S units, and other angiosperms (monocotyledons and legumes) contain all three. These types of lignin can therefore be used as diagnostic units, since they can be identified readily by geochemical analysis to aid palaeoenvironmental interpretation (De Leeuw *et al.*, 1995).

covers the aerial parts of all terrestrial plants, be so well preserved when all of the rest of the leaf is turned to coal during fossilization?

Initially, the presence of highly resistant biomacromolecules, such as cutan, in plant cuticle was believed to be a necessary prerequisite to the preservation of cuticle in the plant fossil record (Van Bergen *et al.*, 1995). However, more recent research has demonstrated that this is not the case (Mosle *et al.*, 1997). The original chemical make-up of fossil plant cuticle is apparently not as important as the chemical alteration of the macromolecules within the cuticle, which occurs during the process of diagenesis (rock formation) (Collinson *et al.*, 1998; Stankiewicz *et al.*, 1998b). During rock formation, certain residues of macromolecules present in leaf cuticle are stabilized into a highly decay-resistant form and it is this process that aids the preservation of plant fossil cuticle in the fossil record. Understanding this process helps to explain the extraordinary preservation of plant cuticle in rocks that are as old as 400 million years.

These studies also have important implications in the interpretation of the plant fossil record (Tegelaar *et al.*, 1991). For instance, does the absence of a particular taxon from the fossil record imply that it had not yet evolved? Or was it absent because of a low preservation or fossilization potential? Water lilies (*Nymphaea*), for example, possess a lignin–cellulose complex in their seed coats. As this is a very resistant biomacromolecule, it is believed to contribute to their seeds being common in the fossil record (Van Bergen *et al.*, 1996, 1997). The seed coats of another genus of water lily (*Nelumbo*), on the other hand, are devoid of lignin and the absence of this resistant biomacromolecule is believed to be the main reason for the absence of fossil *Nelumbo* seeds in older sediments. This bias, due to differences in fossilization potential of two different, but related, genera of water lilies, which also share similar ecologies, must be taken into account when interpreting their evolutionary lineage in the fossil record. The late appearance of *Nelumbo* fossil seeds, in this case, does not indicate that this genus evolved relatively recently, but was, rather, most probably present in the same environments as *Nymphaea* but, due to significant differences in its seed walls, had a low fossilization potential. It is interesting to note, however, that *Nelumbo* seeds are known for their remarkable longevity in modern lake sediments and seeds as old as 1200 years have been germinated successfully. The chemical requirements, in this case, for the long dormancy of *Nelumbo* seeds are not therefore the same as those required for excellent preservation in the plant fossil record (Van Bergen *et al.*, 1997).

Biomarkers in the reconstruction of long-term environmental change

The use of the fossil biomolecular record to reconstruct long-term environmental change has tended to focus more on the biomarker record contained in ocean rather than terrestrial sediments. This is because marine records have a greater potential in terms of providing a long, continuous record through geological time.

One of many examples (reviewed by De Leeuw *et al.*, 1995) of ancient biomolecules derived from plant material, proving to be of particular use in the reconstruction of long-term environmental change, are lipids of prymnesiophyte algae (unicellular biflagellate algae, known as coccoliths). Laboratory results have revealed that many marine algae can biosynthetically tailor their lipid composition to match environmental conditions of stress, such as changing sea-surface temperatures, by changing the molecular composition of the lipid bilayer and thus maintaining the fluidity of their membranes (Hardwood and Russell, 1984). These marine phyotoplankton tend to be extremely well preserved in marine sediments and their molecular composition readily measured. Comparison of the molecular composition of their ancient lipid bilayers with standards obtained from present-day populations in regions of known oceanic temperatures, have thus enabled temperature curves to be reconstructed over geological time (Brassell *et al.*, 1986).

Other examples of how biomolecules which are present in marine organic sediment can be used as 'palaeoenvironmental indicators' include chromans of algae and cyanobacteria as indicators of salinity, and long chain *n*-alkanes as indicators of input by terrestrial vegetation into marine sediments. Certain fatty acids and triterpanes are indicative of seagrasses and mangrove-type vegetation respectively, whereas transformed biomacromolecules such as cutans, sporopollenin, and lignins may indicate either freshwater or terrestrial plant input to marine organic sediments (De Leeuw *et al.*, 1995). Often, multiple analyses of many of these biomarkers can be used to infer the proximity of marine sediment to land, which is particularly useful in the reconstruction of palaeo-shore lines.

9.7 Stable carbon isotopes ($\delta^{13}C$) and the fossil plant record

Isotopes are forms of elements that have the same atomic number but a different atomic mass. There are two stable isotopes of carbon, namely ^{13}C and ^{12}C, and the ratio of these two isotopes (the stable carbon isotopic composition, $\delta^{13}C$) in

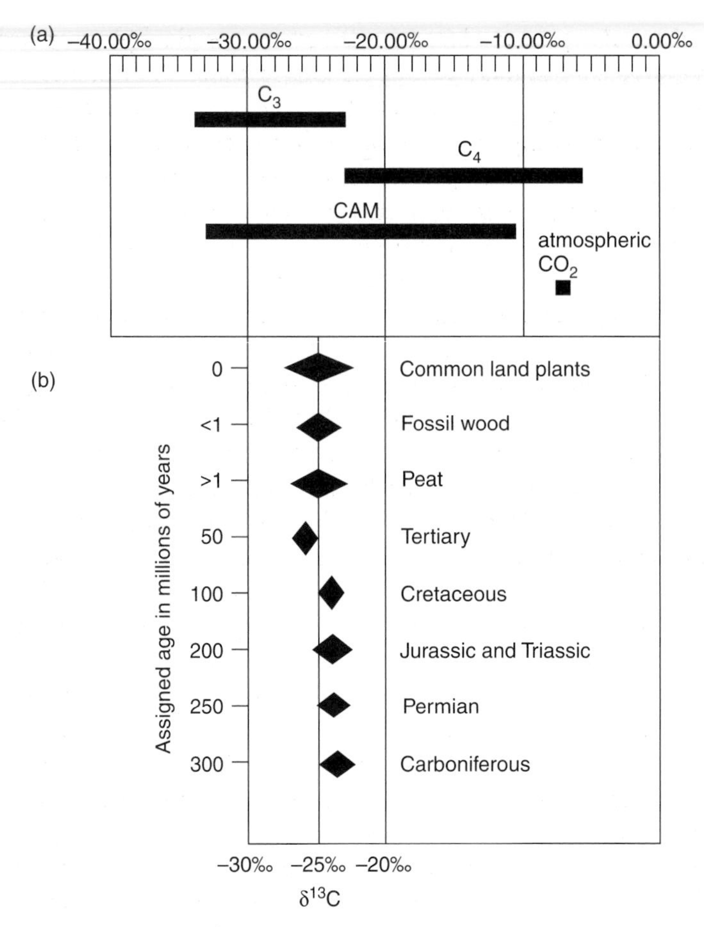

Figure 9.12 (a) The ranges of $\delta^{13}C$ values of modern C_3, C_4, and CAM plants. (b) The ranges of $\delta^{13}C$ values of common land plants, fossil wood, peat, and Tertiary to Carboniferous coal (redrawn from Jones, 1993).

plant tissue ranges from between –23‰ and –30‰ in plants with C_3 photosynthetic pathways, –10‰ to –18‰ for plants with a C_4 photosynthetic pathway, and between –12‰ and –30‰ for CAM plants (Figure 9.12) (see Chapter 7). The stable carbon isotopic composition of all plants is much lower than that of the atmosphere (1.1‰) because plants preferentially take up the lighter isotope, ^{12}C, for photosynthesis, resulting in a lower ratio of ^{13}C to ^{12}C. The stable carbon isotopic composition of fossil plant organs (such as leaves, wood, fruits) and tissues (such as cellulose, cuticle, periderm) remains even less altered than many of the most resistant biomolecules and biomacromolecules following fossilization. Jones and Chaloner (1991) have estimated that even high-temperature charcoalification does not drastically alter the stable carbon isotopic composition of original plant material. Furthermore, it has been suggested that, once plant material is preserved in the fossil record, original plant carbon (such as coal) retains the approximate $\delta^{13}C$ of the parent plant whole tissue, regardless of the fossil's age, phylogeny, or burial history (Degens, 1967). Differences in the stable carbon isotopic composition do occur, however, between different tissues and biomolecules of the same plant. Thus, when utilizing fossil plant stable carbon isotopic ratios in long-term environmental reconstructions, the same tissues (cuticle) or even biomacromolecules (lignin etc) should ideally be used, rather than comparing results from different tissues.

The carbon isotopic composition of fossil plants has been used to investigate a range of different environmental and biological phenomena, from the first appearance and evolution of the three photosynthetic pathways (C_3, C_4, and CAM; see Chapter 7) (e.g. Namburdiri *et al.*, 1978; Bocherens *et al.*, 1994) and changes in ecosystem composition (e.g. Cerling *et al.*, 1993), to long-term changes in the isotopic composition of the atmosphere (e.g. Gröcke *et al.*, 1999; Arens and Jahren, 2000). This method has also been employed to detect changes in concentrations of both carbon dioxide (e.g. Cerling, 1992; Ekart *et al.*, 1999) and oxygen in ancient atmospheres (e.g. Karhu and Holland, 1996). Carbon isotope ratios can also provide signals of the salinity or moisture content of the soils in which plants lived in the past before fossilization (Nyguyen Tu *et al.*, 1999). Temperature changes (Beerling and Jolley, 1998) and physiological processes related to leaf gas exchange and photosynthesis have also been inferred from the $\delta^{13}C$ of fossil plants (McElwain *et al.*, 1999). The potential of stable carbon isotopic investigations using fossil plants is therefore certainly very broad (reviewed by Bocherens *et al.*, 1994; Beerling, 1997; Gröcke, 1998). The main limitation associated with their use, however, is in trying to decipher which of the myriad of biological and environmental factors are responsible for any trends in plant $\delta^{13}C$ detected in the fossil record.

Summary

1. The potential of ancient DNA was seen as far reaching in evolutionary research. This was not only at the level of the whole organism (e.g. identification of extinct species, the verification of previous fossil identifications, and in testing classification of well-studied fossil groups), but also at the molecular level, in the temporal measurement of DNA diagenesis, sequence evolution, and calibration of molecular clocks.

2. Strict protocols must be followed in the extraction and amplification of ancient DNA in order to avoid contamination. Successful extractions of plant DNA have been obtained from herbarium specimens, charred seeds and cobs, mummified seed, anoxic fossils, and amber.

3. The nature of the evolutionary questions being asked mainly determines which genome to target. In plants there is the choice of three genomes to target: mitochondrial DNA (mt DNA), plastid DNA (cpDNA), and nuclear DNA (DNA).

4. Examples of current research where ancient DNA has been extracted from plant tissue include ancient cpDNA sequence from *Magnolia* leaves (17–20 Ma), ancient cpDNA sequence from *Taxodium* fossils (17–20 Ma), and ancient cpDNA from the extinct tree *Hymenaea protera* (35 Ma).

5. There is, however, a fair amount of controversy surrounding the extraction of ancient and fossil DNA and the authenticity of a number of studies.

6. Concerns are expressed regarding the degradation, amplification, and contamination of ancient DNA.

7. Experimentation indicates that theoretically no fragments greater than 150 base pairs should remain after approximately 2700 years and no purines after 4 million years.

8. If the numbers of base pairs of fossil DNA recovered are small, large errors will occur in the amplification process.

9. There is also an increased probability of contamination from the micro-organisms that have been associated with the fossil over time, and therefore a risk of amplifying the DNA from these organisms rather than the fossil of interest.

10. One area where ancient DNA is proving to be an invaluable tool, however, is in the study of the more 'recent' past. There have been numerous successful extractions of ancient DNA from samples that are less than 100 000 years old, and these are starting to provide an important tool for archaeological research.

11. Studies of other fossil plant biomolecules, biomacromolecules, and chemical constituents indicate that the chemical make-up of many of the more resistant fossil plant tissues has remained highly conserved through geological time. The ability to isolate recognizable chemical constituents has far-reaching potential for interpretation of the fossil record, including identification of morphologically indistinct plant remains, information on the plant fossilization process, and in the reconstruction of long-term environmental change.

12. Stable isotopes ($\delta^{13}C$) in plant tissue are also proving to be extremely important tools in the study of long-term environmental and biological change, providing information ranging from the first appearance of the three different photosynthetic pathways and changes in ecosystem composition, to long-term changes in the isotopic composition of the atmosphere.

10 Evolutionary theories and the plant fossil record

Molecular and morphological studies of both living and fossil plants have made huge inroads in our understanding of evolution since Darwin proposed the theory of natural selection (Darwin, 1859). However, the majority of macro-evolutionary theory, which seeks to define and explain the course and tempo of large-scale patterns in evolution, has been based almost exclusively on evidence from the animal kingdom, with only a handful of notable exceptions (cf. Stebbins, 1950, 1974; Grant, 1963; Chaloner, 1967, 1970; Banks, 1968, 1970; Knoll, 1984, 2001; Niklas *et al.*, 1985; Traverse, 1988; Bennett, 1997; Niklas, 1997). This final chapter attempts to redress this balance by outlining a number of the main evolutionary theories, which are very much part of current and lively debate (e.g. Dawkins, 1990; Gould, 1998), and examining them in the light of the plant fossil record described throughout this book.

10.1 Evolutionary theories

Phyletic gradualism (Darwin, 1859)

The theory that is now termed 'phyletic gradualism' was proposed originally by Charles Darwin back in 1859. It is based on the principle that evolutionary change results from descent with modification and is accomplished by means of a process termed 'natural selection'. Species are therefore temporary stages in continuing gradual change. Ecological interactions visible today are the ultimate driving force behind macroevolution. Phyletic gradualism thus supports the view that speciation is a gradual and continuous process and 'if variations useful to any organic being do occur, individuals thus characterised will have the best chance of being preserved in the struggle for life; and from the strong principle of inheritance they will tend to produce offspring similarly characterized' (Darwin, 1859). A diagrammatic representation of this theory (the only diagram in Darwin's book) indicates a cone of increasing diversity (Figure 10.1a), broadening out from a single trunk at the base to numerous branches at the top, thus representing continuously increasing morphology and diversity through time. Inherent in this theory is the assumption that the present ecosystem is the most

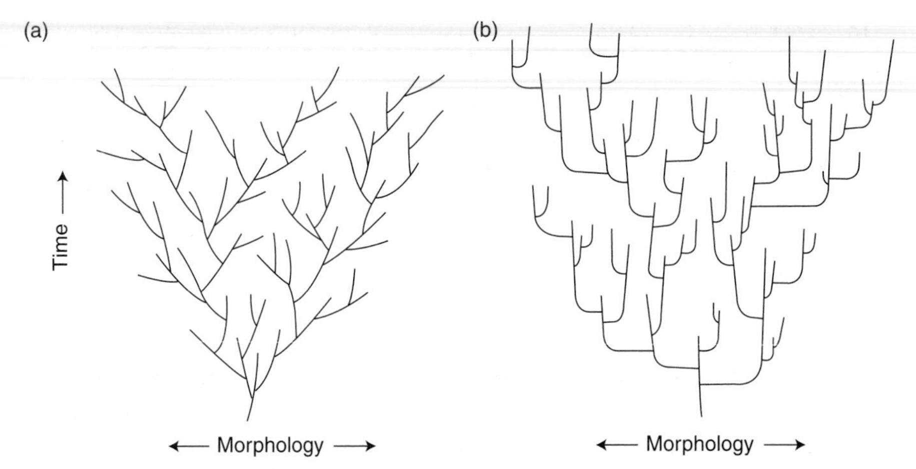

Figure 10.1 Diagrammatic representation of (a) Phyletic gradualism (*sensu* Darwin, 1859), where speciation is a gradual and continuous process. This is indicated by a cone of increasing diversity broadening out from a single trunk at the base to numerous branches at the top, thus representing continuously increasing morphology and diversity through time. (b) Punctuated equilibrium (*sensu* Eldredge and Gould, 1972), where there are relatively rapid periods of speciation interspersed by long periods where there appears to be little morphological change or speciation, referred to as stasis. This is indicated by a cone of increasing diversity, but with vertical branches representing periods of stasis, interspersed by horizontal lines representing periods of rapid change (redrawn from Stanley, 1989).

diverse and morphologically adapted in geological time (Bennett, 1997). With increasing scientific knowledge, especially with regard to processes of isolation (both geographic and genetic) and genetic inheritance, many embellishments have been added to this theory (see, for example, Mayr, 1954, 1963; Dawkins, 1976, 1986, 1989).

Punctuated equilibrium (Eldredge and Gould, 1972)

The theory of punctuated equilibrium was first proposed by Eldredge and Gould in 1972. It was based on their observation that the geological record does not indicate a long sequence of continuous intermediate species from the parental to the new (descendant) species (which would be expected if phyletic gradualism was occurring), but rather, there are long periods where there appears to be little morphological change or speciation, followed by periods of relatively rapid change. Historically, this pattern of abrupt speciation in the geological record had been attributed to gaps in the fossil record. Eldredge and Gould argued, however, that this assumption was unsafe and rather that the 'breaks' could represent periods of rapid speciation. Thus they envisaged a phyletic line that was usually in stasis disturbed only rarely by rapid and episodic events of speciation (Price, 1996). A diagrammatic representation of the punctuated equilibrium theory (Figure 10.1b) thus indicates a cone of increasing diversity, but this time with vertical branches representing periods of stasis, interspersed by horizontal lines representing periods of rapid change.

In recent years the theory of punctuated equilibrium has been further developed to incorporate the suggestion that mass extinctions (physically or biologically mediated) are the punctuating events leading to the apparent sudden speciation events. It is proposed that mass extinctions effectively wipe the slate clean by clearing away the accumulated evolution of a particular lineage, and thereby allow a new lineage to evolve rapidly to fill the niche (Gould, 1985). Thus Gould would argue that evolutionary time is a hierarchical system of distinct tiers (Table 10.1), with the uppermost tier of mass extinctions representing the ultimate driving mechanism behind macroevolution.

Many others before Eldredge and Gould had noted long periods of stasis in the fossil record, including, for example, Lyell in 1856 (cited in Bennett, 1997) and, indeed, Darwin himself commented that 'it is far more probable that each form remains for long periods unaltered, and then again undergoes modification'. However, the idea of stasis being punctuated by 'rapid periods' of change was new, as was the concept that mass extinctions were the ultimate driving force. The theory of punctuated equilibrium has now become a well-established alternative to the model of phyletic gradualism in the evolutionary debate, and has led some to go as far as to suggest that 'natural selection seems to provide little more than fine material and fine adjustment to large-scale evolution' (Stanley, 1975).

The debate remains lively, however, as to which of these two modes of evolution, gradualism or punctuation and stasis, predominates in the fossil record. A recent survey of 58 studies investigating whether speciation in the fossil record predominantly conformed to phyletic gradualism or punctuated equilibrium concluded that 'paleontological evidence overwhelmingly supports a view that speciation is sometimes gradual and sometimes punctuated and that no one mode characterizes this very complicated process in the history of life' (Erwin and Anstey, 1995). It is interesting to note that not a single study included in this comprehensive review was based on the plant fossil record. How does the fossil plant record fit into either of these two evolutionary theories? The existence of 'living fossils', such as *Ginkgo biloba*, which show remarkable morphological 'stasis' since the Jurassic (~200 Ma), would seem to support partially the theory of punctuated equilibrium. In terms of the mode of speciation, however, too few studies have addressed this issue in the plant fossil record. There are currently no examples of studies on punctuated equilibrium, using

Table 10.1 Temporal hierarchy of dominant processes controlling evolutionary patterns seen in the geological record (after Gould, 1985)

First tier	Microevolutionary processes, e.g. competition	Evolutionary events of the ecological moment
Second tier	Punctuated events, e.g. orbital variations	Trends with lineages over geological time
Third tier	Mass extinctions	Mass extinction

fossil plants, that stand up to the rigorous set of criteria required (see review in Gould, 2000) to really test the theory.

Irrespective of the mode of speciation, are there any large-scale macroevolutionary patterns observable in the plant fossil record? In a hierarchical scheme (*sensu* Gould, 1985) mass extinctions are the ultimate driving mechanism behind macroevolution in the animal kingdom. It was established in Chapter 8, however, that plants do not undergo mass extinction. If this is the case, what is the pattern and timing of major originations and extinction in the plant fossil record? Does plant evolution as represented in the fossil record indicate a trend of increasing total species diversity through time, or decimation of species numbers followed by renewed diversification? Are major origination events concentrated in certain intervals of time or do they occur evenly spaced throughout the evolutionary history of plants? Are there other driving mechanisms of large-scale patterns of plant evolution that are discernible? And how do theories of macroevolution tie in with what is known of the molecular record—do the forces that determine microevolution have any bearing on macroevolution?

10.2 Patterns of evolutionary change in the plant fossil record

Surprisingly few studies have examined data from the plant fossil record in terms of overall evolutionary patterns through geological time. However, those records that have attempted a synthesis, based on species-level compilations of originations and extinctions, broadly agree on the overall patterns of plant diversification (Chaloner, 1967, 1970; Banks, 1968, 1970; Knoll *et al.*, 1979; Niklas and Pratt, 1980; Niklas *et al.*, 1983; Lidgard and Crane, 1988; Crane and Lidgard, 1989; Edwards and Davies, 1990; Tiffney and Niklas, 1990; Kovach and Batten, 1993; Niklas and Tiffney, 1994; Niklas, 1997; Lupia *et al.*, 1999, 2000).

From the first appearance of terrestrial vascular plants approximately 430 million years ago (Silurian), the plant fossil record demonstrates a trend of increasing total species diversity through geological time (Niklas *et al.*, 1983; Niklas and Tiffney, 1994). Four major plant groups, each with common structural characteristics and/or reproductive strategies have successively dominated the terrestrial flora (Figure 10.2). The first major group, includes the earliest vascular plants, which originated approximately 430 million years ago (Silurian). These were superseded by the next two major groups or 'reproductive grades', the pteridophytes and gymnosperms, which originated in the late Devonian and lower Carboniferous, respectively, around 390 Ma and 340 Ma. The pteridophytes include the spore-producing plants (lycopsids, filicopsids, sphenopsids, progymnosperms) and the gymnosperms include the earliest seed plants (cordaites and pteridiosperms), conifers, cycads, bennettites, Gnetales,

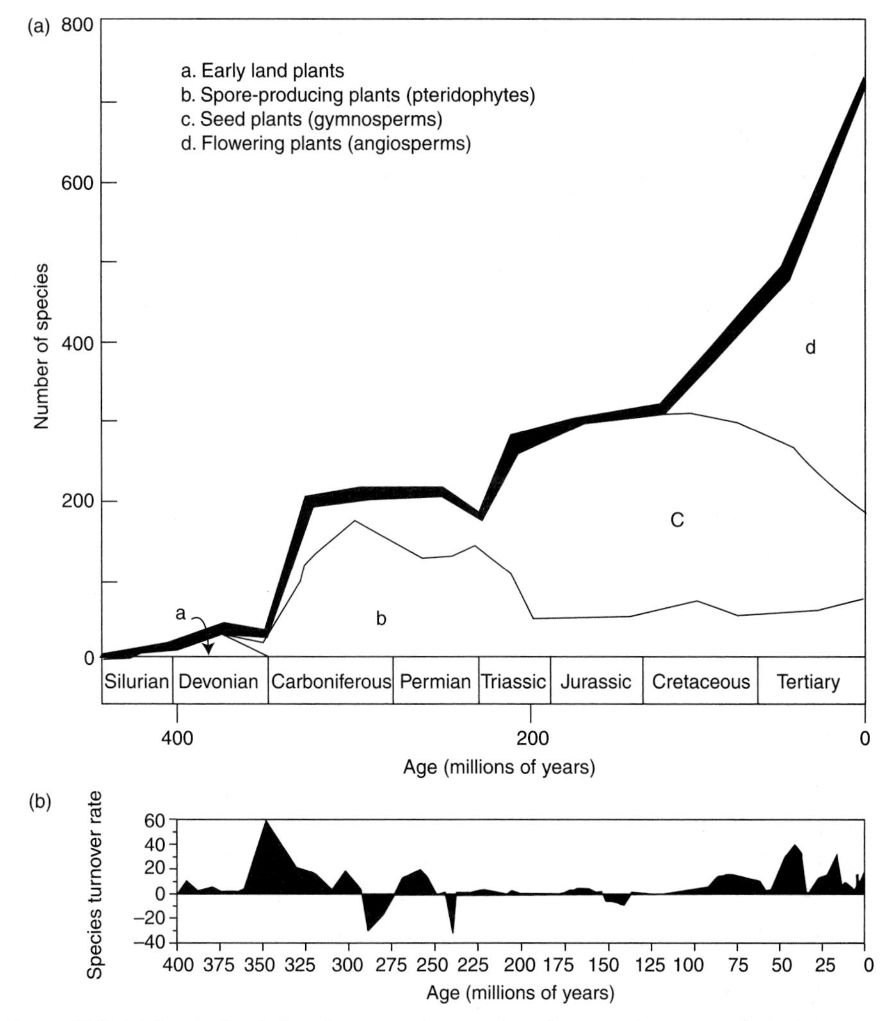

Figure 10.2 (a) Species-level diversity curves for vascular plants species over geological time. Also indicated is their classification into the four plant major groups (redrawn from Niklas *et al.*, 1983). (b) Rates of plant species turnover (originations minus extinctions) through geological time (redrawn from Niklas, 1997).

and ginkgos. The last major reproductive innovation and radiation was that of the angiosperms in the Cretaceous, from approximately 120 Ma. With the adaptive radiation of each new reproductive grade, a gradual decline in the previous grade occurred (Figure 10.2). There is, as yet, no evidence for large-scale reductions in total species-level diversity in the plant fossil record comparable to that seen in the faunal record and classified as mass extinction (with the possible exception of the Permian–Triassic boundary, ~250 Ma; see Chapter 8).

Various studies on rates of change in total species-level diversity through geological time are also in broad agreement (e.g. Niklas *et al.*, 1985; Crane and

Lidgard, 1989; Kovach and Batten, 1993; Niklas, 1997). These compilations provide a means to investigate the tempo of plant evolution, and indicate that following the origination of each new reproductive group or 'evolutionary flora', each group radiated, as revealed by an overall rise in total species number (Figure 10.2) and increased species turnover rate (where the fossil record indicates more originations over extinctions). It is interesting to note that the evolution of grasses, approximately 60 million years ago, also occurred during a period of increased species turnover. With the radiation of each new group (i.e. early vascular plants, pteridophytes, gymnosperms, angiosperms), the speciation rates were initially very high and then decreased to a low baseline (Figure 10.2). Thus the plant fossil record appears to indicate periods of rapid speciation and associated increases in species diversity, followed by a plateauing and then a decline (Niklas *et al.*, 1983; Niklas and Tiffney, 1994; Niklas, 1997). Furthermore, although the radiation of each new 'evolutionary flora' resulted in incremental increases in total species number, the diversity of the group being replaced (e.g. the gymnosperms by the angiosperms) then declined. The only radiation that does not demonstrate this decrease to a low baseline is the most recent (the angiosperms). It could be argued, however, that this low baseline has yet to be seen in the angiosperms, since overall species number are still increasing and have yet to plateau.

At the broadest scale, therefore, the overall pattern of plant evolution appears to be concentrated in distinct intervals in geological time rather than evenly spread throughout. Similarly, increases in morphological complexity and total plant species diversity have occurred during relatively short intervals of geological time, interspersed by longer periods of relative stasis (in terms of major evolutionary innovations) and plateaus in species diversity. The peaks of plant originations seen in the plant fossil record, do not, however, appear to match those seen in the animal record, which tend to follow mass extinction events. Furthermore, the peaks of plant extinction do not coincide with at least four of the 'big five' mass extinctions events (see Chapter 8).

10.3 The mechanisms driving evolutionary change

Is it possible therefore to identify specific climatic/environmental events that may be responsible for major evolutionary changes seen in the plant fossil record? And at what scale (time and space) do these events have to be, in order to bring about major evolutionary change in the plant kingdom?

In the literature, the idea that changes in the environment may account for evolutionary change has warranted much discussion. Regardless of whether these changes are gradual or punctuated, geologists (in particular) have long argued that changes in the environment are largely responsible for the radiation and extinction of taxa (Knoll, 1991). This is also consistent with recent studies of islands, which show that speciation is accelerated by chronic physical or biological disturbances (Niklas, 1997). However, it is only recently that these ideas have been fully incorporated into the theory of punctuated equilibrium, in particular through the work of Vrba (1985, 1993) and Bennett (1990, 1997) to develop the so-called 'turnover-pulse' hypothesis.

The turnover-pulse hypothesis builds on the ideas put forward by Eldredge and Gould (1972) to suggest that evolution is normally conservative and speciation does not occur unless forced by changes in the physical environment. This hypothesis therefore supports the concept of punctuated equilibrium, but takes it one step further to suggest that periods of speciation are largely a function of changes in the physical environment, especially climatic change, possibly of tectonic origin, but most probably of astronomical origin (Vrba, 1985, 1993) (Figure 10.3). The model also proposes that extinction may result directly from climate change or from ecological interactions among newly associated species. The question therefore arises as to whether periods of evolutionary change in the plant record can be related to evidence in the geological record for periods (pulses) of changing environmental conditions. Is there a set of environmental conditions particularly associated with the periods of increased originations over extinctions (that is, turnover) in the plant fossil record? Also, can changes in the physical environment bring about evolutionary change both at the macro- and microevolutionary level?

At the broadest scale, factors responsible for bringing about major environmental and/or climatic change include orbital variations and plate tectonics.

Orbital variations

Over geological time, calculations indicate that variations in the Earth's axial tilt, precession, and changes in the eccentricity of the elliptical orbit around the sun have produced differences both in the amount of solar radiation received by the Earth and also in the latitudinal and seasonal distribution of this radiation (Milankovitch, 1930). The cyclical nature of these variations, known as Milankovitch cycles, has resulted in large and significant astronomical forcing of the climate at intervals of approximately 400, 100, 41, and 23–19 thousand years (Imbrie and Imbrie, 1979). Evidence from both ocean and terrestrial records indicates that these cycles have extended as least as far back as the Jurassic (approximately 200 million years ago), but they have probably occurred throughout geological time (Shackleton and Opdyke, 1976; Herbert and Fischer, 1986; Herbert and D'Hondt, 1990). The reduction of incoming solar radiation induced by variations in the Earth's orbit triggers global climatic change, leading to cold-stage conditions and the formation of major ice caps at the poles during some intervals, including the Quatnernary (Bradley, 1999).

Therefore the Milankovitch cycles have paced numerous warm/cold stages throughout geological history, including, for example, the glacial–interglacial cycles of the Quaternary (the last 1.8 million years), and the pacing effect of

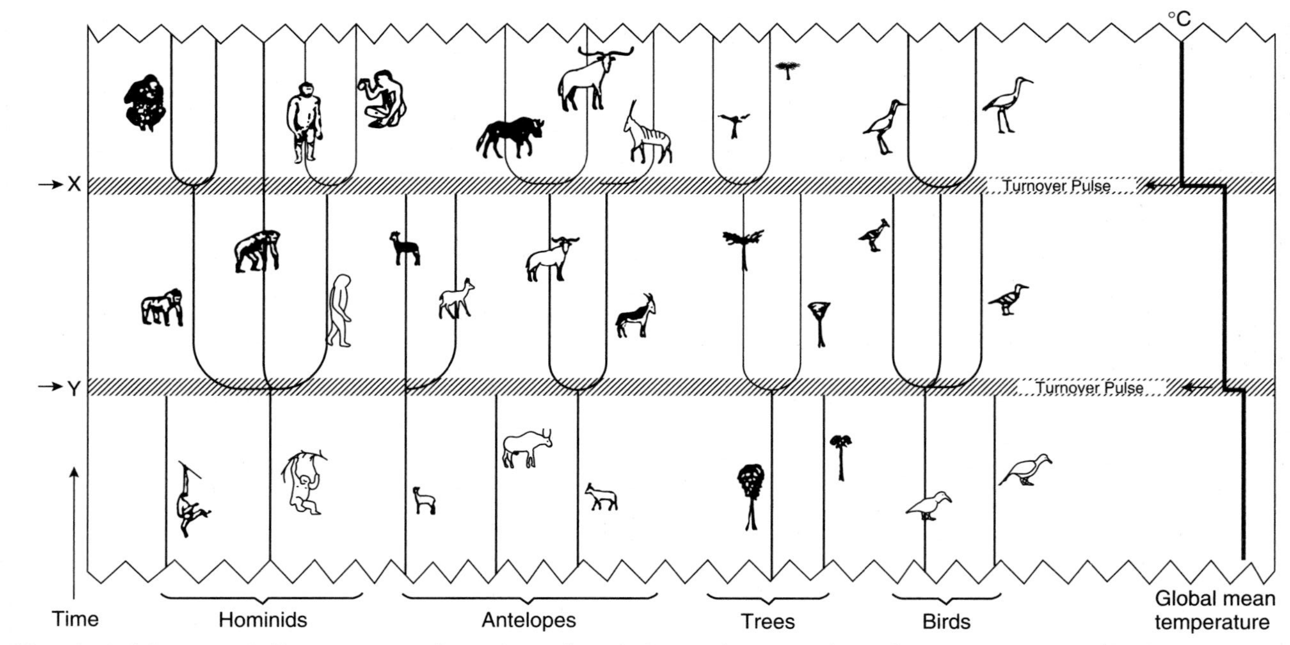

Figure 10.3 Hypothetical diagram to indicate turnover pulses at times of major climatic change x and y resulting in speciations and extinctions in groups as different as hominids, antelopes, trees, and birds (redrawn from Vrba, 1993).

these Milankovitch cycles has been proposed as the pulsing mechanism behind punctuated equilibrium (Bennett, 1990, 1997). It is suggested that the cumulative long-term effect of communities becoming split up and isolated during the cold stages, followed by recombination in warm stages will ultimately lead to speciation and evolution of new lineages. Small adaptive changes may accumulate during ecological time, but most of these changes will become lost as communities are restored as a result of climatic changes on Milankovitch timescales. Speciation will thus occasionally take place in populations which have become isolated because of perpetual environmental change at Milankovitch timescales (Table 10.2).

Some of the most detailed fossil evidence for the impact of Milankovitch cycles upon terrestrial vegetation is contained in pollen records. A record from Bogota in Colombia (Hooghiemstra, 1989, 1995; Hooghiemstra and Sarmiento, 1991) spanning the past 3.5 million years, for example, clearly indicates a strong vegetational response to climatic changes associated with Milankovitch forcing of the climate (Figure 10.4). The vegetation on the high plain of Bogota fluctuated between mixed Andean forest (containing, for example, species such as *Podocarpus, Hedyosmum, Weinmannia*, and *Myrica*) during the interglacials, and subparamo (e.g. Ericaceae, *Hypericum*, Compositae) or grassparamo (e.g. Gramineae, *Valeriana*, Caryophyllaceae) during the glacials. A sequence from a crater lake in Hungary, deposited between 3–2.7 million years ago and containing a detailed chronology in well-defined yearly layers within the sediment, indicates clear cycles in the vegetation at intervals of approximately 100, 41, and 23–19 thousand years. The pollen record from this Hungarian site indicates that in response to orbital variations in insolation, vegetation changed from a forest dominated by subtropical and temperate species to one dominated by those types presently found in the northern latitude boreal forests (Willis *et al.*, 1999a,b) (Figure 10.5). Thus during the cold stages, boreal forest extended south to fill the niche occupied by subtropical and temperate species during the warm stages.

Table 10.2 Temporal hierarchy of dominant processes controlling evolutionary patterns seen in the geological record (from Bennett, 1990, 1997)

Tier	Periodicity	Cause	Evolutionary process
First	–	Natural selection	Microevolutionary change within species
Second	20 000–100 000 years	Orbital forcing	Disruption of communities, loss of accumulated change
Third	–	Isolation	Speciation
Fourth	~26 Ma	Mass extinctions	Sorting of species

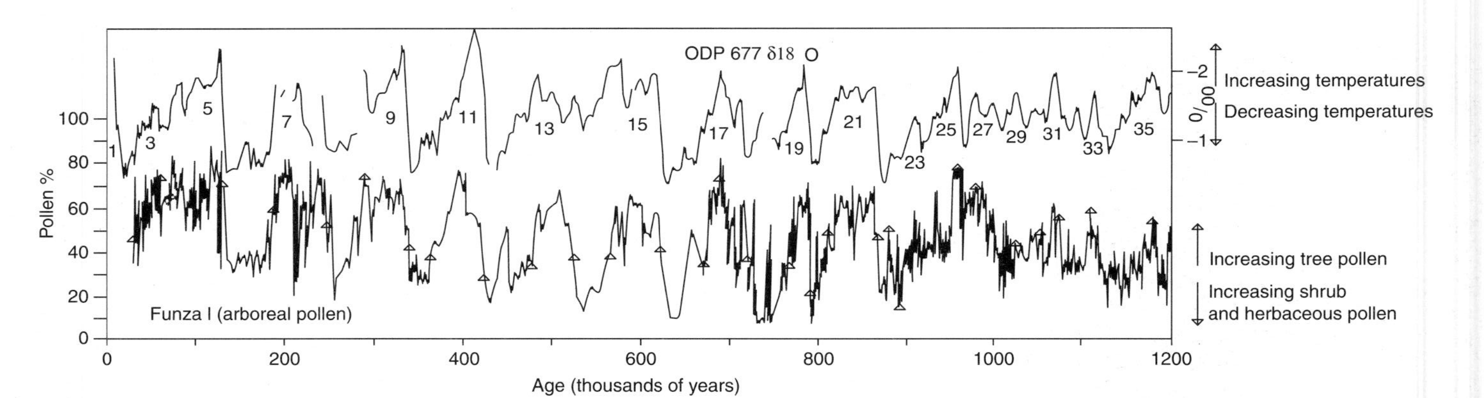

Figure 10.4 Graphical correlation of the tree pollen record spanning the past 1.2 Ma from Funza, on the High Plain, Bogota, with a proxy temperature record obtained from the $\delta^{18}O$ record from deep ocean core ODP677 (Shackleton *et al.*, 1990; redrawn from Hooghiemstra, 1995). The increase in tree pollen represents mixed Andean forest (containing, for example, species such as *Podocarpus*, *Hedyosmum*, *Weinmannia*, and *Myrica*; the increase in shrub and herbaceous pollen representing subparamo (e.g. Ericaceae, *Hypericum*, Compositae) or grassparamo (e.g. Gramineae, *Valeriana*, Caryophyllaceae).

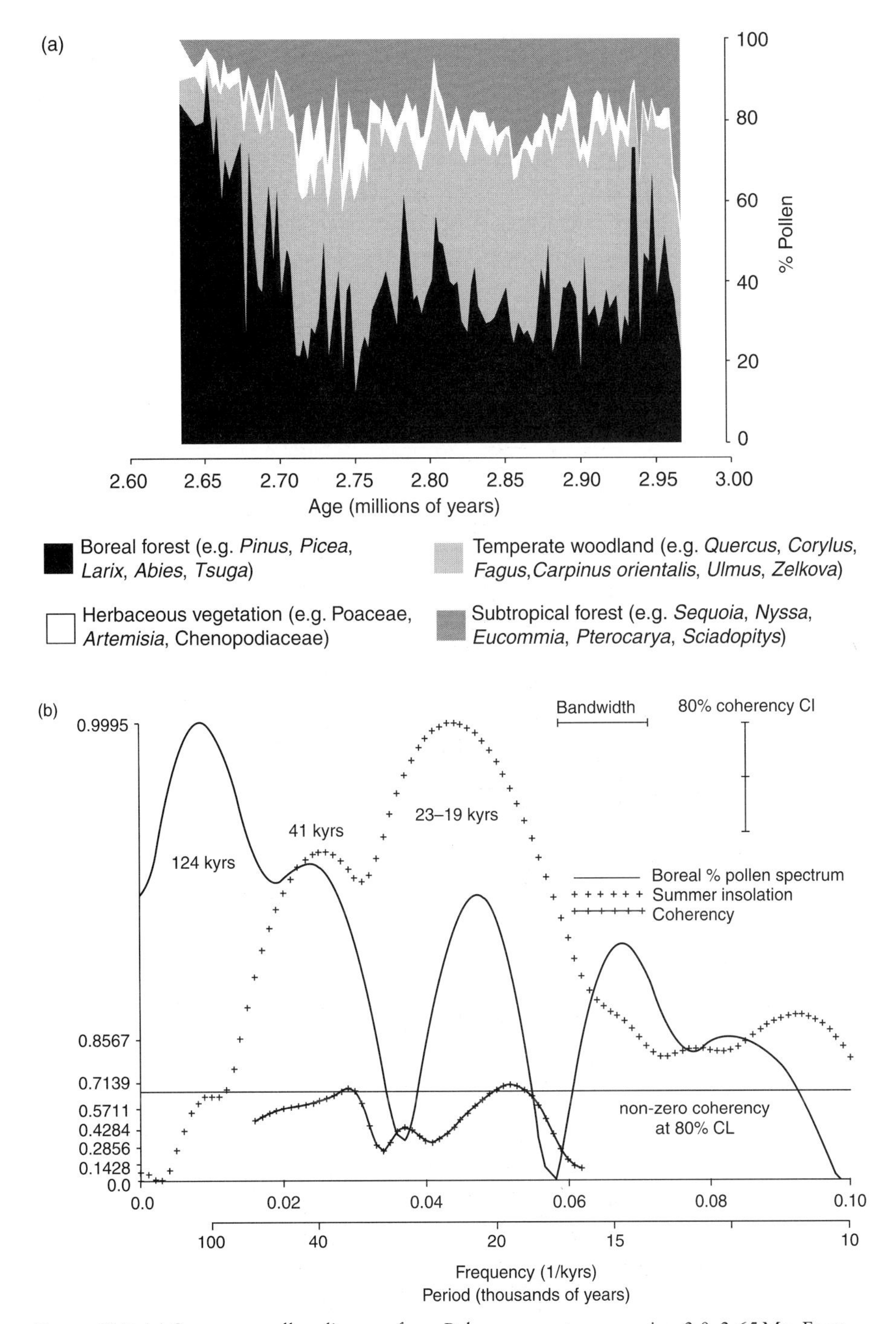

Figure 10.5 (a) Summary pollen diagram from Pula maar crater, spanning 3.0–2.65 Ma. Four predominant vegetation formations can be recognized in this diagram, namely boreal forest, temperate woodland, herbaceous vegetation, and subtropical vegetation. (b) Spectral analysis of this data set (thus indicating the predominant cycles in the data) revealed that during 3.0–2.65 Ma the vegetation alternated between one dominated by boreal forest to one dominated by subtropical temperate woodland, and that the periodicity of these fluctuations was at ~100, 41, and 21–19 kyr, corresponding closely with climatic change at Milankovitch frequencies. (Redrawn from Willis *et al.*, 1999a, b.)

The overall biogeographical effects of these alternating cold/warm stages has therefore been a global redistribution of vegetation. During the cold stages (glacials) temperate vegetation becomes isolated into regions where micro-environmentally favourable conditions exist (refugia). It then re-expands from these regions during the warm stages (interglacials). On a broad global scale, regions such as the Tropics provided important cold-stage refugial areas during the cold stages of the Quaternary (Bush, 1994; Colinvaux *et al.*, 2000) and, on a more localized scale, in Europe for example, temperate vegetation became isolated into small refugial areas in south and south-eastern Europe (Figure 10.6) (Bennett *et al.*, 1991; Willis, 1996; Willis and Whittaker, 2000). In fact, during the 1.8 million years of the Quaternary, plant communities have spent more time in isolation than in the integrated communities of the interglacials (Willis, 1996).

Many of the regions that are thought to have been large (e.g. Amazon; Bush, 1994) or small refugial areas (e.g. southern Europe; Bennett *et al.*, 1991) during the glacial periods appear to have the greatest present-day species diversity (Willis and Whittaker, 2000), thus suggesting that long-term isolation has possibly led to diversification and speciation within these regions. Molecular evidence adds further support to this suggestion (Hewitt, 2000). Examination of extant populations of the silver fir (*Abies alba*) (Konnert and Bergmann, 1995), oak (*Quercus robur*) (Hewitt *et al.*, 1985), and beech (*Fagus sylvatica*) (Demesure *et al.*, 1996; Taberlet and Bouvet, 1994) in central, southern, and northern Europe, for example, indicates that there is significant genetic

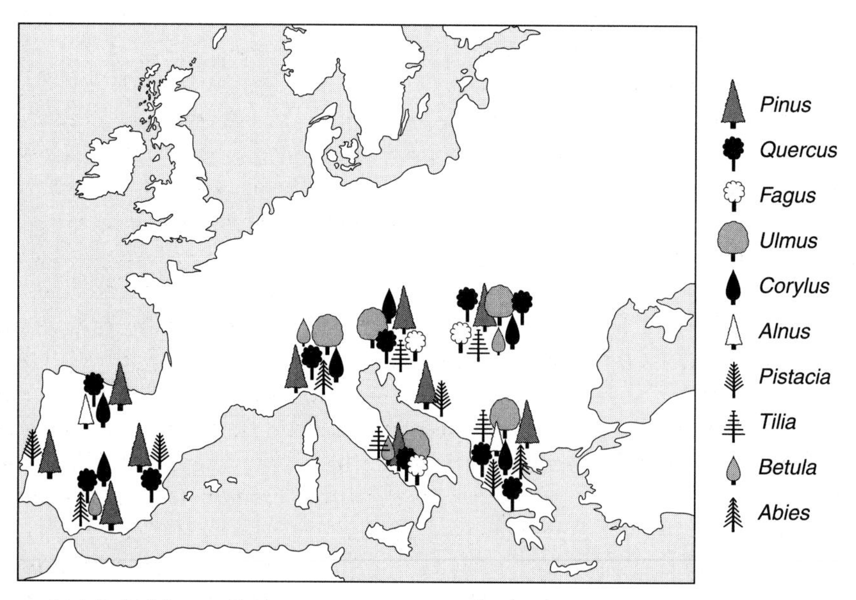

Figure 10.6 Refugial areas for temperate tree taxa in the three southern peninsulas of southern Europe during the last full glacial (~100 000–15 000 years ago). Evidence from microfossils, macrofossils, and macrofossil charcoal data (redrawn from Willis, 1996).

differentiation between populations relating to their regions of isolation during the last full glacial (between approximately 100 and 15 thousand years ago). In the study of *Abies alba*, the genetic characteristics of populations were studied throughout central and southern Europe (Konnert and Bergmann, 1995). The results indicated that *Abies alba* was genetically differentiated to a high degree in five postulated regions, and can be identified as populations with alleles characteristic of the Balkans, northern Italy, southern Italy, southern France, and southern Spain (Figure 10.7a). These regions correspond closely with palaeoecological evidence for the refugial areas of *Abies alba* during the last full glacial. However, evidence from this study also indicates that in only two of these regions have the populations remained isolated during the current warm stage, namely in Spain and southern Italy. In comparison, populations from the other three regions have colonized central and eastern Europe, creating a large mixed gene pool (Figure 10.7b). From an evolutionary viewpoint, therefore, it is only those populations that remain isolated during the warm stages that will carry the genetic differentiation on into the next glacial period (Willis, 1996).

In terms of isolation, redistribution of communities, and punctuated events, evidence from the fossil record would, therefore, appear to suggest that orbitally induced variations in insolation leading to climatic variations at Milankovitch frequencies are possibly one of the driving mechanisms behind punctuated equilibrium (Vrba, 1985, 1995; Bennett, 1990, 1997). However, it is not so easy to demonstrate from the fossil record that this process of isolation and redistribution leads to speciation and thus micro- and macroevolution (Willis and Whittaker, 2000; Hewitt, 2000).

Considering the long-term effects of isolation and redistribution through multiple glacial/interglacial cycles during the Quaternary, the palaeoecological evidence is ambiguous. The sequence from Hungary (described above), for example, indicates that the overall long-term effect of the continuos warm/cold stages in the 320 000-year interval covered by this section appears to be extinction rather than speciation (Willis *et al.*, 1999a,b). Similarly if the 3.5 Ma sequence from Bogota is examined, there is a similar overall reduction in species number with time. In fact, throughout the northern hemisphere, the overall trend of the Quaternary has been one of plant extinction (Van der Hammen *et al.*, 1971; Tallis, 1991; Coxon and Waldren, 1997) (Figure 10.8). But it is also questionable whether this really represents extinction, since the majority of plants that have disappeared are still living, but in more southerly latitudes (Figure 10.8) and there are certainly no mass extinctions. It is also questionable, therefore, whether a long-enough time-slice has been considered. The major radiations in the plant fossil record occurred over millions of years, whereas Quaternary glaciations barely provide continuous isolation for 2 million years.

But what happens over longer timescales? Is there increased speciation in relation to orbital oscillations at Milankovitch frequencies, or do these effects become lost in subsequent interglacials? Geological evidence suggests that the impact of orbital variations on patterns of global climate has varied

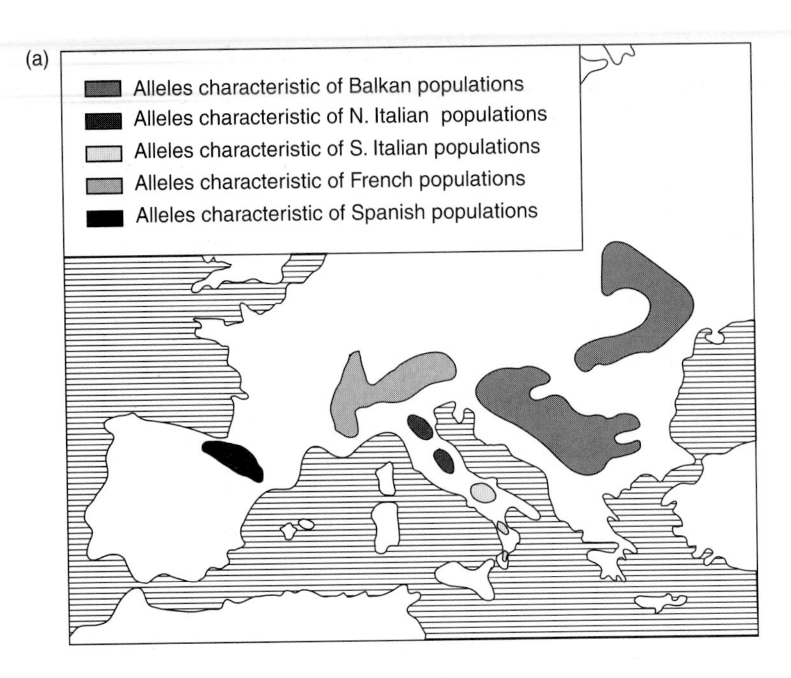

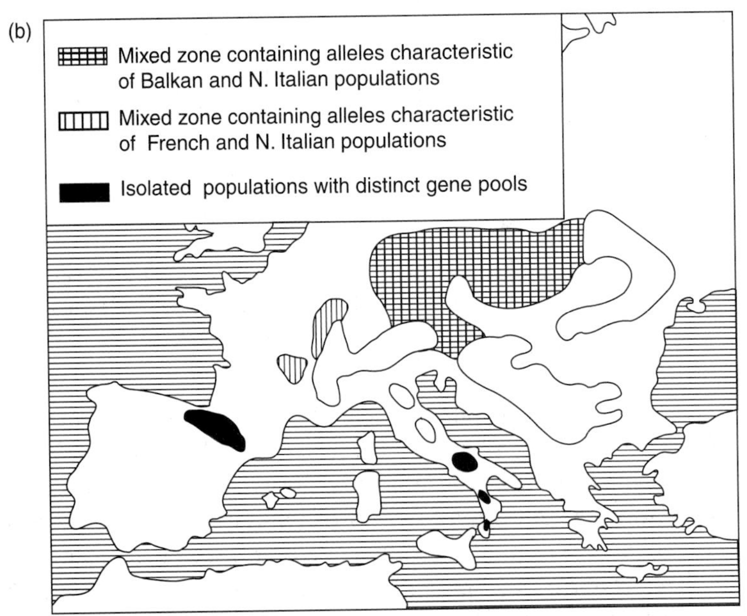

Figure 10.7 (a) The genetic characteristics of populations of *Abies alba* throughout central and southern Europe indicated that *Abies alba* is genetically differentiated to a high degree in five postulated regions and can be identified as populations with alleles characteristic of the Balkans, northern Italy, southern Italy, southern France, and southern Spain. (b) Evidence from this study also indicates that there is a large mixed gene pool in central and eastern Europe, and that in only two of these regions (Spain and southern Italy) have the populations remained isolated during the current warm stage. (Redrawn from Konnert and Bergmann, 1996.)

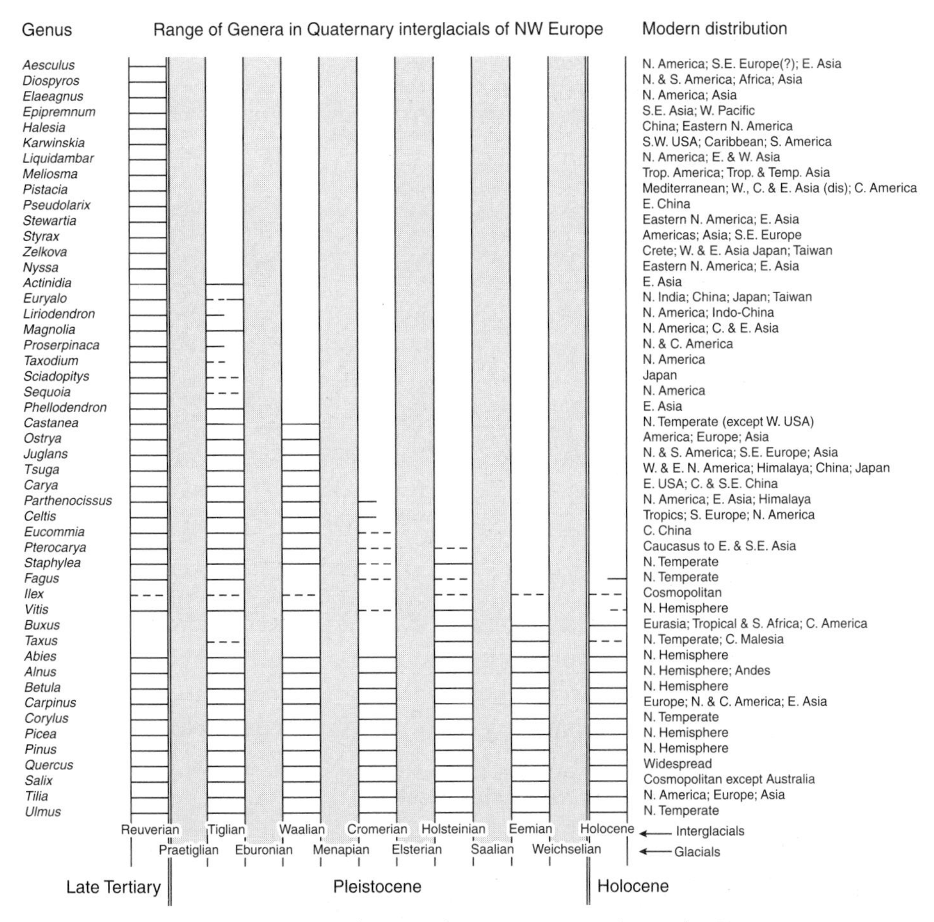

Figure 10.8 Tree taxa to disappear from north-western Europe during the Quaternary. The present-day distribution of these species is also indicated. Glacials are shaded. Dashed lines indicate a discontinuous presence during the interglacial (Adapted from Van der Hammen *et al.*, 1971; Tallis, 1992; Coxon and Waldren, 1997.)

considerably. At certain periods of the Earth's history, for example, there was no build-up of ice at the poles, whereas the impact of global climate change was much more severe at other times. The relative impact of the orbital oscillations upon global climate is related to a number of different climatic/geological phenomena (Frakes *et al.*, 1992). At the broadest scale, the Earth's climatic history has been categorized into two modes: 'icehouse' and 'greenhouse' (Figure 10.9) (Fischer, 1984). The switch between icehouse/greenhouse mode is on a cycle of approximately 30 million years (Fischer and Arthur, 1977). Thus, the climatic changes apparently controlled by orbital variations at Milankovitch frequencies (i.e. at periods of 400, 100, 41, and 23–19 thousand years) are in effect, nested within this larger climatic framework. In theory, therefore, it ought to be those times of global 'icehouse' mode, when the effects of orbital variations would have been most pronounced, that should relate to times of increased isolation and, associated with this, speciation and evolution of new taxa.

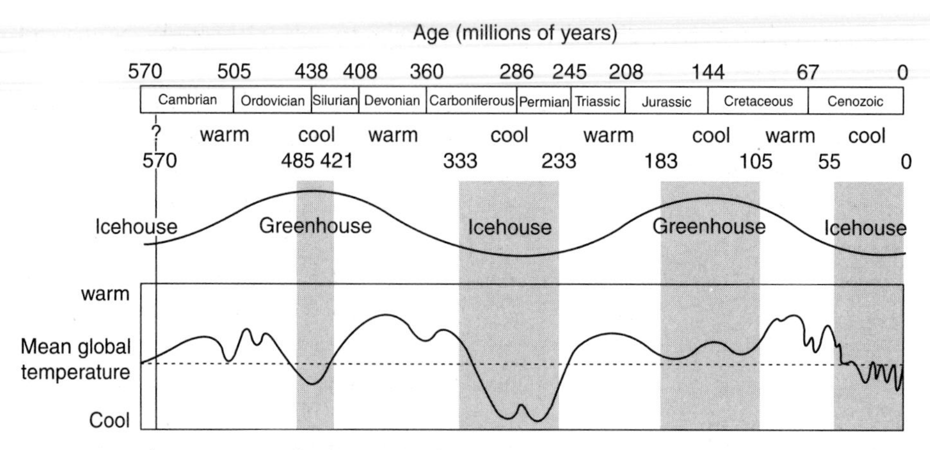

Figure 10.9 Climatic megacycles during the Phanerozoic (~570 million years) (redrawn from Huggett, 1997). Sources for greenhouse/icehouse data from Fischer (1981, 1984); warm and cool modes from Frakes *et al.* (1992); generalized temperature curve Martin (1995).

However, during these icehouse phases, the plant fossil record would still appear to suggest that the predominant trend is for a reduction in turnover and relative plateauing in total global species diversity, rather than an increase. During the Carbo-Permian glaciation (~300–280 Ma), for example, where there is clear geological evidence for extensive ice caps at the poles (Frakes *et al.*, 1992) (Figure 4.27), the plant fossil record indicates that a larger number of extinctions over originations occurred and species turnover was at its lowest throughout geological history (Niklas, 1997). Therefore even during an extensive period of glaciation (~20 Ma) there appears to be no evidence for increased speciation due to cold-stage isolation.

The fossil evidence would therefore appear to suggest that plants do respond to Milankovitch cyclicity. However, at the broadest scale, isolation and redistribution during the multiple glacial/interglacial cycles lead to a reduction in species numbers as more extinctions occur than originations. Therefore, what is observed on Milankovitch timescales could be no more than what occurs on much shorter (i.e. ecological) timescales. In a sense this is what is being suggested in the turnover pulse hypothesis of Bennett (1990, 1997) because it is predicted that in a tiered system the second tier will be orbital forcing of the climate, leading to disruption of communities and loss of accumulated change, and only rarely will the third tier, that of speciation through geographically isolated populations created by the perpetual environmental changes of Milankovitch timescales, come into play (Table 10.2). In addition, it is suggested on the broadest timescale (in this case 26 million years is suggested), a fourth tier, that of mass extinctions, will be the ultimate driving mechanism though the sorting of species created in the third tier. But as discussed previously, this tier is 'borrowed' from the animal record, and since evidence from the plant fossil record suggests that there are no comparable mass extinctions, resulting in significant declines in plant species diversity, another fourth tier must be sought (Willis and Bennett, 1995).

So what is the uppermost tier driving plant evolution? If the assumption is still to be made that the ultimate driving mechanism behind plant evolution is large-scale environmental change, then maybe a consideration of the scale of abiotic extrinsic stress (*sensu* DiMichele *et al.*, 1987) is needed. Small-scale extrinsic abiotic stress occurs when populations/species can migrate, adapt, become concentrated into refugial areas, and, following the removal of this adverse situation, species can usually re-emerge to fill their previously occupied niche. Thus it could be argued that glacial/interglacial cycles cause small-scale extrinsic abiotic stress. Evidence from the fossil record suggests that plant populations/species either migrate, adapt, or become concentrated into refugial areas and when warm-stage conditions return, species re-emerge to fill their previously occupied niche. Small-scale extrinsic abiotic stress is therefore a two-way process, with populations responding to climatic change by shifts in distribution when the conditions are unfavourable and returning to the region when conditions once again improve. The same is not true, however, of large-scale extrinsic abiotic stress.

Large-scale extrinsic abiotic stress is where populations cannot migrate fast enough, or the aerial extent of the stress exceeds their dispersal capacity (DiMichele *et al.*, 1987). This type of stress tends to be unidirectional in that the change is invariably a threshold event from which there is no return. Examples of such environmental changes that may bring about such thresholds, at least for the plant kingdom, are changing continental positions resulting from tectonic activity, and associated environmental and climatic change.

Plate tectonics

Plate boundary reorganizations involving collisions, rifting, and changing patterns of intraplate stress have occurred throughout Earth history and each 'pulse' of activity will have influenced almost all aspects of the global environment (Partridge *et al.*, 1995). Environmental changes associated with plate boundary reorganizations include the physical effects of plates moving together or apart, resulting for example, in the opening or closing of seaways, large-scale tectonic uplift, isolation or recombination of continents, changing sea-levels, and climatic change (Ruddiman, 1997). Climate change associated with plate boundary reorganizations results from alterations to the major patterns of ocean current and atmospheric circulation, due to changes in the position of continental plates and orogenic belts associated with tectonic uplift (thus influencing continental patterns of wind and rainfall), and from changing atmospheric concentrations.

Geological evidence from continental flood-basalts, continental drift rates, sea-floor spreading, and subduction rates and orogenic events, indicate that global plate spreading and subduction may have been episodic, with relatively short periods of activity separated by long periods of inactivity. These short periods of activity are termed 'pulsation tectonics' (Sheridan, 1987, 1997; Huggett, 1997). The proposed mechanisms behind the pulsation tectonics model are described elsewhere (reviewed by Sheridan, 1987). Of interest here is the timing and environmental impact of these 'pulses'. Five periods of increased tectonic activity (measured from sea-floor spreading rates) have been recognized in the geological record between approximately 460–430 Ma, 375–350 Ma, 300–260 Ma, 170–160 Ma, and 120–80 Ma (Figure 10.10). These pulses were associated with increased rates of sea-floor spreading, enhanced volcanic activity, and mantle degassing. These combined tectonic changes would have resulted in enhanced greenhouse gas emissions (in particular increased atmospheric CO_2) (McElwain *et al.*, 1999) and eustatic sea-level rise (Berner *et al.*, 1983; Arthur *et al.*, 1985; Sheridan, 1987, 1997; Larson, 1990). The overall effect of these changes on global climate would have been to increase global temperatures and also humidity (increased humidity related to the greater areas of shallow sea), thus leading to increased evaporation and moisture-holding capacity of the atmosphere.

If the timing of these periods of pulsation tectonics (*sensu* Sheridan, 1987, 1997) are compared to changes in the plant fossil record, some interesting patterns start to emerge (Figure 10.10).

460–430 Ma (Mid-Ordovician–Silurian)

The first identified period of rapid plate spreading, between approximately 460 and 430 million years ago, resulted in the closing of the Iapetus ocean, a large expanse that separated Laurentia (North America) from Baltica (Europe) and Gondwana (South America, Africa, Antarctica, and Australia). This period relates to a time in the plant fossil record when there is the first evidence for the terrestrialization of land by non-vascular plants. Evidence from fossil spores and cuticles without stomata are indicative of the presence of bryophyte-type plants (Graham *et al.*, 2000) dating as far back as 460 million years ago, and true vascular plants such as *Cooksonia* by the late Silurian (430 Ma) (see Chapter 3).

375–350 Ma (Mid-Devonian–early Carboniferous)

The next period relates to a time known as the late Devonian–Carboniferous Acadian orogeny, when Baltica collided against Laurentia and the Avalon volcanic arc terrane collided with Canada. This was accompanied by the rapid spreading of a new oceanic crust between Baltica and Gondwana. In the plant fossil record this is a time when there was rapid evolution and expansion of the lycopsids, sphenopsids, progymnosperms (both arborescent and herbaceous forms), and the two earliest seed plant groups, cordaites and pteridosperms (see Chapter 4).

300–260 Ma (late Carboniferous–early Permian)

This was another period of plate coalescence, this time resulting in the creation of Pangea from the collision between Gondwana and Kazakhstan, and Siberia

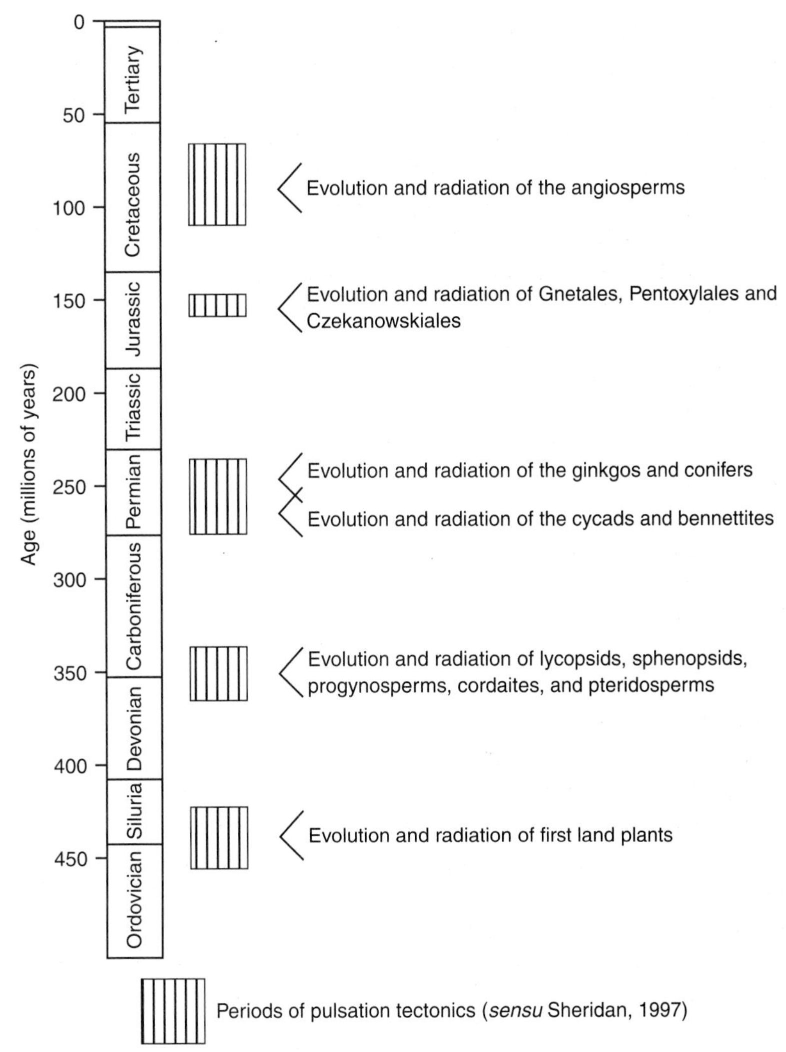

Figure 10.10 Periods of pulsation tectonics (from Sheridan, 1997). Also indicated are the evolution and radiation of major plant groups during these times (described in Chapters 3–7).

and Kazakhstan (Rowley *et al.*, 1985; Briggs, 1995). In the plant fossil record from 280 Ma (end of the Carbo-Permian glaciation) there was evolution and major radiation of the gymnosperms, starting with first appearance of cycads and bennettites at approximately 280 million years ago, ginkgos at 270 million years ago, and the conifers from 260 million years ago or earlier (Chapter 5).

170–160 Ma (middle Jurassic)

Between 170 and 160 million years ago the initial break-up of Pangea occurred, with the rapid spreading of the 'proto Atlantic' oceanic crust, parting North

America from Africa/South America (Gondwana). In the plant fossil record, unlike all the other 'pulses', no new major reproductive grades originated at this time, although there was a notable addition to the gymnosperms of the Pentoxylales and Czekanowskiales (for detailed description see Stewart and Rothwell, 1993), along with equivocal evidence for Gnetales. All three families have been suggested as possible precursors to the angiosperms (Hughes, 1994) (see Chapter 6 for a full discussion). Among the conifers, Taxales also make their first appearance at this time.

120–80 Ma (Cretaceous (Aptian–Campanian))

This final 'pulse' is related to the time of a major break-up of Pangea and rapid sea-floor spreading episodes in the Atlantic and Pacific. These resulted in formation of the South Atlantic Ocean margins, the Indian Ocean margins (east of Madagascar and between Australia and Antarctica), and the northern Atlantic Ocean margins between Europe, Canada, and Greenland. The first unequivocal evidence for the appearance and initiation of a massive radiation of the angiosperms occurred between approximately 120 and 80 Ma (see Chapter 6).

The five observed periods of 'tectonic pulses' can therefore be split into those that resulted in coalescence of the continents and those that resulted in the break-up. Regardless of whether these 'pulses' represented coalescence or break-up of continents, these episodes appear to correspond to times of increased rates of species turnover and the evolution of major groups in the plant fossil record. In contrast, times of relative tectonic stability correspond to times of fewer major originations in the plant fossil record. Pulses of fast plate spreading have shorter duration than intervals of slow spreading. Likewise, periods of low origination rate in the plant record are the norm, interspersed by periods of 'rapid' origination and turnover.

It could be argued that the past 65 million years is the exception to this rule, since the plant fossil record indicates increasing total species diversity and high turnover at intervals during the Tertiary (Figure 10.2; see Chapter 7), even though in terms of the pulsation tectonics it has been a period of relative stability, with slow sea-floor spreading rates. The counter argument to this, however, is that there has been relatively little evolutionary change in the past 65 million years. Angiosperms have remained dominant throughout and no new major 'evolutionary' or reproductive grade has emerged (as yet) to outcompete them. The evolution of grasses sometime in the early Tertiary ($\sim$65–50 Ma), arid-adapted angiosperms from approximately 50 Ma (Eocene), and the C_4 photosynthetic pathways approximately 16 million years ago (see Chapter 7), if viewed within a wider evolutionary framework, should be seen as no more than 'lower-order' evolutionary adjustments, similar to those in the second and third tiers in an evolutionary hierarchy (Gould, 1985; Bennett, 1990, 1997; Vrba, 1985, 1995).

10.4 Why should plant evolution be related to periods of increased continental plate movement?

The question therefore arises as to why periods of increased tectonic activity and plate spreading are temporally correlated with times of major evolutionary change in the plant fossil record. There are various environmental impacts that need to be taken into consideration. First, there is the question of whether, as a consequence of increased plate spreading, the amount of available land area, as opposed to the amount of ocean, influences plant evolution. Second, and closely associated with the previous point, is the question of whether biogeographic isolation of continents, rather than the actual area available, has a direct impact on plant evolution. Third, is the consideration that changing climatic conditions and/or atmospheric composition associated with increased tectonic activity has had a direct and/or indirect impact on plant evolution. These will all be considered in the following sections.

Amount of available land area?

Modern ecological studies often stress the importance of total land area on biotic diversity (Tiffney and Niklas, 1990) and in the case of the fossil faunal record the implications of changing continental area have been widely discussed (Briggs, 1995). A study on ammonite appearances and extinctions in relation to the area of continents covered by the sea between the Triassic and late Tertiary (Kennedy and Cobban, 1977; Hallam, 1983), for example, suggests that both the total number of taxa and appearance of new taxa are positively correlated with total oceanic area. The greater global area covered by water, the greater the potential for speciation. But what is potentially a good 'evolutionary environment' for marine organisms, might be the total opposite for terrestrial organisms, since an increase in sea level may, in certain circumstances, relate to a decrease in available land area, and vice versa.

Attempts to correlate land area to diversity in the plant fossil record for the 12 geological stages from the Devonian to middle Miocene of the northern hemisphere (~410–10 Ma) revealed a strong positive correlation between land area and plant diversity (Tingley and Knoll, 1990). It is, however, difficult to determine whether this is associated with a pronounced correlation between land area and total number of uplands, or purely land area alone. Thus increased diversity may result from the increased area available or as a result of topographical diversity created by a greater number of habitats over a smaller latitudinal area, resulting in increased isolation from biotic exchange through topographical barriers. In addition, both total land area and upland areas strongly correlate with geological age (i.e. there are increasingly greater areas of land and also mountains as the Earth gets older). It is concluded, therefore, that this correlation is probably indicative of a long-term trend of increased area of continental plate correlating with increased species numbers through time, rather than punctuated events of evolution corresponding to periods of increased land area resulting from continental plate movements (Tiffney and Niklas, 1990).

Number of separate land masses?

There is much evidence from the animal record (both marine and terrestrial) to suggest that the greater the number of land masses, the greater the diversity of organisms (Kurten, 1967, 1969; Valentine and Moores, 1970). Thus the fragmentation leads to isolated populations and, over time, this leads to the evolution of new species (allopatric speciation). Again, what is good in evolutionary terms for the terrestrial record might be the total opposite for the ocean record, and vice versa. As two continents approach each other and eventually become joined, for example, the terrestrial fauna and flora may exhibit convergence because of the creation of a corridor, whereas the marine organisms may exhibit divergence, forming new species because of separation and isolation (Hallam, 1983).

There are numerous examples where a causal mechanism has been invoked between number of species and continental fragmentation. For example, it has been suggested that without break-up of the global landmass into continents and islands, the Earth would have only half of its present biodiversity (Holmes, 1998). However, attempts to correlate the dispersion of land masses through geological time with land plant diversity suggest that there is no significant correlation (Tiffney and Niklas, 1990). Thus the number of land plants shows no correlation with increasing numbers of separate land masses. This evidence therefore appears to be at odds with observations of the present global diversity of plants and animals. But the scale of evolutionary change examined must also be remembered, since the changes in land plant diversity being compared by Tiffney and Niklas (1990) are the broadest-scale changes—those in effect attributable to an uppermost tier in an evolutionary model (*sensu* Gould, 1995). Lower-order evolutionary change has obviously occurred throughout geological time and is still occurring presently, whereby isolation may lead to speciation (allopatry). At the broadest evolutionary scale (uppermost tier), however, separation of land masses at particular points in geological time does not appear to account for the major changes seen in the plant evolutionary record. This is also consistent with the pulsation tectonics model, since although three of the periods of increased tectonic activity relate to continental break-up (i.e. 460–430, 170–160, and 120–80 million years ago), the other two relate to periods of coalescence, thus resulting in the totally opposite biogeographical effect.

Latitudinal distribution of land masses

A comparison of the diversity of species with the latitudinal distribution of land masses must also be considered. It may not be the association between tectonic activity and the size or number of land masses that has influenced large-scale patterns in plant evolution through time, but rather the effect that plate spreading has had on their latitudinal position. Present-day observations on global plant diversity, for example, indicate that there is a negative relationship between latitude and species diversity (Stevens, 1992), with decreasing diversity towards the poles and increasing diversity towards the equator. Whether or not this relates to local extinction associated with more inclement and/or unstable climatic conditions as certain land masses moved towards the poles, or rather greater rates of origination or speciation in areas of more favourable climate or climatic stability, is highly debatable.

Evidence from the geological record suggests some justification for the latter point, both in terms of regions of origination and also correlation between land plant diversity and latitudinal distribution of the land masses. Evidence from the plant fossil record suggests that, with one or two exceptions, the main angiosperm groups originated in the equatorial regions, either in South America–Africa or the general South-East Asia area (see Chapter 6), and then radiated out into higher latitudes (Lidgard and Crane, 1988; Barrett and Willis, 2001). In addition, a comparison of the diversity of pteridophyte species with the latitudinal distribution of land over geological time apparently indicates a positive correlation with area of land in the equatorial region (Tiffney and Niklas, 1990). Thus as area of land at or near the equator increased, so too did the number of pteridophyte taxa. There is also a positive correlation between seed plant diversity and poleward displacement of continental area (Tiffney and Niklas, 1990).

But this still does not provide an explanation as to why major plant groups evolved at particular times in the geological record. Land masses with relatively stable climatic regimes were located in the equatorial regions long before the origination of the main pteridophyte or angiosperm groups. Furthermore, although these regions were probably important centres of origination with regard to both plants (Crane and Lidgard, 1989) and animals (Jablonski, 1993), the times of major innovation in the plant fossil record (Figure 10.10) do not consistently coincide with times of plate movements towards the equator. Furthermore, during much of the Permian and Jurassic, it appears from plant biogeographical analysis that regions of high plant diversity were centered on the temperate latitudes rather than the tropical latitudes which were predominantly low diversity summerwet and desert biomes. Without a more detailed data set, including, for example, clearer evidence for regions of origination for lycopsids, sphenopsids, ginkgoales, and cycadales, to name but a few, along with interpretation of climatic conditions associated with the latitudes at these points in time, such correlations will remain spurious. A similar conclusion was reached by Tiffney and Niklas (1990), who suggested 'that although the latitudinal distribution of land may have profoundly affected diversity in times of steep global temperature gradient, such gradients are so scattered in the history of plant evolution that they may have had little effect on the overall trend of land plant diversity'.

Climatic conditions associated with periods of enhanced tectonic activity?

So far the discussion as to why there may be a possible link between major periods of plant evolution and pulsation tectonics has focused upon the physical changes resulting from the effects of enhanced sea-floor spreading upon the position and aerial extent of land masses. Other factors that must be considered are climate change and the effects of enhanced tectonic activity on atmospheric composition. Plate boundary reorganizations alter the major patterns of ocean current circulation, continental patterns of wind, rainfall, and atmospheric circulation. All will have a direct impact on global climates. However, as noted earlier for the Quaternary, Milankovitch-scale global climatic changes (400, 100, 41, 23–19 kyr), involving increases or decreases in temperature and/or precipitation, while having a profound effect upon the global distribution of vegetation, do not appear to bring about significant evolutionary innovation.

This is also true of the effects of global climatic change on vegetation during the great Carbo-Permian glaciation (~300 Ma). A combination of increased seasonality and decreased precipitation in the tropical everwet biome at this time resulted in major extinction among the dominant lycopsids and sphenopsids, which were replaced by pteridosperms (DiMichele *et al.*, 1996). Refugial lycopsid and sphenopsid populations persisted, however, in the Cathaysian region (China) and the main effects of these climatic changes were extinction and vegetation reorganization rather than origination of major new evolutionary groups. It can be suggested that this is because the effects of major climate change do not cause universal changes in climatic variables over the whole Earth, but rather result in the expansion and/or contraction of different climatic zones. This is evident from the expansion and contraction of the different biomes through time (Figure 10.11) and determined by a variety of factors, including temperature contrasts between the equator and poles, between the continents and oceans, and the rotation of the Earth (Erwin, 1993).

However, one change associated with plate spreading that is global in extent, and therefore in impact, is the effect of enhanced tectonic activity on the concentration of atmospheric carbon dioxide. Atmospheric CO_2, along with water and other essential nutrients, is one of the most important requirements

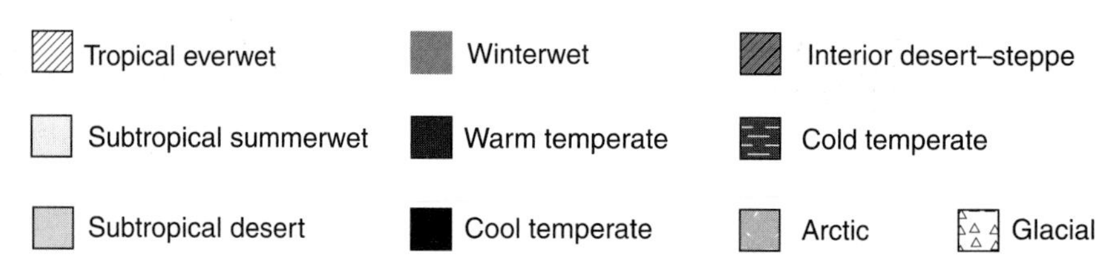

Figure 10.11 *(opposite)* Major changes in the distribution of global vegetation biomes from the early Carboniferous through to the late Miocene.

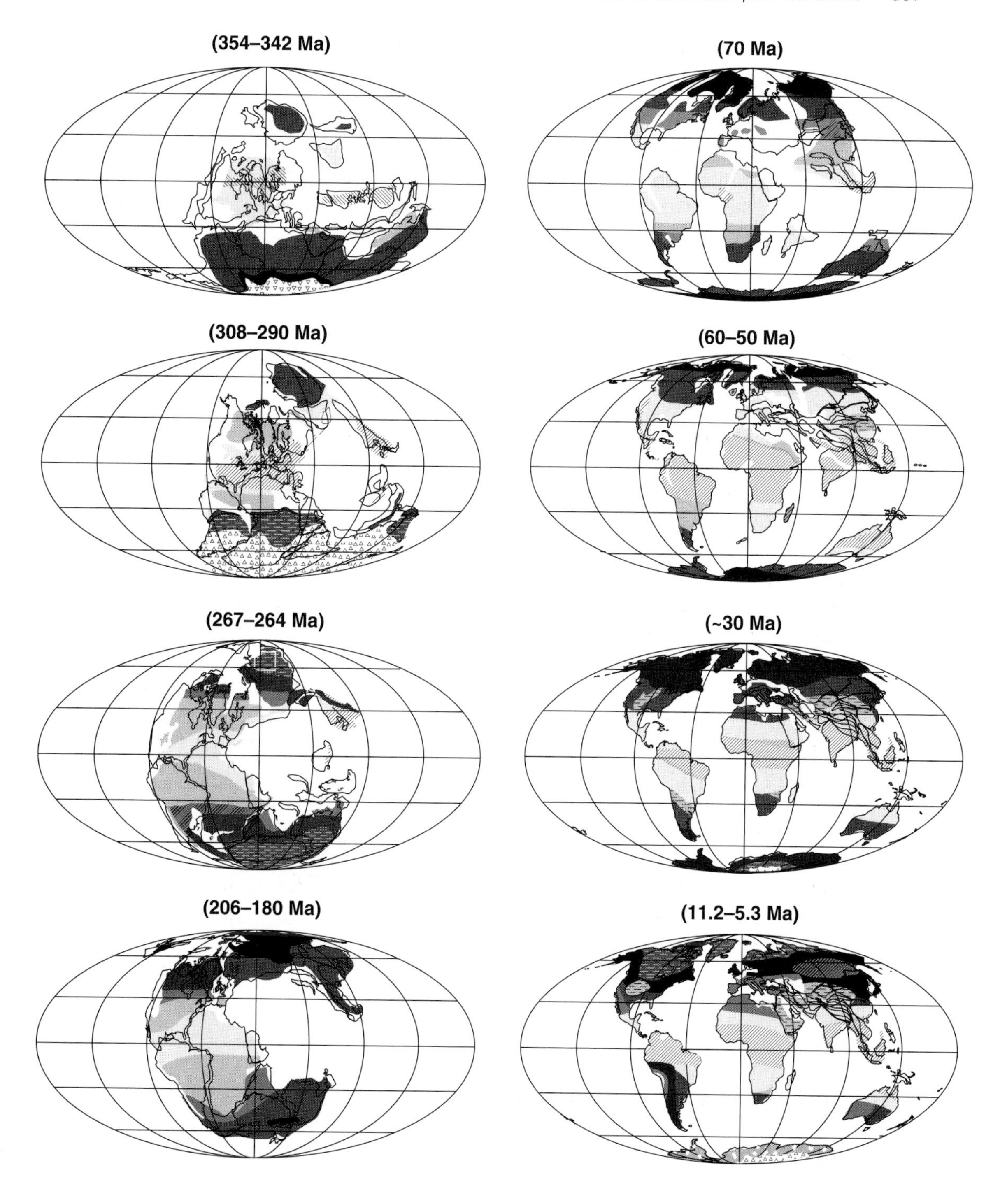
(354–342 Ma)
(70 Ma)
(308–290 Ma)
(60–50 Ma)
(267–264 Ma)
(~30 Ma)
(206–180 Ma)
(11.2–5.3 Ma)

for plant growth and survival. Plants modify their anatomy, morphology, overall architecture, and even biochemical pathways (e.g. C_3 and C_4) to optimize the uptake of carbon dioxide for photosynthesis against water loss. It would not be surprising, therefore, if variations in what is essentially a plant's main food source, that is CO_2, played a role in plant macroevolution.

Atmospheric changes associated with periods of enhanced tectonic activity

There is strong evidence from various geological and palaeobiological indicators of past CO_2 concentrations that atmospheric CO_2 levels increased during episodes of pulsation tectonics. This is most probably due to enhanced rates of seafloor spreading, volcanic activity and mantle outgassing associated with the upwelling of hot plumes of mantle (Sheridan, 1987, 1997; Larson, 1991a, b) (Figure 10.12). The supporting evidence for higher CO_2 concentrations during these 'superplume' events comes from the isotopic composition of fossil soils (Cerling, 1991; Mora *et al.*, 1996; Ekart *et al.*, 1999), changes in stomatal frequency from fossil leaf cuticles (McElwain and Chaloner, 1995; McElwain *et al.*, 1999; Retallack, 2001), and various marine isotopic indicators (Pagani *et al.*, 1999; Pearson and Palmer, 1999). These 'palaeo-CO_2 barometers' are also in good agreement with estimates from carbon cycle models (e.g. Berner, 1991, 1994) indicating that the four major episodes of pulsation tectonics identified by Sheridan are characterized by either anomalously rapid rates of atmospheric CO_2 rise, transient peaks in CO_2 concentration, or generally a trend of increasing atmospheric CO_2.

The first episode, between 460 and 430 Ma (mid-Ordovician–Silurian) is associated with an increase in CO_2 concentration from 4200 to 4500 p.p.m., equivalent to 14 times and 15 times the pre-industrial CO_2 level of 300 p.p.m., respectively (Berner, 1991). The second, between 375 and 350 Ma (mid-Devonian to early Carboniferous), is associated with high CO_2 levels of 3000 p.p.m., following a rapid CO_2 rise from 2400 p.p.m. between 390 and 370 Ma (Berner, 1991). The third proposed episode of pulsation tectonics is coincident with a rapid transient CO_2 peak detected from stomatal frequency changes in fossil plant leaves (Retallack, 2001). A rapid increase from approximately 2200 to just under 4000 p.p.m. and back again occurred during this brief interval in the Jurassic. Finally, the fourth superplume episode, between 120 and 80 Ma in the Cretaceous, is associated with a rapid increase in atmospheric CO_2 from 900 to 1800 p.p.m. (Berner, 1991; Tajika, 1999) or possibly even higher, to more than 4000 p.p.m. (Retallack, 2001).

An intriguing and consistent pattern therefore emerges between the timing of major evolutionary events in the plant fossil record, and fluctuations in the concentration of atmospheric carbon dioxide. It appears that the timing of origination of each successive evolutionary flora, each of which brought about

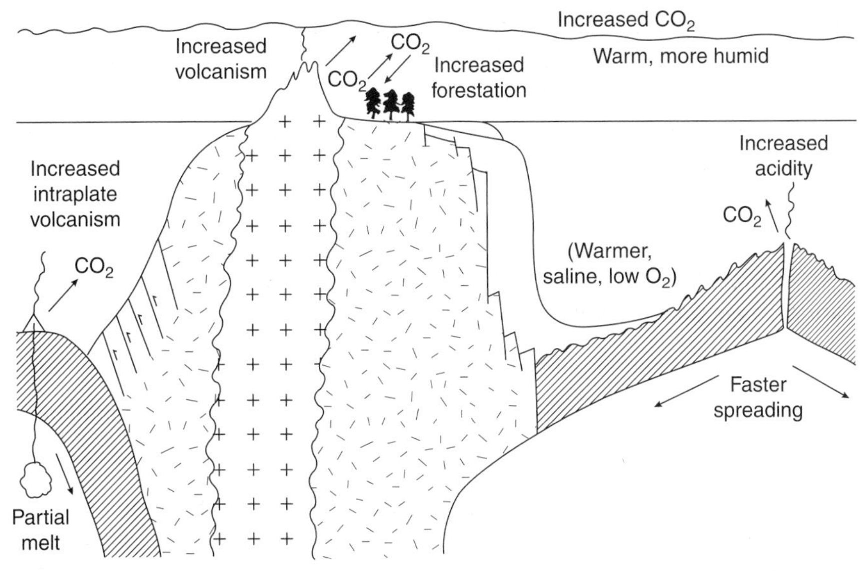

Figure 10.12 Schematic model to demonstrate the stratigraphic/atmospheric effects of periods of faster spreading episodes of the pulsation tectonic cycle model. Proposed changes in vegetation, volcanism, atmospheric CO_2, surface temperature and humidity, ocean temperature, O_2 content, and acidity are indicated (redrawn from Sheridan, 1986).

major reproductive innovations and subsequent increases in global species diversity, was coincident with an episode of rapid CO_2 increase in Earth history. Similarly, the origination of the grasses, although on a lesser scale in terms of evolutionary events compared with the evolution of angiosperms as a whole, is coincident with rapid transient CO_2 changes. Grasses evolved sometime in the late Cretaceous (60 Ma) and late Palaeocene (55 Ma), both of which are characterized by short CO_2 peaks of around 4000 p.p.m. according to recent stomatal and palaeosol data (Figure 7.4) (Ekart *et al.*, 1999; Retallack, 2001).

Further exploration of this intriguing relationship demonstrates that the rate of origination of fossil plant species within each of the three reproductive grades (that is pteridophytes, gymnosperms, and angiosperms) in the fossil record is significantly correlated with both the absolute CO_2 concentration and the rate of CO_2 change (Figure 10.13). It is interesting to note, in the case of the angiosperms, that although they originated during an episode of anomalously rapid CO_2 change (two times greater than normal background rates), their subsequent radiation and diversification occurred under progressively declining CO_2 concentrations throughout the Cretaceous (Figure 10.14). The overall trend of declining CO_2 concentration through the Cretaceous differentially influenced each grade (Lupia *et al.*, 2000), favouring angiosperms over gymnosperms, over pteridophytes, respectively.

It is well recognized from studies of modern plants that angiosperm photosynthetic responses are more optimized to low CO_2 concentrations than those of

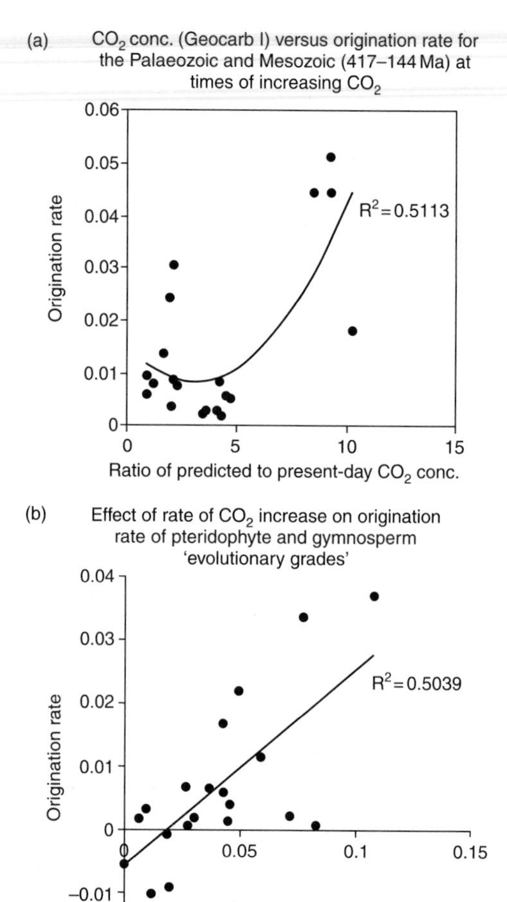

Figure 10.13 Plot of the rate of origination of fossil plant species against absolute CO_2 concentration and rate of CO_2 change during the Palaeozoic and Mesozoic. Fossil plant data from Niklas (1997) and atmospheric CO_2 data from Geocarb I (Berner, 1991). Results indicate that origination rates of fossil plants increased significantly with both the absolute CO_2 concentration (a) and the rate of CO_2 change (b). Origination rate is measured as the change in species diversity over time (stage) over the total standing diversity per stage.

gymnosperms and pteridophytes, respectively (Beerling and Woodward, 1997). Such an observation led Robinson (1994) to hypothesize that 'CO_2 starvation' played an important role in the evolution of photosynthetically efficient angiosperm herbs and grasses in the Tertiary. CO_2 starvation has also been proposed as a mechanism driving the evolution of plants with C_4 photosynthetic pathways in the Tertiary (Ehleringer and Monson, 1993), and the evolution of true leaves (megaphylls) in the Devonian (Beerling *et al.*, 2000). It is highly probable, therefore, that declining CO_2 concentrations through the Cretaceous provided a competitive physiological edge to the angiosperms. Along with their

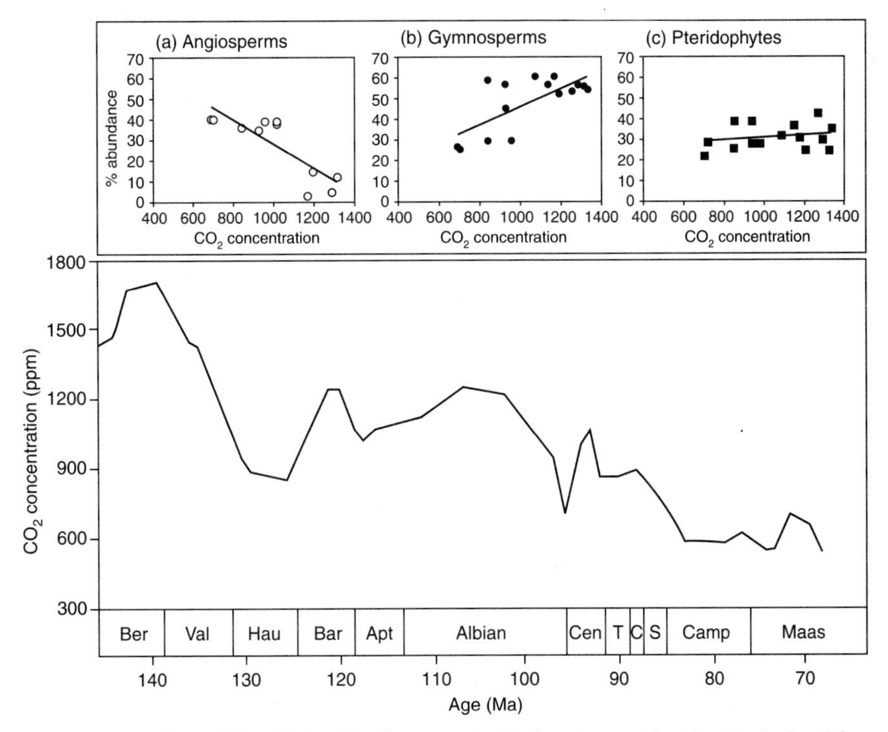

Figure 10.14 (a), (b), (c) The relationships between the % abundance of fossil pollen in fossil floras of angiosperms, gymnosperms, and pteridophytes, respectively, with atmospheric CO_2 change during the Cretaceous (~ last 140 million years). The lower graph shows the estimated CO_2 change through the Cretaceous, based on a long-term carbon cycle model from Tajika *et al.* (1999). Results indicate that the overall trend of declining CO_2 concentration through the Cretaceous differentially influenced each grade, favouring angiosperms over gymnosperms over pteridophytes, respectively. The % abundance data is from Lupia *et al.* (1999).

superior reproductive strategy, this would contribute to the competitive replacement of less physiologically optimized gymnosperm and pteridophyte groups through the Cretaceous, leading to their rapid radiation and ultimate rise to dominance in nearly every biome by the Tertiary.

Although the differential effects of 'CO_2 starvation' on plant physiology can be readily explained, such reasoning provides no hints as to why major evolutionary innovations in the plant fossil record, such as the evolution of seed plants (gymnosperms), coincide with times of increasing CO_2 concentration, or why rates of origination are positively correlated with rates of CO_2 change. Is this correlation an artefact, or could there be a possible mechanism as to why rapid rates of CO_2 increase accelerate or promote natural selection?

Experimental investigations of individual plant responses to elevated CO_2 concentrations have shown a variety of well-documented physiological effects. For instance, many species show increased water use efficiency (dry matter production per unit water lost), increased nutrient use efficiency (tissue C/N ratio) (Woodward *et al.*, 1991), and enhanced photosynthesis when grown in high CO_2 concentrations. The extent and activity of symbiotic associations with mycorrhizal fungi and N_2-fixing bacteria also increased for a number of species with elevated CO_2. Aspects of plant reproductive biology, such as flowering phenology (timing of flowering), seed number, size, and nutrient content may be altered when plants are grown in elevated CO_2 (Bazzaz, 1990). High carbon dioxide concentrations can also modify the effects of different plant stresses or limiting factors for photosynthesis. For example, an increase in atmospheric CO_2 concentration can compensate for low light, which limits photosynthesis, and ameliorate the effects of high levels of salt stress through indirect physiological effects (Lemon, 1983; Bazzaz, 1990).

The most important feature of all of these experimental findings, however, is that there are no rules or a uniform direction of responses to elevated CO_2 for all species. Elevated carbon dioxide does not result in universally enhanced photosynthesis, increased water-use efficiency, and improved seed quality. In contrast, some plant species show no change or reduced photosynthesis under elevated CO_2 concentrations. Furthermore, even the plant taxa which initially show enhanced rates of photosynthesis, and hence higher productivity, tend to reach a threshold CO_2 level, beyond which further productivity gains cannot be attained. Other essential factors for growth and repair, such as nitrogen, become limiting in natural ecosystems, and certain species begin to show negative physiological and/or reproductive responses (Mackowiak and Wheeler, 1996; Reuveni and Bugbee, 1997).

When experimental investigations of plant responses to elevated CO_2 are undertaken on populations and communities (rather than individuals) enhanced competitive interaction between species for other resources, such as light, water, and nitrogen are observed (Lemon, 1983; Bazzaz, 1990). Studies have also indicated that elevated CO_2 greatly enhances interplant competition when they are grown in dense stands rather than individually (Bazzaz *et al.*, 1995). This has therefore led Bazzaz and colleagues to conclude that 'the overall effects of increasing CO_2 on the intensity of selection in plant populations may be surprisingly large, potentially resulting in substantial changes in the genetic diversity of plant populations'. The suggestion that increasing CO_2 levels can act to intensify selection and genetic drift in density-dependent plant populations may hold one of the many keys to the observed positive correlation between rates of origination and atmospheric CO_2 concentration in the plant fossil record. Enhanced genetic diversity within populations due to elevated CO_2 concentrations may facilitate speciation/origination.

However, factors responsible for the observed increase in speciation during times of increasing atmospheric CO_2 may not only have been due to inter- and intraspecific competition, but also due to the direct effect of atmospheric CO_2 on plant generation times. Recent studies on tropical forest dynamics (Phillips and Gentry, 1994) have shown that, over the past 40 years, the age of tree mortality has decreased, i.e. trees are dying younger (Figure 10.15). This they attribute to

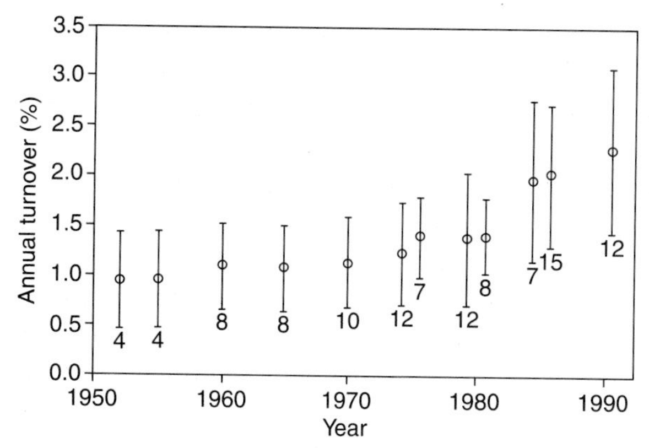

Figure 10.15 Tree turnover rates (measured by tree mortality and recruitment) at 40 tropical forest sites during the past 40 years (redrawn from Phillips and Gentry, 1994).

the recent anthropogenic increases in atmospheric CO_2 (see also Phillips and Sheil, 1997). Greenhouse experiments have also demonstrated that increased CO_2 could decrease the average generation time for plants (Omer and Hovarth, 1983). Decreased generation time will increase turnover rate and thus increase the probability of genetic mutations becoming fixed in a population.

The compounding effects of possible global climatic change and greenhouse warming on plant evolution during pulsation episodes must also be considered. Fossil plant responses to a fourfold CO_2 increase and global warming across the Triassic–Jurassic (~205 Ma) boundary may offer some important clues to the combined effects of increasing CO_2 and temperatures on plant turnover (McElwain *et al.*, 1999). The combined effects of temperature and CO_2 in this case had a detrimental effect on the leaf temperatures of large-leaved species, causing overheating and most probably protein denaturation and ultimately death, which contributed to a 95% species-level turnover across the Triassic–Jurassic boundary. Species with small and dissected leaf shapes, which enabled more efficient cooling, are hypothesized to have evolved, whereas those with large leaves prone to overheating either went extinct or migrated to cooler climates (McElwain *et al.*, 1999).

Large-scale patterns in plant evolution over geological time may well have been influenced, therefore, by a combination of direct effects of CO_2 on inter-plant competition, genotypic diversity, reproductive biology, and generation time (Figure 10.16). The fact that changes in atmospheric CO_2 are global, not local, is another point in favour of this theory; the opportunities for plant migration do not arise, thereby increasing the evolutionary and selection pressures on plants. This contrasts with migration during, for example, glacial–interglacial cycles (Bennett, 1997). In terms of extrinsic abiotic stress (*sensu* DiMichele *et al.*, 1985), increasing levels of atmospheric CO_2 can therefore be considered as 'large-scale', since it will provide a level of abiotic stress which

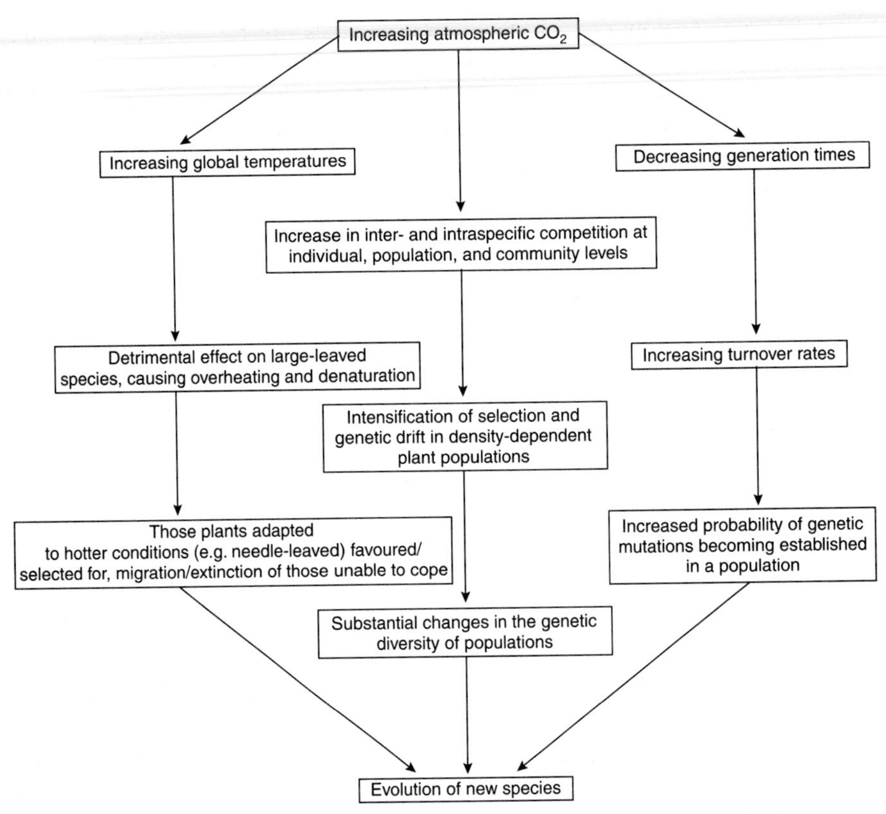

Figure 10.16 The three suggested mechanisms that may account for major periods of plant evolution occurring during times of increasing atmospheric CO_2.

exceeds dispersal capacity and from which plant populations cannot escape. Therefore changing levels of atmospheric CO_2 may well provide the uppermost tier in a hierarchical model of plant evolution.

10.5 Conclusions

In conclusion, evidence from the plant fossil record indicates a broadening spectrum of diversity and morphological complexity through time (e.g. Niklas *et al.*, 1980, 1983, 1985; Lidgard and Crane, 1988, 1990; Kovach and Batten, 1993). The plant fossil record suggests that, on the broadest scale, major evolutionary change and innovation was concentrated into relatively short intervals in geological time, such as the late Devonian and early Carboniferous, followed by long intervals of relatively little evolutionary innovation. Most of the major reproductive grades declined in total species diversity following the evolution of a new grade, although this has yet to happen with the angiosperms.

The macroevolutionary timescale of plants does not follow that of animals (Traverse, 1988). There is no evidence for the mass extinction events seen in both the oceanic and terrestrial faunal records during at least four of the 'big five' extinction events. Biological interactions including adaptation, competition, and coevolution must have played a role in terrestrial diversification (Tiffney and Niklas, 1990; Niklas, 1997) and, in particular, they are thought to have been responsible for the decline in species diversity of incumbent grades or clades of plants as new ones evolved (Niklas *et al.*, 1985). However, these interactions do not appear to be responsible for the major episodes of speciation observed in plant fossil record.

If the ultimate causes of evolutionary change in the plant fossil record are organized into a hierarchical scheme of causality, then an alternative uppermost tier to mass extinctions (proposed for the animal record; Gould, 1995) must be sought, since there is little evidence for mass extinction in the plant fossil record. Also, unlike the animal record, pulses of origination do not tend to follow mass extinction events during their so-called 'recovery phase'.

There is a close chronological relationship between major evolutionary change in the plant fossil record and proposed pulses of global plate spreading and increased tectonic activity (Sheridan, 1987, 1997). It is therefore suggested that various physical/climatic parameters associated with these tectonic pulses may provide the answer to what is driving plant evolution. In particular, we propose that the pulses of increased atmospheric CO_2 might well have been crucial to plant evolution, by providing a global extrinsic abiotic stress from which plants could not escape. A sobering thought for the future.

References

Aitken, M. J. (1985) Thermo-luminescence dating—past progress and future-trends. *Nuclear Tracks and Radiation Measurements* **10**, 3–6.

Aitken, M. J. (1990) *Science-based dating in archaeology*. London: Longman.

Aitken, M. J. (1998) *An introduction to optical dating*, First edn. Oxford: Oxford University Press.

Algeo, T. J. and Scheckler, S. E. (1999) Terrestrial–marine teleconnections in the Devonian: links between the evolution of land plants, weathering processes, and marine anoxic events. *Philosophical Transactions of the Royal Society of London, Series B* **353**, 113–130.

Alvarez, L. W., Alvarez, W., Asaro, F. and Michel, H. V. (1980) Extraterrestrial cause for the Cretaceous–Tertiary extinction. *Science* **208**, 1095–1108.

Alvin, K. and Chaloner, W. G. (1970) Parallel evolution in leaf veination: an alternative view of angiosperm origins. *Nature* **226**, 662–663.

Anders, E. (1989) Pre-biotic organic matter from comets and asteroids. *Nature* **342**, 255–257.

Andreae, O. M. and Jaeschke, W. A. (1992) Exchange of sulphur between biosphere and atmosphere over temperate and tropical regions. In: Howard, R.W. (Ed.) *Sulphur cycling on the continents*, pp. 27–66. London: Wiley.

Andrews, H. N. (1961) *Studies in paleobotany*. New York: Wiley.

Arber, E. A. N. and Parkin, J. (1907) On the origins of angiosperms. *Journal of the Linnean Society of Botany* **38**, 29–80.

Arber, E. A. N. and Parkin, J. (1908) Studies on the evolution of angiosperms. The relationships of angiosperms to the Gnetales. *Annals of Botany* **22**, 489–515.

Archangelsky, S. (1990) Plant distribution in Gondwana during the Late Paleozoic. In: Taylor T. N. and Taylor E. L. (Eds) *Antarctic paleobotany: its role in the reconstruction of Gondwana*, pp. 102–117. New York: Springer.

Archangelsky, S. and Cuneo, R. (1990) Polyspermophyllum, a new Permian gymnosperm from Argentina, with considerations about the Dicranophyllales. *Review of Palaeobotany and Palynology* **63**, 117–135.

Archibald, J. D. and Bryant, L. J. (1990) Differential Cretaceous/Tertiary extinctions of nonmarine vertebrates; evidence from northeastern Montana. *Geological Society of America Special Paper* **247**, 549–562.

Archibold, O. W. (1995) *Ecology of world vegetation*, First edn. London: Chapman & Hall.

Arens, N. C. and Jahren, H. (2000) Carbon isotope excursion in atmospheric CO_2 at the Cretaceous–Tertiary boundary: evidence from terrestrial sediments. *Palaios* **15**, 314–322.

Arthur, M. A., Dean, W. E. and Claypool, G. E. (1985) Anomalous ^{13}C enrichment in modern marine organic carbon. *Nature* **315**, 216–218.

Ash, S. (1986) Fossil plants and the Triassic–Jurassic boundary. In: Padian, K. (Ed.) *The beginning of the age of dinosaurs*, pp. 21–30. Cambridge: Cambridge University Press.

Austin, J. J., Ross, A. J., Smith, A. B., Fortey, R. A. and Thomas, R. H. (1997a) Problems of reproducibility – does geologically ancient DNA survive in amber-preserved insects? *Proceedings of the Royal Society London, B* **264**, 467–474.

Austin, J. J., Smith, A. B. and Thomas, R. H. (1997b) Palaeontology in a molecular world: the search for authentic ancient DNA. *Trends in Ecology and Evolution* **12**, 303–306.

Axelrod, D. I. (1952) A theory of angiosperm evolution. *Evolution* **6**, 60.

Axelrod, D. I. (1966) Origin of deciduous and evergreen habits in temperate forests. *Evolution* **20**, 1–15.

Axelrod, D. I. (1975) Evolution and biogeography of Madrean–Tethyan sclerophyll vegetation. *Annals of the Missouri Botanical Garden* **62**, 280–334.

Bada, G., Wang, X. S., Poinar, H. N., Pääbo, S. and Poinar, G. O. Jr. (1994) Amino acid racemization in amber-entombed insects: implications for DNA preservation. *Geochimica Cosmochimica Acta* **58**, 3131–3135.

Bada, J. L., Wang, X. S. and Hamilton, H. (1999) Preservation of key biomolecules in the fossil record: current knowledge and future challenges. *Philosophical Transactions of the Royal Society of London, B* **354**, 77–87.

Bakker, R. T. (1978) Dinosaur feeding behaviour and the origin of flowering plants. *Nature* **274**, 661–663.

Bakker, R. T. (1986) *The dinosaur heresies: new theories unlocking the mystery of the dinosaurs and their extinction.* New York: William Morrow.

Bambach, R. K., Scotese, C. R. and Ziegler, A. M. (1980) Before Pangea: the geographies of the Paleozoic world. *American Scientist* **68**, 26–38.

Banks, H. (1968) The early history of land plants. In: Drake E. T. (Ed.) *Evolution and environment (Symposium – 100th anniversary foundation of the Peabody Museum of Natural History)*, pp. 73–107. New Haven: Yale University Press.

Banks, H. (1970) *Evolution and plants of the past.* Belmont: Wadsworth Publishing.

Banks, H. (1980) Floral assemblages in the Siluro-Devonian. In: Dilcher, D. L. and Taylor, T. N. (Eds) *Biostratigraphy of fossil plants*, pp. 1–24. Stroudsberg, Pa: Dowden, Hutchinson & Ross.

Banks, H. P., Leclercq, S. and Hueber, F. M. (1975) Anatomy and morphology of *Psilophyton dawsonii* sp.n., from the Late Lower Devonian of Quebec (Gaspe) and Ontario, Canada. *Palaeontographica America* **8**, 77–127.

Barghoorn, E. S. and Tyler, S. A. (1965) Microorganisms from the Gunflint Chert. *Science* **147**, 563–577.

Barnosky, C. W. (1984) Late Miocene vegetational and climatic variations inferred from a pollen record in northwest Wyoming. *Science* **223**, 49–51.

Barrett, P. M. (2000) Evolutionary consequences of dating the Yixian formation. *Trends in Ecology and Evolution* **15**, 99–103.

Barrett, P. M. and Willis, K. J. (2001) Did dinosaurs invent flowers? Dinosaur–angiosperm coevolution revisited. *Biological Reviews of the Cambridge Philosophical Society* **76**, 411–447.

Barron, E. J. and Washington, W. M. (1984) The role of geographic variables in explaining paleo-climates: results from Cretaceous climate model sensitivity studies. *Journal of Geophysical Research* **89**, 1267–1279.

Bateman, R. M., Crane, P. R., DiMichele, W. A., Kenrick, P., Rowe, N. P. and Speck, T. (1998) Early evolution of land plants: phylogeny, physiology, and ecology of the primary terrestrial radiation. *Annual Review of Ecology and Systematics* **29**, 263–292.

Bazzaz, F. A. (1990) The response of natural ecosystems to the rising global CO_2 levels. *Annual Review of Ecology and Systematics* **21**, 167–196.

Bazzaz, F. A., Jasienski, M., Thomas, S. C. and Wayne, P. (1995) Microevolutionary responses in experimental populations of plants to CO_2-enriched environments – parallel results from two model systems. *Proceedings of the National Academy of Sciences of the United States of America* **92**, 8161–8165.

Beck, C. B. (1960) Connection between *Archaeopteris* and *Callixylon. Science* **131**, 1524–1525.

Beck, C. B. (1962) Reconstructions of *Archaeopteris* and further considerations of its phylogenetic position. *American Journal of Botany* **49**, 373–382.

Beerling, D. J. (1997) Interpreting environmental and biological signals from the stable isotopic composition of fossilized organic and inorganic carbon. *Journal of the Geological Society, London* **154**, 303–309.

Beerling, D. J. and Jolley, D. W. (1998) Fossil plants record an atmospheric $^{12}CO_2$ and temperature spike across the Palaeocene–Eocene transition in NW Europe. *Journal of the Geological Society, London* **155**, 591–594.

Beerling, D. J. and Woodward, F. I. (1997) Changes in land plant function over the Phanerozoic: reconstructions based on the fossil record. *Botanical Journal of the Linnean Society* **124**, 137–153.

Beerling, D. J., Osborne, C. P. and Chaloner, W. G. (2001) Evolution of leaf-form in land plants linked to atmospheric CO_2 decline in the Late Palaeozoic Era. *Nature* **410**, 352–354.

Bell, P. R. (1992) *Green plants: their origin and diversity*, First edn. Cambridge: Cambridge University Press.

Bennett, K. D. (1990) Milankovitch cycles and their effects on species in ecological and evolutionary time. *Paleobiology* **16**, 11–21.

Bennett, K. D. (1997) *Evolution and ecology: the pace of life*, First edn. Cambridge: Cambridge University Press.

Bennett, K. D., Tzedakis, P. C. and Willis, K. J. (1991) Quaternary refugia of north European trees. *Journal of Biogeography* **18**, 103–115.

Benton, M. J. (1986) The late Triassic tetrapod extinction events. In: Padian, K. (Ed.) *The beginning of the age of the dinosaurs*, pp. 303–320. Cambridge: Cambridge University Press.

Benton, M. J. (1989) Mass extinctions among tetrapods and the quality of the fossil record. *Philosophical Transactions of the Royal Society of London, B* **325**, 369–386.

Benton, M. J. (1990) End-Triassic. In: Briggs, D. E. G. and Crowther, P. R. (Eds) *Paleobiology: a synthesis*, pp. 239–251. Oxford: Blackwell.

Berner, R. A. (1991) A model for atmospheric CO_2 over phanerozoic time. *American Journal of Science* **291**, 339–375.

Berner, R. A. (1993) Paleozoic atmospheric CO_2: Importance of solar radiation and plant evolution. *Science* **261**, 68–70.

Berner, R. A. (1994) GEOCARB II: A revised model for atmospheric CO_2 over Phanerozoic time. *American Journal of Science* **294**, 56–91.

Berner, R. A. (1995) Chemical weathering and its effect on atmospheric CO_2 and climate. *Reviews of Mineralogy* **31**, 565–583.

Berner, R. A. (1997) The rise of plants and their effect on weathering and atmospheric CO_2. *Science* **276**, 544–546.

Berner, R. A. (1998) Sensitivity of Phanerozoic atmospheric CO_2 to paleogeographically induced changes in land temperature and surface runoff. In: Crowley, T. J. and Burke, K. C. (Eds) *Tectonic boundary conditions for climatic reconstructions*, pp. 251–261. Oxford: Oxford University Press.

Betancourt, J. L. M. P. S., van Devender, T. R. and Martin, P. S. (1990) *Packrat middens. The last 40 000 years of biotic change.* Tucson: University of Arizona Press.

Bewley, J. D. and Black, M. (1982) *Physiology and biochemistry of seeds, volume 2: Viability, dormancy and environmental control.* Berlin: Springer.

Blackmore, S. and Barnes, S. H. (1987) Embryophyte spore walls: origin, development and homologies. *Cladistics* **3**, 185–195.

Bocherens, H., Friis, E. M., Mariotti, A. and Pedersen, K. J. (1994) Carbon isotopic abundance in Mesozoic and Cenozoic fossil plants: palaeoecological interpretations. *Lethaia* **26**, 347–358.

Bockelie, J. F. (1994) Plant roots in core. In: Donovan, S. K. (Ed.) *The palaeobiology of trace fossils*, pp. 177–199. Chichester: John Wiley & Sons.

Boucot, A. J., Xu, C. and Scotese, C. R. (2001) Atlas of lithologic indicators of climate. *Geological Society of America Special Publication* (in press).

Boulter, M. C., Spicer, R. A. and Thomas, B. A. (1988) Patterns of plant extinction from some paleobotanical evidence. In: Larwood, G. P. (Ed.) *Extinction and survival in the fossil record*, pp. 1–36. Oxford: Clarendon Press.

Bower, F. O. (1935) *Primitive land plants*. London: Macmillan.

Bradley, R. S. (1999) *Paleoclimatology. Reconstructing climates of the Quaternary*. London: Academic Press.

Brand, U. (1989) Global climate changed during the Devonian–Mississippian: stable isotope biogeochemistry of brachiopods. *Palaeogeography, Palaeoclimatology, Palaeoecology* **75**, 311–329.

Brashier, C. K. (1968) Vascularization of cycad leaflets. *Phytomorphology* **18**, 35–43.

Brassell, S. C., Eglinton, G., Marlowe, I. T., Pflaumann, U. and Sarnthein, M. (1986) Molecular stratigraphy: a new tool for climatic assessment. *Nature* **320**, 129–133.

Bremer, K. (2000) Early Cretaceous lineages of monocot flowering plants. *Proceedings of the National Academy of Sciences of the United States of America* **97**, 4707–4711.

Brenchley, P. J. (1990) End Ordovician. In: Briggs, D. E. G. and Crowther, P. R. (Eds) *Paleobiology: a synthesis*, pp. 181–184. Oxford: Blackwell.

Brenner, G. J. (1996) Evidence for the earliest stage of angiosperm pollen evolution: a paleoequatorial section from Israel. In: Taylor, D. W. and Hickey, L. J. (Eds) *Flowering plant origin, evolution and phylogeny*, pp. 91–115. New York: Chapman & Hall.

Briggs, D. and Walters, S. M. (1984) *Plant variation and evolution*. Cambridge: Cambridge University Press.

Briggs, D. E. G. (1999) Molecular taphonomy of animal and plant cuticles: selective preservation and diagenesis. *Philosophical Transactions of the Royal Society of London, B* **354**, 7–17.

Briggs, D. E. G. and Eglinton, G. (1994) Chemical traces of ancient life. *Chemistry in Britain* **30**, 907–912.

Briggs, J. C. (1995) *Global biogeography*, First edn. Amsterdam: Elsevier Science.

Brocks, J. J., Logan, G. A., Buick, R. and Summons, R. E. (1999) Archaean molecular fossils and the early rise of eukaryotes. *Science* **285**, 1033–1036.

Brown, J. H. and Lomolino, M. V. (1998) *Biogeography*, Second edn. Sunderland, Massachusetts: Sinauer Associates.

Brown, T. A. (1999) How ancient DNA may help in understanding the origin and spread of agriculture. *Philosophical Transactions of the Royal Society of London, B* **354**, 89–98.

Brown, T. A. and Brown, K. A. (1994) Ancient DNA: using molecular biology to explore the past. *BioEssays* **16**, 719–726.

Buggisch, W. (1972) Zur geologie und geochemie der Kellwasserkalke und ihrer begleitenden sedimente (Unteres Oberdevon). *Abhabdlungen des Hessischen Landesamt fur Bodenforschung* **62**, 1–68.

Buggisch, W. (1991) The global Frasnian–Famennian 'Kellwasser Event'. *Geologische Rundschau* **80**, 49–72.

Bush, M. B. (1994) Amazonian speciation – a necessarily complex model. *Journal of Biogeography* **21**, 5–17.

Butterfield, N. J. (2000) *Bangiomorpha pubescens* n. gen., n. sp.: implications for the evolution of sex, multicellularity, and the Mesoproterozoic/ Neoproterozoic radiation of eukaryotes. *Paleobiology* **26**, 386–404.

Butterfield, N. J., Knoll, A. H. and Swett, K. (1988) Exceptional preservation of fossils in Upper Proterozoic shale. *Nature* **334**, 424–426.

Butterfield, N. J., Knoll, A. H. and Swett, K. (1990) A bangiophyte red alga from the Proterozoic of arctic Canada. *Science* **250**, 104–107.

Butterfield, N. J., Knoll, A. H. and Swett, K. (1994) Paleobiology of the Neoproterozoic Svanbergfjellet Formation, Spitzbergen. *Fossils and Strata* **34**, 75–82.

Caccavari, M. A. (1996) Analysis of the South American fossil pollen record of Mimosoideae (Leguminosae). *Review of Palaeobotany and Palynology* **94**, 123–135.

Caldeira, K. and Rampino, M. R. (1991) The Mid-Cretaceous super plume, carbon-dioxide, and global warming. *Geophysical Research Letters* **18**, 987–990.

Cande, S. C. and Kent, D. V. (1992) A new geomagnetic polarity time scale for the Late Cretaceous and Cenozoic. *Journal of Geophysical Research* **97**, 13 917–13 951.

Canfield, D. E. and Teske, A. (1996) Late Proterozoic rise in atmospheric oxygen concentration inferred from phylogenetic and sulphur-isotope studies. *Nature* **382**, 127–132.

Cantrill, D. J. and Nichols, G. J. (1996) Taxonomy and palaeoecology of Early Cretaceous (Late Albian) angiosperm leaves from Alexander Island, Antarctica. *Review of Palaeobotany and Palynology* **92**, 1–28.

Carson, H. L. (1982) Speciation as a major reorganisation of polygenic balances. In: Barigozzi, C. (Ed.) *Mechanisms of speciation*, pp. 411–433. New York: Alan R. Liss.

Cattolico, R. A. (1986) Chloroplast evolution in algae and land plants. *Tree* **1**, 64–66.

Cerling, T. E. (1992) Use of carbon isotopes in paleosols as an indicator of the P(CO_2) of the paleoatmosphere. *Global Biogeochemical Cycles* **6**, 307–314.

Cerling, T. E. and Quade, J. (1993) Stable carbon and oxygen isotopes in soil carbonates. *Geophysical Monographs* **78**, 231.

Cerling, T. E., Wang, Y. and Quade, J. (1993) Global ecological changes in the late Miocene: expansion of C4 ecosystems. *Nature* **361**, 345.

Cerling, T. E., Harris, J. M., MacFadden, B. J., Leakey, M. G., Quade, J., Eisenmann, V. and Ehleringer, J. R. (1997) Global vegetation change through the Miocene/Pliocene boundary. *Nature* **389**, 153–158.

Chaloner, W. G. (1967) Spores and land plant evolution. *Review of Palaeobotany and Palynology* **1**, 89–93.

Chaloner, W. G. (1970) The rise of the first land plants. *Biological Reviews* **45**, 353–377.

Chaloner, W. G. (1999) Plant and spore compression in sediments. In: Jones, T. P. and Rowe, N. P. (Eds) *Fossil plants and spores*, pp. 36–41. London: The Geological Society.

Chaloner, W. G. and Creber, G. T. (1988) Fossil plants as indicators of late Palaeozoic plate positions. In Audley-Charles, M. G. and Hallam, A. (Eds.) *Gondwana and Tethys Geological Society Special Publication* 37, 201–210.

Chaloner, W. G. and Creber, G. T. (1990) Do fossil plants give a climate signal? *Journal of the Geological Society London* **147**, 343–350.

Chaloner, W. G. and Hallam, A. (1994) *Evolution and extinction: Proceedings of a Joint Symposium of the Royal Society and the Linnean Society, 9–10 November 1989*. Cambridge: Cambridge University Press.

Chaloner, W. G. and McElwain, J. C. (1997) The fossil plant record and global climatic change. *Review of Palaeobotany and Palynology* **95**, 73–82.

Chaloner, W. G. and Sheerin, A. (1979) Devonian macrofloras. In: House, M. R., Scrutton, C. T. and Bassett, M. G. (Eds) *The Devonian system*, pp. 145–161. Special papers in Palaentology 23.

Chandler, M. E. J. (1964) *The lower Tertiary floras of southern England IV*. London: British Museum (Natural History).

Chang, S. (1994) The planetary setting of prebiotic evolution. In: Bengston, S. (Ed.) *Early life on Earth*, pp. 10–23. New York: Columbia University Press.

Chapman, D. J. (1985) Geological factors and biochemical aspects of the origin of land plants. In: Tiffney, B. H. (Ed.) *Geological factors and the evolution of plants*, pp. 23–45. New Haven: Yale University Press.

Chapman, R. L. and Buchheim, M. A. (1991) Ribosomal RNA gene sequences: analysis and significance in the phylogeny and taxonomy of green algae. *Critical Reviews in Plant Sciences* **10**, 343–368.

Chase, C. G., Gregory-Wodzicki, K. M., Parrish, J. T. and DeCelles, P. G. (1998) Topographic history of the Western Cordillera of North America and controls on climate. In: Crowley, T. J. and Burke, K. C. (Eds) *Tectonic boundary conditions for climate reconstructions*, pp. 73–100. Oxford: Oxford University Press.

Chernov, Y. I. (1985) *The living tundra*. Cambridge: Cambridge University Press.

Christopher, R. A. (1979) Normapolles and triporate pollen assemblages from the Raritan and Magothy Formations (Upper Cretaceous) of New Jersey. *Palynology* **3**, 73–122.

Chyba, C., Thomas, P., Brookshaw, L. and Sagan, C. (1990) Cometary delivery of organic molecules to the early Earth. *Science* **249**, 366–373.

Cleal, C. J. and Thomas, B. A. (1991) Carboniferous and Permian paleogeography. In: Cleal, C. J. (Ed.) *Plant fossils in geological investigation; the Palaeozoic*, pp. 154–181. New York: Ellis Horwood.

Cleal, C. J. and Thomas, B. A. (1999) *Plant fossils*. Woodbridge: The Boydell Press.

Clement-Westerhof, J. A. (1988) Morphology and phylogeny of Paleozoic conifers. In: Beck, C. B. (Ed.) *Origin and evolution of gymnosperms*, pp. 298–337. New York: Columbia University Press.

Cole, M. M. (1986) *The savannas: biogeography and geobotany*. London: Academic Press.

Colinvaux, P. A., De Oliveira, P. E. and Bush, M. B. (2000) Amazonian and neotropical plant communities on glacial time-scales: the failure of the aridity and refuge hypotheses. *Quaternary Science Reviews* **19**, 141–169.

Collinson, M. E. (1990) Plant evolution and ecology during the early Cainozoic diversification. *Advances in Botanical Research* **17**, 1–98.

Collinson, M. E. (1992) Vegetational and floristic changes around the Eocene/Oligocene boundary in western and central Europe. In: Prothero, D. R. and Berggren, W. A. (Eds) *Eocene–Oligocene climate and biotic evolution*, pp. 437–450. Princeton, NJ: Princeton University Press.

Collinson, M. E. (2000) Cainozoic evolution of modern plant communities and vegetation. In: Culver, S. J. and Rawson, P. F. (Eds) *Biotic response to global change*, pp. 223–243. Cambridge: Cambridge University Press.

Collinson, M. E., Boutler, M. C. and Holmes, P. L. (1993) Magnoliophyta ('Angiospermae'). In: Benton, M. J. (Ed.) *The fossil record* 2, pp. 809–841. London: Chapman & Hall.

Collinson, M. E., Van Bergen, P. V., Scott, A. C. and De Leeuw, J. W. (1994) The oil generating potential of plants from coal and coal generating strata through time: a review with new evidence from Carboniferous plants. *Geological Society Special Publication* **77**, 31–70.

Collinson, M. E., Scott, A. C., Finch, P. and Wilson, R. (1998) The preservation of plant cuticle in the fossil record: a chemical and microscopical investigation. *Ancient Biomolecules* **2**, 251–265.

Compton, J. S. and Mallinson, D. J. (1996) Geochemical consequences of increased Late Cenozoic weathering rates and the global CO_2 balance since 100 Ma. *Paleoceanography* **11**, 431–446.

Condie, K. C. (1997) *Plate tectonics and crustal evolution*, Fourth edn. New York: Pergamon Press.

Condie, K. C. and Sloan, R. E. (1998) *Origin and evolution of Earth*, First edn. New Jersey: Prentice Hall.

Copper, P. (1986) Frasnian–Famennian mass extinction and cold-water oceans. *Geology* **14**, 835–839.

Courtillot, V. (1990) A volcanic eruption. *Scientific American* **263**, 53–60.

Coxon, P. and Clayton, G. (1999) Light microscopy of fossil pollen and spores. In: Jones, T. P. and Rowe, N. P. (Eds) *Fossil plants and spores: modern techniques*, pp. 47–52. London: Geological Society.

Coxon, P. and Waldren, S. (1997) Flora and vegetation of the Quaternary temperate stages of NW Europe: evidence for large-scale range changes. In: Huntley B. E., Cramer W., Morgan A. V., Prentice H. C. and Allen J. R. M. (Eds) *Past and future rapid environmental changes: the spatial and evolutionary responses of terrestrial biota*, pp. 103–117. Berlin: Springer-Verlag.

Crane, P. R. (1985) Phylogenetic analysis of seed plants and the origins of angiosperms. *Annals of the Missouri Botanical Garden* **72**, 716–793.

Crane, P. R. (1987) Vegetational consequences of the angiosperm diversification. In: Friis, E. M., Chaloner, W. G. and Crane, P. R. (Eds) *The origins of angiosperms and their biological consequences*, pp. 105–144. Cambridge: Cambridge University Press.

Crane, P. R. and Kenrick, P. (1997) Diverted development of reproductive organs: a source of morphological innovation in land plants. *Plant Systematics and Evolution* **206**, 161–174.

Crane, P. R. and Lidgard, S. (1989) Angiosperm diversification and paleolatitudinal gradients in cretaceous floristic diversity. *Science* **246**, 675–678.

Crane, P. R., Friis, E. M. and Pedersen, K. R. (1995) The origin and early diversification of angiosperms. *Nature* **374**, 27–33.

Crepet, W. L. (1996) Timing in the evolution of derived floral characteristics: Upper Cretaceous (Turonian) taxa with tricolpate and tricolpate-derived pollen. *Review of Palaeobotany and Palynology* **90**, 339–359.

Crepet, W. L. and Feldman, G. D. (1991) The earliest remains of grasses in the fossil record. *American Journal of Botany* **78**, 1010–1014.

Crepet, W. L. and Friis, E. M. (1987) The evolution of insect pollination in angiosperms. In: Friis, E. M., Chaloner, W. G. and Crane, P. R. (Eds) *The origin of angiosperms and their biological consequences*, pp. 181–201. Cambridge: Cambridge University Press.

Cross, A. T. and Phillips, T. L. (1990) Coal-forming plants through time in North America. *International Journal of Coal Geology* **16**, 1–46.

Crowley, T. J. (1998) Significance of tectonic boundary conditions for paleoclimate simulations. In: Crowley, T. J. and Burke, K. C. (Eds) *Tectonic boundary conditions for climate reconstructions*, pp. 3–21. Oxford: Oxford University Press.

Crowley, T. J. and North, G. R. (1991) *Paleoclimatology*. Oxford: Oxford University Press.

Crowley, T. J., Mengel, J. G. and Short, D. A. (1987) Gondwanaland's seasonal cycle. *Nature* **329**, 803–807.

Cuneo, R. (1989) Phytogeography and paleoecology of late Paleozoic floras from southern South America and their relationships with other floral regions. *International Geological Congress, Abstracts* **28** (**1**), 351(Abstract).

Daghlian, C. P. (1981) A review of the fossil record of monocotyledons. *The Botanical Review* **47**, 517–555.

Dalrymple, G. B. (1991) Modern radiometric methods: how they work. In: Dalrymple, G. B. (Ed.) *The age of the Earth*, p. 79. Stanford: Stanford University Press.

Darwin, C. (1859) *The origin of species*, Penguin Classics 1985 edn. London: Penguin Books.

Davis, J. I. (1995) A phylogenetic structure for the monocotyledons, as inferred from chloroplast DNA restriction site variation, and a comparison of measures of clade support. *Systematic Botany* **20**, 503–527.

Dawkins, R. (1976) *The selfish gene*. Oxford: Oxford University Press.

Dawkins, R. (1986) *The blind watchmaker*. Harlow, England: Longman Scientific & Technical.

Deamer, D. W. (1993) Prebiotic conditions and the first cells. In: Lipps, J. H. (Ed.) *Fossil prokaryotes and protists*, pp. 11–18. Boston: Blackwell Scientific.
Deamer, D. W., Mahon, E. H. and Bosco, S. (1994) Self-assembly and function of primitive membrane structures. In: Bengston, S. (Ed.) *Early life on Earth*, pp. 107–123. New York: Columbia University Press.
Degens, E. T. (1967) Biogeochemistry of stable carbon isotopes. In: Eglinton, G. and Murphy M. T. (Eds) *Organic geochemistry: methods and results*, pp. 303–329. New York: Longman.
deJesus, M. D., Tabatabai, F. and Chapman, D. J. (1989) Taxonomic distribution of copper–zinc superoxide dismutase in green algae and its phylogenetic importance. *Journal of Phycology* **25**, 767–772.
De Leeuw, J. W. and Largeau, C. (1993) A review of macromolecular organic compounds that comprise living organisms and their role in kerogen, coal and petroleum formation. In: Engel, M. H. and Macko, S. A. (Eds) *Organic geochemistry*, pp. 23–72. New York: Plenum.
De Leeuw, J. W., Frewin, N. L., Van Bergen, P. V. and Collinson, M. E. (1995) Organic carbon as a palaeoenvironmental indicator in the marine realm. In: Bosence, D. W. J. and Allison, P. A. (Eds) *Marine palaeoenvironmental analysis from fossils*, pp. 43–71. London: Geological Society Press.
Delevoryas, T. and Hope, C. R. (1976) More evidence for a slender growth habit in Mesozoic cycadophytes. *Review of Palaeobotany and Palynology* **21**, 93–100.
Delwiche, C. F., Graham, L. E. and Thomson, N. (1989) Lignin-like compounds and sporopollenin in *Coleochaete*, an algal model for land plant ancestry. *Science* **245**, 399–401.
Demesure, B., Comps, B. and Petit, R. J. (1996) Chloroplast DNA phylogeography of the common beech (*Fagus sylvatica* L.) in Europe. *Evolution* **50**, 2515–2520.
Denton, G. H. and Armstrong, R. L. (1969) Miocene–Pliocene glaciations in southern Alaska. *American Journal of Science* **267**, 1121–1142.
DeSalle, R. (1994) Implications of ancient DNA for phylogenetic studies. *Experientia* **50**, 543–550.
DeSalle, R., Gatesby, J., Wheeler, W. and Grimaldi, D. (1992) DNA sequences from a fossil termite in Oligo-Miocene amber and their phylogenetic implications. *Science* **257**, 1933–1936.
Des Marais, D., Strauss, H., Summons, R. E. and Hayes, J. M. (1992) Carbon isotope evidence for the stepwise oxidation of the Proterozoic environment. *Nature* **359**, 605–609.
Dettmann, M. E. (1992) Structure and floristics of Cretaceous vegetation of southern Gondwana: implications for angiosperm biogeography. In: Pons, D. and Broutin, J. (Eds) *Organisation internationale de paleobotanique, Ivème conference: resumçès des communications*, p. 47. Villeneuve d'Asca: Organisation Française de Paleobotanique.
Dettmann, M. E. and Jarzen, D. M. (1990) The Antarctic Australian rift-valley – Late Cretaceous cradle of northeastern Australasian relicts. *Review of Palaeobotany and Palynology* **65**, 131–144.
Dettmann, M. E., Pocknall, D. T., Romero, E. J. and Zamaloa, M. D. C. (1990) *Nothofagidites* Erdtman ex Potonié, 1960: a catalogue of species with notes on the paleographic distribution of *Nothofagus*. B1 (southern beech). *New Zealand Geological Survey Paleontological Bulletin* **60**, 1–79.
DeVries, P. J., Simons, R. J. and VanBeem, A. P. (1983) Sporopollenin in the spore wall of *Spirogyra. Acta Botanica Neerlandica* **32**, 25–28.
Dilcher, D. L. (1989) The occurrence of fruits with affinities to Ceratophylaceae in lower and mid-Cretaceous sediments. *American Journal of Botany* **76**, 162.
Dilcher, D. and Crane, P. R. (1984) In pursuit of the 1st flowers. *Natural History* **93**, 56–61.

Dilcher, D. L. and Crane, P. R. (1984) Archaeanthus: an early angiosperm from the Cenomanian of the western interior of North America. *Annals of the Missouri Botanical Gardens* **71**, 351–783.

DiMichele, W. A., Phillips, T. L. and Olmstead, R. G. (1987) Opportunistic evolution: abiotic environmental stress and the fossil record of plants. *Review of Palaeobotany and Palynology* **50**, 151–178.

DiMichele, W. A., Hook, R. W., Beerbower, J. R., *et al.* (1992) Paleozoic terrestrial ecosystems. In: Behrensmeyer, J. D. *et al.* (Eds) *Evolutionary paleoecology of terrestrial plants and animals*, pp. 205–235. Chicago: University of Chicago Press.

DiMichele, W. A., Pfefferkorn H. W. and Phillips T. L. (1996) Persistence of Late Carboniferous tropical vegetation during glacially driven climatic sea-level fluctuations. *Palaeogeography, Palaeoclimatology, Palaeoecology* **125**, 105–128.

Dobruskina, I. A. (1987) Phytogeography of Eurasia during the Early Triassic. *Palaeogeography, Palaeoclimatology, Palaeoecology* **58**, 75–86.

Doyle, J. A. and Donoghue, M. J. (1986) Seed plant phylogeny and the origins of angiosperms: an experimental cladistic approach. *The Botanical Review* **52**, 321–431.

Doyle, J. A. and Donoghue, M. J. (1987) The origin of angiosperms: a cladistic approach. In: Friis, E. M., Chaloner, W. G. and Crane, P. R. (Eds) *The origins of angiosperms and their biological consequences*, pp. 17–49. Cambridge: Cambridge University Press.

Doyle, J. A. and Hickey, L. J. (1976) Pollen and leaves from Mid-Cretaceous Potomac Group and their bearing on early angiosperm evolution. In: Beck, C. B. (Ed.) *Origin and early evolution of angiosperms*, pp. 139–206. New York: Columbia University Press.

Doyle, J. A., Donoghue, M. J. and Zimmer, E. A. (1994) Intergration of morphological and ribosomal RNA data on the origin of angiosperms. *Annals of the Missouri Botanical Garden* **81**, 419–450.

Drinnan, A. N. and Crane, P. R. (1990) Cretaceous paleobotany and its bearing on the biogeogrpahy of austral angiosperms. In: Taylor, T. N. and Taylor, E. L. (Eds) *Antarctic paleobiology: Its role in the reconstruction of Gondwana*, pp. 192–219. New York: Springer-Verlag.

Drinnan, A. N., Crane, P. R., Friis, E. M., Pedersen, K. R. (1990) Lauraceous flowers from the Potomac Group (mid Cretaceous) of eastern North America. *Botanical Gazette* **151**, 370–380.

Dunlop, D. J. and Özdemir, Ö. (1997) Magnetism in nature. In: Dunlop, D. J. and Özdemir, Ö. (Eds) *Rock magnetism: fundamentals and frontiers*, pp. 1–15. Cambridge: Cambridge University Press.

Durante, M. V. (1995) Reconstruction of late Paleozoic Climate changes in angarland according to phytogeographic data. *Stratigraphy and Geological Correlation* 3(2), 123–133.

Dwyer, E., Gregoire, J.-M. and Pereira, J. M. C. (2000) Climate and vegetation as driving factors in global fire activity. In: Innes, J. L., Beniston, M. and Verstraete, M. M. (Eds) *Biomass burning and its inter-relationships with the climate system*, pp. 171–191. Dordrecht: Kluwer Academic Publishers.

Dyer, B. D. and Obar, R. A. (1994) *Tracing the history of eukaryotic cells: the enigmatic smile*, First edn. New York: Columbia University Press.

Dyson, W. G. and Herbin, G. A. (1968) Studies on plant cuticular waxes. IV. Leaf wax alkanes as a taxonomic discriminant for cypresses grown in Kenya. *Phytochemistry* **7**, 1339–1344.

Edwards, D. (1982) Fragmentary non-vascular plant microfossils from the Late Silurian of Wales. *Botanical Journal of the Linnean Society* **84**, 223–256.

Edwards, D. (1993) Cells and tissues in the vegetative sporophytes of early land plants. *New Phytologist* **125**, 225–247.

Edwards, D. and Berry, C. (1991) Silurian and Devonian. In: Cleal, C. J. (Ed.) *Plant fossils in geological invesitgations*, pp. 117–148. New York: Ellis Horwood.

Edwards, D. and Davies, M. S. (1990) Interpretations of early land plant radiations: 'facile adaptionist guesswork' or reasoned speculation? In: Taylor, P. D. and Larwood, G. P. (Eds) *Major evolutionary radiations*, pp. 351–376. Oxford: Oxford University Press.

Edwards, D. and Fanning, U. (1985) Evolution and environment in the Late Silurian–Early Devonian – the rise of the pteridophytes. *Philosophical Transactions of the Royal Society of London, Series B* **309**, 147.

Edwards, D. and Selden, P. A. (1991) The development of early terrestrial ecosystems. *Botanical Journal of Scotland* **46**, 337–366.

Edwards, D., Feehan, J. and Smith, D. G. (1983) A late Wenlock flora from Co. Tipperary, Ireland. *Botanical Journal of the Linnean Society* **86**, 19–36.

Edwards, D., Duckett, J. G. and Richardson, J. B. (1995) Hepatic characters in the earliest land plants. *Nature* **374**, 635–636.

Edwards, D. S. (1980) Evidence for the sporophytic status of the Lower Devonian plant *Rhynia gwynne- vaughanii* Kidston and Lang. *Review of Palaeobotany and Palynology* **29**, 177–188.

Edwards, D. S. (1986) *Aglaophyton major*, a non-vascular land-plant from the Devonian Rhynie Chert. *Botanical Journal of the Linnean Society* **93**, 173–204.

Eggert, D. A. (1961) The ontogeny of Carboniferous arborescent Lycopsida. *Palaeontographica* **108B**, 43–92.

Eglinton, G. and Logan, G. A. (1991) Molecular preservation. *Philosophical Transactions of the Royal Society of London, Series B* **333**, 315–328.

Ehleringer, J. R. and Monson, R. K. (1993) Evolutionary and ecological aspects of photosynthetic pathway variation. *Annual Review of Ecology and Systematics* **24**, 411–439.

Ekart, D. P., Cerling, T. E., Montanez, I. P. and Tabor, N. J. (1999) A 400 million year carbon isotope record of pedogenic carbonate: implications for paleoatmospheric carbon dioxide. *American Journal of Science* **299**, 805–827.

Eldredge, N. and Gould, S. J. (1972) Punctuated equilibria: an alternative to phyletic gradualism. In: Schopf, T. J. (Ed.) *Models in paleobiology*, pp. 82–243. San Francisco: Freeman, Cooper & Company.

Elick, J. M., Driese, S. G. and Mora, C. I. (1998) Very large plant and root traces from Early to Middle Devonian: implications for early terrestrial ecosystems and atmospheric $p(CO_2)$. *Geology* **26**, 143–146.

Endress, P. K. (1987) The early evolution of the angiosperm flower. *Tree* **2**, 300–304.

Eriksson, P. G. and Cheney, E. S. (1992) Evidence for the transition to an oxygen-rich atmosphere during the evolution of red beds in the lower Proterozoic sequences of southern Africa. *Precambrian Research* **54**, 257–269.

Erwin, D. H. (1990) The end-Permian mass extinction. *Annual Review of Ecology and Systematics* **21**, 69–91.

Erwin, D. H. (1993) *The great Paleozoic crisis: life and death in the Permian*, First edn. New York: Columbia University Press.

Erwin, D. H. and Anstey, R. L. (1995) New approaches to speciation in the fossil record. Columbia University Press, New York.

Evershed, R. P. (1993) Biomolecular archaeology and lipids. *World Archaeology* **25**, 74–93.

Falcon-Lang, H. (1999) The Early Carboniferous (Asbian-Brigantian) seasonal tropical climate of northern Britain. *Palaios* **14**, 116–126.

Farrimond, P. and Eglinton, G. (1990) The role of organic compounds and the nature of source rocks. In: Briggs, D. and Crowther, P. R. (Eds) *Paleobiology: a synthesis*, pp. 217–222. Oxford: Blackwell.

Fischer, A. G. (1984) Biological innovations and the sedimentary record. In: Holland, H. and Trendall A. F. (Eds) *Patterns of change in Earth evolution*, pp. 145–157. Berlin: Springer-Verlag.

Fischer, A. G. and Arthur, M. A. (1977) Secular variations in the pelagic realm. In: Cook, H. E. and Enos H. E. (Eds) *Deep water carbonate environments, special publication 25*, pp. 19–50. Tulsa, Oklahoma: Society of Economic Paleontologists and Mineralogists.

Florin, R. (1938) Die Koniferen des Oberkarbons und des Unteren Perms. *Palaeontographica* **B85**, H1, 1–62.

Florin, R. (1951) Evolution in cordaites and conifers. *Acta Horti Bergiani* **15**, 285–388.

Flower, B. P. and Kennett, J. P. (1994) The Middle Miocene climatic transition: east Antarctica ice sheet development, deep ocean circulation and global carbon cycling. *Palaeogeography, Palaeoclimatology, Palaeoecology* **108**, 537–555.

Fowell, S. J., Cornet, B. and Olsen, P. E. (1994) Geologically rapid Late Triassic extinctions; palynological evidence from the Newark Supergroup. In: Klein, G. D. (Ed.) *Pangea; paleoclimate, tectonics, and sedimentation during accretion, zenith and breakup of a supercontinent*, pp. 197–206. Boulder: Geological Society of America.

Frakes, L. A., Francis, J. E. and Syktus, J. I. (1992) *Climate modes of the Phanerozoic: the history of the Earth's climate over the past 600 million years.* Cambridge: Cambridge University Press.

Frederick, S. E., Gruber, P. J. and Tolbert, N. E. (1973) The occurrence of glycolate dehydrogenase and glycolate oxidase in green plants. *Plant Physiology* **52**, 318–323.

Friedman, W. E. (1996) Biology and evolution of the Gnetales. *International Journal of Plant Sciences* **157**, 1–220.

Friis, E. M., Crane, P. R. and Pedersen, K. R. (1986) Floral evidence for Cretaceous chloranthoid angiosperms. *Nature* **320**, 163–164.

Friis, E. M., Crane, P. R. and Pedersen, K. R. (1986) Reproductive structures of Cretaceous Platanaceae. *Biol. Skr* **41**, 5–45.

Friis, E. M. and Crepet, W. L. (1987) Time of appearance of floral features. In: Friis, E. M., Chaloner, W. G. and Crane, P. R. (Eds) *The origin of angiosperms and their biological consequences*, pp. 145–179. Cambridge: Cambridge University Press.

Friis, E. M., Chaloner, W. G. and Crane, P. R. (1987) *The origins of angiosperms and their biological consequences.* Cambridge: Cambridge University Press.

Friis, E. M., Pedersen K. R. and Crane, P. R. (1995) *Appomattoxia ancistrophora* gen et sp-nov, a new Early Cretaceous plant with similarities to *Circaeaster* and extant Magnoliidae. *American Journal of Botany* **82**, 933–943.

Friis, E. M., Pedersen, K. R. and Crane, P. R. (1999) Early angiosperm diversification: the diversity of pollen associated with angiosperm reproductive structures in Early Cretaceous floras from Portugal. *Annals of the Missouri Botanical Garden* **86**, 259–296.

Futuyma, D. J. (1998) *Evolutionary biology*, Third edn. Massachusetts: Sinauer Associates.

Galtier, J. and Phillips, T. L. (1999) The acetate peel technique. In: Jones, T. P. and Rowe, N. P. (Eds) *Fossil plants and spores*, pp. 67–70. London: The Geological Society.

Gandolfo, M. A., Nixon, K. C., Crepet, W. L., Stevenson, D. W. and Friis, E. M. (1998) Oldest known fossils of monocotyledons. *Nature* **394**, 532–533.

Garratt, M. J., Tims, J. D., Richards, R. B., Chambers, T. C. and Douglas, J. G. (1984) The appearance of *Baragwanathia* (Lycophytina) in the Silurian. *Botanical Journal of the Linnean Society* **89**, 355–358.

Gensel, P. G. (1982) A new species of *Zosterophyllum* from the Early Devonian of New Brunswick. *American Journal of Botany* **69**, 651–669.

Gensel, P. G. and Andrews, H. N. (1984) *Plant life in the Devonian.* New York: Praeger.

Gensel, P. G. and Andrews, H. N. (1987) The evolution of early land plants. *American Scientist* **75**, 478–489.

Germeraas, J. H., Hopping, C. A. and Muller, J. (1968) Palynology of tertiary sediments from tropical areas. *Review of Palaeobotany and Palynology* **6**, 189–348.

Gifford, E. M. and Foster, A. S. (1989) *Morphology and evolution of vascular plants*, Third edn. New York: Freeman.
Gillespie, J. H. (1991) *The causes of molecular evolution*. New York: Oxford University Press.
Golenberg, E. M. (1991) Amplification and analysis of Miocene plant fossil DNA. *Philosophical Transactions of the Royal Society of London, Series B* **333**, 419–427.
Golenberg, E. M. (1994) DNA from plant compression fossils. In: Herrman, B. and Hummel, S. (Eds) *Ancient DNA*, pp. 237–257. Berlin: Springer-Verlag.
Golenberg, E. M. (1999) Isolation, identification and authentication of DNA sequences derived from fossil material. In: Jones, T. P. and Rowe, N. P. (Eds) *Fossil plants and spores: modern techniques*, pp. 156–161. London: Geological Society.
Golenberg, E. M., Giannasi, D. E., Clegg, M. T., Smiley, C. J., Durbin, M., Henderson, D. and Zurawski, G. (1990) Chloroplast DNA sequence from a Miocene *Magnolia* species. *Nature* **344**, 656–658.
Goloubinoff, P., Pääbo, S. and Wilson, A. C. (1993) Evolution of maize inferred from sequence diversity of an *Adh2* gene segment from archaeological specimens. *Proceedings of the National Academy of Sciences of the United States of America* **90**, 1997–2001.
Golubic, S. (1976) Organisms that build stromatolites. In: Walter, M. R. (Ed.) *Stromatolites*, pp. 113–126. Amsterdam: Elsevier.
Goodwin, A. M. (1991) *Precambrian geology*. London: Academic Press.
Gould, S. J. (1985) The paradox of the first tier: an agenda for paleobiology. *Paleobiology* **11**, 2–12.
Gould, S. J. (1989) *Wonderful life*. London: Penguin Books.
Gould, S. J. (1995) A task for paleobiology at the threshold of majority. *Paleobiology* **21**, 1–14.
Gould, S. J. (1998) Gulliver's further travels: the necessity and difficulty of a hierarchical theory of selection. *Philosophical Transaction of the Royal Society of London B* **353**, 307–314.
Graham, A. (1999) *Late Cretaceous and Cenozoic history of North American vegetation*. Oxford: Oxford University Press.
Graham, L. E. (1990) Meiospore formation in charophycean algae. In: Blackmore, S. and Knox, A. (Eds) *Microspores: evolution and ontogeny*, pp. 43–54. London: Academic Press.
Graham, L. E. (1993) *Origin of land plants*. New York: Wiley.
Graham, L. E. and Delwiche, C. F. (1992) The occurrence and phylogenetic significance of a surface layer on thalli of *Coleochaete*. *American Journal of Botany* **72**, 102.
Graham, L. E., Graham, J. M., Downs, J. and Gerwing, J. (1992) *Colechaete* and the periphyton ecology of a northern oligotrophic lake. *Journal of Phycology* **28**, 8.
Graham, L. K. E., Wilcox, L. W. and Kenrick, P. (2000) The origin of alternation of generations in land plants: a focus on matrotrophy and hexose transport. *Philosophical Transaction of the Royal Society of London B* **355**, 757–767.
Grant, V. (1963) *The origin of adaptations*. New York: Columbia University Press.
Grant, V. (1971) *Plant speciation*. New York: Columbia University Press.
Gray, J. (1985) The microfossil record of early land plants: advances in understanding of early terrestrialisation, 1970–1984. *Philosophical Transactions of the Royal Society of London, Series B* **309**, 167–195.
Gray, J. (1993) Major Paleozoic land plant evolutionary bio-events. *Palaeogeography, Palaeoclimatology, Palaeoecology* **104**, 153–169.
Gray, J. and Shear, W. (1992) Early life on land. *American Scientist* **80**, 444–456.
Gray, J., Massa, D. and Boucot, A. J. (1982) Caradocian land plant microfossils from Libya. *Geology* **10**, 197–201.
Greenwood, D. R. (2000) Early Paleogene warm climates and vegetation in southeastern Australia. *Geological Society of Australia Abstracts* **59**, 192.

Grimaldi, D. (1999) The co-radiations of pollinating insects and angiosperms in the Cretaceous. *Annals of the Missouri Botanical Garden* **86**, 373–406.

Gröcke, D. (1998) Carbon isotope analyses of fossil plants as a chemistratigraphic and palaeoenvironmental tool. *Lethaia* **31**, 1–13.

Gröcke, D., Hesselbo, S. P. and Jenkyns, H. C. (1999) Caron-isotope composition of Lower Cretaceous fossil wood: ocean–atmosphere chemistry in relation to sea-level change. *Geology* **27**, 155–158.

Grotzinger, J. P. and Knoll, A. H. (1999) Stromatolites in Precambrian carbonates: evolutionary mileposts or environmental dipsticks? *Annual Review of Earth and Planetary Sciences* **27**, 313–358.

Groves, A. T. and Rackham, O. (2001) *The nature of Mediterranean Europe: an ecological history*. New Haven: Yale University Press.

Groves, R. H. and DiCastri, F. (1991) *Biogeography of Mediterranean invasions*, First edn. Cambridge: Cambridge University Press.

Gubeli, A. A., Hochuli, P. A. and Wildi, W. (1984) Lower Cretaceous turbiditic sediments from the Rif chain (northern Morocco) – palynology, stratigraphy and palaeogeographic setting. *Geologische Rundschau* **73**, 1081–1114.

Guo, S. (1993) The evolution of the Cenozoic tropical monsoon climate and monsoon forests of southwestern China. In: Jablonski, N. G. and Chak-Lam, S. (Eds) *Evolving landscapes and evolving biotas of east Asia since the mid Tertiary*, pp. 123–136. Hong Kong: Centre of Asian Studies, University of Hong Kong.

Hallam, A. (1983) Plate tectonics and evolution. In: Bendall, D. S. (Ed.) *Evolution from molecules to men*, pp. 367–386. Cambridge: Cambridge University Press.

Hallam, A. (1989) The case for sea-level change as a dominant causal factor in mass extinctions of marine invertebrates. *Philosophical Transactions of the Royal Society of London, Series B* **325**, 653–655.

Hallam, A. (1992) *Phanerozoic sea level changes*. New York: Columbia University Press.

Hallam, A. (1994) *An outline of Phanerozoic biogeography*, First edn. Oxford: Oxford University Press.

Hallam, A. and Wignall, P. B. (1997) *Mass extinctions and their aftermath*. Oxford: Oxford University Press.

Halstead, L. B. (1990) Cretaceous–Tertiary (Terrestrial). In: Briggs, D. E. G. and Crowther, P. R. (Eds) *Palaeobiology: a synthesis*, pp. 203–207. Oxford: Blackwell Science.

Han, T. and Runnegar, B. (1992) Megascopic eukaryotic algae from the 2.1-billion-year-old Negaunee iron-formation, Michigan. *Science* **257**, 232–235.

Haq, B. U., Hardenbol, J. and Vail, P. R. (1987) Chronology of fluctuating sea levels since the Triassic. *Science* **235**, 1156–1167.

Hardwood, J. R. and Russell, N. J. (1984) *Lipids in plants and microbes*. London: George Allen & Unwin.

Harland, W. B., Armstrong, R. L., Cox, A. V., Craig, L. E., Smith, A. G. and Smith, D. G. (1990) *A geologic time scale*, Second edn. Cambridge: Cambridge University Press.

Harris, T. M. (1932) The fossil flora of Scoresby Sound, East Greenland. Part 2: description of seed plants *Incertae sedis* together with a discussion of certain cycadophytic cuticles. *Meddelelser om Grønland* **85**, 4–7.

Harris, T. M. (1937) The fossil flora of Scoresby Sound, East Greenland. Part 5: Stratigraphic relations of the plant beds. Kobenhavn: *Medd. Grønland* **112**, 1–11.

Harris, T. M. (1979) *The Yorkshire Jurassic Flora. V. Coniferales*. British Museum (Natural History), London, 163p.

Harris, T. M. (1964) *The Yorkshire Jurassic Flora II. Caytoniales, Cycadales, Pteridosperms*. British Museum of Natural History, London.

Harris, T. M. (1969) *The Yorkshire Jurassic Flora III. Bennettitales*. British Museum of Natural History, London.

Harris, T. M. and Millington, W. (1974) *The Yorkshire Jurassic Flora. IV. 1. Ginkgoales.* British Museum (Natural History), London.
Harrison, T. M., Yin, A. and Ryerson, F. J. (1998) Orographic evolution of the Himalaya and Tibetan plateau. In: Crowley, T. J. and Burke, K. C. (Eds) *Tectonic boundary conditions for climate reconstructions*, pp. 39–73. Oxford: Oxford University Press.
Hay, W. W., Behensky, J. F., Barron, E. J. and Sloan II, J. L. (1982) The Triassic–Liassic paleoclimatology of the proto-central North Atlantic rift system. *Palaeogeography, Palaeoclimatology, Palaeoecology* **40**, 13–30.
Hayes, J. M. (1994) Global methanotrophy at the Archean–Proterozoic transition. In: Bengston, S. (Ed.) *Early life on Earth*, pp. 220–236. New York: Columbia University Press.
Hemsley, A. R., Barrie, P. L., Scott, A. C. and Chaloner, W. G. (1994) Studies of fossil and modern spore and pollen wall biomacromolecules suing ^{13}C solid state NMR. *NERC Special Publication* **94**, 15–19.
Herbert, T. D. and D'Hondt, S. L. (1990) Precessional climate cyclicity in Late Cretaceous–Early Tertiary marine sediments: a high resolution chronometer of Cretaceous–Tertiary boundary events. *Earth and Planetary Science Letters* **99**, 263–275.
Herbert, T. D. and Fischer, A. G. (1986) Milankovitch climatic origin of Mid-Cretaceous black shale rythms in central Italy. *Nature* **321**, 739–743.
Herendeen, P. S. (1991) Charcoalified angiosperm wood from the Cretaceous of eastern North America and Europe. *Review of Palaeobotany and Palynology* **70**, 225–239.
Herendeen, P. S. and Crane, P. R. (1995) The fossil history of the monocotyledons. In: Rudall, P. J., Cribb, P. J., Cutler, D. F. and Humphries, C. J. (Eds) *Monocotyledons: systematics and evolution*, pp. 1–21. Kew, London: Royal Botanic Gardens.
Herngreen, G. F. W. and Chlonovo, A. F. (1981) Cretaceous microfloral provinces. *Pollen et Spores* **xxiii**, 441–555.
Herrman, B. and Hummel, S. (1994) Introduction. In: Herrman, B. and Hummel, S. (Eds) *Ancient DNA*, pp. 1–13. Berlin: Springer-Verlag.
Hickey, L. J. and Doyle, J. A. (1977) Early Cretaceous fossil evidence for angiosperm evolution. *Botanical Review* **43**, 3–104.
Hiesel, R., Haeseler, A.V. and Brennicke, A. (1994) Plant mitochondrial nucleic acid sequences as a tool for phylogenetic analysis. *Proceedings of the National Academy of Sciences of the United States of America* **91**, 634–638.
Hill, C. R. and Crane, P. R. (1982) Evolutionary cladistics and the origin of angiosperms. *Systematic Association* **21**, 269–361.
Hill, R. S. and Scriven, L. J. (1995) The angiosperm-dominated woody vegetation of Antarctica: a review. *Review of Palaeobotany and Palynology* **86**, 175–198.
Hills, L. V., Klovan, J. E. and Swett, A. R. (1974) *Juglans eocinerea* n.sp., Beaufort Formation (Tertiary), southwestern Banks Island, arctic Canada. *Canadian Journal of Botany* **52**, 65–90.
Hoffman, A. (1989a) Mass extinctions: the view of a sceptic. *Journal of the Geological Society, London* **146**, 21–35.
Hoffman, A. (1989b) Changing palaeontological views on mass extinction phenomena. In: Donovan, S. K. (Ed.) *Mass extinctions: processes and evidence*, pp. 1–18. New York: Columbia University Press.
Holland, H. D. (1984) Earth's earliest biosphere – its origin and evolution. *American Scientist* **72**, 391–392.
Holland, H. D. (1994) Early Proterozoic atmospheric change. In: Bengston, S. (Ed.) *Early life on Earth*, pp. 237–244. New York: Columbia University Press.
Holland, H. D. and Beukes, X. (1990) A paleoweathering profile from Griqualand west South Africa: evidence for a dramatic rise in atmospheric oxygen between 2.2 and 1.9 b.y.b.p. *American Journal of Science* **290**, 1–34.

Holmes, W. B. K. (1998) The Triassic vegetation of eastern Australia. *Journal of African Earth Sciences* **27**, 115–116.

Holser, W. T. and Magaritz, M. (1987) Events near the Permian–Triassic boundary. *Modern Geology* **11**, 115–280.

Honey, J. G., Harrison, J. A., Prothero, D. R. and Stevens, M. S. (1998) Camelidae. In: Janis, C. M., Scott, K. M. and Jacobs, L. L. (Eds) *Evolution of Tertiary mammals of North America*, pp. 439–463. Cambridge: Cambridge University Press.

Hooghiemstra, H. (1989) Quaternary and Upper Pliocene glaciations and forest development in the tropical Andes: evidence from a long high-resolution pollen record from the sedimentary basin of Bogota, Colombia. *Palaeogeography, Palaeoclimatology, Palaeoecology* **72**, 11–26.

Hooghiemstra, H. (1995) Environmental and paleoclimatic evolution in Late Pliocene–Quaternary Colombia. In: Vrba, E. S., Denton, G. H. and Partridge, T. C. (Eds) *Paleoclimate and evolution, with emphasis on human origins*, pp. 249–261. Yale: Yale University Press.

Hooghiemstra, H. and Sarmiento, G. (1991) Long continental pollen record from a tropical intermontane basin: Late Pliocene and Pleistocene history from a 540-meter core. *Episodes* **14**, 107–115.

Horrell, M. A. (1991) Phytogeography and palaeoclimatic interpretation of the Maastrichtian. *Palaeogeography, Palaeoclimatology, Palaeoecology* **86**, 87–138.

Hsu, K. J. (1983) *The Mediterranean was a desert.* Princeton: Princeton University Press.

Hsu, K. J., Montadert, L., Bernoulli, D., Cita, M. B., Erikson, A., Garrison, R. E., Kidd, R. B. *et al.* (1977) History of Mediterranean salinity crisis. *Nature* **267**, 399–403.

Hueber, F. M. (1992) Thoughts on the early lycopsids and zosterophylls. *Annals of the Missouri Botanical Garden* **79**, 474–499.

Huggett, R. J. (1997) *Enviromental change: the evolving ecosphere*, First edn. London: Routledge.

Hughes, N. F. (1994) *The enigma of angiosperm origins*, First edn. Cambridge: Cambridge University Press.

Hughes, N. F. and McDougall, A. B. (1987) Records of angiospermid pollen entry into the English Early Cretaceous. *Review of Palaeobotany and Palynology* **50**, 255–272.

Hughes, N. F. and McDougall, A. B. (1994) Search for antecedents of Early Cretaceous monosulcate columellate pollen. *Review of Palaeobotany and Palynology* **83**, 175–183.

Imbrie, J. and Imbrie, K. P. (1979) *Ice ages: solving the mystery.* London: Macmillian.

Ingrouille, M. (1992) *Diversity and evolution of land plants.* London: Chapman & Hall.

Jablonski, D. (1986a) Background and mass extinctions: the alternation of macroevolutionary regimes. *Science* **231**, 129–133.

Jablonski, D. (1986b) Causes and consequences of mass extinctions. In: Elliot, D. K. (Ed.) *Dynamics of extinctions*, pp. 183–229. New York: John Wiley.

Jablonski, D. (1993) The tropics as a source of evolutionary novelty through geological time. *Nature* **364**, 142–144.

Jablonski, D. (1995) Extinctions in the fossil record. In: Lawton, J. and May, R. M. (Eds) *Extinction rates*, pp. 25–44. Oxford: Oxford University Press.

Jacobs, B. F., Kingston, J. D. and Jacobs, L. L. (1999) The origin of grass dominated ecosystems. *Annals of the Missouri Botanical Garden* **86**, 590–644.

Jacobs, J. A. (1994) *Reversals of the Earth's magnetic field*, Second edn. Cambridge: Cambridge University Press.

Janis, C. M. (1984) The use of fossil ungulate communities and indicators of climate and environment. In: Brenchley, P. (Ed.) *Fossils and climate*, pp. 85–104. New York: John Wiley & Sons.

Janis, C. M. (1993) Tertiary mammal evolution in the context of changing climates, vegetation and tectonic events. *Annual Review of Ecology and Systematics* **24**, 467–500.

Janis, C. M., Damuth, J. and Theodor, J. M. (2000) Miocene ungulates and terrestrial primary productivity: where have all the browsers gone? *Proceedings of the National Academy of Sciences of the United States of America* **97**, 7899–7904.

Jarzen, D. M. (1978) Some Maestrichtian palynomorphs and their phytogeographical and paleoecological implications. *Palynology* **2**, 29–38.

Jones, M. (2001) *The molecule hunt.* London: Penguin.

Jones, M. and Brown, T. A. (2000) Agricultural origins: the evidence of modern and ancient DNA. *The Holocene* **10**, 769–777.

Jones, T. P. (1993) 13C enriched lower Carboniferous fossil plants from Donegal, Ireland: Carbon isotope constraints on taphonomy, diagenesis and palaeoenvironments. *Review of Palaeobotany and Palynology* **81**, 53–64.

Jones, T. P. and Chaloner, W. G. (1991) Fossil charcoal, its recognition and palaeoatmosphere significance. *Palaeogeography, Palaeoclimatology, Palaeoecology (Global and Planetary Change Section)* **97**, 35–50.

Karhu, J. A. and Holland, H. D. (1996) Carbon isotopes and the rise of atmospheric oxygen. *Geology* **24**, 867–870.

Kasting, J. F. (1988) Runaway and moist greenhouse atmospheres and the evolution of Earth and Venus. *Icarus* **74**, 472–494.

Kasting, J. F., Zahnle, K. J. and Walker, J. C. G. (1983) Photochemistry of methane in the Earth's early atmosphere. *Precambrian Research* **20**, 121–148.

Kellman, M. and Tackaberry, R. (1997) The history of tropical environments. In: Kellman, M. and Tackaberry, R. (Eds) *Tropical environments: the functioning and management of tropical ecosystems*, pp. 7–26. London: Routledge.

Kennedy, W. J. and Cobban, W. A. (1977) The role of ammonites in biostratigraphy. In: Kauffman E. G. and Hazel J. E. (Eds) *Concepts and methods of biostratigraphy*, pp. 309–320. Stroudsburg: Dowden, Hutchinson & Ross.

Kennett, J. P. (1977) Cenozoic evolution of Antarctic glaciation, the Circum-Antarctic Ocean, and their impact on global paleoceanography. *Journal of Geophysical Research* **82**, 3843–3859.

Kennett, J. P. (1995) A review of polar climatic evolution during the Neogene, based on the marine sediment record. In: Vrba, E. S., Denton, G. H., Partridge, T. C. and Burckle, L. H. (Eds) *Paleoclimate and evolution*, pp. 49–64. New Haven: Yale University Press.

Kennett, J. P. and Stott, L. D. (1991) Abrupt deep-sea warming, palaeoceanographic changes and benthic extinctions at the end of the Palaeocene. *Nature* **353**, 225–229.

Kennett, J. P. and Stott, L. D. (1995) Terminal Paleocene mass extinction in the deep sea; association with global warming. In: Stanley S. M. (Ed.) *Effects of past global change on life*, pp. 94–107. Washington: National Academy Press.

Kenrick, P. (1999a) Botany: the family tree flowers. *Nature* **402**, 358–359.

Kenrick, P. (1999b) Opaque petrifaction techniques. In: Jones, T. P. and Rowe, N. P. (Eds) *Fossil plants and spores*, pp. 87–91. London: The Geological Society.

Kenrick, P. (2001) Turning over a new leaf. *Nature* **410**, 309–310.

Kenrick, P. and Crane, P. R. (1991) Water-conducting cells in early fossil and land plants – implications for the early evolution of tracheophytes. *Botanical Gazette* **152**, 335–356.

Kenrick, P. and Crane, P. R. (1997a) *The origin and early diversification of land plants.* Washington: Smithsonian Institution Press.

Kenrick, P. and Crane, P. R. (1997b) The origin and early evolution of land plants. *Nature* **389**, 33–39.

Kerp, H. and Krings, M. (1999) Light microscopy of cuticles. In: Jones, T. P. and Rowe, N. P. (Eds) *Fossil plants and spores*, pp. 52–56. London: The Geological Society.

Kimura, M. (1983) *The neutral theory of molecular evolution.* Cambridge: Cambridge University Press.

Kimura, M. (1991) Recent development of the neutral theory viewed form the Wrightian tradition of theoretical population genetics. *Proceedings of the National Academy of Sciences of the United States of America* **88**, 5969–5973.

Kivilaan, A. and Bandurski, R. S. (1981) The one-hundred year period for Dr. Beal's seed viability experiment. *American Journal of Botany* **68**, 1290–1292.

Knaus, M. J., Upchurch, G. R. Jr. and Gillespie, W. H. (2000) *Charbeckia macrophylla* gen. et. sp. nov. from the Lower Mississippian Price (Pocono) Formation of southeastern West Virginia. *Review of Palaeobotany and Palynology* **111**, 71–92.

Knoll, A. H. (1984) Patterns of extinction in the fossil record of vascular plants. In: Nitecki, M. H. (Ed.) *Extinctions*, pp. 21–68. Chicago: The University of Chicago Press.

Knoll, A. H. (1985) Patterns of evolution in the Archean and Proterozoic eons. *Palaeobiology* **11**, 53–64.

Knoll, A. H. (1991) Environmental context of evolutionary change: an example from the end of the Proterozoic eon. In: Warren, L. and Koprowski, H. (Eds) *New perspectives on evolution*, pp. 77–85. Pennsylvania: John Wiley & Sons.

Knoll, A. H. (1992) The early evolution of eukaryotes: a geological perspective. *Science* **256**, 622–627.

Knoll, A. H. (1994a) Neoproterozoic evolution. In: Bengston, S. (Ed.) *Early life on Earth*, pp. 439–449. New York: Columbia University Press.

Knoll, A. H. (1994b) Proterozoic and Early Cambrian protists: evidence for accelerating evolutionary tempo. *Proceedings of the National Academy of Sciences of the United States of America* **91**, 6743–6750.

Knoll, A. H. (1996) Breathing room for early animals. *Nature* **382**, 111–112.

Knoll, A. H. (2001) Patterns of extinction in the fossil record of vascular plants. In: Nitecki, M. H. (Ed.) *Extinctions*, pp. 21–68. Chicago: University of Chicago Press.

Knoll, A. H. and Barghoorn, E. S. (1977) Archean microfossils showing cell division from the Swaziland system of South Africa. *Science* **198**, 396–398.

Knoll, A. H. and Lipps, J. H. (1993) Evolutionary history of prokaryotes and protists. In: Lipps, J. H. (Ed.) *Fossil prokaryotes and protists*, First edn, pp. 19–31. Boston: Blackwell Scientific Publications.

Knoll, A. H. and Niklas, K. J. (1987) Adaptation, plant evolution, and the fossil record. *Review of Palaeobotany and Palynology* **50**, 127–149.

Knoll, A. H., Niklas, K. J. and Tiffney, B. T. (1979) Phanerozoic land-plant diversity in North America. *Science* **206**, 1400–1402.

Kodner, R. B. and Graham, L. E. (2001) High-temperature, acid-hydrolyzed remains of *Polytrichum* (Musci, Polytrichaceae) resemble enigmatic Silurian–Devonian tubular microfossils. *American Journal of Botany* **88**, 462–466.

Konnert, M. and Bergmann, F. (1995) The geographical distribution of genetic variation of silver fir (*Abies alba, Pinaceae*) in relation to its migration history. *Plant Systematics and Evolution* **196**, 19–30.

Kovach, W. L. and Batten, D. J. (1993) Diversity changes in Lycopsid and aquatic fern megaspores through geologic time. *Paleobiology* **19**, 28–42.

Kranz, H. D. and Huss, V. A. R. (1996) Molecular evolution of pteridophytes and their relationship to seed plants: evidence from complete 18S rRNA gene sequences. *Plant Systematics and Evolution* **202**, 1–11.

Kranz, H. D., Miks, D., Siegler, M. L., Capesius, I., Sensen, C. W. and Huss, V. A. R. (1995) The origin of land plants – phylogenetic relationships among charophytes, bryophytes, and vascular plants inferred from complete small-subunit ribosomal-RNA gene sequences. *Journal of Molecular Evolution* **41**, 74–84.

Krassilov, V. A. and Dobruskina, I. A. (1995) Angiosperm fruit from the Lower Cretaceous of Israel and origins in rift valleys. *Paleontological Journal* **29**, 110–115.

Kutzbach, J. E. and Gallimore, R. G. (1989) Pangaean climates – megamonsoons of the megacontinent. *Journal of Geophysical Research – Atmospheres* **94**, 3341–3357.

Labandeira, C. C. (1997) Permian pollen eating. *Science* **277**, 1422–1423.

Labandeira, C. C. (1998) The role of insects in Late Jurassic to Middle Cretaceous ecosystems. *Bulletin of the New Mexico Museum of Natural History and Science* **14**, 105–124.

Labandeira, C. C. and Sepkoski, J. J. (1993) Insect diversity in the fossil record. *Science* **261**, 310–315.

Labandeira, C. C., Dilcher, D. L., Davis, D. R. and Wagner, D. L. (1994) Ninety-seven million years of angiosperm–insect association: paleobiological insights into the meaning of coevolution. *Proceedings of the National Academy of Sciences of the United States of America* **91**, 12 278–12 282.

Lakhanpal, R. N. (1970) Tertiary floras of India and their bearing on the historical geology of the region. *Taxon*, **19**, 675–694.

Lang, W. H. and Cookson, I. C. (1935) On the flora, including vascular land plants, associated with *Monograptus,* in rocks of Silurian Age, from Victoria, Australia. *Philosophical Transactions of the Royal Society of London, Series B* **224**, 421–449.

Lange, R. T. (1982) Australian Tertiary vegetation. In: Smith J. M. B. (Ed.) *A history of Australasian vegetation*, pp. 44–89. Sydney: McGraw-Hill.

Langley, C. H. and Fitch, W. M. (1974) An examination of the constancy rate of molecular evolution. *Journal of Molecular Evolution* **3**, 161–177.

Larcher, W. (1995) *Physiological plant ecology*, Third edn. Berlin: Springer.

Larson, R. L. (1991a) Geological consequences of superplumes. *Geology* **19**, 963–966.

Larson, R. L. (1991b) Latest pulse of Earth: evidence for a mid-Cretaceous superplume. *Geology* **19**, 547–550.

Lawver, L. A. and Gahagan, L. M. (1998) Opening of Drake Passage and its impact on Cenozoic ocean circulation. In: Crowley, T. J. and Burke, K. C. (Eds) *Tectonic boundary conditions for climate reconstructions*, pp. 212–223. Oxford: Oxford University Press.

Lemon, E. R. (1983) *CO_2 and plants.* Boulder, Colorado: Westview Press.

Leopold, E. B. (1967) Late Cenozoic patterns of plant extinction. In: Martin, P. S. and Wright, H. E. Jr. (Eds) *Pleistocene extinctions*, pp. 203–246. New Haven: Yale University Press.

Leopold, E. B. and Denton, M. F. (1987) Comparative age of grasslands and steppe east and west of the northern Rocky Mountains. *Annals of the Missouri Botanical Garden* **74**, 841–867.

Leopold, E. B., Liu, G. and Clay-Poole, S. (1992) Low-biomass vegetation in the Oligocene? In: Prothero, D. R. and Berggren, W. A. (Eds) *Eocene–Oligocene climatic and biotic evolution*, pp. 399–420. Princeton, NJ: Princeton University Press.

Les, D. H., Garvin, D. K. and Whimpee, C. F. (1991) Molecular evolutionary history of ancient aquatic angiosperms. *Proceedings of the National Academy of Sciences of the United States of America* **88**, 10 119–10 123.

Lewin, R. (1997) *Patterns in evolution.* New York: Scientific American Library.

Li, H.-M., Sun, X. -J. and Walker, D. (1984) The potential of paleobotany in the explanation of China's plant geography. In: Whyte, R. O. (Ed.) *The evolution of the east Asian environment*, pp. 433–447. Hong Kong: Centre of Asian Studies, University of Hong Kong.

Li, W.-B. and Liu, Z.-S. (1994) The Cretaceous palynofloras and their bearing on stratigraphic correlation in China. *Cretaceous Research* **15**, 365.

Li, W. -H., Gouy, M., Wolfe, K. H. and Sharp, P. M. (1989) Angiosperm origins. *Nature* **342**, 131–132.

Lidgard, S. and Crane, P. R. (1988) Quantitative analyses of the early angiosperm radiation. *Nature* **331**, 344–346.

Lidgard, S. and Crane, P. R. (1990) Angiosperm diversification and Cretaceous florisitc trends: a comparison of palynofloras and leaf macrofloras. *Paleobiology* **16**, 77–93.

Lindahl, T. (1993) Instability and decay of the primary structure of DNA. *Nature* **362**, 709–715.

Lindahl, T. (1997) Facts and artifacts of ancient DNA. *Cell* **90**, 1–3.

Lipps, J. H. (1993a) *Fossil prokaryotes and protists*. Boston: Blackwell Scientific.

Lipps, J. H. (1993b) Introduction to fossil prokaryotes and protists. In: Lipps, J. H. (Ed.) *Fossil prokaryotes and protists*, pp. 1–76. Boston: Blackwell Scientific.

Lister, A. M., Kaldwell, M., Kaagan, L. M., Jordan, W. C., Richards, M. B. and Stanley, H. F. (1999) Ancient and modern DNA in a study of horse domestication. *Ancient Biomolecules* **2**, 267–280.

Litke, R. (1968) Pflanzenreste aus dem Untermiozaen in Nordwestsachsen. Plant remains from the Lower Miocene of northwestern Saxony. *Palaeontographica* **123**, 173–182.

Lockheart, M. J. (1997) Isotope compositions and distributions of individual compounds as indicators for environmental conditions: comparisons between contemporary and *Clarkia* fossil leaves. PhD thesis, University of Bristol.

Loconte, H. (1996) Comparisons of alternative hypotheses for the origin of angiosperms. In: Taylor, D. W. and Hickey, L. J. (Eds) *Flowering plant origin, evolution and phylogeny*, pp. 267–285. New York: Chapman & Hall.

Loconte, H. and Stevenson, D. W. (1991) Cladistics of the Magnoliidae. *Cladistics* **7**, 267–296.

Logan, G. A., Boon, J. J. and Glinton, G. (1993) Structural biopolymer preservation in Miocene leaf fossils from the Clarkia site, northern Idaho. *Proceedings of the National Academy of Sciences of the United States of America* **90**, 2246–2250.

Logan, G. A., Smiley, C. J. and Eglinton, G. (1995) Preservation of fossil leaf waxes in association with their source tissues, Clarkia, northern Idaho. *Geochimica Cosmochimica Acta* **59**, 751–763.

Looy, C. V., Brugman, W. A., Dilcher, D. L. and Visscher, H. (1999) The delayed resurgence of equatorial forest after the Permian Triassic ecological crisis. *Proceedings of the National Academy of Sciences of the United States of America* **96**, 13 857–13 862.

Loper, D. E., McCartney, K. and Buzyna, G. (1988) Mantle plumes and the periodicity of magnetic field reversals, climate and mass extinctions. *Journal of Geology* **96**, 1–15.

Lowe, D. R. (1994) Early environments: constraints and opportunities for early evolution. In: Bengston, S. (Ed.) *Early life on Earth*, pp. 25–35. New York: Columbia University Press.

Lowe, J. J. and Walker, M. J. C. (1997) *Reconstructing Quaternary environments*, Second edn. London: Longmans.

Lowenstein, J. and Scheuenstuhl, G. (1991) Immunological methods in molecular paleontology. *Philosophical Transactions of the Royal Society of London, Series B* **333**, 375–380.

Luo, Z. (1999) A refugium for relicts. *Nature* **400**, 24–25.

Lupia, R., Lidgard, S. and Crane, P. R. (1999) Comparing palynological abundance and diversity: implications for biotic replacement during the Cretaceous angiosperm radiation. *Paleobiology* **25**, 305–340.

Lupia, R., Crane, P. R. and Lidgard, S. (2000) Angiosperm diversification and mid-Cretaceous environmental change. In: Culver, S. J. and Rawson, P. F. (Eds) *Biotic response to global change: the last 245 million years*, pp. 207–222. Cambridge: Cambridge University Press.

MacFadden, B. J. (1998) Equidae. In: Janis, C. M., Scott, K. M. and Jacobs, L. L. (Eds) *Evolution of Tertiary mammals of North America*, pp. 537–560. Cambridge: Cambridge University Press.

MacFadden, B. J. and Cerling, T. E. (1994) Fossil horses, carbon isotopes and global change. *Trends in Ecology and Evolution* **9**, 485.

Mackowiak, C. L. and Wheeler, R. M. (1996) Growth and stomatal behaviour of hydroponically cultured potato (*Solanum tuberosum* L.) at elevated and super-elevated CO_2. *Journal of Plant Physiology* **149**, 205–210.

Magallon, S., Crane, P. R. and Herendeen, P. S. (1999) Phylogenetic pattern, diversity and diversification of eudicots. *Annals of the Missouri Botanical Garden* **86**, 297–372.

Mai, D. H. (2000) Palm trees in the past – palaeoclimatological and palaeoecological indicators. *GFF Special Issue on Early Paleogene Warm Climates and Biosphere Dynamics* **122** (1), 97–99.

Mallard, L. D. and Rogers, J. J. W. (1997) Relationship of avalonian and cadomian terranes to grenville and pan–African events. *Journal of Geodynamics* **23**, 197–221.

Manchester, S. R. (1999) Biogeographical relationships of North America Tertiary floras. *Annals of the Missouri Botanical Garden* **86**, 472–523.

Manhart, J. R. and Palmer, J. D. (1990) The gain of two chloroplast tRNA introns marks the green algal ancestors of land plants. *Nature* **345**, 268–270.

Marchant, H. J., Pickett-Heaps, J. D. and Jacobs, K. (1973) An ultrastructural study of zoosporogenesis and the mature zoospore of *Klebsormidium flaccidum. Cytobios* **8**, 95–107.

Martin, R. E. (1995) Cyclic and secular variation in microfossil biomineralisation: clues to the biogeochemical evolution of Phanerozoic oceans. *Global and Planetary Change*, **11**, 1–23.

Martin, R. E. (1998) *One long experiment: scale and process in Earth history*. New York: Columbia University Press.

Martin, W., Gierl, A. and Saedler, H. (1989) Molecular evidence for pre-Cretaceous angiosperm origins. *Nature* **339**, 46–48.

Maslin, M. A., Haug, G. H., Sarnthein, M. and Tiedemann, R. (1996) The progressive intensification of northern hemisphere glaciation as seen from the North Pacific. *Geol. Rundsch.* **85**, 452–465.

Mattox, K. R. and Stewart, K. D. (1984) Classification of the green algae: a concept based on comparative cytology. In: Irvine, D. E. G. and John, D. M. (Eds) *Systematics of the green algae*, pp. 29–72. London: Academic Press.

Mayr, E. (1954) Change of genetic environment and evolution. In: Huxley, J., Hardy, A. C. and Ford, E. B. (Eds) *Evolution as a process*, pp. 157–180. London: George Allen & Unwin.

Mayr, E. (1963) *Animal species and evolution.* Cambridge, MA: Harvard University Press.

McElwain, J. C. (1998) Do fossil plants signal palaeo-atmospheric CO_2 concentration in the geological past? *Philosophical Transactions of the Royal Society of London, Series B* **353**, 1–15.

McElwain, J. C. and Chaloner, W. G. (1995) Stomatal density and index of fossil plants track atmospheric carbon-dioxide in the Paleozoic. *Annals of Botany* **76**, 389–395.

McElwain, J. C., Beerling, D. J. and Woodward, F. I. (1999) Fossil plants and global warming at the Triassic–Jurassic boundary. *Science* **285**, 1386–1390.

McGhee, G. R. (1990) Frasnian–Famennian. In: Briggs, D. E. G. and Crowther, P. R. (Eds) *Paleobiology: a synthesis*, pp. 184–187. Oxford: Blackwell.

McGhee, G. R. (1996) *The Late Devonian extinction.* New York: Columbia University Press.

McIver, E. E. and Basinger, J. F. (1994) Early Tertiary floral evolution in the Canadian high arctic. *Annals of the Missouri Botanical Garden* **86**, 523–546.

McKay, D. S., Gibson E. K. Jr., Thomas-Keprta, K. L., Vali, H., Romanek, C. S., Clemett, S. J., Chillier, X. D. F., Maechling, C. R. and Zare, R. N. (1996) Search for life on Mars: possible relic biogenic activity in Martian meteorite ALH84001. *Science* **273**, 924–930.

McLaren, D. J. (1983) Bolides and biostrategraphy. *Geological Society of America Bulletin* **94**, 318–324.

Melville, R. (1983) Glossopteridae, Angiospermidae and the evidence for angiosperm origin. *Botanical Journal of the Linnean Society* **86**, 279–323.

Mendelson, C. V. (1993) Acritachs and Prasinophytes In: Lipps, J. E. (Ed) *Fossil prokaryotes and protists*, pp. 77–104. Boston: Blackwell Scientific Publications.

Meyen, S. V. (1982) The Carboniferous and Permian floras of Angaraland (a synthesis). *Biological Memoirs* **7**, 1–109.

Meyen, S. V. (1987) Geography of macroevolution in higher-plants. *Zhurnal Obshchei Biologii* **48**, 291–309.

Meyer, H. W. and Manchester, S. R. (1997) The Oligocene Bridge Creek flora of the John Day formation, Oregon. *University of California Publications in Geological Sciences* **141**, 1–197.

Meyer-Berthaud, B., Scheckler, S. E. and Wendt, J. (1999) *Archaeopteris* is the earliest known modern tree. *Nature* **398**, 700–701.

Milankovitch, M. (1930) Mathematische Klimalehre und astronomische Theorie der Klimaschwankungen. In: Köppen, W. and Geiger, R. (Eds) *Handbuch der Klimatologie I (A)*, pp. 1–176. Berlin: Gebrüder Borntraeger.

Miller, C. N. (1982) Current status of Paleozoic and Mesozoic conifers. *Review of Palaeobotany and Palynology* **37**, 99–114.

Miller, C. N. (1988) The origin of modern conifer families. In: Beck, C. B. (Ed.) *Origin and evolution of gymnosperms*, pp. 448–486. New York: Columbia University Press.

Miller, K. G., Fairbanks, R. G. and Mountain, G. S. (1987) Tertiary oxygen isotope synthesis, sea level history and continental margin erosion. *Paleoceanography* **2**, 1–19.

Miller, S. L. (1992) The prebiotic synthesis of organic compounds as a step towards the origin of life. In: Schopf, J. M. (Ed.) *Major events in the history of life*, pp. 1–28. Boston: Jones and Bartlett.

Mimura, M. R. M., Salatino, M. L. F., Salatino, A. and Baumgratz, J. F. A. (1998) Alkanes from foliar epicuticular waxes of *Huberia* species: taxonomic implications. *Biochemical Systematics and Ecology* **26**, 581–588.

Mishler, B. D. and Churchill, S. P. (1985) Transition to a land flora: phylogenetic relationships of the green algae and bryophytes. *Cladistics* **1**, 305–328.

Mora, C. I., Driese, S. G. and Colarusso, L. A. (1996) Middle to Late Paleozoic atmospheric CO_2 levels from soil carbonate and organic matter. *Science* **271**, 1105–1107.

Morley, R. J. (2000) *Origin and evolution of tropical rain forests*. Chichester: John Wiley & Sons.

Morley, R. J. and Richard, K. (1993) Gramineae cuticle: a key indicator of late Cenozoic climatic change in the Niger Delta. *Review of Palaeobotany and Palynology* **77**, 119–127.

Mosle, B., Finch, P., Collinson, M. E. and Scott, A. C. (1997) Comparison of modern and fossil plant cuticles by selective chemical extraction monitored by flash pyrolysis–gas chromatography–mass spectrometry and electron microscopy. *Journal of Analytical and Applied Pyrolysis* **40–41**, 585–597.

Muller, J. (1981) Fossil pollen records of extant angiosperms. *Botanical Review* **47**, 1–142.

Muller, J. (1984) Significance of fossil pollen for angiosperm history. *Annals of the Missouri Botanical Garden* **71**, 419–433.

Namburdiri, E. M. V., Tidwell, W. D., Smith, B. N. and Herbert, N. P. (1978) A C4 plant from the Miocene. *Nature* **276**, 816–817.

Naveh, Z. and Vernet, J-L. (1991) The palaeohistory of the Mediterranean biota. In *Biogeography of Mediterranean Invasions*. Edited by R. H. Groves and F. Di Castri. Cambridge: Cambridge University Press.

Niklas, K. J. (1976) The chemotaxonomy of *Parka decipiens* from the lower Old Red Sandstone, Scotland (UK). *Review of Palaeobotany and Palynology* **21**, 205–217.

Niklas, K. J. (1983) The influence of Palaeozoic ovule and cupule morphologies on wind pollination. *Evolution* **37**, 968–986.

Niklas, K. J. (1986) Evolution of plant shape: design constraints. *Tree* **1**, 67–72.

Niklas, K. J. (1994) *Plant allometry: the scaling of form and process*. Chicago: Chicago University Press.

Niklas, K. J. (1997) *The evolutionary biology of plants*. Chicago: The University of Chicago Press.

Niklas, K. J. and Pratt, L. M. (1980) Evidence for lignin-like constituents in Early Silurian (Llandoverian) plant fossils. *Science* **209**, 396–397.

Niklas, K. J. and Tiffney, B. H. (1994) The quantification of plant biodiversity through time. *Philosophical Transactions of the Royal Society of London, Series B* **345**, 35–44.

Niklas, K. J., Tiffney, B. H. and Knoll, A. H. (1983) Patterns in vascular land plant diversification. *Nature* **303**, 614–616.

Niklas, K. J., Tiffney, B. H. and Knoll, A. H. (1985a) Patterns in Vascular land plant diversification: an analysis at species level. In *Phanerozoic diversity patterns: profiles in macroevolution.* Princeton: Princeton University Press, pp. 97–128.

Niklas, K. J., Brown R. M. Jr. and Santos, R. (1985b) Ultrastructure and cytochemistry of Miocene angiosperm leaf tissues. In: Smiley, C. J. (Ed.) *Late Cenozoic history of the Pacific Northwest*, pp. 143–160. San Francisco: Pacific Division of the American Association for the Advancement of Science.

Nikolov, N. and Helmisaari, H. (1992) Silvics of the circumpolar boreal forest tree species. In: Shugart, H. H., Leemans, R. and Bonan, G. B. (Eds) *A systems analysis of the global boreal forest*, pp. 13–84. Cambridge: Cambridge University Press.

Nip, M., Brinkhuis, H., De Leeuw, J. W., Schenck, P. A. and Holloway, P. J. (1986) Analysisof modern and fossil plant cuticles by Curie point Oy-Gc and Curie point Py-GC-MS: recognition of a new, highly aliphatic and resistant bipolymer. *Organic Geochemistry* **10**, 769–778.

Nisbet, E. G. and Fowler, C. M. R. (1999) Archaean metabolic evolution of microbial mats. *Proceedings of Royal Society London B* **266**, 2375–2382.

Nisbet, E. G. and Sleep, N. H. (2001) The habit and nature of early life. *Nature* **409**, 1083–1091.

Nisbet, E. G., Cann, J. R. and van Dover, C. L. (1995) Origins of photosynthesis. *Nature* **373**, 479–480.

Nishida, H. and Nishida, M. (1988) *Protomonimia kasai-nakajhongii* gen. et sp. nov.: a permineralised magnolialean fructification from the mid-Cretaceous of Japan. *Botanical Magazine (Tokyo)* **101**, 397–426.

Nyguyen Tu, T. T., Bocherens, H., Mariotti, A., Baudin, F., Pons, D., Broutin, J., Derenne, S. and Largeau, C. (1999) Ecological distribution of Cenomanian terrestrial plants based on $^{13}C/^{12}C$ ratios. *Palaeogeography, Palaeoclimatology, Palaeoecology* **145**, 79–93.

O'Donoghue, K., Clapham, A., Evershed, R. P. and Brown T. (1996) Remarkable preservation of biomolecules in ancient radish seeds. *Proceedings of the Royal Society of London B*, **263**, 541–547.

Officer, C. B. and Drake, C. L. (1985) Terminal Cretaceous environmental events. *Science* **227**, 1161–1167.

Olson, J. S. (1985) Cenozoic fluctuations in biotic parts of the global carbon cycle. In: Sundquist, E. T. and Broecker, W. S. (Eds) *The carbon cycle and atmospheric CO_2: natural variations Archean to present*, pp. 377–396. Washington: American Geophysical Union.

Olsen, P. E., Shubin, N. H. and Anders, P. E. (1987) New Early Jurassic tetrapod assemblages constrain Triassic-Jurassic tetrapod extinction event. *Science*, **237**, 1025–1029.

Omer, L. and Hovarth, S. M. (1983) Elevated carbon dioxide concentrations and whole plant senescence. *Ecology* **64**, 1311–1314.

Otto-Bliesner, B. L. (1998) Effects of tropical mountain elevations on the climate during the past: climate model experiments. In: *Tectonic Boundary Conditions for Climate Reconstructions*, T. J. Crowley and K. C. Burke (Eds), Oxford Monographs on Geology and Geophysics, Chapter 5, 100–115. Oxford.

Pääbo, S. (1989) Ancient DNA: extraction, characterization, molecular cloning and enzymatic amplification. *Proceedings of the National Academy of Sciences of the United States of America* **86**, 1939–1943.

Pääbo, S., Higuchi, R. G. and Wilson, A. C. (1988) Polymerase chain reaction reveals cloning artifacts. *Nature* **334**, 387–388.

Pagani, M., Freeman, K. H. and Arthur, M. A. (1999) A Late Miocene atmospheric CO_2 concentration and expansion of C_4 grasses. *Science* **285**, 876–879.

Palmer, J. D. and Herbon, L. J. (1988) Plant mitochondrial DNA evolves rapidly in structure but slowly in sequence. *Journal of Molecular Evolution*, **28**, 87–97.

Parrish, J. M., Parrish, J. T. and Ziegler, A. M. (1986) Permian-Triassic paleogeography and paleoclimatology and implications for Therapsid distribution. In: Hotton, N., MacLean, P. D., Roth, J. J. and Roth, E. C. (Eds) *The ecology and biology of mammal-like reptiles*, pp. 109–131. Washington, DC: Smithsonian Institution Press.

Parrish, J. T. (1993) Climate of the supercontinent pangea. *Journal of Geology* **101**, 215–233.

Parrish, J. T. (1998) *Interpreting pre-quaternary climate from the geologic record*. New York: Columbia University Press.

Partridge, T. C. (1997) Late Neogene uplift in eastern and southern Africa and its paleoclimatic implications. In: Ruddiman, W. F. (Ed.) *Tectonic uplift and climate change*, pp. 63–86. New York: Plenum Press.

Partridge, T. C., Bond, G. C., Hartnady, J. H., deMenocal, P. B. and Ruddiman, W. F. (1995) Climatic effects of Late Neogene tectonism and volcanism. In: Vrba, E. S., Denton, G. H., Partridge, T. C. and Burckle, L. H. (Eds) *Paleoclimate and evolution*, pp. 8–23. New Haven: Yale University Press.

Pearson, L. C. (1995) *The diversity and evolution of plants*, First edn. Boca Raton: CRC Press.

Pearson, P. N. and Palmer, M. R. (1999) Middle Eocene seawater pH and atmospheric carbon dioxide concentrations. *Science* **284**, 1824–1825.

Petit, J. R., Jouzel, J., Raynaud, D., Barkov, N. I., *et al.* (1999) Climate and atmospheric history of the past 420,000 years from the Vostok ice core, Antarctica, *Nature* **399**: 429–435.

Phillips, T. L., Peppers, R. A. and DiMichele, W. A. (1985) Stratigraphic and interregional changes in Pennsylvanian coal-swamp vegetation: Environmental inferences. *International Journal of Coal Geology* **5**, 43–109.

Phillips, O. L. and Gentry, A. H. (1994) Increasing turnover through time in tropical forests. *Science* **263**, 954–958.

Phillips, O. L. and Sheil, D. (1997) Forest turnover, diversity and CO_2. *Trends in Ecology and Evolution* **12**, 404.

Phillips, T. L., and DiMichele, W. A. (1999) Coal ball sampling and quantification. In: Jones, T. P. and Rowe, N. P. (Eds) *Fossil plants and spores*, pp. 206–209. London: The Geological Society.

Pickering, K. T. (2000) The Cenozoic world. In: Culver, S. J. and Rawson, P. F. (Eds) *Biotic response to global change: the last 145 million years*, pp. 20–34. Cambridge: Cambridge University Press.

Pickett-Heaps, J. D. (1975) *Green algae. Structure, reproduction and evolution in selected genera*. Sunderland, MA: Sinauer.

Pigg, K. B. and Trivett, M. L. (1994) Evolution of the Glossopterid gymnosperms from Permian Gondwana. *Journal of Plant Research* **107**, 461–477.

Poinar, G. O. Jr. (1994) The range of life in amber: significance and implications in DNA studies. *Experientia* **50**, 536–542.

Poinar, G. O. Jr., Höss, M., Bada, G. and Pääbo, S. (1996) Amino acid racemization and the preservation of ancient DNA. *Science* **272**, 864–866.

Poinar, H. N., Cano, R. J. and Poinar, G. O. Jr. (1993) DNA from an extinct plant. *Nature* **363**, 677.

Pole, M. (1999) Latest Albian–earliest Cenomanian monocotyledonous leaves from Australia. *Botanical Journal of the Linnean Society* **129**, 177–186.

Pole, M. S. and Macphail, M. K. (1996) Eocene *Nypa* from Regatta Point Tasmania. *Review of Palaeobotany and Palynology* **92**, 55–67.

Price, P. W. (1996) *Biological evolution*, First edn. Fort Worth: Saunders College Publishing, Harcourt Brace College Publishers.

Pryer, K. M., Schneider, H., Smith, A. R., Cranfill, R., Wolf, P. G., Hunt, J. S. and Sipes, S. D. (2001) Horsetails and ferns are a monophyletic group and the closest living relatives to seed plants. *Nature* **409**, 618–622.

Qiu, Y.-L., Cho, Y., Cox, C. and Palmer, J. D. (1998) The gain of three mitochondrial introns identified liverworts as the earliest land plants. *Nature* **394**, 671–674.

Qiu, Y.-L., Lee, J., Bernasconi-Quadroni, F., Soltis, D. E., Soltis, P., Zanis, M., Zimmer, E. A., Chen, Z., Savolainen, V. and Chase, M. W. (1999) The earliest angiosperms. *Nature* **402**, 404–407.

Ramanujam, C. G. K., Rothwell, G. W. and Stewart, W. N. (1974) Probable attachment of the *Dolerotheca. American Journal of Botany* **61**, 1057–1066.

Rampino, M. R. and Stothers, R. B. (1988) Flood basalt volcanism during the past 250 million years. *Science* **241**, 663–668.

Rampino, M. R. and Stothers, R. B. (1984) Terrestrial mass extinctions, cometary impacts and the sun's motions perpendicular to the galactic plane. *Nature* **308**, 709–712.

Ramussen, R. (2000) Filamentous microfossils in a 3,235 million-year-old volcanogenic massive sulphide deposit. *Nature* **405**, 676–679.

Raubeson, L. A. and Jansen, R. K. (1992) Chloroplast DNA evidence on the ancient evolutionary split in vascular land plants. *Science* **255**, 1697–1699.

Raup, D. M. (1979) Size of the Permo-Triassic bottleneck and its evolutionary implications. *Science* **206**, 217–218.

Raup, D. M. (1988) Extinction in the geologic past. In: Osterbrock, D. E. and Raven, P. H. (Eds) *Origins and extinctions*, pp. 109–119. New Haven: Yale University Press.

Raup, D. M. and Sepkoski, J. J. (1982) Mass extinctions in the marine fossil record. *Science* **215**, 1501–1503.

Raven, J. A. (1993) The evolution of vascular plants in relation to quantitative functioning of dead water-conducting cells and stomata. *Biological Reviews of the Cambridge Philosophical Society* **68**, 337–363.

Raven, J. A. (1995) The early evolution of land plants: aquatic ancestors and atmospheric interactions. *Botanical Journal of Scotland* **47**, 151–175.

Raven, J. A. and Spicer, R. A. (1996) The evolution of crassulacean acid metabolism. In Crassulacean Acid Metabolism. *Ecological Studies* **114**, 360–385.

Raven, J. A. and Sprent, J. I. (1989) Phototrophy, dizotrophy and palaeoatmospheres: biological catalysis and the H, C, N and O cycles. *Journal of the Geological Society, London* **146**, 161–170.

Raven, P. H., Evert, R. F. and Eichhorn, S. E. (1992) *Biology of plants*, Fifth edn. New York: Worth Publishers.

Raymo, M. E. and Ruddiman, W. F. (1992) Tectonic forcing of late Cenozoic climate. *Nature* **359**, 117–122.

Raymo, M. E., Ruddiman, W. F. and Froelich, P. N. (1988) Influence of late Cenozoic mountain building on ocean geochemical cycles. *Geology* **16**, 649–653.

Raymond, A. (1985) Floral diversity, phytogeography, and climatic amelioration during the Early Caroniferous (Dinantian). *Paleobiology* **11**, 293–309.

Raymond, A. (1987) Palaeogeographic distribution of Early Devonian plant traits. *Palaios* **2**, 113–132.

Raymond, A. and Metz, C. (1995) Laurussian land-plant diversity during the Silurian and Devonian: mass extinction, sampling bias, or both? *Paleobiology* **21**, 74–91.

Raymond, A., Parker, W. C. and Barrett, S. F. (1985) Early Devonian phytogeography. In: Tiffney, B. H. (Ed.) *Geological factors and the evolution of plants*, pp. 129–168. New Haven: Yale University Press.

Rea, D. K. and Schrader, H. (1985) Late Pliocene onset of glaciation: ice-rafting and diatom stratigraphy of North Pacific DSPD cores. *Palaeogeography, Palaeoclimatology, Palaeoecology* **49**, 313–325.

Rea, D. K., Zachos, J. C., Owen, R. M. and Gingerich, P. D. (1990) Global change at the Paleocene–Eocene boundary: climatic and evolutionary consequences of tectonic events. *Palaeogeography, Palaeoclimatology, Palaeoecology* **79**, 117–128.

Read, J. and Francis, J. E. (1992) Responses of the Southern Hemisphere tree species to a prolonged dark period and their implications for high latitude Cretaceous and Tertiary floras. *Palaeogeography, Palaeoclimatology, Palaeoecology* **99**, 271–290.

Rees, P. M., Gibbs, M. T., Ziegler, A. M., Kutzbach, J. E. and Behling, P. J. (1999) Permian climates: evaluating model predictions using global paleobotanical data. *Geology* **27**, 891–894.

Rees, P. M., Ziegler, A. M. and Valdes, P. J. (2000) Jurassic phytogeography and climates: new data and model comparisons. In: Hueber, F. M., Macleod, K. G. and Wing, S. L. (Eds) *Warm climates in Earth history*, pp. 297–318. Cambridge: Cambridge University Press.

Reid, G. C., McAfee, J. R. and Crutzen, P. J. (1978) Effects of intense stratospheric ionisation events. *Nature* **275**, 489–492.

Renne, P. R., Zichao, Z., Richards, M. A., Black, M. T. and Basu, A. R. (1995) Synchrony and causal relations between Permian–Triassic boundary crises and Siberian flood volcanism. *Science* **269**, 1413–1416.

Retallack, G. J. (1980) Late Carboniferous to Middle Triassic megafossil floras from Sydney Basin. In: Herbert, C. and Helby, R. (Eds). A Guide to the Sydney Basin. *Geological Survey of New South Wales Bulletin* **26**, 385–430.

Retallack, G. J. (1985) Fossil soils as grounds for interpreting the advent of large plants and animals on land. *Philosophical Transactions of the Royal Society of London, Series B* **309**, 105.

Retallack, G. J. (1986) The fossil record of soils. In: Wright, V. P. (Ed.) *Paleosols: their recognition and interpretation*, pp. 1–57. Oxford: Blackwell Scientific.

Retallack, G. J. (1992) Paleosols and changes in climate across the Eocene/Oligocene boundary. In: Prothero, D. R. and Berggren, W. A. (Eds) *Eocene–Oligocene climatic and biotic evolution*, pp. 382–398. Princeton, NJ: Princeton University Press.

Retallack, G. J. (1994) Were the Ediacaran fossils lichens? *Paleobiology* **20**, 523–544.

Retallack, G. J. (1995). Permian–Triassic life crisis on land. *Science* **267**, 77–80.

Retallack, G. J. (1999) Postapocalyptic greenhouse paleoclimate revealed by earliest Triassic paleosols in the Sydney Basin, Australia. *Geological Society of America Bulletin* **111**, 52–70.

Retallack, G. J. (2001) A 300-million-year record of atmospheric carbon dioxide from fossil plant cuticules. *Nature* **411**, 287–290.

Retallack, G. J. and Dilcher, D. L. (1981) Arguments for a glossopterid ancestry of angiosperms. *Palaeobiology* **7**, 54–67.

Reuveni, J. and Bugbee, B. (1997) Very high CO_2 reduces photosynthesis, dark respiration and yield in wheat. *Annals of Botany* **80**, 539–546.

Rice, C. M. (1994) A Devonian auriferous hot spring system, Rhynie, Scotland. *Journal of the Geological Society, London* **152**, 229–250.

Richardson, J. B. and McGregor, D. C. (1986) Silurian and Devonian spore zones of the Old Red Sandstone continent and adjacent regions. *Geological Survey of Canada Bulletin* **364**, 1–79.

Riding, R. (1994) Evolution of algal and cyanobacterial calcification. In: Bengston, S. (Ed.) *Early life on Earth*, pp. 427–437. New York: Columbia University Press.

Rieseberg, L. H. (1995) The role of hybridization in evolution: old wine in new skins. *American Journal of Botany* **82**, 944–953.

Rieseberg, L. H. and Soltis, D. (1987) Flavonoids of fossil Miocene *Platanus* and its extant relatives. *Biochemical Systematics and Ecology* **15**, 109–112.

Riou-Nivert, P. (1996) *Les Résineux: Tome I Connaissance et reconnaissance*. Paris: Institut pour le développement forestier.

Rivera, M. C. and Lake, J. A. (1992) Evidence that eukaryotes and eocyte prokaryotes are immediate relatives. *Science* **257**, 74–76.

Roberts, N. (1998) *The Holocene*, Second edn. Oxford: Blackwell.

Robinson, J. M. (1991) Phanerozoic atmospheric reconstructions: a terrestrial perspective. *Palaeogeography, Palaeoclimatology, Palaeoecology* **97**, 51–62.

Robinson, J. M. (1994) Speculations on carbon-dioxide starvation, Late Tertiary evolution of stomatal regulation and floristic modernization. *Plant Cell and Environment* **17**, 345–354.

Rogers, J. J. W. (1996) A history of continents in the past three billion years. *Journal of Geology* **104**, 91–107.

Rogers, S. O. and Bendich, A. J. (1985) Extraction of DNA from milligram amounts of fresh, herbarium and mummified plant tissues. *Plant Molecular Biology* **5**, 69–76.

Rollo, F., Venanzi, F. M. and Amici, A. (1994) DNA and RNA from ancient plant seeds. In: Herrman, B. and Hummel, S. (Eds) *Ancient DNA*, pp. 218–235. Berlin: Springer-Verlag.

Romero, E. J. (1986) Paleogene phytography and climatology of South America. *Annals of the Missouri Botanical Garden* **73**, 449–461.

Romero, E. J. and Archangelsky, S. (1986) Early Cretaceous angiosperm leaves from southern South America. *Science* **234**, 1580–1582.

Rothwell, G. W. and Serbet, R. (1994) Lignophyte phylogeny and the evolution of spermatophytes – a numerical cladistic analysis. *Systematic Botany* **19**, 443–482.

Rowley, D. B., Raymond, A., Parrish, J. T., Lottes, A. L., Scotese, C. R. and Ziegler, A. M. (1985) Carboniferous paleogeographic, phytogeographic, and paleoclimatic reconstructions. *International Journal of Coal Geology* **5**, 7–42.

Ruddiman, W. F. (1997) Tropical Atlantic terrigenous fluxes since 25 000 yrs B.P. *Marine Geology* **136**, 189–207.

Ruddiman, W. F. and Kutzbach, J. E. (1989) Forcing of the Late Cenozoic northern hemisphere climate by plateau uplift in southern Asia and the American West. *Journal of Geophysical Research* **94**, 18 409–18 427.

Ruddiman, W. F. and Raymo, M. E. (1988) Northern Hemisphere climate regimes during the past 3 Ma: possible tectonic connections. *Philosophical Transactions of the Royal Society of London, Series B*, **318**, 411–430.

Ruddiman, W. F., Prell, W. L. and Raymo, M. E. (1989) Late Cenozoic uplift of southern Asia and the American West: rationale for general circulation modelling experiments. *Journal of Geophysical Research* **94**, 18 379–18 391.

Ruddiman, W. F., Kutzbach, J. E. and Prentice, I. C. (1997) Testing the climatic effects of orography and CO_2 with general circulation and biome models. In: Ruddiman, W. F. (Ed.) *Tectonic uplift and climate change*, pp. 203–233. New York: Plenum Press.

Runnegar, B. (1994) Discovery of megascopic fossils resembling *Grypanian spiralis* in 2.1 Ga old banded iron formations in Northern Michigan. In: Bengston, S. (Ed.) *Early life on Earth*, pp. 287–298. New York: Columbia University Press.

Saito, T., Yamanoi, T. and Kaiho, K. (1986) End-Cretaceous devastation of terrestrial flora in the boreal Far East. *Nature* **323**, 253–255.

Schatz, G. E., Lowry, P. P. II and Ramisamihantanirina, A. (1998) *Takhtajania perrieri*. *Nature* **391**, 133–134.

Scheckler, S. E. (1985), Origins of the coal swamp biome: Evidence from the southern Appalachians. *Geological Society of America, Abstract with programs* **17(2)**, 134.

Scheckler, S. E. (1986) Geology, floristics and paleoecology of Late Devonian coal swamps from Appalachian Laurentia (U.S.A.). *Annals de la Société géologique de Belgique* **109**, 209–222.

Schidlowski, M. (1983) Evolution of photoautotrophy and early atmospheric oxygen levels. *Precambrian Research* **20**, 319–335.

Schindewolf O. H. (1963) Neocatastrophism? *Zeitschrift der Deutschen Geologischen Geseuschaft*, **114**, 430–445.

Schlegel, M. (1994) Molecular phylogeny of eukaryotes. *Trends in Ecology and Evolution* **9**, 330–335.

Schopf, J. M. (1975) Modes of fossil preservation. *Review of Palaeobotany and Palynology* **20**, 27–53.

Schopf, J. M. (1976) Morphologic interpretations of fertile structures in *Glossopteris* gymnosperms. *Review of Palaeobotany and Palynology* **21**, 25–64.

Schopf, J. M. (1992) The oldest fossils and what they mean. In: Schopf, J. M. (Ed.) *Major events in the history of life*, pp. 29–63. Boston, MA: Jones and Bartlett.

Schopf, J. M. (1993) Microfossils of the early Archean apex chert: new evidence of the antiquity of life. *Science* **260**, 640–650.

Schopf, J. W. (1994) Disparate rates, differing fates: tempo and mode of evolution changed from the Precambrian to the Phanerozoic. *Proceedings of the National Academy of Sciences of the United States of America* **91**, 6735–6742.

Schopf, J. W. and Packer, B. M. (1987) Early Archean (3.3-billion to 3.5-billion-year-old) microfossils from Warrawoona Group, Australia. *Science* **237**, 70–73.

Scotese, C. R. and McKerrow, W. S. (1990) Revised world maps and introduction. In: McKerrow, W. S. and Scotese, C. R. (Eds) *Palaeozoic palaeogeography and biogeography*, pp. 1–21. London: Blackwell, for the Geological Society of London.

Scott, A. C. (1987) *Coal and coal-bearing strata: recent advances.* Oxford: Blackwell Scientific.

Scott, A. C. and Chaloner, W. G. (1983) The earliest fossil conifer from the Westphalian B of Yorkshire. *Proceedings of the Royal Society London, Series B* **220**, 163–182.

Scott, A. C. and Galtier, J. (1996) A review of the problems in the stratigraphical, palaeoecological and palaeobibgeographical interpretation of Lower Carboniferous (Dinantian) floras of Western Europe (Montpellier, France). *Review of Palaeobotany and Paynology* **90**, 141–153.

Scott, A. and Rex, G. (1985) The formation and significance of carboniferous coal balls. *Philosophical Transactions of the Royal Society of London, Series B* **311**, 123–137.

Scott, A. C., Stephenson, J. and Chaloner, W. G. (1992) Interaction and coevolution of plants and arthropods during the Palaeozoic and Mesozoic. *Philosophical Transactions of the Royal Society of London, Series B* **335**, 129–165.

Sepkoski, J. J. (1986) An overview of Phanerozoic mass extinctions. In: Jablonski, D. and Raup, D. M. (Eds) *Pattern and process in the history of life*, pp. 277–295. Berlin: Springer-Verlag.

Shackleton, N. J. and Opdyke N. D. (1976) Oxygen-isotope and paleomagnetic stratigraphy of Pacific core V28–239 Late Pliocene to Latest Pleistocene. *Geological Society of America Memoir* **145**, 449–464.

Shackleton, N. J., Hall, M. A. and Pate, D. (1995) Pliocene stable isotope stratigraphy of site 846. In: Pisias, N. G., Janecek, T. R., Palmer-Julson, A. and van Andel, T. H. (Eds) *Proceedings of the Ocean Drilling Program, Scientific Results, volume 138*, pp. 337–355. Texas: College Station.

Shen-Miller, J., Mudgett, M. B., Schopf, J. W., Clarke, S. and Berger, R. (1995) Exceptional seed longevity and robust growth: ancient sacred lotus from China. *American Journal of Botany* **82**, 1367–1380.

Sheridan, R. E. (1987) Pulsation tectonics as the control of long-term stratigraphic cycles. *Paleoceanography* **2**, 97–118.

Sheridan, R. E. (1997) Pulsation tectonics as a control on the dispersal and assembly of supercontinents. *Journal of Geodynamics* **23**, 173–196.

Sidow, A. and Thomas, W. (1994) A molecular evolutionary framework for eukaryotic model organisms. *Current Biology* **4**, 596–603.

Sidow, A., Wilson, A. C. and Pääbo, S. (1991) Bacterial DNA in *Clarkia* fossils. *Philosophical Transactions of the Royal Society of London, Series B* **333**, 429–433.

Singh, A. (1988) History of aridland vegetation and climate: a global perspective. *Biological Reviews of the Cambridge Philosophical Society* **63**, 156–198.

Sirois, L. (1992) The transition between boreal forest and tundra. In: Shugart, H. H., Leemans, R. and Bonan, G. B. (Eds) *A systems analysis of the global boreal forest*, pp. 196–215. Cambridge: Cambridge University Press.

Sleep, N. H. and Zahnle, K. (1998) Refugia from asteroid impacts on early Mars and the early Earth. *Journal of Geophysical Research* **103**, 28 529–28 544.

Sloan, L. C. and Barron, E. J. (1992) Paleogene climatic evolution: a climate model investigation of the influence of continental elevation and sea-surface temperature upon continental climate. In: Prothero, D. R. and Berggren, W. A. (Eds) *Eocene–Oligocene climate and biotic evolution*, pp. 202–217. Princeton, NJ: Princeton University Press.

Smit, J. (1990) Meteorite impact, extinctions and the Cretaceous–Tertiary boundary. *Geol. Miijnboum* **69**, 187–204.

Smith, A. B. and Littlewood, D. T. J. (1994) Paleontological data and molecular phylogenetic analysis. *Paleobiology* **20**, 259–273.

Soltis, P., Soltis, D. and Smiley, C. J. (1992) An RBCL sequence from a Miocene *Taxodium* (bald cypress). *Proceedings of the National Academy of Sciences of the United States of America* **89**, 449–451.

Soltis, P. S., Soltis, D. E. and Chase, M. W. (1999) Angiosperm phylogeny inferred from multiple genes as a tool for comparative biology. *Nature* **402**, 402–404.

Speck, T. and Vogellehner, D. (1988) Biophysical examinations of bending stability of various stele types and the upright axes of early 'vascular' land plants. *Botanica Acta* **101**, 262–268.

Spicer, R. A. (1989) Plants at the K/T boundary. *Philosophical Transactions of the Royal Society of London, Series B* **325**, 291–305.

Spicer, R. A. (1990) Reconstructing high latitude Cretaceous vegetation and climate: Arctic and Antarctic compared. In: Taylor, T. N. and Taylor, T. (Eds) *Antarctic paleobiology: its role in reconstructing Gondwana*, pp. 27–36. New York: Springer–Verlag.

Spicer, R. A. and Parrish, J. M. (1990) Late Cretaceous–Early Tertiary palaeoclimates of northern high latitudes: a quantitative view. *Journal of the Geological Society, London* **147**, 329–341.

Srivastava, A. K. (1991) Evolution tendancy in the venation pattern of Glossopteridales. *Geobios* **24**, 383–386.

Stankiewicz, B. A., Poinar, H. N., Briggs, D. E. G., Evershed, R. P. and Poinar, G. O. Jr. (1998a) Chemical preservation of plants and insects in natural resins. *Proceedings of the Royal Society London B* **256**, 641–671.

Stankiewicz, B. A., Scott, A. C., Collinson, M. E., Finch, P., Mosle, B., Briggs, D. E. G. and Evershed, R. P. (1998b) Molecular taphonomy of arthropod and plant cuticles from the Carboniferous of North America: implications for the origin of kerogen. *Journal of the Geological Society, London* **155**, 453–462.

Stanley, S. M. (1975) A theory of evolution above the species level. *Proceedings of the National Academy of Sciences of the United States of America* **72**, 646–650.

Stanley, S. M. (1988) Paleozoic mass extinctions: shared patterns suggest global cooling as a common cause. *American Journal of Science* **288**, 334–352.

Stanley, S. M. (1989) *Earth and life through time*. New York: W.H. Freeman.

Stebbins, G. L. (1950) *Variation and evolution in plants*. New York: Columbia University Press.

Stebbins, G. L. (1959) The role of hybridization in evolution. *Proceedings of the American Philosophical Society* **103**, 231–251.

Stebbins, G. L. (1971) *Chromosomal evolution in higher plants*. London: Edward Arnold.

Stebbins, G. L. (1974) *Flowering plants: evolution above the species level*. Cambridge, MA: Harvard University Press.

Stebbins, G. L. (1981) Coevolution of grasses and herbivores. *Annals of the Missouri Botanical Garden* **68**, 75–86.

Stevens, C. H. (1977) Was development of brackish oceans a factor in Permian extinctions? *Geology* **88**, 133–138.

Stevens, G. C. (1992) The elevational gradient in altitudinal range: an extension of Rapport's Latitudinal Rule to altitude. *American Naturalist* **140**, 893–911.

Stevenson, D. W. and Loconte, H. (1995) Cladistic analysis of monocot families. In: Rudall, P. J., Cribb, P. J., Cutler, D. F. and Humphries, C. J. (Eds) *Monocotyledons: systematics and evolution*, pp. 543–578. Kew, London: Royal Botanic Gardens.

Stewart, W. N. and Rothwell, G. W. (1993) *Paleobotany and the evolution of plants*, Second edn. Cambridge: Cambridge University Press.

Stokes, S. (1999) Luminescence dating applications in geomorphological research. *Geomorphology* **29**, 153–171.

Stokes, S. (2000) Luminescence dating methods. In: Thomas, D. S. G. and Goudie, A. (Eds) *Dictionary of physical geography*, Third edn. Oxford: Blackwell Scientific.

Stone, A. C. and Stoneking, M. (1999) Analysis of ancient DNA from a prehistoric Amerindian cemetery. *Philosophical Transactions of the Royal Society of London, Series B* **354**, 153–159.

Strutz, H. C. and Thomas, L. K. (1964) Hybridisation and introgression between *Cowania* and *Purshia*. *Evolution* **18**, 183–195.

Summons, R. E., Jahnke, L. L., Hope, J. M. and Logan, G. A. (1999) 2-Methylhopanoids as biomarkers for cyanobacteria oxygenic photosynthesis. *Nature* **49**, 554–557.

Sun, G., Dilcher, D. L., Zheng, S. and Zhou, Z. (1998) In search of the first flower: a Jurassic angiosperm, *Archaefructus*, from northeast China. *Science* **282**, 1692–1695.

Surlyk, F. (1990) Cretaceous–Tertiary (Marine). In: Briggs, D. E. G. and Crowther, P. R. (Eds) *Palaeobiology: a synthesis*, pp. 198–203. Oxford: Blackwell Science.

Swisher, C. C. I., Wang, Y.-Q., Wang, X.d-L. and Wang, Y. (1999) Cretaceous age for the feathered dinosaurs of Liaoning, China. *Nature* **400**, 58–61.

Taberlet, P. and Bouvet, J. (1994) Mitochondrial DNA polymorphism, phylogeography, and conservation genetics of the brown bear *Ursus arctos* in Europe. **225**, 195–200.

Tajika, E. (1999) Carbon cycle and climate change during the Cretaceous inferred from a biogeochemical carbon cycle model. *The Island Arc* **8**, 293–303.

Tallis, J. H. (1991) *Plant community history: long-term changes in plant distribution and diversity*, First edn. London: Chapman & Hall.

Tanai, T. (1961) Neogene floral change in Japan. *Journal of the Faculty of Science, Hokkaido University* **10**, 119–398.

Tanai, T. (1972) Tertiary history of vegetation in Japan. In: Graham, A. (Ed.) *Floristics and paleofloristics of Asia and eastern North America*, pp. 235–255. Amsterdam: Elsevier.

Taylor, D. W. and Hickey, L. J. (1990) An aptian plant with attached leaves and flowers – implications for angiosperm origin. *Science* **247**, 702–704.

Taylor, D. W. and Hickey, L. J. (1992) Phylogenetic evidence for the herbaceous origin of angiosperms. *Plant Systematics and Evolution* **180**, 137–156.

Taylor, D. W. and Hickey, L. J. (1996) Evidence for and implications of an herbaceous origin for angiosperms. In: Taylor, D. W. and Hickey, L. J. (Eds) *Flowering plant origin, evolution and phylogeny*, pp. 232–266. New York: Chapman & Hall.

Taylor, D. W. and Millay, A. M. (1979) Pollination biology and reproduction in early seed plants. *Review of Palaeobotany and Palynology* **27**, 329–355.

Taylor, T. N. (1999) The ultrastructure of fossil cuticle. In: Jones, T. P. and Rowe, N. P. (Eds) *Fossil plants and spores*, pp. 113–115. London: The Geological Society.

Taylor, T. N. and Taylor, E. L. (1993) *The biology and evolution of fossil plants*, First edn. Englewood Cliffs: Prentice-Hall.

Taylor, T. N., Hass, H., Remy, W. and Kerp, H. (1995) The oldest fossil lichen. *Nature* **378**, 244.

Tegelaar, E. W., Kerp, H., Visscher, H., Schenck, P. A. and De Leeuw, J. W. (1991) Bias of the paleobotanical record as a consequence of variations in the chemical composition of higher vascular plants. *Palaeobiology* **17**, 133–144.

Tegelaar, E. W., Hollman, G., Van de Vegt, P., De Leeuw, J. W. and Hollowat, P. J. (1995) Chemical characterisation of the periderm tissue of some angiosperm species: recognition of an insouble, non-hydrolyzable, aliphatic biomacromolecule (suberan). *Organic Geochemistry* **23**, 239–250.

Terry, K. D. and Tucker, W. H. (1968) Biological effects of supernovae. *Science* **159**, 421–423.

Thomas, B. A. (1986) The biochemical analysis of fossil plants and its use in taxonomy and systematics. In: Spicer, R. A. and Thomas, B. A. (Eds) *Systematic and taxonomic approaches in palaeobotany*. Oxford: Clarendon Press.

Thomas, B. A. and Spicer, R. A. (1987) *The evolution and palaeobiology of land plants*, First edn. London: Croom Helm.

Thomasson, J. R. (1987) Fossil grasses: 1820–1986 and beyond. In: Soderstrom, T. R., Hilu, K. W., Campbell, C. S. and Barkworth, M. E. (Eds) *Grass systematics and evolution.* Washington, DC: Smithsonian Institution Press.

Thusu, B., van der Eem, J. G. L. A., El-Mehdawi, A. and Bu-Argoub, F. (1988) Jurassic–Early Cretaceous palynostratigraphy in north-east Libya. In: El-Arnauti, A., Owens, B. and Thusu, B. (Eds) *Subsurface palynology of northeast Libya*, pp. 171–213. Benghazi: Garyounis University Publications.

Tiffney, B. H. (1984) Seed size, dispersal syndromes, and the rise of angiosperms: evidence and hypothesis. *Annals of the Missouri Botanical Garden* **71**, 551–576.

Tiffney, B. H. and Niklas, C. J. (1985) Clonal growth in land plants: a paleobotanical perspective. In: Jackson, J. B. C., Buss, L. W. and Cook, R. E. (Eds) *Population biology and evolution of clonal organisms*, pp. 35–66. New Haven: Yale University Press.

Tiffney, B. H. and Niklas, K. J. (1990) Continental area, dispersion, latitudinal distribution, and topographic variety: a test of correlation with terrestrial plant diversity. In: Ross, R. M. and Allmon, W. D. (Eds) *Causes of evolution: a paleontological perspective*, pp. 76–102. Chicago: The University of Chicago Press.

Tims, J. D. and Chambers, T. C. (1984) Rhyniophytina and Trimerophytina from the early land flora of Victoria, Australia. *Palaeontology* **27**, 265–279.

Traverse, A. (1988a) Plant evolution dances to a different beat. *Historical Biology* **1**, 277–301.

Traverse, A. (1988b) *Paleopalynology*, First edn. Boston: Unwin Hyman.

Troughton, J. H., Card, K. A., Hendy, C. H. (1974) Photosynthetic pathways and carbon isotope discrimination by plants. *Carnegie Institute of Washington Year Book* **73**, 768–780.

Truswell, E. M. (1990) Cretaceous and Tertiary vegetation of Antarctica: a palynological perspective. In: Taylor, T. N. and Taylor, E. L. (Eds) *Antarctic paleobiology*, pp. 71–88. New York: Springer-Verlag.

Tschudy, R. H. and Tschudy, B. D. (1986) Extinction and survival of plant life following the Cretaceous/Tertiary boundary event, Western Interior, North America. *Geology* **14**, 667–670.

Upchurch, G. R. Jr. and Wolfe, J. A. (1987) Mid-Cretaceous to early Tertiary vegetation and climate: evidence from fossil leaves and woods. In: Friis, E. M., Chaloner, W. G. and Crane, P. R. (Eds) *The origins of angiosperms and their biological consequences*, pp. 75–105. Cambridge: Cambridge University Press.

Upchurch, G. R. Jr., Otto-Bliesner, B. L. and Scotese, C. R. (1999) Terrestrial vegetation and its effect on climate during the latest Cretaceous. In: Barrera, E. and Johnson, C. C. (Eds) *Evolution of the Cretaceous ocean-climate system*, pp. 406–426. Boulder, Colorado: Geological Society of America.

Vakhrameev, V. A. (1991) *Jurrasic and Cretaceous floras and climates of the Earth*. Cambridge: Cambridge University Press.

Valentine, J. W. and Moores, E. M. (1970) Plate-tectonic regulation of faunal diversity and sea level: a model. *Nature* **228**, 657–659.

Van Bergen, P. V. (1999) Pyrolysis and chemolysis of fossil plant remains: applications to palaeobotany. In: Jones, T. P. and Rowe, N. P. (Eds) *Fossil plants and spores*, pp. 143–149. London: The Geological Society.

Van Bergen, P. V., Collinson, M. E., Hatcher, P. G. and De Leeuw, J. W. (1993) Lithological control on the state of preservation of fossil seed coats of water plants. *Advances in Organic Geochemistry* **22**, 683–702.

Van Bergen, P. V., Collinson, M. E., Briggs, D. E. G., De Leeuw, J. W., Scott, A. C., Evershed, R. P. and Finch, P. (1995) Resistant biomacromolecules in the fossil record. *Acta Botanica Neerlandica* **44**, 319–342.

Van Bergen, P. V., Collinson, M. E. and De Leeuw, J. W. (1996) Characterisation of the insoluble constituents of propagule walls of fossil and extant water lilies: implications for the fossil record. *Ancient Biomolecules* **1**, 55–81.

Van Bergen, P. V., Hatcher, P. G., Boon, J. J., Collinson, M. E. and De Leeuw, J. W. (1997) Macromolecular composition of the propagule wall of *Nelumbo nucifera. Phytochemistry* **45**, 601–610.

Van der Hammen, T. (1983) Paleoecology of tropical South America. In: Prance, G. T. (Ed.) *Biological diversification in the Tropics*, pp. 60–66. New York: Columbia University Press.

Vaughan, T. A. (1986) *Mammalogy*, Third edn. Philadelphia: Saunders.

Visscher, H. and Brugman, W. A. (1981) Ranges of selected Palynomorphs in the Alpine Triassic of Europe. *Review of Palaeobotany and Palynology* **34**, 115–128.

Visscher, H., Brinkhauis, H., Dilcher, D., Elsik, D., Eshet, W. C., Looy, C. V., Rampino, M. R. and Traverse, A. (1996) The terminal Paleozoic fungal event: evidence of terrestrial ecosystem destabilization and collapse. *Proceedings of the National Academy of Sciences of the United States of America* **93**, 2155–2158.

Von Wettstein, R. R. (1907) *Handbuch der systematischen Botanik, II. Band.*, Leipzig: Deuticke.

Vrba, E. S. (1985) Environment and evolution: alternative causes of the temporal distribution of evolutionary events. *South African Journal of Science* **81**, 229–236.

Vrba, E. S. (1992) Mammals as a key to Evolutionary Theory. *Journal of Mammalogy*, 73, 1–28.

Vrba, E. S. (1993) Turnover-pulses, the Red Queen, and related topics. *American Journal of Science* **293-A**, 418–452.

Vrba, E. S. (1995) On the connections between paleoclimate and evolution. In: Vrba, E. S., Denton, G. H. and Partridge, T. C. (Eds) *Paleoclimate and evolution, with emphasis on human origins*, pp. 24–45. New Haven: Yale University Press.

Walker, J. W., Brenner, G. J. and Walker, A. G. (1983) Winteraceous pollen in the Lower Cretaceous of Israel: early evidence of magnolialean angiosperm family. *Science* **220**, 1273–1275.

Walker, J. W. and Walker, A. G. (1984) Ultrastructure of Lower Cretaceous angiosperm pollen and origin and early evolution of flowering plants. *Annals of the Missouri Botanical Garden* **71**, 521.

Walsh, M. M. (1992) Microfossils from early Archean Onverwacht Group, Barberton Mountain land, South Africa. *Precambrian Research* **54**, 271–293.

Walter, H. (1986) *Vegetation of the Earth*. Berlin: Springer-Verlag.

Walter, M. R. (1977) Interpreting stromatolites. *American Scientist* **65**, 563–571.

Walter, M. R. (1994) Stromatolites: the main geological source of information on the evolution of the early benthos. In: Bengston, S. (Ed.) *Early life on Earth*, pp. 278–286. New York: Columbia University Press.

Webb, S. D., Hulbert, R. C. and Lambert, W. D. (1995) Climatic implications of large-herbivore distributions in the Miocene of North America. In: Vrba, E. S., Denton, G. H., Patridge, T. C. and Burckle, L. H. (Eds) *Paleoclimate and evolution*, pp. 91–108. New Haven: Yale University Press.

Weishampel, D. B. and Norman, D. B. (1989) Vertebrate herbivory in the Mesozoic; jaws, plants and evolutionary metrics. *Special Papers of the Geological Society of America* **238**, 87–100.

Wheeler, E. A. and Baas, P. (1993) The potentials and limitations of dicotyledonous wood anatomy for climatic reconstructions. *Paleobiology* **19**, 487–498.

White, M. E. (1990) *The greening of Gondwana*, First edn. Princeton: Princeton University Press.

Wieland, G. R. (1915) *American fossil Cycads*. Washington DC: Carnegie Institute.

Wignall, P. B. and Hallam, A. (1992) Anoxia as a cause of the Permian/Triassic extinction: facies evidence from northern Italy and the western United States. *Palaeogeography, Palaeoclimatology, Palaeoecology* **93**, 21–46.

Wignall, P. B. and Twitchett, R. J. (1996) Oceanic anoxia and the end Permian mass extinction. *Science* **272**, 1155–1158.

Wilf, P., Labandeira, C. C., Kress, W. J., Staines, C. L., Windsor, D. M., Allen, A. L. and Johnson, K. R. (2000) Timing the radiations of leaf beetles: hispines on gingers from the latest Cretaceous to recent. *Science* **289**, 291–294.

Williams, C. A. (1986) An oceanwide view of Palaeogene plate tectonic events. *Palaeogeography, Palaeoclimatology, Palaeoecology* **57**, 3–25.

Williams, M. A. J., Dunkerley, D. L., DeDeckker, P., Kershaw, A. P. and Stokes, T. J. (1993) *Quaternary Environments*, First edn. London: Edward Arnold.

Williams, M. A. J., Dunkerley, D., De Deker, P., Kershaw, P. and Chapell, J. (1998) *Quaternary Environments*, Second edn. London: Arnold.

Willis, K. J. (1996) Where did all the flowers go? The fate of temperate European flora during glacial periods. *Endeavour* **20**, 110–114.

Willis, K. J. and Bennett, K. D. (1995) Mass extinction, punctuated equilibrium and the fossil plant record. *Trends in Ecology and Evolution* **10**, 308–309.

Willis, K. J. and Whittaker, R. (2000) The refugal debate. *Science* **287**, 1406–1407.

Willis, K. J., Kleczkowski, A., Briggs, K. M. and Gilligan, C. A. (1999a) The role of sub-Milankovitch climatic forcing in the initiation of the Northern Hemisphere glaciation. *Science* **285**, 568–571.

Willis, K. J., Kleczkowski, A. and Crowhurst, S. J. (1999b) 124 000-year periodicity in terrestrial vegetation change during the late Pliocene epoch. *Nature* **397**, 685–688.

Wing, S. L. (1998) Tertiary vegetation of North America as a context for mammalian evolution. In: Janis, C. M., Scott, K. M. and Jacobs, L. L. (Eds) *Evolution of Tertiary mammals of North America*, pp. 37–66. Cambridge: Cambridge University Press.

Wing, S. L. and Boucher, L. D. (1998) Ecological aspects of the Cretaceous flowering plant radiation. *Annual Review of Earth and Planetary Sciences* **26**, 379–421.

Wing, S. L. and Tiffney, B. H. (1987) The reciprocal interaction of angiosperm evolution and tetrapod herbivory. *Review of Palaeobotany and Palynology* **50**, 179–210.

Wing, S. L., Hickey, L. J. and Swisher, C. C. (1993) Implications of an exceptional fossil flora for Late Cretaceous vegetation. *Nature* **363**, 342–344.

Wing, S. L., Sues, H. D., Tiffney, B. H., Stucky, R. K., Weishampel, D. B., Spicer, R. A., Jablonski, D., Badgley, C. E., Wilson, M. V. H. and Kovach, W. L. (1992) Mesozoic and Early Cenozoic terretrial ecosystems. In: Behrensmeyer, A. K., Damuth, J. D., DiMichele, W. A., Potts, R., Sues, H. D. and Wing, S. L. (Eds) *Terrestrial ecosystems through time*, pp. 327–416. Chicago: University of Chicago Press.

Wnuk, C. (1996) The development of floristic provinciality during the Middle and Late Paleozoic. *Review of Palaeobotany and Palynology* **90**, 6–40.

Woese, C. R. (1994) There must be a prokaryote somewhere. *Microbiological Reviews* **58**, 1–9.

Woese, C. R. and Fox, G. E. (1977) Phylogenetic structure of the prokaryotic domain: the primary kingdoms. *Proceedings of the National Academy of Sciences of the United States of America* **74**, 5088–5090.

Woese, C. R., Kandler, O. and Wheelis, M. L. (1990) Towards a natural system of organisms: proposal for the domains Archaea, Bacteria, and Eucarya. *Proceedings of the National Academy of Sciences of the United States of America* **87**, 4576–4579.

Wolfe, J. A. (1975) Some aspects of plant geography of the northern hemisphere during the Late Cretaceous and Tertiary. *Annals of the Missouri Botanical Garden* **62**, 264–279.

Wolfe, J. A. (1985) Distribution of major vegetation types during the Tertiary. In: Sundquist, E. T. and Broecker, W. S. (Eds) *The carbon cycle and atmospheric* CO_2*: natural variations Archean to present*, pp. 357–375. Washington, DC: American Geophysical Union.

Wolfe, J. A. (1987) An overview of the origins of modern vegetation and flora of the northern Rocky Mountains. *Annals of the Missouri Botanical Garden* **74**, 785–803.

Wolfe, J. A. (1991) Palaeobotanical evidence for a June 'impact winter' at the Cretaceous/Tertiary boundary. *Nature* **352**, 420–423.

Wolfe, J. A. (1992) Climatic, floristic and vegetational changes near the Eocene/Oligocene boundary in North America. In: Prothero, D. R. and Berggren, W. A. (Eds) *Eocene–Oligocene climatic and biotic evolution*, pp. 421–436. Princeton, NJ: Princeton University Press.

Wolfe, J. A. and Upchurch G. R. Jr. (1986) Vegetation, climate and floral changes at the Cretaceous–Tertiary boundary. *Nature* **324**, 148–152.

Wolfe, J. A. and Upchurch G. R. Jr. (1987) Leaf assemblages across the Cretaceous–Tertiary boundary in the Raton Basin, New Mexico and Colorado. *Proceedings of the National Academy of Sciences of the United States of America* **84**, 5096–5100.

Wolfe, K. H., Wen-Hsiung, L. and Sharp, P. M. (1987) Rates of nucleotide substitution vary greatly among plant mitochondrial, chloroplast and nuclear DNAs. *Proceedings of the National Academy of Sciences of the United States of America* **89**, 449–451.

Wolfe, K. H., Gouy, M., Yang, Y.-W., Sharp, P. M. and Li, W.-H. (1989) Date of the monocot–dicot divergence estimated from chloroplast DNA sequence data. *Proceedings of the National Academy of Sciences of the United States of America* **86**, 6201–6205.

Woodward, F. I., Thompson, G. B. and McKee, I. F. (1991) The effects of elevated concentrations of carbon-dioxide on individual plants, populations, communities and ecosystems. *Annals of Botany* **67**, 23–38.

Woolbach, W. S., Lewis, R. S. and Anders, E. (1985) Cretaceous extinctions: evidence for wildfires and search for meteoric material. *Science* **230**, 167–170.

Worsley, T. R., Nance, R. D. and Moody, J. B. (1986) Tectonic cycles and the history of the Earth's biogeochemical and paleoceanographic record. *Paleoceanography* **1**, 233–263.

Wright, J. D. (1998) Role of the Greenland–Scotland Ridge in Neogene climate changes. In: Crowley, T. J. and Burke, K. C. (Eds) *Tectonic boundary conditions for climate reconstructions*. Oxford: Oxford University Press.

Wright, R. and Cita, M. B. (1979) Geodynamics and biodynamics effects of the Messinian salinity crisis in the Mediterranean. *Palaeogeography, Palaeoclimatology, Palaeoecology* **29**, 215–222.

Xiong, J., Fischer A. G., Inoue, W. M., Nakahara, K. and Bauer, C. E. (2000) Molecular evidence for the early evolution of photosynthesis. *Science* **289**, 1724–1730.

Yapp, C. and Poths, H. (1992) Ancient atmospheric CO_2 pressures inferred from natural goethites. *Nature* **355**, 342–344.

Yin-Long, Q., Bernasconi-Quadroni, F., Soltis, D. E., Soltis, P. S., Zanis, M., Zimmer, E. A., Chen, Z., Savolainen, V. and Chase, M. W. (1999) The earliest angiosperms: evidence from mitochondrial, plastid and nuclear genomes. *Nature* **402**, 404–407.

Zavada, M. S. (1984) Angiosperm origins and evolution based on dispersed fossil pollen ultrastructure. *Annals of the Missouri Botanical Garden* **71**, 444–463.

Ziegler, A. M., Bambach, R. K., Parrish, J. T., *et al.* (1981) Paleozoic biogeography and climatology. In: Niklas, K. J. (Ed.) *Paleobotany, paleoecology and evolution*, vol. 2, Praeger Scientific.

Ziegler, A. M. (1990) Phytogeographic patterns and continental configurations during the Permian period. In: McKerrow, W. S. and Scotese, C. R. (Eds) *Palaeozoic, Palaeogeography and Biogeography*, pp. 363–379. London: Geological Society.

Ziegler, A. M., Raymond, A., Gierlowski, T. C., Horrell, M. A., Rowley, D. B. and Lottes, A. L. (1987) Coal, climate and terrestrial productivity: the present and Early Cretaceous compared. In: Scott, A. C. (Ed.) *Coal and coal-bearing strata*, pp. 25–49. Geological Society of America Special Publication 32.

Ziegler, A. M., Parrish, J., Yao, E. D. *et al.* (1993) Early Mesozoic phytogeography and climate. In: *Palaeoclimates and heir modelling with special reference to the Mesozoic Era. Philosophical Transactions of the Royal Society of London*, Ser. B, 341.

Zimmerman, W. (1952) Main results of the 'telome theory'. *Palaeobotanist* **1**, 456–470.

Zuckerkandl, E. and Pauling, L. (1965) Evolutionary divergence and convergence of proteins. In: Bryson, V. and Vogel, H. J. (Eds) *Evolving genes and proteins*, pp. 97–166. New York: Academic Press.

Glossary

Abscission The loss of structures from the stem of the plant following the growth of a layer of specialized cells. Usually applied to the loss of leaves from plants with seasonal change in climate.

Adventitious roots Roots which grow from an unusual position, e.g. from the leaf or stem.

Aerial The above ground part of the plant.

Allele One of two or more forms of a gene.

Amber A fossilized tree resin.

Amino acid Organic acids from which proteins are built.

Ammonoid(s) Extinct aquatic mollusk of the order Cephalopoda that were abundant in the Mesozoic era (225–65 Ma).

Annual(s) Plants that have a life cycle that is completed in a year.

Anoxic An environment that is devoid of free oxygen or has very low levels of free oxygen.

Anthropogenic Process that is directly or indirectly influenced by humans.

Apical meristem Region at tip of each shoot and root that contains cells that are continually dividing to produce new shoot and root tissue respectively.

Archipelago Group of islands.

Base substitution An alteration in DNA resulting in a change in a single base pair at a particular site on the DNA.

Bauxite Soil deposits consisting mainly of the ores of aluminium oxides which are particularly prevalent in tropical and subtropical climates with seasonal precipitation.

Benthic Usually ascribed to aquatic organisms that live on or near the ocean floor but can also be used for deep dwelling river or lake-bed organisms.

Binding site The region on the surface of the enzyme that holds the substrate.

Biomacromolecule Biologically produced molecule of high molecular mass, e.g. sporopollenin, cutan.

Biomarker An organic compound or molecule which is transformed into a more stable product with a higher preservation potential during the process of rock formation.

Biomolecule Any molecule that is involved in the metabolic processes and maintenance of living organisms, e.g. lipids, proteins, and nucleic acids.

Biopolymer (see also biomacromolecule) Biologically produced compound of high molecular weight made up of repeating chemical structures, e.g. sporopollenin, cutan.

Biome Biological subdivision of the Earth's surface that reflects the ecological and physiognomic character of the vegetation.

Biotic degradation Degradation resulting from the activities of living organisms.

Bisaccate A pollen grain with two air sacs.

Bolide impact Impact on Earth by one large or a shower of smaller meteorites with devastating environmental impacts.

Boreal (taiga) Cold temperate regions of the northern hemisphere incorporating the northern coniferous dominated forests and taiga.

Buttress root Root that is asymmetrically thickened on its upper side to provide extra support for a tree. Common in many tropical trees.

C3 plant A plant which produces phosphoglyceric acid (containing three carbon atoms) as the first product of photosynthesis.

C4 plant A plant which produces oxaloacetic acid (containing four carbon atoms) as the first product of photosynthesis.

Calcrete (caliche) A hardened layer of calcium or magnesium carbonate that accumulates in the soil profile usually above the water table, in semi-arid regions.

CAM Plants that have a variation of the C4 pathway where crassulacean acid metabolism is used to fix atmospheric CO_2. Characteristic of most succulent plants.

Carbonaceous Deposits containing abundant organic matter.

Carbonate A group of minerals containing the anion CO_3^-, the most common of which are calcite and dolomite.

Carboxylation Incorporation of carbon dioxide into a substrate molecule.

Cellulose A polysaccharide composed solely of glucose units linked by glycosidic bonds.

Charcoalification Process by which plant fragments after exposure to high temperatures in the presence of limited oxygen are converted to a carbonaceous material.

Chemosystematics The use of biochemical characteristics of whole organisms, organs or tissues in systematics.

Chemotaxonomy The use of biochemical characteristics as a tool in the classification of organisms.

Chloroplast Chlorophyll containing organelles found in plant cells and essential to photosynthesis.

Chronology A term referring to the measurement of time.

Clade A monophyletic group.

Coccolith Unicellular marine algae covered in calcium carbonate platelets.

Continental drift A hypothesis put forward to account for various features of the Earth's surface, e.g. the distribution of the continents and ocean basins.

Cutan A highly resistant biopolymer, which is an important chemical component of plant cuticle.

Cuticle General term for the waxy coating made up of highly resistant biopolymers, covering the outer walls of all above ground parts of all land plants. The cuticle provides protection and prevents excessive water loss from plant surfaces.

Cutin An important chemical component of plant cuticle, which is less resistant to decay than cutan and is therefore not detected as often as cutan in the fossil record.

Cyanobacteria Blue-green algae.

Dentition The type, arrangement and number of teeth in an organism.

Depauperization Becoming impoverished.

Depurination The main chemical reaction in the breakdown and fragmentation of DNA involving hydrolysis of the deoxyribose-adenine or guanine bond.

Diagenesis The chemical and physical processes influencing sediment before and during rock formation.

DNA amplification The amplification of a DNA strand using a pre-existing strand as a template.

Domesticates Plants and animals that have become genetically modified through the direct or indirect activities of humans making them more suited to the needs of people and/or the environment.

Dorsiventral Having distinct upper and lower sides.

Drip-tip An extension of the leaf apex thought to channel excess water from the leaf — leaves with drip-tips are abundant in tropical and everwet climates.

Embryo A young sporophytic plant.

Endosperm Storage tissue in the seeds of most angiosperms.

Endosymbiosis Symbiosis in which one symbiont lives inside the other.

Enzyme A biological catalyst.

Ephemeral therophyte A plant that survives unfavourable conditions in the form of a seed.

Epicuticular Upon or above the cuticle.

Epidermal The outermost layer or layers of cells of the leaf and of young stems and roots.

Epiphyte A plant that grows upon another plant (e.g. for support or anchorage) but is not parasitic on it.

Ester Organic compounds formed by reaction between alcohols and acids.

Eubacteria One of the three primary kingdoms of living organisms encompassing all typical bacteria.

Eustatic Changes in the global sea level resulting in an increase or decrease in the exposure of continental plates — leading to so-called eustatic uplift.

Evaporite deposit A mineral, sediment or sedimentary rock that forms by chemical precipitation of salty water by evaporation.

Exine Outer part of the wall of a pollen grain composed of the resistant biopolymer sporopollenin.

Extant Living — the opposite to extinct.

Extraction buffer Chemical medium used to extract DNA strands from cell material.

Fauna All members of the animal kingdom.

Flagellates Threadlike projections arising from the surface of motile unicellular algae, bacteria and fungi.

Flood basalt Flood basalts are one type of large igneous province that characterise the Earth's surface and have been formed at various times in the geological past —they represent the largest eruptions of lava on Earth, with known volumes of individual lava flows exceeding 2000 cubic kilometers extending over areas of more than a million square kilometres.

Foraminifera Single celled animals that inhabit nearly all marine environments.

Formations A unit of classification of vegetation defined on the basis of distinctive physiogonomy rather than distinctive speices composition. Formations are broadly equivalent with biomes.

Fructification Seed or spore-bearing structure.

Fruiting axes The branch or part of the plant that bears the fruit.

Galactose An aldohexose sugar commonly found in plants.

Gamete A cell or nucleus that undergoes sexual fusion to form a zygote.

Gametophyte The gamete-producing generation in the life cycle of a plant, i.e. the embryo sac in angiosperms and the prothallus in ferns.

Gas chromatogram (GC) The column of paper on which some or all the constituents of a mixture have been absorbed and separated.

Generalist Plant or animal that can exist in a broad range of environments.

Genetic loci A specific place on a chromosome where a gene is located.

Genetic drift The random fluctuations of gene frequencies in a population such that the genes of offspring are not a perfectly representative sampling of the parental genes.

Genetic substitution See base substitution.

Genome The total genetic constitution of an organism.

Genotype The recognition of an organism by its genetic makeup.

Genotypic Where individuals all share the same genetic constitution.

Geophyte A plant that perennates by organs or renewal buds such as corms or rhizomes buried below the soil surface.

Glaciated Previously or currently covered in continental ice.

Glycolytic A pathway or enzyme present in fungi and plants involving the stepwise anaerobic degradation of glucose to produce CO_2 and ethanol as end products.

Gondwana Supercontinent thought to have existed in the southern hemisphere over 200 million years ago which consisted of South America, Africa, Arabia, Madagascar, India and the East Indies.

Grade A group of organisms which do not share a common ancestor.

Grassland Regions where grass is the dominant vegetation.

Green algae Green-pigmented algae commonly found in aquatic or moist terrestrial habitats.

Greenhouse gases Gases in the atmosphere (e.g. CO_2, CH_4) that trap radiation re-emitted by the Earth resulting in increasing global temperatures.

Growth ring Rings in the cross-section of woody stems that usually represent the xylem formed in one year of growth.

Guard cell Pair of specialized epidermal cells that surround a pore or stoma and enable regulation of gases and moisture into and out of a plant.

Haploid Having one set of chromosomes.

Humic acid A complex mixture of an organically produced brown coloured acid present in vegetated soils.

Hydrophobic Having a low affinity for water.

Hydrophilic Having a high affinity for water.

Hypersaline Extremely saline.

Ice rafting Where break-off of large pieces of ice from major ice-sheets (ice-sheet calving) results in extensive rafts of ice in the oceans.

Inaperturate A term used to describe pollen grains that have no pores or furrows in the grain wall.

Inflorescence A flower cluster.

Integument Outermost layer of tissue covering the ovule that develops into the seed coat.

Intercalary Growth tissue that is not restricted to the apical meristem.

Interglacial An episode of thermal improvement separating cold glacial episodes, when climatic conditions were similar to, or warmer than, those of today.

Intine Inner layer of the wall of pollen grain or spore.

Lamina Flattened blade-like portion of the leaf.

Lanceolate Narrowing or tapering of the leaf at both ends resulting in 'spear-shaped' leaf.

Leaf mat A build-up of leaves on the ground surface or in waterways following leaf abscission.

Leaf mine(s) A trace or trail produced where animals, particularly insects, mine into the leaf between the upper and lower epidermis resulting in visible damage.

Liana Woody climbing plant with very long stems (up to 70 m) that grow from ground level to the canopy of trees. Very common in tropical forests.

Lignin Biomolecules made up of complex carbohydrate polymers and found in plant cells and tissues with a support function, e.g. in water conducting cells such as xylem.

Limonite A group of hydrous iron oxides formed by direct precipitation from marine or freshwater and by oxidation of iron-rich minerals.

Lipid A water-soluble hydrocarbon found in plant cells and tissues that plays an important role in energy storage, the structure of membranes and control of water loss from the plant.

Lithological Classification of a sequence according to the general characteristics of rock types contained within it.

Loess Wind-blown sediment of predominantly silt-sized particles that is derived from aridlands and/or outwash surfaces such as glacial outwash plains.

Macrofossil Fossils usually larger than approximately 0.5 cm and therefore visible with the naked eye.

Macrophyllous A term generally applied to angiosperm leaves with leaf lengths greater than 12 cm.

Mangrove swamp Highly productive tropical and subtropical ecosystem of shallow rooted trees and shrubs located on the tidal muddy swamps around tropical coastlines.

Mantle The part of the Earth's interior that underlies the crust and surrounds the core. The relatively fluid outermost layer of the mantle allows the movement of the Earth's tectonic plates.

Megafossil A very large fossil or subfossil such as a tree trunk, too large to be manipulated by hand and normally found *in situ*.

Megasporophyll(s) Leaf-like structures which bear megaspores or ovules.

Megaspore A spore which on germination develops into a female gametophyte.

Microspore Smaller spore of a heterosporous plant which on germination develops into a male gametophyte.

Metamorphic The process by which rocks are altered in their mineralogy, texture and internal structure due to changes in temperature, pressure and/or chemical environment.

Methane Atmospheric trace gas that has a strong greenhouse influence.

Microfossil A small fossil (e.g. pollen, spores) usually less than 0.5 cm in diameter and thus requiring microscopic analysis for identification.

Microphyllous A term generally applied to angiosperm leaves with leaf lengths less than 8 cm or smaller.

Micropyle An opening in the integument (seed coat) through which the pollen usually enters.

Microsporophyll A modified leaf that bears the microsporangia.

Mitochondria A structure within the cytoplasm of plant cells that carries out aerobic respiration.

Milankovitch cycle Mathematic theory proposed by Milankovitch in 1937 to explain the periodicity and timing of the Earth's orbit around the sun. Now used extensively to explain the timing of glacial-interglacial cycles in Earth's history.

Monocolpate A term usually applied to pollen and spores that have a single furrow on the wall of the grain.

Monoecious Having male and female reproductive organs separated in different floral structures on the same plant.

Monolete A term usually applied to spores which bear a linear scar on their proximal surface.

Monophyletic Applied to a group of taxa which share a common single ancestor.

Monopodial branching A plant with single main axis from which secondary shoots arise.

Monosaccate A pollen grain with one air sac usually extending all around the grain.

Motile Having the ability to swim such as in some algal and fungal spores that swim to the gamete.

mtDNA A circular ring of DNA found in the mitochondria.

Mycorrhiza(e) A symbiotic association between a fungus and the roots of vascular plants.

Notophyllous A term generally applied to angiosperm leaves with leaf lengths between 8 and 12 cm.

Nucleus Large body found in cytoplasm of all eukaryotic plant cells that contains the genetic material.

Organelles Minute membrane bound structures within plant cells that have a particular function, e.g. mitochondria.

Orogeny An episode of mountain building.

Orographic effect Impact of newly formed mountain ranges on environmental and climatic conditions.

Ovule Unfertilised seed.

Ovuliferous scale A scale that bears ovules and then seeds in the female cone of conifers.

Palmate A compound leaf with four or more leaflets arising from a single point, e.g. the leaves of horse chestnut *Aesculus hippocastanum.*

Palynology The study of fossil pollen and spores used to reconstruct long-term vegetation dynamics.

Palynoflora(s) Fossil floral assemblages made up of fossil pollen and spores.

Pampas South American grassland.

Papillae Projections usually from epidermal cells—often looking like minute hairs on leaves. Important in protecting plants from sunlight and excessive water loss.

Perennial Plant that lives for many years either surviving inclement seasons in underground storage (e.g. bulbs) or perennating organs (e.g. rhizomes) or persisting above ground sometimes with loss of foliage (e.g. deciduous trees).

Perianth Structure that protects the developing reproductive parts of the flower.

Periderm Outer protective layers in the stems of woody plants consisting of a corky tissue and an inner secondary cortex layer.

Petiole Leaf stalk.

Phenotype The recognition of an organism by its observable (morphological) characteristics.

Phloem Tissue that conducts food material (products of photosynthesis) in vascular plants away from leaves to other parts of the plant.

Photorespiration Light dependent respiratory process in plants where glycolic acid is produced in chloroplasts and is subsequently oxidised in peroxisomes.

Phycobilins Water-soluble pigments that occur in red algae and cyanobateria.

Phylogenetic Describing a system of classification of organisms in an attempt to show their evolutionary history.

Phylogeny Evolutionary history of an organism or a group of related organisms.

Phytoplankton Plankton that obtain their energy from photosynthesis. Consists mainly of microscopic algae such a diatoms and dinoflagellates.

Phytolith Microscopic intracellular bodies formed of biogenic opaline silica in certain plant cells, mainly leaves, stems and inflorescences. They often have a characteristic size and shape that can be used to identify the family, genus or occasional species from which they came.

Pinna(e) A primary division or leaflet of a compound leaf or frond.

Pith Tissue occupying the center of the stem or root.

Plastids Organelles found in the majority of plant cells that are responsible for processes such as food manufacture and storage e.g. chloroplasts.

Polymerase chain reaction (PCR) Technique used to replicate a fragment of DNA so as to produce many copies of a particular DNA sequence.

Polyphyletic With evolutionary descent from more than one group. A polyphyletic group does not contain an ancestral species common to all taxa.

Polyporate Descriptive term applied to pollen grains that have numerous pores distributed over their surface.

Pore A circular opening in the wall of pollen grains or an elliptical opening in the epidermis surrounded by two guard cells.

Propagule Collective term for propagative plant organs e.g. fruits, seeds and spores.

Protista Single-celled organisms with both plant and animal characteristics. Can be further subdivided into prokaryotes (no separation of genetic material from rest of cell) and eukaryotes (genetic material in a defined nucleus).

Proxies Fossil data sets (organic and inorganic) that are used in the reconstruction of past environmental variables.

Purine An organic nitrogenous base that gives rise to a group of important derivatives, notably adenine and guanine that occur in nucleotides and nucleic acids.

Pyrimidines An organic nitrogenous base that gives rise to a group of important derivatives, notably uracil, thiamine and cytosine that occur in nucleotides and nucleic acids.

Pyretic Fossil preservation in which cellular spaces have been impregnated by iron pyrite.

Rachis The main axis of a compound leaf or frond.

Radiation Evolutionary pattern of rapid appearance of many diverse descendants of one linage.

Radiometric dating Dating techniques that use the process of radioactive decay as the basis for measurement of periods of time.

Red algae Red-pigmented algae most of which live in Tropical seas.

Resin Sticky substance consisting of mainly polymerized acids, esters and terpenoids that is exuded by certain plants particularly when wounded. Particularly prevalent in conifers.

Resin canal A longitudinal channel in the secondary xylem and leaves of many gymnosperms.

Reticulate venation Net-like pattern of the veins in the leaf blade.

Rhizome Underground stem that grows horizontally, and through branching, acts as an agent of vegetative propagation.

Ribosome(s) Protoplasmic particles that act as the sites for the assembly of amino acids into the polypeptide chains of protein molecules.

Rosette Form of plant that has leaves radiating outwards from a short stem at soil level.

Sauropod Large four-limbed saurischian dinosaurs that were lizard-hipped with extremely long necks and tails, small heads and elephantine bodies.

Savannah Tropical grassland with some scattered trees or shrubs.

Sclerophyll A type of vegetation which is typically scrub but also woodland, the trees and shrubs of which possess small, hard and thick, leathery leaves. This vegetation type is well adapted to the dry summer months of the winterwet (Mediterranean) climate.

Sea-floor spreading Formation and lateral displacement of oceanic crust by upwelling and melting of the mantle beneath mid-ocean ridges.

Seasonality With distinctive seasons.

Secondary growth Increase in diameter of a plant resulting from the formation of secondary tissue e.g. secondary xylem, secondary phloem.

Sedimentary Unconsolidated material that has accumulated on the Earth's surface.

Sedimentary sequences Layers of sedimentary rocks recording accumulation over time.

Seed coat Outer layer of the seed developed from the integuments of the ovule.

Selfing Self-pollination or self-fertilization.

Sexine The outer layer of the exine in pollen grains.

Shale(s) A sedimentary rock with dominant grain size of mud grade.

Solar radiation Electromagnetic radiation emitted by the Sun.

Speciation The formation of a new species.

Spikelet The basic unit of a grass inflorescence consisting of a short axis, two bracts and one or more florets.

Spinose margin Margin that is spiny in appearance.

Sporangium Structure in which the spores are produced.

Sporophyte Spore producing stage of plant in the alternation of generations.

Sporopollenin The main biopolymer of which spore and pollen exine is composed. It is highly resistant to decay.

Stamen Pollen-producing organ in a flower.

Stasis An evolutionary term used to describe periods of little or no apparent change.

Stegosaur Bird-hipped dinosaur (ornithischian) characterised by bony plates and spines that extended along the back.

Stele The vascular tissue of the stem or root also known as the vascular cylinder.

Stellar arrangement The arrangement of the vascular tissue in the stem or root.

Steppe Vegetation dominated by grasslands in temperate regions where the precipitation: evaporation ratio is too low to enable trees to grow.

Stigma The receptive tip of the carpel, which receives the pollen at pollination and on which the pollen grain germinates.

Stomata Pores in the epidermis of above ground parts of plants, which enables gaseous exchange between the internal tissues and the atmosphere.

Stomatal density The number of stomatal pores per mm^2 leaf surface area.

Striated Pollen with multi-parallel grooves and ridges giving the grain a striped appearance.

Subduction A term used to describe the process when an ocean plate collides with a continental plate, causing the oceanic plate (which is carrying the heavier oceanic crust) to slide beneath the continental crust and melt into the mantle.

Succulent A plant that conserves water by storing it in large cells in swollen stems and leaves.

Symbiotic A relationship between individuals in which both species benefit.

Taiga see Boreal

Tannin Complex organic compounds commonly found in leaves, bark and unripe fruit.

Taphonomy All bioliogical, physical and chemical process influencing the preservation of fossils (e.g. decay, burial, transport, disarticulation).

Tap-root The primary root of a plant from which arise smaller lateral branches.

Tectonic Processes that relate to the deformation, uplift, subsidence and lateral movement of the Earth's crust.

Tectum Outer layer of exine in pollen grain wall.

Temperate Presently encompasses most of the latitudes between the tropics and polar circles in both hemispheres. The climate of the temperate regions are characterised by moderate temperatures (i.e. not excessive or extreme) with alternating long warm summers and short cold winters.

Temporal range Range through time.

Tetrapod Four-limbed animal.

Thalli Generalized term for simple plant body of nonvascular plants.

Thermophylic Organisms that grow best at temperatures of 50°C or higher.

Thrust faulting A process of rock deformation when rock on one side of a fracture (fault) plane is thrust up and over the rock on the other side of the fracture (fault) plane.

Thylakoid One of the membranous layers making up the grana in the chloroplast of plant cells.

Tonoplast The membrane surrounding the vacuole in plant cells.

Topography The altitudinal variation in the Earth's surface.

Tracheid(s) Elongated water conducting cells.

Tracheophyte(s) Vascular plants.

Tricolpate Pollen grain with three furrows arranged equidistally about the grain.

Triporate Pollen grain with three pores arranged equatorially at 120° apart.

Tropical Presently encompasses the latitudes between 23–27°N and 23–27°S. The climate of the tropical regions are characterised by high temperature, humidity and rainfall throughout most of the year.

Tundra A treeless circumpolar region predominantly found in the northern hemisphere north of the Arctic Circle.

Ultrastructure Structural details of cells only visible with an electron microscope.

Understorey In a forest, the vegetation between the canopy and the ground story — usually composed of shade tolerant trees of moderate height.

Uni-ovulate Containing one ovule.

Uniparentally Reproduction and inheritance from a single parent, as in vegetative reproduction.

Vacuole A fluid filled cavity within the cytoplasm of plant cells.

Venation The pattern formed by the veins on the leaf.

Vessel members Tubelike structure of the xylem composed on elongated cells joined end to end and connected by perforations. The function of vessel members are to transport water and mineral through the plant body. They are found in nearly all flowering plants and some ferns.

Whorl A circle of flower parts or leaves.

Xeric Having very little moisture.

Xeromorphic Plants that have structural or functional adaptations to prevent water loss by evaporation.

Xerophytic Plants that live in a dry habit often showing xeromorphic or succulent adaptations.

Xylem Tissue found in vascular plants; the principle function of which is the translocation of water and solutes.

Zoospores Independently motile spores found among protists, some fungi and algae.

Zygote The diploid cell resulting from the fusion of the male and female gametes.

Appendices

Appendix 1: Sources of plant drawings, Figure 4.26

(a) *Sigillaria* reconstruction redrawn from Tidwell (1998). Bark details from *Sigillaria* sp. (Field Museum specimen PP 32608).

(b) *Lepidodendron* reconstruction redrawn from Eggert (1961). Bark details from *Lepidodendron velttreimii* (Field Museum specimen PP16349).

(c) *Stigmaria* reconstruction adapted from Meyen (1987). Bark details from *Stigmaria* sp. (Field Museum specimen PP25342).

(d) Details of *Lepidodendropsis* from *Lepidodendropsis vandegrachtii* (Field Museum specimens PP45122, PP45123).

(e) Adapted from *Sphenophyllum cuneifolium* (Meyen, 1987).

(f) *Calamites* reconstruction and details of *Asterophyllites* leaves adapted from Tidwell (1998), Meyen (1987).

(g) Stylized pteridosperm frond adapted from reconstructions of *Archaeopteridium tschermackii* (Rowe, 1992) and line tracing of *Charbeckia* gen nov. (Knaus *et al.*, 2000).

(h) *Tomiodendron* reconstruction redrawn from *Tomiodendron kemorviense* in Meyen (1987).

(i) Immature *Tomiodendron* redrawn from Meyen (1987).

(j) Detail of *Ursodendron* redrawn from Meyen (1987).

Appendix 2: Sources of plant drawings, Figure 4.27

(a) *Medullosa noei* reconstruction redrawn from Stewart and Delevorayas (1956).

(b) See Appendix 1(b).

(c) *Psaronius* reconstruction redrawn from Morgan (1959).

(d) Cordaites tree adapted from a whole plant model (FMNH # CSB74112) and an original reconstruction from Cridland (1964).

(e) See Appendix 1(f).

(f) *Lebachia piniformis* redrawn from Florin (1944) with details from Field Museum specimen P1608.

(g) *Botrychiopsis plantiana* adapted from White (1994).

(h) Rhacopterid adapted from White (1994).

Appendix 3: Sources of plant drawings, Figure 5.10

(a) Stylized *Baiera* leaf adapted from Harris (1974).

(b) Stylized *Sphenobaiera* leaf adapted from Harris (1974).

(c) *Lebachia piniformis* redrawn from Florin (1944) with details from Field Museum specimen *Walchia piniformis* (P1608).

(d) Line drawing of *Ullmannia frumentaria* (P1613).

(e) Cordaites tree adapted from a whole-plant model (FMNH # CSB74112) and an original reconstruction from Cridland (1964).

(f) Line drawing of *Gigantoclea guizhouensis* (PP34440).

(g) *Medullosa noei* reconstruction redrawn from Stewart and Delevorayas (1956).

(h) Adapted from line drawing of *Sphenophyllum cuneifolium* (Meyen, 1987).

(i) Line drawing of *Plagiozamites oblongifolius* (UP1083).

(j) Line drawing of *Callipteris conferata* (P1607).

(k) *Glossopteris* reconstruction redrawn from Gould and Delevoryas (1977).

(l) Equisetites adapted from Tidwell (1998).

(m) *Lepidodendron* reconstruction redrawn from Eggert (1961).

Appendix 4: Sources of plant drawings, Figure 5.14

(a) Stylized cycad based on living *Zamia floridana.*

(b) *Otozamites* sp. modified from Harris (1969).

(c) *Ptilophyllum* sp. modified from Harris (1969).

(d) *Zamites* sp. modified from Harris (1969).

(e) Reconstruction of *Williamsonia sewardiana* redrawn from Sahni (1932).

(f) *Brachyphyllum* redrawn from Harris (1979).

(g) *Pagiophyllum* redrawn from Harris (1979).

(h) *Pseudocycas* redrawn from Florin (1933).

(i) *Anomozamites* redrawn from Harris (1969).

(j) *Nilssonia* redrawn from Harris (1964).

(k) *Lindleycladus* leaves redrawn from Harris (1979).

(l) Stylized *Baiera* leaf adapted from Harris (1974).

(m) Stylized *Sphenobaiera* leaf adapted from Harris (1974).

(n) Reconstructed Mesozoic *Ginkgo* tree modified from Thomas and Cleal (1998).

Appendix 5: Sources of plant drawings, Figure 6.21

(a) Stylized leaf redrawn from Wolfe (1985).

(b) Generalized unnamed Cretaceous flower redrawn from Basinger and Dilcher (1980).

(c) Stylized leaf redrawn from Wolfe (1985).

(d) *Sabalites* palm redrawn from Tidwell (1998).

(e) *Nypa* palm adapted from Collinson (1983) and living *Nypa* specimens.

(f) *Cyathea* tree fern courtesy of K. Pryer (Pryer *et al.*, 2000).

(g) *Dicksonia* tree fern stylized from living specimens.

(h) *Myrica torreyi* redrawn from Dorf (1942).

(i) *Ficus planicostata* redrawn from Dorf (1942).

(j) *Dryophyllum subfalcatum* redrawn from Dorf (1942).

(k) *Viburnum marginatum* redrawn from Dorf (1942).

(l) *Grewiopsis saportana* redrawn from Dorf (1942).

(m) Reconstruction of *Araucaria* tree adapted from Thomas and Cleal (1998).

(n) Reconstruction of cheirolepidaceous tree adapted from Thomas and Cleal (1998).

(o) *Podocarpus* leaves modelled from many examples in Salmon (1980).

(p) Generalized from fossil whole-tree reconstructions and modern trees.

(q) Stylized microphyllous leaves redrawn from Wolfe (1985).

Appendix 6: Sources of plant drawings, Figure 7.6

(a) Line drawing of *Metasequoia occidentalis* (Field Museum specimen PP9636).

(b) *Corylites* sp. leaf adapted from Kvaček (1994).

(c) *Trochodendroides* sp. leaf redrawn from Kvaček (1994) and Basinger *et al.* (1999).

See Appendix 5 for other sources.

Appendix 7: Sources of plant drawings, Figure 7.8

(a) Line drawing of *Metasequoia occidentalis* (Field Museum specimen PP9636).

Index

Page numbers in **bold** refer to figures and tables.